Christianity in Roman Pannonia

An evaluation of Early Christian finds and sites from Hungary

Dorottya Gáspár

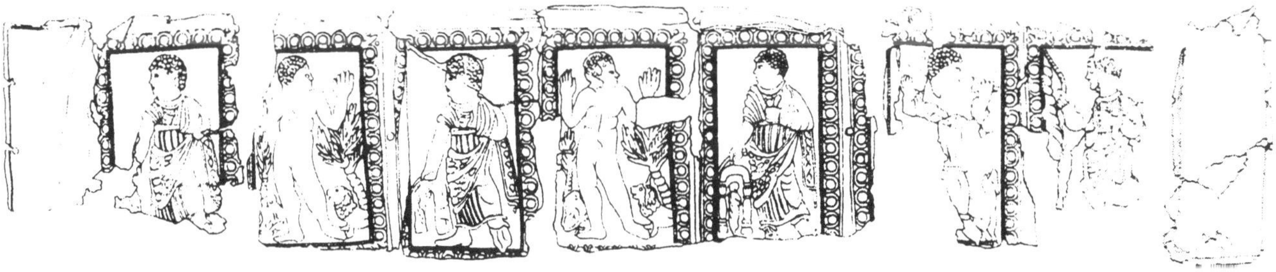

BAR International Series 1010
2002

Published in 2016 by
BAR Publishing, Oxford

BAR International Series 1010

Christianity in Roman Pannonia

ISBN 978 1 84171 288 8

BAR Publishing is the trading name of British Archaeological Reports (Oxford) Ltd. British Archaeological Reports was first incorporated in 1974 to publish the BAR Series, International and British. In 1992 Hadrian Books Ltd became part of the BAR group. This volume was originally published by Archaeopress in conjunction with British Archaeological Reports (Oxford) Ltd / Hadrian Books Ltd, the Series principal publisher, in 2002. This present volume is published by BAR Publishing, 2016.

Printed in England

BAR
PUBLISHING

BAR titles are available from:

BAR Publishing
122 Banbury Rd, Oxford, OX2 7BP, UK
EMAIL info@barpublishing.com
PHONE +44 (0)1865 310431
FAX +44 (0)1865 316916
www.barpublishing.com

BONAE MEMORIAE

MAGISTRI ANDREAS MÓCSY

DEDI DEDICAVI

CONTENTS

I. Acknowledgements

First, I wish to express my sincere gratitude to everyone who assisted me with my work. My thanks are due to several archaeologists, draughtspeople, photographers, colleagues, institutiones and museums, as well as to the administration of the Archaeological Institute. In this passage, I particularly emphasize the colleagues and institutes who helped me with things, which by nature, cannot be listed in form of notes. The others will not remain anonymous as I shall mention them by name in the appropriate places.

I should like to thank the Archaeological Institute of Hungarian Academy of Sciences for the material support they provided me with. I was one of their scholars when I completed the first version of this manuscript early in 1994.

My thanks are due to Dr István Torma, the senior member of the Archaeological Institute, for his precise reading regarding topographical questions. After considering and accepting his comments I could improve my manuscript. I am also grateful to Dr Branka Migotti for her cooperation by which I could become more acquainted with Croatian literature and also with the archaeological materials of the region.

My very special thanks are also due to Dr Rajka Makjanić for her help as a publisher and a friend.

Finally, I would like to remark that owing to a number of difficulties the book could only now be published, so a certain revision of the formulations dating from the year 1994 was necessary. The basis for the revision was the latest literature which extended the range of things to be catalogued and also forced me to formulate my concept even more accurately.

Dorottya Gáspár
Hungarian text March 1999, English translation 2000.

LIST OF FIGURES

168. Magyaregregy, finger ring, 28.II.a. **a:** photo, **b:** drawing. Based on Severin Katalog.
169. Mórichida-Kisárpás, casket mount, 29.I.a. Photo: SUGAR Lajos.
170. Mucsfa, roof tile, 30.II.a. Based on THOMAS 1982b.
171. Nagyberki-Szalacska, bronze lamp, 31.I.a. Based on IIL.
172. Pécs, general plan. Based on FÜLEP 1984. **a:** map of the graves found around the cathedral, **b:** map of the graves inside the present centre of town.
173. Pécs, one of the 13 graves, 32.I.a. Based on KOLLER 1804.
174. Glass finds of 32.I.a. Based on KOLLER 1804.
175. Pécs, tomb III, 32.II.a. Based on GOSZTONYI 1943.
176. Entrance of 32.II.a. Based on FÜLEP 1984.
177. Pécs, tomb IV, 32.II.b. Based on GOSZTONYI 1943.
178. Entrance of 32.II.b. from the outside (south). Based on FÜLEP 1984.
179. Niche in the western wall, 32.II.b. Based on FÜLEP 1984.
180. Northern wall of 32.II.b. Based on FÜLEP 1984.
181. Pécs, tomb VI, 32.II.d. Based on GOSZTONYI 1943.
182. Pécs, general plan of the excavation of tombs VIII-IX. 32.II.f-g. Based on TÖRÖK Gy. 1942.
183. 32.II.f. from the entrance. Based on TÖRÖK Gy 1942.
184. 32.II.f. from the north. Based on TÖRÖK Gy 1942.
185. Northern wall of 32.II.f. with the place of burial. Based on TÖRÖK Gy 1942.
186. Southern wall of 32.II.g. with the entrance. Based on TÖRÖK Gy 1942.
187. Northern side of 32.II.g. Based on TÖRÖK Gy 1942.
188. Bronze buckle and pottery fragment found on the floor of 32.II.f. Based on TÖRÖK Gy 1942.
189. Fresco fragments found above 32.II.f-g. Based on TÖRÖK Gy 1942.
190. Pécs, tomb X, **a:** pieces of mosaic floor found near 32.II.h. **b:** reconstruction. Based on FÜLEP 1984.
191. Pécs, painted tomb I, 32.I.b. **a:** ground plan**, b:** section, based on HENSZLMANN 1876. The measurements can be found here.
192. Pécs, painted tomb I, 32.I.b. **a:** ground plan, together with the ground plan of the 32.I.c. **b:** sectional, **c:** reconstruction based on GOSZTONYI 1943.
193. Pécs, painted tomb I, 32.I.b. **a:** Section, based on SZŐNYI O 1907. The description of the tomb follows the marks of this drawing. **b:** drawing of frescos, based on GOSZTONYI s.a.
194. Fresco on the entrance (southern) wall of 32.I.b. Based on FÜLEP 1984.
195. Fresco of the northern wall of 32.I.b. Based on SZŐNYI O 1907.
196. Frescos in the joining of the northern and eastern walls of 32.I.b. Based on SZŐNYI O 1907.
197. Frescos in the joining of the western and southern walls of 32.I.b. Based on SZŐNY O 1907.
198. Ceiling fresco of 32.I.b: portrait in medallion. Based on FÜLEP 1984.
199. Western wall of 32.I.b. Based on FÜLEP 1984.
200. 'Mary's' scene in the fresco of 32.I.b. **a:** based on FÜLEP 1984. **b:** based on BALOGH 1936.
201. Western wall of the 32.I.b. Three youths in the furnace. Based on FÜLEP 1984.
202. Western wall of the 32.I.b. Noah in the Ark. Based on FÜLEP 1984.
203. Eastern wall of the 32.I.b. 'Eve'. Based on FÜLEP 1984. **a:** full-length figure, **b:** head.
204. Part of the fresco of 32.I.b.: Jonah's story. Based on FÜLEP 1984.
205. Scene of 'the Magi' of 32.I.b. Based on SZŐNYI O 1907. **a:** drawing, **b:** head of one of 'the Magi'.
206. Drawing of a portrait on the southeast part of the ceiling of 32.I.b. Based on SZŐNYI O 1907.
207. Fresco fragment in the antechamber of 32.I.b. Candelabrum in the southeast corner. Based on SZŐNYI O 1907.
208. Rome, Domitilla catacomb, Martyr Petronilla leads Veneranda. Based on TESTINI 1966.
209. Rome, part of the fresco in the hypogeum. Based on TESTINI 1966.
210a-b. Or ha-Ner, fresco: two portraits. Based on Encyclopedia 1978 (TSRAFRIR)
211. Or ha-Ner, fresco: series of portraits in medallions. Based on Encyclopedia 1978 (TSRAFRIR)
212. *Venus pudica*, bronze statue, H: 25 cm (Cf. Note 721.).
213. Pécs, painted tomb II, 32.I.c. Based on GOSZTONYI 1943. a: Cross-section oriented E-W, b: cross-section oriented N-S, c: ground plan of the tomb. The superstructure is only indicated nearby, d: ground plan of the superstructure, the tomb inside is only indicated. e, f, g: reconstruction of the superstructure.
214. Photo of the grave in 32.I.c. Based on FÜLEP 1984.
215. Measurements of 32.I.c. indicated on a drawing; ground plan. Based on FÜLEP/FETTER 1969-1970.
216. Measurements of 32.I.c. indicated on a drawing; section of the eastern wall. Based on FÜLEP/FETTER 1969-1970.
217. Measurements of 32.I.c. indicated on a drawing; section of the western wall. Based on FÜLEP/FETTER 1969-1970.
218. Measurements of 32.I.c. indicated on a drawing; section of the northern wall. Based on FÜLEP/FETTER 1969-1970.
219. Measurements of 32.I.c. indicated on a drawing; section of the southern wall. Based on FÜLEP/FETTER 1969-1970.
220. Measurements of 32.I.c. indicated on a drawing; picture of the southern outside wall. Based on FÜLEP/FETTER 1969-1970.
221. Drawing of the frescos of 32.I.c. Based on GOSZTONYI 1942. **a:** entrance (southern) side, **b:** northern side, **c:** western side, **d:** eastern side.
222. Fresco on the northern wall of 32.I.c. Based on FÜLEP 1984.
223a, b. Entrance and footing of 32.I.c. Based on FÜLEP 1984.

282. Ságvár, cella III, 36.II.c. Based on BURGER 1966.
283. Ságvár, buckle, 36.II.d. Based on BURGER 1966. **a:** photo. **b:** drawing.
284. Ságvár, brick with incised pattern, 36.II.e. Based on BURGER 1966. **a:** photo. **b:** drawing.
285. Ságvár, buckle. Based on BURGER 1966. **a:** photo. **b:** drawing.
286. Keszthely-Fenékpuszta, finger ring. Based on BARKÓCZI 1968.
287. Ságvár, brick with incised pattern, 36.II.f. Based on BURGER 1966. **a:** photo. **b:** drawing.
288. Darova (Daruvár), brick. Based on NAGY L 1938.
289. Sárisáp, general plan: 17: site of Roman settlement, 18: site of the so-called grave-chapel. Based on MRT 5, 19.
290. Sárisáp, ground plan of the so-called grave-chapel, 37.II.a. Based on MRT 5, 19.
291. Somlóvásárhely, fibula, 40.I.a. Based on TÓTH E 1994a. **a:** photo. **b:** drawing, **c:** detail.
292. Somogyszil, pottery with incised pattern, 41.II.a. Drawn by DUKAY Bernadett.
293. Somogyszil, pottery with incised pattern, 41.II.b. Drawn by DUKAY Bernadett.
294. Somogyszil, pottery with incised pattern, 41.II.c. **a:** drawing by DUKAY Bernadett. **b:** photo: Mrs. GŐZSY.
295. Somogyszil, pottery with incised pattern, 41.II.d. Drawn by DUKAY Bernadett.
296. Somogyszil, pottery with incised pattern, 41.II.e. Drawn by DUKAY Bernadett.
297. Somogyszil, pottery with incised pattern, 41.II.f. Drawn by DUKAY Bernadett.
298. Somogyszil, **a:** ichi smoothed in the bottom of an earthenware bowl, 41.II.g. Photo: Mrs. GŐZSY. **b:** Esztergom, altar. Based on MRT 5.
299. Somogyszil, pottery with incised pattern, 41.II.h. Photo: Mrs. GŐZSY.
300. Somogyszil, pottery with incised pattern, 41.II.i. Drawn by DUKAY Bernadett.
301. Somogyszil, pottery with incised pattern, 41.II.j. Photo: Mrs. GŐZSY.
302. Somogyszil, X smoothed in the bottom of an earthenware bowl, 41.II.k. Photo: Mrs. GŐZSY Gáborné.
303. Tác, drawings of earthenware incisions. Based on GABLER/KOCZTUR 1976.
304. Sopron, general plan of the excavation of the forum. Based on GÖMÖRI 1986.
305. Sopron, jug fragment with inscription, 42.II.a. Based on GÖMÖRI 1986.
306. Sopron, brick sherd, 42.II.b. Based on GÖMÖRI. **a:** photo. **b:** drawing.
307. Sopron, marble slab, 42.II.c. Photo by ADORJÁN Attila.
308. Sümeg, general plan: 23: site of the villa and the so-called early Christian basilica, 24: site of Roman settlement, 25: site of Roman finds from the surface. Based on MRT 3, 54.
309. Sümeg, building I (villa), 43.I.a. Based on THOMAS 1964.
310. Sümeg, building II (so-called early Christian basilica), 43.II.a. Based on THOMAS 1964.
311. Parallels to building I. Based on RIVET 1969.
312. Ground plan of the Orlandovci villa. Based on SARNOWSKI 1978.
313. Ground plan of the Pforzheim villa. Based on SARNOWSKI 1978.
314. Szabadszállás, general plan. Based on H.-TÓTH E 1969.
315. Szabadszállás, glass cup, 44.II.a. Based on H.-TÓTH E 1971.
316. Part of 44.II.a. The first medallion. Based on H.-TÓTH E 1971.
317. Part of 44.II.a. The woman figure between the first and the second medallions. Based on H.-TÓTH E 1971.
318. Detail of 44.II.a. Second medallion. Based on H.-TÓTH E 1971.
319. Detail of 44.II.a. Satyr between the medallions. Based on H.-TÓTH E 1971.
320. Detail of 44.II.a. Third medallion. Based on H.-TÓTH E 1971.
321. Detail of 44.II.a. Third figure between the medallions. Based on H.-TÓTH E 1971.
322. Superficies of 44.II.a. Drawing by Dr H.-TÓTH E.
323. Analogues of the shape of the cup. **a:** Bátaszék-Kövesdpuszta, based on PÉTERFI 1993. **b:** Ságvár, grave 32, based on BURGER 1966.
324. Százhalombatta, general plan: 6: fort, 7: vicus belonging to the fort, 8: southern cemetery, 9: northern cemetery, 23: watchtower, Based on MRT 7, 27.
325. Százhalombatta-Dunafüred, bone tessera, 45.I.a. Based on MÓCSY 1955.
326. Székesfehérvár, screen, 46.I.a. as it was found, based on NAGY L 1938.
327. Szekszárd, vas diatretum, 47.II.b. Based on Severin Katalog.
328. Szekszárd, sarcophagus, 47.II.a. Based on Severin Katalog.
329. Szentendre, general plan. Based on MRT 7, 28.
330. Szentendre, fortress and its environs, general plan. Based on MRT 7, 28.
331. Szentendre, casket mount, 48.I.a. Photo: KÁDAS Tibor.
332. Szentendre, cemetery, 48.I.b. Based on NAGY L 1938. I-II. cemetery chapel.
333. Szentkirályszabadja-Romkút, comprehensive drawing of the buildings, 49.II.a. **a:** based on THOMAS 1964. **b:** based on MRT 2, 44.
334. Szilágypuszta-Malomalja, bronze sheet, 50.II.a. Based on THOMAS 1964.
335. Szombathely, reconstruction of the 15th century conditions: 1. castle and the Virgin Mary Church, 2. Franciscan Church, 3. Saint Martin's Church. Based on KISS G/ TÓTH E 1993.
336. Szombathely's topography and periods according to earlier knowledge. Based on BUOCZ 1967.
337. Link between Savaria and Arrabona. Based on GRAF 1936.
338. Link between Savaria and Bassiana. Based on TÓTH E 1977b.
339. Savaria's town wall and gates. Based on MEDGYES 1997.
340. Szombathely, reconstructed places of the *passion*'s events based on TÓTH E 1973. 1. Imperial palace, 2. Sopron gate, 3. Saint Martin's supposed native house and supposed Christian cemetery, 4. place of Quirinus' martyrdom, 5. western gate, 6. amphitheatre, 7. southern gate.

II. Introduction

The first five Christian centuries, as a limited period for research are not without any difficulties. Does the end of the 5th century mean a real break either in the whole of the Christian world or only in Pannonia? On the other hand, should we draw the chronological boundary at some other date? In a certain respect, the *constitutio* of Mediolanum forms an obvious division, but as it brought important changes in the life of the Christians, it cannot mark an end of an era. But the survey divides this period into two branches, namely to the period before Constantine as ancient Christianity and the period after Constantine as early Christianity.

Proclamation of Christianity as the state religion cannot be regarded as a temporal boundary either, but only as a new situation[1]. This rising or raising of Christianity achieved and could have achieved certain effects, which must be examined in Pannonia as well. Because of these consequences, the period in question cannot be considered complete at the end of the 4th century. Furthermore, the year 476 indicating the fall of the Roman Empire in the West also suggests that the first five centuries should be considered as a unit. That is why I have taken the first five centuries as a unit as well. Despite of this it was necessary sometimes to have a look at the events that took place during the 6th century. The fall of Sirmium in the year 568 would perhaps justify considering the first six centuries as a unit; however, I had two reasons for not doing so. The first reason was that the decline in Sirmium had begun earlier, in fact at the beginning of the 5th century when the praefectura was transferred[2], and the Avars only completed the decline in the 6th century. Secondly, the 6th century was characterized far more by the Migration and much less by the Roman life over the whole Pannonia - except for some towns and other certain areas.

On the other hand, I ought to explain why the first century belongs to this temporal unit. The organization of the province of Pannonia was the result of a process[3] that took place in the course of the first century A.D. The names of the provinces were changed: Illyricum Superius and Inferius, the two parts of Illyricum, became Dalmatia and Pannonia[4]. As I share the opinion of scholars[5] who thought that no special organization of the provinces was undertaken as a result of the expansion of Christianity, I am leaving aside the question of when and how Pannonia became an organized province. As for my subject, Emona[6] is important, because this town was annexed to Italy towards the end of the 1st century A.D[7]. Poetovio, however, was annexed to Noricum during the reign of Diocletian, as a Christian town went on belonging to Pannonia.[8]

Joseph Lortz's Church History[9] has helped me to decide whether Altertum or Völkerwanderungszeit is more precise for the both, the period and space. This book is not a usual history of Christianity, but a history of ideas and thoughts. Joseph Lortz explains what the Church history[10] means theologically. When we talk about history, we witness the temporal character of the thinking. This kind of thinking belongs to the modern human being, and it supposes that above all we have the past. If we think about the Church in this special way, it wins special significance 'weil die Kirche es auch und zwar wesentlich mit unwandelbaren Elementen zu tun hat.'[11] The Church is Christ's mystical body; it is Christ, who lives on, consequently something 'divine', the object of belief. That is why it can hardly be comprehensible to mind of man, 'doch vermag er genügend tief in ihr Wesen und Wirken einzudringen, um von ihr eine wissenschaftliche Darstellung zu geben.' With the aid of history of Christianity, we do so. First moment of this history is the embodiment of Jesus Christ, i.e. the divine Logos, the Word of God appeared in the temporal history.[12] Now, I shall attempt to illustrate this fact in archaeology. The idea and thoughts will be expressed by the *interpretatio Christiana*, while the body, the living organism as the sensible world interprets depending on time, space, and its own culture. I call this technique of interpretation *interpretatio Pannoniaca*, which must have had a Christian and a non-Christian version. The above mentioned dependence does not allow the scholar to lay aside the realities, but it rather urges one to be extremely disciplined; 'Die Geschichte der Kirche kann man nicht aus Ideen, auch nicht aus geoffenbarten, ableiten; man muß sie in Treue und Entsagung an dem, was einmal ohne unser Zutun wurde und war, ablesen.'[13] According to Joseph Lortz, if we insist on discipline then history of Christianity helps us to get to know and to comprehend the deeper meaning of the Church. Considering his warnings and viewpoints, I have worked out my method, which I am going to describe systematically. I have always been insisting on empirical evidence in order to get real conclusions. I do not wish anything else by my hypotheses than to comment on the

---

[1] It seems useful to mention the fact that it was Theodosius († 395), who proclaimed Christianity as the state religion.

[2] Constantine created four *praefecturae*. Pannonia belonged to the region of the *praefectura Italiae*, which also included Illyricum. Constantius II changed this arrangement. He joined Pannonia to the diocesis of Macedonia. The name of the new praefectura was *praefectura Illyrici* whose capital was Sirmium. Although Julian restored the conditions at the reign of Constantine I, at the end of the 4th century the arrangement by Constantius II was valid again. This is the *praefectura*, which was later transferred from Sirmium.

[3] Mócsy 1974: 31-52.

[4] Nagy T 1961a: 135; MÓCSY 1974: 34.

[5] Barton 1975: 12; NOETHLICHS 1983: 1-8.

[6] Plesničar-Gec et al.: Old Christian Centre in Emona. Katalogi in monografije, 21 (Emona IV) Ljubljana 1983; PLESNIČAR-GEC 1976; PLESNIČAR-GEC 1983; PLESNIČAR-GEC 1990.

[7] Mócsy 1974: 94.

[8] In Pannonia, boundaries were sometimes changed rather complicatedly. See Nagy T 1939: 200-216. Poetovio is discussed on page 206 and it is stated that the church affiliation of Poetovio preserved the civil administrative boundaries that were valid before the reign of Diocletian.

[9] Lortz 1962

[10] Dagmar Stutzinger argued so. Stutzinger 1983: 13.

[11] Lortz 1962: 3.

[12] Lortz 1962: 3.

[13] Lortz 1962: 6.

tendencies that are implicit in the material. I do not intend to declare facts by these hypotheses, but rather I would like to demonstrate the uncertainties and unsolved problems to those who deal or are going to deal with these themes.

The unity of the Church. 'Die Einheit der Kirche besagt nicht, ... daß die Heiden nicht zur einen Kirche gehörten.'[14] By speaking of this, Joseph Lortz refers to St Augustine who writes that there are a lot of us who do not belong to us, but they are of us; and many belong to us who are indeed external.

In this single sentence, Augustine could demonstrate the connection between Pagans and Christians, and at the same time the fundamental distinction between them. Pagans and Christians have a common ground like a fundamental unity, which concerns the thesis and the persisting on thesis. Generally, we speak about polytheism and various ancient religions, behind which we find the same belief, the mutual unity. Ancient authors, philosophers adumbrated several times this unity called *tormentum infiniti*, the longing for eternity, and longing for the Paradise, indicating briefly the knowledge above the knowledge,[15] which may be a paraphrase of St Paul's quotation: 'nunc cognosco ex parte, tunc autem cognoscam sicut et cognitus sum.'[16] The revelation which the man has received *in illo tempore* or *in principio* when he was created has been preserved by the man living in the Paradise or – in terms of mythology – by the people of the Golden Age. People have always treasured some reminiscence of life in

for which they have been longing without stopping. This reminiscence and desire has achieved an identical knowledge which has enabled also pagans to think in a Christian mentality and to make their utensils by remembering the world of 'ideas'. Moreover, the pictures on these utensils can be interpreted in Christian way.[17] It has involved *interpretatio Christiana*.[18] The present-day influence caused the rigorous distinction of paganism and Christianity and it ascribes, first, to material conception as we see the history today, but there are scholars who speak about historical designation mythically. This mythical view differentiated between the fallen and the redeemed humanity, it never neglected the fact that fallen human being was also God's creature, so he could not be pagan in the absolute sense. The symbols used by paganism can be found in Christian art, as well.[19]

'Christlich besagt Wahrheit und Liebe, beides in untrennbarer Einheit. Nur die, von der Liebe, also von der Begeisterung befruchtete Erkenntnis führt ins Innerste der Dinge. Aber die liebende Erkenntnis darf nur der Wirklichkeit gelten. Für die Erkenntnis der Wahrheit (und dies gerade auch in der Kirchengeschichte) sind also nötig: Begeisterung *und* Kritik, Liebe *und* Wahrhaftigkeit; die Gesamthaltung muß sein: nüchterne Begeisterung.'[20]

Each religion and denomination has some emotional aspects, but as long as they can control the emotions by love and justice, the quiet is preserved, so 'pagan' and Christian live quietly side by side. The statement that pagan gods are idols insulted love and justice and by doing so, the emotions escaped. St Augustine's quoted sentence - 'many belong to us, who are indeed external' - is valid for the one who first spoke out this.

Pagan worship cannot be pure idolatry because it subsists on the revelation of *in illo tempore* and *in principio*. A pagan, an uneducated believer who prays and has difficulties with invocation, so he says 'sive mas sive femina'[21] or si deus, si dea[22] instead of the name of his God, nevertheless is led in life by love and justice and can hardly be called an idolater. He and others like him 'do not belong to us, but they are of us.' The *mos maiorum* is not at all an idol; it is an adherence to the primordial one, whose honour was carefully guarded.[23] Lortz tells us not to examine the Roman Empire only by side of the persecutions.[24] To turn on the persecutions we have to mention that Romans had contraries not only with Christians. Everyone who - symbolically speaking - was not a worshipper of SOL was regarded as an enemy in the Roman Empire. The Roman Empire failed to judge the Christians, because they were thought, continuing the symbol, to be worshippers of LUNA.[25] Christian attitude is different not only because of the frequently mentioned monotheism, but also because of the linear time with Christas the centre.[26] The cyclic time used before assured the prosperity for the whole year by the celebration of the Emperor's birthday, and his accession to the throne[27] and also by the *navigatio Isidis*[28], the ritual that assured wealth, fertility, peace, and prosperity of the *temporum felicitas* every year. When the sacrifice for the welfare of the emperor was skipped, the

[14] LORTZ 1962: 8.

[15] I do not mean the heretical Gnostic sense when I wrote the above expression. The expression itself was mentioned by Dr András László. Lecture 'Oriental and Western Religions' October 4th, 1993.

[16] 1 Cor. 13.12

[17] Brendel 1977. The mosaic in Villa Albani

[18] ELIADE deals with these pictures. I quote here the following: 'Les images, les symbols, les myths, ne sont pas des creations irresponsables de la psyché; ils répondent à une nécessité et remplissent une fonction: metre à nu les plus secretes modalités ded l'étre.' (1952: 13-14.) Eliade explains that the pictures and symbols come from the lost Paradise and that primordial man is the archetype. (1952: 14-15).

[19] For example, the two lines form a right angle, and the point where they cut each other is a centre, the axis mundi, where cosmic levels are penetrable. For Christians these two lines also form a four-armed cross with a centre where God dwells. This cross-shaped symbol was often misinterpreted, explaining why non-Christian finds were considered Christian ones.

[20] LORTZ 1962: 9.

[21] VERSNEL 1981: 17.

[22] Appel 1909 No. 22 (Macrobius), 51 (Cato).

[23] Considering *mos maiorum* and *crimen laesae maiestatis* we can read about the persecution of Christians in a work by Torben Christensen: CHRISTENSEN 1981

[24] LORTZ 1962: 31.

[25] The differences between the worshippers of SOL and those of LUNA are well known. One of these differences appears in the use of the lunar and the solar calendars. For instance, the Sabine Numa Pompilius introduced the lunar calendar, Julius Caesar, however, carried out his calendar reform for expressing his honour to SOL. He intended to follow Romulus' example, that is why he renewed some ancient habits. SUET. Caes. XX, 1-2; XL.

[26] ELIADE 1993: 151-198. For similarities and differences between Romans and Christians GÁSPÁR 1994: 163-172.

[27] Feriale Duranum. FINK 1971: 422-427. No. 117. P. Dur. 54.

[28] ALFÖLDI 1937: 42-58.

prosperity, the *temporum felicitas* was endangered. This attitude was *lease maiestatis* as regards the emperor and disturbance of the daily life and habits, disturbance of the yearly festivals, prayers and offering as regards the common people. This behaviour meant the disturbance of unity[29] all over the Roman Empire. The habitants of the Roman Empire really were not pagan, we ourselves labelled them as pagans; they were only *gentes ante Christum*. Just because of this, they were right to take offence when accused of idolatry.[30] Here to quote Géza Alföldy again, who says the following in connection with the conquests: 'We cannot find excuses for the brutality of the Roman Empire ... however, its birth was not characterized ... by mass-murders, ... not to mention the dreadful deeds of modern dictatorships.'[31] This statement is also true in a religious aspect - the brutality of Rome cannot be compared with modern cruelties, because in that context the old brutality would seem as a tender caress.[32]

We, the succeeding generations, invented the word persecutions. The Roman Empire only knew the legally allowed and forbidden religion, they differentiated between *licita* and *illicita*. Judaism was licita; Christianity was illicita. It was not advisable to practise the *religio illicita*. It counted as a disturbance or as something causing disturbances. We can find more information about this matter in written sources than in archaeological ones. However, this matter must be debated out also from archaeological aspects, because archaeology keeps silent about it. This silence resulted in the widely shared opinion that Christianity spread as late as the 3rd century. The *ante* world and Christianity had one common feature, the *disciplina arcana*, which was required by mysteries as well as by Christians. In this respect the 3rd century, the coincidence between the spreading of the Mithras religion and the strong emergence of Christianity is very important. Furthermore, it is also a fact that Christian buildings are quite frequently situated above or near Mithraism. Could these facts imply a *modus vivendi* between the ancient world and Christianity? Could the *modus vivendi* encourage both the ancient world and Christians to tolerate each other? Could they recognize that the secrecy was the same for both? Similarly, the situation in Dura-Europos where the synagogue, the Christian house-church and the mithraeum existed quietly at the same time, might have appeared anywhere else, even in Pannonia. Of course, there have always been some implacable people, even in Roman time, who showed intolerance by emphasizing the differences and in this way concealing that they knew a lot.[33] The meaning of the term *christianus* becomes visible in behaviour[34] and not in objects.

Further on, Christians consist of diverse believers; for example, the Montanist and the Orthodox differed from each other but for the Romans, they both were the same Christians.[35] It is obvious that Christians used utensils in the same way as pagans did, in fact they must have used the same utensils, but these objects were not always signed by Christian symbols, but rather by common ones, which were given a Christian meaning by the *interpretatio Christiana*. Moreover, just like the pagans who did not know anything about that, we lack the proof, at least in some cases, whether to interpret the symbol as Christian or as pagan. It was not always important or advisable for Christians to make Christian representation clear. I do not think that the reason of the change in the 3rd century was the sudden expansion of Christianity, but rather the period of peace, which started at the beginning of the reign of Gallienus in 260.[36] In our modern times more people attend the church at religious peace and freedom. Everybody who earlier had had fear or had been cautious or cowardly came out of their shelters. This might have been the same in ancient times. The peace, which lasted for 40 years 'Christianised' the most timid as well, or at least apparently.[37]

Summing up, no Christian archaeological finds can be expected from earlier times (from the 1st, the 2nd centuries and from the beginning of the 3rd century), but this does not imply the lack of Christians. A slow, but steady spreading can be presumed and not a sudden increase at a certain time. This is the reason why I do not

[29] We may read about this unity in a contribution of Géza Alföldy: '...the people united by the Imperium adjusted themselves to the Roman culture and way of life and not vica-versa. They ascribed to the supremacy of a Greek and Roman culture that provided an intellectual basis for the Empire, and also established ... such technical achievements as the road network, houses with central heating, baths, and drainage. All these were connected with *mos maiorum*, which was based upon the traditional Roman values and the world of ideas and behaviour hallowed by the Roman religion; it was the single guideline followed by everyone who wanted to succeed in that social and political system until the spread of Christianity.'ALFÖLDY 1992: 17.

[30] It might be useful to mention that not only pictures and statues could be idols, but everything else that people substitute for God, e.g. money, career, power, etc.

[31] ALFÖLDY 1992: 14.

[32] Let me indicate briefly some features of the view of contemporary historians, especially their harsh criticism of the past. The consequence of their method is that our era does not know, or does not want to know, the ideal; it will not know the great elders, forefathers and saints. These historians insist on searching only for mistakes, failures, and sins, which are emphasized with arrogant criticism. If they discover something good, they will trample it underfoot. If they find a saint, they thrust him or her into a vague world of rumours. Finally, when their method has entangled everything in its false criticism, it states triumphantly that this modern era is the most perfect and that modern people are excellent so far. This approach has already reached into Church history, where it has become the propaganda of disbelief. This attitude has started, and resulted in the unpredictable series of false conclusions.

[33] Here I mean Tertullianus, who must have been initiated in to the mysteries of Mithras. I do not intend to deal with the relationship between Christians and believers in Mithras; I have only mentioned it because comparing their terminology and symbols we find they have some similarities. Therefore, we have to suggest that the symbols may be interpreted in different ways. Cf. SCHÜTZE A: Mithras-Mysterien und Urchristentum. Stuttgart 1972.

[34] The so-called pagans experienced the behaviour first. This could not be interpreted so easily for the pagans. See WILKEN 1984: 1-47.

[35] Let me cite only the Montanists by way of example. I do not intend to simplify the heresies as it is not my task to deal with them here; the aim of mentioning this one is to clarify my subject.

[36] I dare state that Christianity did not spread suddenly and unexpectedly, but it did appear to a larger extent during this time. I agree with others that a period of peace was needed, but they also think that Christianity started to spread during this period.

[37] See the opinion of Libanios quoted by Kurt SELIGMANN in German: Das Weltreich der Magie. 500 Jahre Geheime Kunst. Mit einem Nachwort von Prof. Dr G. F. HARTLAUB. Eltville am Rhein 1988: 98.

regard the 3rd century as a separate era, but rather as a continuation into the beginning of the 4th century, until the reign of Constantine I. A new epoch begun with the Constantine's reign. Hiding stopped but other problems emerged. The *interpretatio Christiana* did not cease although, there was no more need for it. The clearly Christian representations, the biblical scenes appeared at the same time.

Summarizing these points, we can state that the 3rd century brought important changes, which were caused not by multiplying the number of Christians but by acknowledging the Christian religion.

Antiquity or Migration Period? This question can rightly be raised in Pannonia regarding the 5th, but especially the 6th century.[38]

'Das kirchengeschichtliche Geschehen tritt uns zunächst als ein buntes Vielerlei auf vielen Schauplätzen in verschiedenen Zonen und Zeiten entgegen.' The life of the church 'und damit ihre Geschichte, vielfältig, nicht nur im Sinne des erwähnten Vielerlei, sondern als Entfaltung von verschiedenen Strukturschichten. Von dieser Sicht aus kann man die Kirchengeschichte aufgliedern ...'[39] Lortz uses special and temporal division here. In the present case, the space is Pannonia, the time is the first five centuries of the antiquity, where the 5th century is highlighted.

The area of Pannonia has been populated since prehistoric times. The Romans 'conquered' the Celts and Illyrians who were living here. Although they accepted the Roman culture, pre-Roman elements could be found even in the 5th century. The Romanised Celtic and Illyrian population as well as Romanised or Hellenised settlers from elsewhere became Christians in Pannonia or arrived in Pannonia already as Christians. They clearly belong to the antiquity in spite of their various cultures and habits. The different objects made by them show Christian elements and traces of Romanisation and Hellenisation even in the 5th century. Consequently, they can rightly be joined into antiquity. On the other hand, the tribes of the Migration Period were not romanised even if they were Christians.[40] However, the 5th century cannot be called Migration Period just because these tribes appeared. Some tribes who moved in Roman provinces during supremacy of the Roman Empire are simple barbarian wanderers. However, the Romans found here in the 6th century are the descendants of the Romans in the Migration Period. These surviving Romans represent the continuity in the new epoch, named Migration Period. At this time, two[41] populations with different cultures live side by side in the same age, but their material remains can clearly be separated.[42]

---

[38] Reading Tibor Nagy's article, we discover traces of the Migration Period even in the 4th century. NAGY T 1961b: 591-596.

[39] LORTZ 1962: 10-11.

[40] Numerous Pannonian examples may be seen in the catalogue: Germanen 1988.

[41] Those keeping Roman culture and civilization represented the continuity, and all the others formed a different category if we compare them to the Romans.

[42] It is enough to read one book to demonstrate this: Germanen 1988.

III. THE SYSTEM AND METHOD

The Early Christianity and the Christians are an important part of the ancient history. They have played notable roll not only in the Rome itself, but also all over the Roman Empire, to which Pannonia belonged. They have spread geographically; therefore, they, their behaviour, and their utensils always depended on their own time, space and environment. Consequently, if we intend to study the Church history independently of above, it may happen to us that we apply our own way of thinking and our own modern culture, and our own political and economic values to the studied period. It is very probable that this kind of the method always results in a false concept. To avoid this fault, I have created and employed my own system. In order to reach my aim I endeavoured to follow Lortz's viewpoints, which convinced me that we must never neglect the customs, the thinking, the intellectual life, and the culture of population, who became acquainted Christian deliberately thinking, and the 'Kingdom of God'. Therefore, the Christian history of Pannonia is dependent on the political, cultural, civil and economic history of Pannonia. Keeping this viewpoint, we may hope that the distinction between the Christian milieu and non-Christian milieu can be recognized; further, that the features and criteria for the identification of the Christian utensils, capitals, objects, spaces, etc. can be achieved, and finally that the history of Christianity in Pannonia can be adumbrated. Up to now, the research has rolled a straw-stack as an avalanche, which has been produced in the past six decades, by the superficiality in the works, the unevenness of research, and the emotionally influenced research; this avalanche has been overshadowing the clearness and accurateness of the monograph written by Tibor Nagy in 1939. First, the shadow should disappear and the straw-stack should be trussed; then it can be started to systematize the little bit torn, sometimes hardly perceivable plants in the clear field. My chosen and developed method might match this work.

After accepting the principle 'Begeisterung und Kritik' I decided to revise whether all finds and sites were virtually Christian, which had been thought to be Christian before. I have understood that there is no love without justice or at least there is no Christian love without justice. Therefore, by 'Liebe und Wahrhaftigkeit' and with steadfast severity, I deprived of their Christian titles all the finds and sites whose non-Christian character could be proved or at least their Christian character became uncertain. Simultaneously, the Liebe und Wahrhaftigkeit encouraged me to revise whether the neglected and 'außer stehen' finds and sites were rightly ranged. Topographical system seemed to be suitable for this revising and analysing. Thus, we can learn more about the sites, their geographical situation and stratigraphy, the archaeological context of the finds. In each case and every site, I shall give an account of the research, and a description of the finds. I shall analyse the features of the sites and finds, and based on this analysis, I shall divide them into two groups. Christian and Jewish sites and finds came into the first group – 'Group I'. The sites and finds, once considered as Christian, but actually non-Christian came into the second group – 'Group II'. The doubtful or ambiguous sites and finds also came into the 'Group II'. The fact that Jewish sites and finds came into 'Group I' does not mean the equalization of Christian belief and Jewish one.[43] Above-mentioned descriptions are followed by my own argument and opinion. That is why, some repeating inevitably occurs in the final consequences although the two[44] evaluations actually complement each other.

The system of the catalogue. Every site maintains an index number. This index number marks the spot of the site on the map. The sites are arranged according to their names in the alphabetical order. The description of the research is omitted at sites, which were not excavated. The two groups are marked by I. and II. In each group, a letter marks the objects. Hence, three data are needed for the identification of an object: the number of the site, the mark of the group and its own letter, e.g. 4.I.c. The catalogue does not contain sites, which used to belong to Pannonia, but are not parts of Hungary[45]

In addition, I have also adopted some other methods developed by scholars, who deal with philology, sociology or philosophy. Their detailed description became known earlier.[46]

---

[43] As for evaluation, Jews and Christians are connected by the fact that Romans considered Christians a Jewish sect. What actually separates them is that they themselves knew the difference very well. The difference was not the same as modern anti-Semitism, neither in its content nor expression.

[44] There is evaluation after each item, and an evaluation of the whole at the end. The latter represents not only summaries of the earlier ones.

[45] It would have been unnecessary to give the same detailed explanation of these parts as about the Hungarian ones, because a monograph has recently been published about the territories in question: BARTON 1975; Christentum 1994; Catalogue 1994; MIGOTTI 1990; MIGOTTI 1997; NOLL 1954; NOLL 1974-1975.

[46] GÁSPÁR 1989: 5-22. This paper concerns the analysis of the topic's and authors' expressions, which do not always convey information. It is also about the spread of Christianity, the possibilities of its mission, and about the subjective and objective conditions of that mission. In the course of sociological research I developed an experiment to find the ratio of Christians to the whole population in the province, in a determined period. I also raised the question of whether we could take into account the liberating of slaves from a viewpoint of wealth and Christianity.

IV. HISTORY OF RESEARCH

In the 18th century, Daniel Farlatus engaged in research of Christianity in Pannonia. His work resulted in a monograph, most of which has not yet been published; the manuscript is located in Venice.[47] Farlatus travelled around the area of Illyricum in the 18th century and read all the archive material on the subject in the Vatican, even that, which was later burnt. Thus, we have to regard his work as the most important source even if we disagree with some of his statements or our predecessors did so. One of these statements is the question of the episcopacy of Sirmium. Farlatus believed that the episcopacy existed quite early and that in fact it was founded during the fifties. Another author, Stephanus Salagius, was of the same opinion.[48] J. Zeiller, who regarded the sources about the episcopacies in the Danube region during the first century as legends, rejected the statements referring to the episcopacy.[49] At the beginning of the 20th century, Lajos Balics' work on the history of Christianity in Hungary, resulted in a book, which deals with the early period as well.[50]

Tibor Nagy produced an admirable and still useful handbook about the early Christianity in Pannonia.[51] Zoltán Kádár studied Christian art and iconography.[52] Many papers have appeared, which deal with the Christian Pannonia based on the archaeological material, sources, iconography or stratigraphy of the excavation. The scholars write about their interpretations and even reinterpretations. Their attitudes towards this subject and its comprehensions, rejections or approvals them have already been summed up.[53] However, I have to emphasize the critical summary by András Mócsy, which motivated me to continue and fulfil the work of mine.[54] His critical remarks exactly pointed out the main problems, the uncertainties, and drew up the attention the neglected questions. There are no criteria, according to which buildings, architectural elements, inscriptions and any objects or finds could clearly be identified as Christian. We do not know either, how the existence of a Christian object influenced the Christianization of the province. We cannot tell the precise chronology for the early Christian centuries as regards the temporal changes. We cannot make conclusive statements at the present stage of the research. Therefore, the task is to find the criteria, if there are any, or to prove their non-existance.

The antecedent research can summarize the Christian history of Pannonia as follows.

It was presumed that the first Christian communities were formed about the middle of the 3rd century, especially in the southern part of Pannonia. During the reign of Diocletian, some Christian communities were already known. Although the earliest names are Greek, it is still more probable that the first arriving Christians came from Aquileia[55] and Ravenna. There are no written data proving the existence of Christianity in the present region of Transdanubia. Archaeological evidence is also rare and ambiguous. Some scholars assert that there are no archaeological data on Christians in this area at all; others, however, enumerate some objects without considering that their Christian use might be uncertain. Historians believe that the Christian religion spread during the short interval between the reign of Valerianus (253-260) and beginning the persecution of Diocletian (303). This period saw the construction of buildings, tombs and burial-chapels, and the emergence of Christian communities. This opinion is widely accepted in Hungary, although the burial-chapels found in Pannonia are of a later date. Writers on Christian Pannonia have frequently neglected the investigation into the Christian or un-Christian character of the sites and finds. That is why some sites are only presumed Christian, fig. 1. To understand the location of towns and network of roads, see map, fig 2.[56] At the present stand of research, there is a lot of doubt regarding the towns or places of Episcopal sees. Apart from the work of Tibor Nagy,[57] there exist hardly any analysis or records to what is actually meant by the word bishop in this early time of Pannonia.[58] The scholars – because of disinterest or custom - use the title 'bishop' in its modern term. However, the bishop in the antiquity, and the contemporary bishop are not the same, especially not from the 1st to the 3rd centuries.[59] Henceforth the bishop meant the priest whose position was acknowledged by a certain community which showed obedience to him, and called him either episcopus or presbyteros. These two words were interchangeable. Their meaning matches to 'overseer' or 'elder'. But we meet other terms as well, e.g. teachers, ministers, evangelists. The consecration cannot be proved by archaeological

[47] Located in a Jesuit convent. Some volumes, in which he deals also with Pannonia, have already been published; FARLATUS 1751.

[48] [SZALAGYI] 1784

[49] ZEILLER 1918: 31.

[50] BALICS 1901: 3-158.

[51] NAGY T 1939.

[52] KÁDÁR 1939; KÁDÁR 1940-1941; KÁDÁR 1942; KÁDÁR 1962; KÁDÁR 1965; KÁDÁR 1968a; KÁDÁR 1968b; KÁDÁR 1969; KÁDÁR 1972; KÁDÁR/HORVÁTH/GÉFIN 1981; KÁDÁR/NÉMETH/TOMPOS 1987.

[53] GÁSPÁR 1995: 111-120.

[54] Pannonia 1990: 260-264, 371-373.

[55] A. FERRUA: Dalla Pannonia a Roma. Storia della fine del IV secolo. La civiltà catholica 83 (4) 1937: 129-140.

[56] After MÓCSY 1962: 659-660. Abb. 6.

[57] NAGY T 1939.

[58] Schmaus dealt with the question: SCHMAUS 1986: 494-496. H. Wolff considered the problem from the aspect of history: H. WOLFF 1994: 1-28; For the spread of Christianity and the Christian community see BRATOŽ 1996.

[59] LIETZMANN 1958: 141-224 detailed the question. I quote some passages here: "Die ignatianischen Briefe sind unsere älteste Quelle, die einen monarchischen Episkopat kennt. Dreigliedrig ist der Vorstand der Gemeinde: ein Episkopos steht an der Spitze, unter ihm die Kollegien der Presbyter und der Diakonen. ... Nichts darf ohne den Bischof geschehen, weder Taufe, noch Agape oder Eucharistiefeier ..." p. 175; "Die Reihenfolge der Hierarchie ist: Bischof, Presbyter, Diakon, ... Die Presbyter sitzen im Chor der Kirche, mitten unter ihnen auf einem Thron der Bischof ..." p. 177; He often quotes Ireneus regarding successio episcorporum. "... er (sc. Ireneus) könnte aufzählen *eos qui ab apostolis sunt instituti episcopi*, begnügt sich aber, die *successiones episcoporum* der römischen Kirche anzuführen, und nennt nun einen ἐπισκοπος nach dem anderen. ... Der ἐπισκοπος ist πρεσβύτερος, er gehört dem ältestenkollegium an; aber selbstverständlich ist nicht jeder Presbyter auch ἐπισκοπος." pp. 180-181.

material at all and it cannot always be traced in the written sources either. Nevertheless, the community might have known about the consecration. Therefore, I am going to refer to the persons considered bishops because of their age, authority and nomination as bishops.[60] Later, the two terms had different meanings. The Council of Sardica brought changes because it limited the activity of a bishop to a certain place.

Sámuel Timon[61] lists the following Episcopal sees in Pannonia: Sirmium, Cibalae, Carpi,[62] Curta,[63] Siscia, Brigetio, Mursa, Poetovio "et forsitan Scarbantia." Ladislaus Szörény[i] writes particularly about "sedes episcopales" - Sirmium, Mursa, Singidunum, Cibalae, Carpi, Curta, Savaria, Scarabatiensis, Siscia, Nitrensis.[64] M. Dufrèn writes that "leguntur decretis S. Cornelii Papae subscripsisse inter alios Pannoniae antistites, episcopi Sirmiensis, Cibaliensis, Carpensis, Curtensis, Agrensis,[65] Sisciensis et Brigetiensis."[66] According to Tibor Nagy, the towns Cibalae, Poetovio, Sirmium and Siscia had bishops at the turn of the 3rd and 4th centuries. There were no bishops in Pannonia in earlier periods. On the other hand, after 343 there were bishops in following places: Cibalae, Iovia, Mursa, Poetovio Singidunum, Sirmium, Siscia.[67] From the late 4th century, we can speak about 'Kirchenorganization'; at this time, a new problem arose, which is connected with the administrative boundary. The civil administrative boundaries generally were the same as the ecclesiastical ones. However, here in Pannonia, this situation was altered; after the regional distribution by Diocletian, during the reign of Constantine, Poetovio, which had been part of Pannonia, was incorporated into Noricum,[68] but at the same time Bishop Aprianus, the bishop of Poetovio, signed his name as the bishop of Pannonia.[69] As for civil administration Singidunum was situated in Moesia Superior, however, its bishop belonged sometimes to Moesia and sometimes to Pannonia.[70]

Bishops of different towns: It is obvious from the research by Rudolf Egger that Amantius was the bishop of Iovia.[71] ☞ Kapospula. Bonifacius Gams lists the bishops of Sirmium: Epaenetus (about 50), Ireneus (305), Domnus (325), Eutherius (343-344), Photinus haereticus[72] expulsus (347), Germinius (351-370), Anemius (381), Cornelius (391), Laurentius (404), Methodius (880).[73]

Stephanus Salagius knows about even more bishops in Sirmium: Iustus (503)[74] and Sebastianus (about 580).[75] The Avars had destroyed Sirmium; and Ioannes, the bishop, survived this devastation together with some Christians; he intended to rebuild the collapsed Church.[76]

The bishops of Mursa. According to Salagius, there was an Episcopal see in Mursa before the 4th century; Saint Clement founded it while he was the Bishop of Rome.[77] In 335, Arian Valens was the bishop of Mursa.[78]

The bishops of Cibalae were Donatus and Eusebius. Salagius and his predecessors are of the opinion that Donatus suffered martyrdom in the persecution of Traianus Decius and Eusebius during the reign of Diocletian.[79] Gaius, who was an Arian and appeared together with other Arian bishops like Germinius, Ursacius and Valens, is also considered to have been the bishop of Cibalae.[80]

Paulus is regarded as the bishop of Bassiana I, because his name is found among the names signed in the letter to the bishop of Sirmium. The Arian bishops, Ursacius, and Valens wrote and sent the letter to Sirmium.[81] According to Alföldi, it is accepted Bassiana I is the same as Bacensis civitas.[82]

The first bishop of Siscia is Quirinus, who suffered martyrdom in Savaria. Tibor Nagy published a contribution about the Quirinus' translatio.[83] We have some data about bishop names as Marcus, Constantius.[84] Referring to Zeiller, András Alföldi supposed Ioannes a real historical person.[85] The first bishop of Poetovio was Victorinus,[86]

[60] By the end of the 4th century, a bishop in populated towns did not lead every community. It was the time the authority of bishops started to change.

[61] According to TIMON (1735 caput X.) the places were given at the dicretion of Pope Cornelius (254).

[62] It is the same Carpi or Cirpi (Dunabogdány). PATSCH: Carpi RE III (1899) 1611; HÖRING 1950: 16, 19.

[63] The exact place is not known, PATSCH: Curta RE IV (1901) 1863: "Station der Strasse Poetovio Savaria." HÖRING mentions two Curta oppidums, one on the Danube and the other between Poetovio and Alicanum: HÖRING 1950.

[64] SZÖRÉNY[I] 1734: 25-26. Nitrensis is not ancient.

[65] Agria/Eger is Hungarian, but it was not part of Pannonia. Although this solution is obvious if we consider that Szörényi also mentions Nyitra, we have to point out that the Agram/Zagreb solution is also possible, especially if we take the fact into account that, the other listed episcopacies, except for Brigetio, are all situated in South-Pannonia.

[66] DUFRÈN, 1731: 81.

[67] NAGY T 1939: 206 writes that ecclesiastical and political administrative boundaries were different. Hence, Singidunum could belong ecclesiastically to Pannonia.

[68] MÓCSY 1974: 273.

[69] See a detailed analysis: NAGY T 1939: 205-206.

[70] NAGY T 1939: 206-207.

[71] EGGER 1962a: 57-67. In the course of identification, several places emerged: "Jovia, … known originally as Botivo, …" MÓCSY 1974: 222; "the fortification of Heténypuszta near Gölle (ancient name, Jovía)" MÓCSY 1974: 305. András ALFÖLDI 1938: 165-166. commented on the interpretation of the inscription.

[72] NAGY T 1945: 1-20, analysed the heresy in this paper.

[73] GAMS 1893: 378. Cf. 1432. note.

[74] [SZALÁGYI] III 98. In the days of Pope Symmachus there was a concilium in Rome where Iustus Syrmiensis was also present.

[75] [SZALÁGYI] III 102-105.

[76] [SZALÁGYI] III 105.

[77] [SZALÁGYI] III 113.

[78] [SZALÁGYI] III 121-150. He also writes about the events when he and Ursacius played a very important role.

[79] As a matter of fact, it was Eusebius who died as a martyr during the reign of Traianus Decius or Valerianus.

[80] [SZALÁGYI] III 158-161. The martyr Pollio is also mentioned here, but not among the bishops.

[81] [SZALÁGYI] III 164.

[82] ALFÖLDI 1938: 156.

[83] NAGY T 1946: 244-257.

[84] NAGY T 1946: 244-257.

[85] "Together with Zeiller we would like to suppose that bishops Ioannes and Constantius from Siscia (mentioned in the documents about the councils of Salona in 530 and 532) might be regarded as real historical personalities." ALFÖLDI 1938: 155.

[86] "The scholarly activity of the bishop takes place in the last quarter of the 3rd century." (NAGY T 1939: 31). See for this activity (NAGY T 1939: 31-52. See for Christians in Poetovio MÓCSY 1974: 326).

then Aprianus[87] and Marcus.[88] Tibor Nagy paid a proper attention to Victorinus.[89]

Gams regards Migdonius and Megasius as bishops of Savaria, ☞ Szombathely, however, Salagius refers to their Episcopal see as *"sedes in Illyrico incertae"*.[90] Referring to the papal decree of Cornelius, Salagius also talks about the Episcopal sees in Carpi and Curta.[91] However, the bishops are not known.

The list of bishops completed by Salagius is followed by István Gyárfás[92] followed the list of bishops detailed by Salagius; but his summary is shorter because he gives an analysis of archaeological material. M. Jarak published Christian communities and Episcopal sees.[93]

The above-mentioned scholarly works elaborated the sources, and the material data; with aid of these scholarly works, conclusions can be drawn, some question can be asked or assumptions can be put forward. They can be sometimes linked to theories; the reader will find them in detail in the chapter "Topography" at the suitable site.

Martyrs of Pannonia and the Arian religious controversy.[94] The publication of Tibor Nagy analyses both subjects in a uniquely precise way[95]. Hence, this publication is even now a comprehensive work of the research on early Christianity in Pannonia.

The following were the martyrs of Pannonia: Eusebius, the bishop of Cibalae,[96] Ireneus, the bishop of Sirmium, diaconi Fortunatus and Donatus in Singidunum or in Sirmium.[97] diaconus Demetrius in Sirmium, presbyter Montanus in Singidunum, Miximilla, the wife of Montanus. Pollio, primicerius lectorum in Cibalae. Quirinus, the bishop of Siscia, Victorinus, the bishop of Poetovio. Ursacius, tribunus lanciariorum in Sirmium. Further martyrs in Sirmium were Sabbatia, Agrippinus and Secundus, the virgin Basilla, Anastasia, the seven virgins of Sirmium, the gardener Synerotas, the presbyter Romulus. Silvanus, Venustus and Hermogenes lector were martyrs in Singidunum. The martyrs called Sancti Quattuor Coronati were stone-cavers, by the names of Castorius, Claudius, Nicostratus and Simpronianus, who were executed during the reign of Diocletian because they were not willing to cut out the statue of Aesculapius. *Translationes* created "complicated and mysterious problems the story of the four stonecutters in the hagiographical literature".[98] Their story entitled *Passio sanctorum quattuor coronatorum* came down to us in several manuscripts in Latin and also in a Greek translation from the 12th century.[99] The *Passio* contains some data about Cyril, the bishop dragged from Antiochia to Pannonia, who was kept imprisoned for three years and was flogged several times.[100] The compiler of Martyrologium Hieronymianum listed martyrs of Africa and Antiochia among the Pannonian martyrs. Tibor Nagy intended to separate them, and to range them suitably.[101]

The honour of the martyrs' relics led to *translationes* causing sometimes confusion, which can hardly be disentangled. Besides the above-mentioned stone-cavers of the Sancti Quattuor Coronati the honour of Quirinus in Rome and the honour of Demetrius in Thessaloniki was spread by the *translationes* too. In Pannonia persecutions ended in 311. Galerius issued an edict,[102] which allowed the practise of Christian religion. Thus, it was not the *constitutio Constantini* of Mediolanum, which brought changes in Pannonia, as this province stood under the control of Galerius.

The Arianism was the heaviest heresy in Pannonia. Arius, presbyter in Alexandria, preached doctrines different from the true Christian faith. The main point of the doctrines states that the Son is not of exactly the same nature as the Father, that he is only similar to him. The Son is the first creature, so there was a time when he did not exist. Bishop Alexander required Arius to withdraw his doctrines. As Arius was not willing to do so, he was banished from Alexandria. Arius fled to his fellow-student, to Eusebius of Nicomedia, who became his devoted patron. In the meantime, the doctrines were spreading and they led to the Council of Nicaea, where Domnus, the bishop of Sirmium represented Pannonia. Arius was sent into exile to Illyricum by the Council, which meant the beginning of Arianism there too. The two leading Arians in Pannonia were Ursacius and Valens, the bishops of Singidunum and Mursa. The participation of the two bishops in the Orthodox-Arian disagreement resulted in the 1st and 2nd creeds of Sirmium and the creed of Ariminium.[103] They honestly supported the eager Arian emperor Constantius II who professed not only the Arian doctrines, but also the Arian political concepts. Orthodoxy proclaimed autonomy, which meant that the emperor was just a believer in the Church, and he should be obedient to the bishop of Rome. However, Arians declared that the emperor as the ruler provided by God stood even above the Church.[104] Arianism became strong between 328 and 330 in Pannonia. The Arians gained on the councils, which were held between 328 and 330.[105]

The role of Ursacius and Valens raises the question whether the other bishops in Pannonia were also Arian or whether they adhered to orthodoxy. Some scholars are sure that the Christians in Pannonia joined the Arians. They

[87]NAGY T 1939: 205 refers to the signature of Aprianus: *Aprianus Patabione Pannoniae* illustrating the difference between civil and ecclesiastical administration.

[88] [SZALÁGYI] III 202-204.

[89] NAGY T 1939: 35-52.

[90] [SZALÁGYI] III 223.

[91] [SZALÁGYI] III 235-242.

[92] GYÁRFÁS 1889.

[93] 1994.

[94] For a detailed summary, see JARAK 1994: 158-172; JARAK 1996.

[95] First of all NAGY T 1939.

[96] Albin BALOGH does not accept him as bishop of Cibalae, 1932: 97. n. 23.

[97] NAGY T 1939: 59. Cf. VULIĆ 1931; JARAK 1994: 162-163.

[98] NAGY T 1939: 63.

[99] VULIĆ 1931: 359; BALOGH 1932: 78.

[100] VULIĆ 1931: 361.

[101] NAGY T 1939: 58-61. Cf. Cyril of Antiochia.

[102] NAGY T 1936: 32-60; NAGY T 1937: 111-142.

[103] NAGY T 1936: 32-60; NAGY T 1937: 111-142.

[104] SESTON 1947: 128-131.

[105] However at these councils intrigue rather than theology prevailed. Decisions were taken not in favour of knowledge, but at the will of the majority.

argue that Christianity arrived in Pannonia from the East and that Eastern Christianity was Arian. Moreover, the German tribes who settled here were also Arian. András Alföldi quotes Ambrosius, who thought the heresy to be the reason of the misfortune of Valeria.[106].

Jumping conclusions refused to take into consideration that Germans were not involved in a religious controversy. Their Arianism cannot be held the same as that of the polemical Arians.[107] Germans rebelled against the Romans; and their Arianism constituted the basis of their resistance.

The other heresy in Pannonia is less significant than Arianism. It was the doctrine by Photinus, the bishop of Sirmium, who denied the Holy Trinity. It was an intellectual sect. Both the Arians and the Orthodox Church regarded Photinus as their enemy, so they excommunicated him and his followers. Photinus fled from Sirmium. Later the Emperor Julian restored him to his position.[108] That is why he could strengthen his doctrines in this region. At the Council of Aquileia in 381 it was stated "that the followers of Photinus were still having their meetings in Sirmium. ... Even Pope Innocent had to use his influence with the bishop of the town (401-417) so that he should not endure the sect in Sirmium."[109]

The result of the research into the names is "Nomina et cognomina Christiana" a part of which, although unpublished, was available to me.[110] The following names belong to Pannonia: "Aelia[111], Amantius[112], Aminia[113], Anastasius in Greek[114], Artemidora[115], Aspalia[116], Aulia[117], Aulius[118], Aurelia[119] Aurelius[120] Bonosa[121]"

The 'Nomenclature' has not been ready, so it would be too early to appreciate it in any way.

In a paper, Andrea Horváth deals with the Christian inscriptions.[122] She does not follow the clear method of András Mócsy. Her own ideas have been slightly exceeded by previous research. Therefore, I only mention her paper, but disregard it in the Topography of sites and finds.

Finally, I would like to praise the new catalogue by Branka Migotti[123] who studiesd the Gnostic objects as well, which is unusual for an early Christian subject. I agree with her, but nevertheless, I am not going to follow her example because it would enlarge my catalogue. However, it is necessary to elaborate the Gnostic objects with us too. In addition, her catalogue and her previous publication, the topography[124] made my work easier in many aspects. I intended to unite both the catalogue and the topography. Topography is important because it sets out the context without which the sites and finds would hardy be interpretable.

At last, I have to point to the investigations connected with the archaeological material from the 7th century. András Alföldi emphasized that identical material was found in the cemeteries of Pécs-gyárváros and Keszthely-Fenékpuszta.[125] The finds are the disc-fibulae, which have been a subject of recent scholarly works.

[106] ALFÖLDI 1938: 151, here Ambrosius, but fide refers to 2, pp. 139-140.
[107] See the reasoning about the Arianism of the Germans in NAGY T 1961b: 591-596.
[108] NAGY T 1945: 1-20.
[109] ALFÖLDI 1938 153.
[110] Most of the results of the research into the names by András Mócsy are still unpublished. Barnabás Lőrincz and Ferenc Redő published the first volume: Onomasticon, 1994. Cf. the name Bonosa found on a ring ☞ Keszthely-Fenékpuszta. Cf. MÓCSY et al. 1983. Cf. notes 544, 1406.
[111] RIU 76 ☞ Szombathely
[112] CIL III 11206.
[113] CIL III 10232.
[114] CIL III 10611. ☞ Budapest III
[115] CIL III 10233.
[116] RIU 128 ☞ Pornóapáti
[117] CIL III 10236.
[118] As Aulia
[119] As Aminia
[120] RIU 77, 78, 79, 80, 81 ☞ Szombathely, Aurelia also occurs.
[121] AnnÉp 1968 426. ☞ Keszthely-Fenékpuszta
[122] HORVÁTH A 1997.
[123] MIGOTTI 1997.
[124] MIGOTTI 1990.
[125] ALFÖLDI 1938: 158-161.

V. TOPOGRAPHY

CONCORDANCE, FINDS LISTED IN TOPOGRAPHY, **Figs 3-4.**

1.II.a.	Balatonberény	roof tile
2.I.a.	Bátaszék	fibula
3.II.a.	Becsehely	roof tile
4.I.a.	Budapest III.	pottery fragment
4.I.b.	Budapest III.	marble fragment
4.I.c.	Budapest III.	clay lamp
4.I.d.	Budapest III.	baptistery
4.I.e.	Budapest III.	glass vessel
4.I.f.	Budapest III.	burial-chamber
4.I.g.	Budapest III.	building with apses
4.I.h.	Budapest III.	'double church' or Christian basilica
4.I.i.	Budapest III.	Jewish tombstone of Anastasius
4.I.j.	Budapest III.	Jewish tombstone Ael. Silvanus
4.II.a.	Budapest III.	so-called *cella trichora*
4.II.b.	Budapest III.	cemetery at Vihar Street
4.II.c.	Budapest III.	'cemetery chapel'
4.II.d.	Budapest III.	the ROTAS-SATOR square on a roof tile
4.II.e.	Budapest III.	tombstone, *hic quescunt*
4.II.f.	Budapest III.	tombstone of Fl. Calvena
4.II.g.	Budapest II.	fibula and grave goods
4.II.h.	Pannonia	so-called altar slab
5.II.a.	Csákvár	fibula
6.I.a.	Császár	casket mount
7.II.a.	Csopak	marble fragments of sigma-shaped table
8.I.a.	Dombóvár	roof tile
8.I.b.	Dombóvár	amulet
9.	Dunabogány	as an Episcopal see recorded
10.I.a.	Dunapataj	finger ring
11.I.a.	Dunaszekcső	pendant
11.II.a.	Dunaszekcső	fondo d'oro
12.I.a.	Dunaújváros	casket mount with Orpheus scene
12.I.b.	Dunaújváros	lid, portrayed Peter and Paul on it
12.I.c.	Dunaújváros	casket mount with biblical scenes
12.I.d.	Dunaújváros	same as 12.I.c.
12.I.e.	Dunaújváros	casket mount with relief of Apostles
12.I.f.	Dunaújváros	same as 12.I.c.
12.I.g.	Dunaújváros	casket mount with inscription ¬ESV
12.I.h.	Dunaújváros	casket mount, with the Good Shepherd
12.I.i.	Dunaújváros	silver mirror
12.I.j.	Dunaújváros	glass vessel
12.I.k.	Dunaújváros	standard lamp
12.I.l.	= 11.I.a.	
12.II.a.	Dunaújváros	so-called church
12.II.b.	Dunaújváros	fondo d'oro
12.II.c.	Dunaújváros	votive plate with dedication *deo aeterno*
12.II.d	Dunaújváros	tombstone of Malchias
12.II.e.	Dunaújváros	sarcophagus of M. Aur. Sallumas
12.II.f.	Dunaújváros	tombstone of Aurelia Baracha
13.II.a.	Egyházashollós	marble screen / baluster
14.I.a.	Esztergom	Jewish tombstone of Kassia
15.II.a.	Fejér megye(county)	building inscription
16.II.a.	Felsődörgics	carved stone plate
17.I.a.	Győr	finger ring
17.I.b.	Győr	clay lamp
18.II.a.	Kajár	inscribed roof tile

19.I.a.	Kapospula-Alsóheténypuszta	fibula (?)
19.I.b.	Kapospula-Alsóheténypuszta	hanging lamp
20.II.a.	Kecskemét	finger ring
21.I.a.	Kékkút	openwork clay brick
21.I.b.	Kékkút	basilica I.
21.I.c.	Kékkút	basilica II.
22.I.a.	Keszthely-Fenékpuszta	basilica
22.I.b.	Keszthely-Fenékpuszta	*horreum*
22.I.c.	Keszthely-Fenékpuszta	omega, suspended by chain
22.I.d.	Keszthely-Fenékpuszta	silver spoon
22.I.e.	Keszthely-Fenékpuszta	hairpin
22.I.f.	Keszthely-Fenékpuszta	casket mount, biblical scenes on it
22.II.a.	Keszthely-Fenékpuzta	so-called basilica no. I.
22.II.b.	Keszthely-Fenékpuszta	burial chamber
23.I.a.	Keszthely-Dobogó	spindle
23.II.a.	Keszthely-Dobogó	pottery vessel with incised pattern
24.II.a.	Kisdorog	roof tile
25.II.a.	Kismákfa	'paten', 'chalice', buckle (?)
26.II.a.	Kővágószőlős	so-called mausoleum
27.I.a.	Leányfalu	jug
28.II.a.	Magyaregregy	finger ring
29.I.a.	Mórichida-Kisárpás	casket mount
30.II.a	Mucsfa	roof tile
31.I.a.	Nagyberki-Szalacska	bronze lamp
32.I.a.	Pécs	13 graves
32.I.b.	Pécs	painted tomb I.
32.I.c.	Pécs	painted tomb II.
32.I.d.	Pécs	*cella septichora*
32.I.e.	Pécs	so-called mausoleum
32.I.f.	Pécs	G/4 tomb
32.I.g.	Pécs	tomb XIII.
32.I.h.	Pécs	finger ring
32.II.a.	Pécs	tomb III.
32.II.b.	Pécs	tomb IV.
32.II.c.	Pécs	tomb V.
32.II.d.	Pécs	tomb VI.
32.II.e.	Pécs	so-called tomb VII.
32.II.f.	Pécs	tomb VIII.
32.II.g.	Pécs	tomb IX.
32.II.h.	Pécs	tomb X.
32.II.i.	Pécs	double grave
32.II.j.	Pécs	gem
32.II.k.	Pécs	double grave
32.II.l.	Pécs	*cella trichora*
33.I.a.	Pilismarót	building a. (church)
33.I.b.	Pilismarót	building b.
33.I.c.	Pilismarót	building c.
34.II.a.	Pornóapáti	tombstone of Aspalia
35.II.a.	Pölöske	dagger-sheath
36.I.a.	Ságvár	casket mount with relief of Apostles
36.I.b.	Ságvár	casket mount with relief: raising of Lazarus
36.I.c.	Ságvár	fibula
36.II.a.	Ságvár	so-called cemetery chapel
36.II.b.	Ságvár	*cella* II.
36.II.c.	Ságvár	*cella* III.
36.II.d.	Ságvár	buckle
36.II.e.	Ságvár	brick with incised pattern
36.II.f.	Ságvár	same as 36.II.e.
37.II.a.	Sárisáp	so-called cemetery chapel

38.I.a	Sárszentlőrinc	finger ring
39.I.a.	Soklos	Jewish tombstone of Septimia Maria
40.I.a.	Somlóvásárhely	fibula
41.II.a-k.	Somogyszil	pottery with incensed pattern
42.II.a.	Sopron	jug
42.II.b.	Sopron	brick sherd
42.II.c.	Sopron	fragment of marble slab defined az altar table
43.I.a.	Sümeg	building I. (Christian chapel as *domus ecclesiae*)
43.II.a.	Sümeg	building II.
44.II.a.	Szabadszállás	glass cup
45.I.a.	Százhalombatta	bone tessera
46.I.a.	Székesfehérvár	pillar or mullion
47.II.a.	Szekszárd	sarcophagus
47.II.b.	Szekszárd	*vas diatretum*
48.I.a.	Szentendre	casket mount with biblical scenes
48.I.b.	Szentendre	cemetery
49.II.a.	Szentkirályszabadja-Romkút	'hall church'
50.II.a.	Szilágypuszta	bronze sheet
51.I.a.	Szombathely	tombstone of Aurr. Flavianus et Leonis
51.I.b.	Szombathely	tombstone of Aurelius Iodorus
51.I.c.	Szombathely	tombstone of Aurelius Iustina
51.I.d.	Szombathely	tombstone of Florentinus infans
51.I.e.	Szombathely	grave-inscription of *pictures pelegrini*
51.I.f.	Szombathely	tombstone of L. Maximinus
51.I.g.	Szombathely	clay lamp
51.I.h.	Szombathely	Menas ampulla
51.II.a.	Szombathely	marble screen / baluster
51.II.b.	Szombathely	marble slab (cover *loculi*)
51.II.c.	Szombathely	inscription of Aelia Kalendina
51.II.d.	Szombathely	tombstone of Aurelia Iustina
51.II.e.	Szombathely	tombstone with *carmina*
51.II.f.	Szombathely	tombstone of Namius Quintus
51.II.g.	Szombathely	fragment of tombstone
51.II.h.	Szombathely	roof tile
52.I.a.	Szőny	clay lamp
52.I.b.	Szőny	clay lamp
52.I.c.	Szőny	bronze suspension for a lamp
52.I.d.	Szőny	silver finger ring
52.I.e.	Szőny	bar Kochba coin
52.II.a.	Szőny	so-called augur' staff
52.II.b.	Szőny	inscribed roof tile
52.II.c.	Szőny	votive plate
53.II.a.	Tác	*'basilica maior'*
53.II.b.	Tác	*'basilica minor'*
53.II.c.	Tác	roof tile
54.II.a.	Táp	fragment of architrave
55.I.a.	Tata	finger ring
56.I.a.	Tihany-Sajkód	fibula
57.I.a.	Tokod	pottery sherd
57.I.b.	Tokod	pottery sherd
57.I.c.	Tokod	pottery sherd
57.I.d.	Tokod	pottery sherd
57.I.e.	Tokod	casket mount
58.I.a.	Tótvázsony	bronze tablet
59.II.a.	Ugod-Diópuszta	'church'
60.I.a.	Zalaegerszeg-Kaszaháza	bronze lamp
61.II.a.	Rusovce/Oroszvár	roof tile
62.I.a.	prov. unknown	suspension for lamp

62.I.b.	prov. unknown	suspension for lamp
62.I.c.	prov. unknown	suspension for lamp
62.I.d.	prov. unknown	clay lamp
62.I.e.	prov. unknown	clay lamp
62.I.f.	prov. unknown	clay lamp
62.II.a.	prov. unknown	finger ring
62.I.g.	prov. unknown	clay lamp
62.I.h.	prov. unknown	finger rin
62.II.b.	prov. unknown	gem
62.I.i.	prov. unknown	clay lamp
62.I.j.	prov. unknown	bust of an emperor

Abbreviations in topography

De	deep	L	length
Diam	diameter	Th	thickness
H	heigth	W	width

FINDS AND SITES

1. BALATONBERÉNY

1.II.a. Roof tile, fig. 5.

Provenance: A brick grave with a woman's skeleton. Finds: a string of glassbeads, a bracelet, a glass vessel, a glass beaker, pottery, clay lamp, an iron needle, an iron knife, remains of fruits, coins of Licinius, Maximinus Daia, Constantinus Magnus. Dr László Horváth led the excavation. He himself told me that the engraving had been placed in the grave in a way that the dead person could 'look at it'.[126] **Location:** BMK. **Literature:** SÁGI 1968b: 393. Abb. 1. Only the engraving was published on the cover of Diakonia - Evangélikus Szemle 1988 1.

Károly Sági thinks that the engraving is an early depiction of Christ's cross, but he also suggests that it can be interpreted as an ancient symbol of the Sun. However, he prefers the decision of the cross. Furthermore, he points out that the engraving is related to the bottom sign of the vessel 23.II.a. found in the cemetery in Keszthely-Dobogó. He dates the vessel to the 4th century.

Commentary: The engraving is by no means a cross. That its position was important for the dead person proves no more than its significance in the afterlife. The only question is whether the engraving actually had some Christian features. The intersection of the lines shows that the vertical line was drawn first followed by the second line of the semicircle starting from above, inclining to the right, and crossing the vertical line. Then the semicircle was drawn also starting from the above, inclining to the left, intersecting the vertical line up and down. The semicircle is also inclining to the right. In this way, the engraving cannot even be interpreted as a chi-rho drawing. The engraving itself can be interpreted as a letter Φ at best. There is a cross above the engraving on the left while below also on the left there is an X. However, the whole engraving cannot be considered Christian.[127]

2. BÁTASZÉK-KÖVESDPUSZTA

2.I.a. Fibula, fig. 6.

Provenance: A late Roman cemetery, grave 108, male. Other find: a bronze wire-ring. **Location:** WMM 83.4.46. Literature: PÉTERFI 1993: 68, 82 Taf. XV 108/1, XXIX 1-2; TÓTH E 1994a: 154, B4b type 4.

Zsuzsa Péterfi describes it as follows: 'it was made of a bronze sheet; it is enormous, massive, and hollow. Each of the three bulbs of onion closes in bronze wire and is engaged to the body of the fibula. There is a geometrical niello on the foot and on the back; the foot is chiselled with chi-rho monogram on the bottom. The whole surface of the fibula is gilt. L: 8,5 cm, W: 6,4 cm, H: 4 cm.' The author evaluates it by listing the analogies and dates it to the end of the 4th and the beginning of the 5th centuries. She thinks that the chi-rho monogram means that the user of the fibula was Christian.

I would like to add to this statement that the chi-rho monogram might also convey the meaning that its owner was a Christian soldier. ☞ Kapospula-Alsóheténypuszta

[126] I would like to thank him for this information which assisted my work.

[127] The sign may be interpreted as ΘΕΟΥ ΦΑ ΣΩΤΗΡΟΣ only with difficulties, see BARB 1953: 194-197. If we accept this interpretation, the X and the cross are depictions of stars.

3. BECSEHELY

3.II.a. Roof tile, fig.7.

Location: TGYM (there are tiles without engravings).[128] **Literature:** THOMAS 1975-1976: 105-160 Abb. 134; THOMAS 1977b: 76-77.

The tile was found in a late Roman grave. Edit Thomas stated that the engraving depicted a chi-rho monogram. Bricks from a Roman grave are preserved in the museum, but no engravings or paintings can be seen on them. Therefore, I think that the author misinterpreted damage or some other lines, cf. 51.II.h.

BONYHÁD

The provenance of a bronze chi-rho monogram is not Bonyhád; because the site is unknown, see 62.I.c.

4. BUDAPEST III. (AQUINCUM)

RESEARCH HISTORY

The regional history of Aquincum starts in the 1st century. During the 2nd century, Aquincum became a municipium. However, the municipium was organized not from the population of the canabae, but from a different settlement, which was not far away from the canabae. The leaders of the municipium were the natives who were granted the citizenship by Trajan and Hadrian. At the end of the 2nd century, the town was granted the rank of colonia by Septimus Severus. The rank also included those of a canabae because the municipium and the canabae were united legally. Territorial independence was preserved, but legally the earlier municipium and canabae counted as one town. When Diocletian's administrative reform was enforced, Aquincum became the centre of the dux of dioecesis Valeria.

During the last decade, there were several rescue excavations all over Óbuda, and even all over Budapest. Many new finds came to light, but unfortunately most of them remain unpublished. One could read about more beautiful and rare finds in guides and minor publications.[129] About the details of the excavations, short reports are published in Régészeti Füzetek. The buildings of the legionary fortress are the house of tribunus laticlavius and the schola in the praetentura [its topographical place is Pacsirtamező and Föld Streets, fig. 8/5].[130] The military bath was situated in the praetentura of the fortress [its topographical place is Flórián tér].[131] One of the canabae buildings with frescos and mosaics can be found at 14-16. Folyamőr Street (fig.8/9). In the frescos figures, plants and geometrical patterns could be seen. Fragments of stucco

[128] I wish to express gratitude to my colleague Dr László Horváth for his work and help with the identification of the bricks.

[129] Aquincum-Táborvárosi Múzeum; Aquincumi Fürdőmúzeum; Aquincum-Hercules-villa; Aquincum; SZIRMAI K: Bronzestatuetten aus Aquincum, without year; PÓCZY 1983; PÓCZY/ZSIDI; RB 1986.

[130] KOCSIS RégFüz Ser. 1. No. 33 1979, 57/I/5; KOCSIS 1989; KOCSIS 1990; KOCSIS 1991b; MADARASSY 1991a; GABLER 1991.

[131] KABA RégFüz Ser. 1. No. 33 1979 57/I/5; No. 35 1981 69/6; No. 36 1982 67/2.

were also found here.[132] The area surrounded by Árpád Fejedelem útja, Kulcsár, Serfőző, Lajos Street and Korvin Ottó tér was excavated earlier.[133]. Ágnes B.- Tóth, the leader of the excavation, found the 'building with an apse' and the 'building with a hall'[134], which were partly excavated in 1942 (fig. 8/4.) Further south on the corner of Tímár Street and Fényes Adolf Street 'the group of buildings erected in the 2nd half of the 2nd century was remarkably enlarged during the reign of Severus. There were people living in most of the buildings until the end of the 4th century.'[135] (Fig. 8/26). Further to the southwest an amphitheatre was built in Nagyszombat Street (fig.8/27). Raktár Street was also known earlier because the northern gate of the fortress was situated here and a road started here.[136] Raktár Street and its neighbourhood have been well-known places of discovery for a long time, so I quote the report on the excavations: '... A sarcophagus with a flat lid was discovered. A wall with an average height of 1 metre closely surrounded it. The wall covered a part of the sarcophagus and extended above it. The upper storey of the sepulchre or sepulchral chapel could not be reconstructed any more. There was a female skeleton in the coffin lying to the east and west. The grave goods were as follows: 3 glass vessels, 2 bone hairpins with gilt ends, 2 bone jars, 1 bone awl, 2 iron knives with a horn handle, the remains of a wooden case, 1 string of glass beads and 1 string of amber, 1 coin, 1 clay lamp. The grave dates back to the first half of the 4th century A.D. About two metres away from this sepulchre a lead coffin oriented N-S was found and above it, there was a cover made of tegulae. In the lead coffin there was a one-and-a-half-year old child's skeleton.'[137] In 1960 on the corner of Berend Street and Veder Street 'kam ... der Rest eines Gebäudes mit Apsidenabschluss und Terrazzofussboden zum Vorschein. Seine Länge betrug etwa 8 m, die Breite etwa 5,5 m. Es bestand aus einem einzigen Raum. Möglicherweise war es eine altchristliche Kapelle. Der Eingang befand sich am westlichen Ende der Nordmauer.'[138] On the corner of Raktár Street and Vihar Street some graves were found. They were constructed of tegulae, were plundered and without grave goods'... 'they might have belonged to the so-called Roman cemetery in Raktár Street.'[139] (fig. 8/14). Tibor Nagy also reports on the excavations in Raktár Street: 'At 8 Raktár Street (fig. 8/15), across the neighbourhood in the south and cutting the line of earlier László Street we opened three sections.' After excavating a 90-metre long strip of land 'the remains of a north-south, stone-paved road were found. On the west side of the road, a canal with stone walls was found. The level of the road from the 2nd century was once raised. Under the level of the earlier stone road in the north of the segments, we found a wetter sunk into the earth, with a semicircular part on the northern side. ... It can be regarded as a tannery. It dates back to the end of the 1st and to the first decades of the 2nd century which also gives the terminus a quo for the construction of the stone road.' The 'northern wall' of the legionary fortress was not found in these segments. 'To the southwest where the nursery of the housing estate of Óbuda is going to be built a 45-metre long segment ... was excavated.' A stone N-S oriented wall and traces of large fortress were found.[140] 'To the south in the neighbouring block a 17-metre long and 1,85 wide stonewall was running west- and eastwards. The wall was made of blocks of stone larger than one metre. 'At 7 Raktár Street in the yard (fig.8/16-17) ... The industrial arrangements of the canabae and its buildings with terrazzo-floor were discovered. From the 1st to the 4th centuries, five periods of construction can be differentiated here. Both intact lower periods were rich in finds.'[141] In Miklós tér (fig.8/6) the unknown part of the building called horreum III was uncovered by János Szilágyi in the praetentura of the legionary fortress. Furthermore, the remains of the repairs in the horreum were observed and the traces of the earlier fortress, which had been levelled[142]. Along Bogdáni Road, (fig. 8/10) a cemetery was found where 386 graves were opened up in 1982. This was not the complete cemetery; the rest was discovered in Kaszás dűlő (fig.8.) [Raktárrét, number 19241 in cadastral survey].[143] The cemetery of the canabae was found in Bécsi Road, from number 80 to number 269 (fig. 8/1). The graves date back to the 2nd and 3rd centuries.[144] While the heating system was being installed at 203 Bécsi Road (fig.8/18), 14 late Roman and early Christian stone slabs were discovered. Furthermore, an early Christian burial chamber was also found under the road. It was decorated with chi-rho monogram between two birds and an olive branch.[145] In the area surrounded by Kórház, Vihar, Verőfény and Szél Streets (fig.8/15), excavations uncovered traces of a settlement, which preceded the legionary fortress of the 2nd and 3rd centuries. The fortress-road running east- and westwards was also discovered.[146]

Even reports on earlier excavations mention a burial vault in Bécsi Road. At 98 Bécsi Road (fig.8/23) 'in the northern part of the ground-plot, a 0,90-1 metre wide, L-shaped wall was found. It was built of stones, which were removed. Later on, next to west-eastern part of the

---

[132] SZIRMAI RégFüz Ser. 1. No. 33 1979 57/10; NÉMETH 1991b: 93, 98. Abb. 3.

[133] SZILÁGYI 1968 Plan 1. The ala fort was uncovered in the territory surrounded by the above streets. Here it is fig.8/19.

[134] B.-TÓTH RégFüz Ser. 1. No. 34 1980, 55/2; KÉRDŐ RégFüz Ser. 1. No. 35 1981, 34; NÉMETH 1991b: 93, 99. Abb. 3. The building with hall was interpreted as a pier building, and the building with apse next to it as the bath of the ala fort. Then the two parts were merged. Its function is uncertain.

[135] NÉMETH 1991b: 93, 98. Abb. 3.

[136] ZSIDI RégFüz Ser. 1 No. 34 1980, 55/8.

[137] WELLNER, The corner of Berend Street and Raktár Street. RégFüz Ser. 1 No. 26 1972, 55.

[138] WELLNER BpR 21 1964: 329. I have quoted the short report word by word. Neither a drawing nor a detailed report has been published. Also see: SZILÁGYI 1968: 85.

[139] WELLNER RégFüz Ser. 1. No. 24 1970 37.

[140] The width of this wall is given in the report as 2.05 - 10m

[141] NAGY T RégFüz Ser. 1. No. 24 1970 34.

[142] SZIRMAI RégFüz Ser. 1. No. 37 1983 75/14.

[143] ZSIDI, RégFüz Ser. 1. No. 35. 1981 69/8.

[144] TOPÁL RégFüz Ser. 1. No. 37 1983 75/5.

[145] TOPÁL RégFüz Ser. 1. No. 39 1985 68/2.

[146] H. KÉRDŐ/MADARASSY RégFüz Ser. 1. No. 38 1984 66/12.

removed wall and a little bit above it, a wall of mortar was built. The wall could be followed for 31 metres. The corners were removed. Near the northeastern corner there was a terazzo-floored room measuring 1,98x 2,15m; however, late Roman quadrangular burial vault measuring 11.70x5,60 m was built above the southern wing. A wall divided the tomb into two different-sized parts. The western part was smaller, in which two graves made of stone slabs were found, one on the northern and the other one on the western wall. The former was completely plundered. The cover of the latter, which was carved of one single piece of stone, was broken, and the bones were swept into one of the corners. No grave goods were found in either of them. Among the stone slabs, in secondary use, a fragment of a tombstone of a cornicularius legionis IV Flaviae was also found. Along the southern wall, there was a plundered brick grave. In the larger, eastern part, next to the northern wall, a grave of stone slab stood, but it had been robbed. In order to put together this grave, the front slab of a sarcophagus from the neighbouring cemetery dated back to the 3rd century was also used. According to the inscription, Flavia Paula rested once in the sarcophagus. ... The traces of another terrazzo-floor survived in the rammed floor above the sepulchre at an average height of 40cm. The walls were decorated with painted green and red stripes. Like the buildings found in Ostia and Köln the sepulchre in Bécsi Road must have been a storied building above ground level.'[147]

Of religious beliefs, Mithras' cult, with its five mithraeums, was of great importance in the town.[148] One of them was inside the fortress.[149] On the basis of the scholarly works, there are a few finds, which refer the religious mentality and cults of other gods and goddesses, as well, e.g. the lead disc of the Danubian Rider-god,[150] and some altars with astral symbols.[151] An altar was erected for Adonis. In addition, the cults of Silvanus, Diana, Jupiter, Dionysos, Aesculapius, Hygieia, Lares, Iuno, Mars, Hercules, etc. can also be found.[152] The statue of Jupiter Teutanus [153] and the sanctuaries in the civil town all show that religion was important in Aquincum.[154] Sándor Scheiber is certain that Jews also lived here.[155] There were some Jews among the Syrian immigrants. There are also data proving that at the time of the Marcomannian war, Marcus Aurelius negotiated the support of Jews of the Danube Bend. Some of these Jews might have lived here. The word 'here' might mean the line of the Danube Bend from Aquincum to Solva[156]. The villa at Kaszás dűlő-Csikós Street (fig.8/25) should be mentioned; it was suggested that the wealthy Oriental population of Aquincum settled down and had villas built outside the town walls, but still inside the *territorium* of the town[157].

Among the published finds, there are only few data about early Christianity. I have to point out again that most of the finds are still unpublished although more accurate conclusions could be drawn from them concerning not only early Christianity but religious life in general. Fig. 8[158] is a map showing the sites of the above mentioned excavations.

Lajos Nagy, who was in 1942 of the opinion that the excavations 'finally cleared up the religious conditions of the 4th century in our town', completed a summary about Christianity in Aquincum.[159] In this summary, we can read about a cella trichora, a 'cemetery basilica' found in Vihar Street[160] in 1933. 'A private house was rebuilt in the middle of the 4th century. ... Some objects with Christian features were found in the military town." Lajos Nagy regards the inscription, which was found in the area of Goldberg Factory, because of the word *suscitabit* as a Christian find. He listed also the belt decorated with niello because of the patterns, which he interpreted as crosses.[161] In the course of a detailed analysis, these finds turned out not to have any early Christian characteristics.

Klára Póczy[162] summarized the topography of the sites and the evaluation of the uncovered buildings. Fig. 9a shows the layout of different sites along the Danube. For the identification of streets and squares mentioned as sites, it is necessary to enclose a map of 1934 because the places where the stone carvings, tombstones, statues, and miscellaneous sculptures were discovered can be seen on it. Furthermore, in many cases the sites are recorded according to their old names, which can be identified only on old maps. Fig.10.[163]

4.I.a. Pottery fragment, figs. 11, 8/3

Location: BTM 34.225 (VI). **Literature:** unpublished

It was found 3,20 m deep, in front of an oven at 177 Lajos Street. The thickness of the fragment is 1,1 cm, the length is 5,4 cm. There is a cross cut on it. The shape of the cross originally was an equilateral one, with bar 1,5 cm wide. This cross was altered into a non-equilateral cross by deepening further the vertical bar. However, it was not

147 NAGY T 1943: 374. As a matter of fact, Lajos Nagy reports on some of the ground sepulchres (1939).
148 MÓCSY 1962: 737.
149 KOCSIS 1989.
150 HAMPEL 1903a No. 59; NAGY T 1964: 51.
151 WELLNER 1976.
152 BT 1942: 386-463 (Nagy L); WELLNER 1962; WELLNER 1965; SZILÁGYI 1968: 116-120; SZIRMAI 1972; SZIRMAI 1973b.
153 NAGY T 1961a; NAGY T 1965
154 PÓCZY 1983: [19-23].
155 SCHEIBER 1960: 16.
156 PÓCZY 1989: This article is not only about Jews in Aquincum and in its environs, but also about Jews in whole of Pannonia.

157 ZSIDI 1994: 295.
158 Drawn by Bernadett Dukai. It is necessary to include here the changes of street names in Budapest occurring on the map and in the text.

previous names	present names	
Magyar Lajos	Kis Korona	Óbuda
Zamercev tér	Miklós Street	Óbuda
Korvin Ottó	Pacsirtamező	Óbuda
Korvin Ottó tér	Szentlélek tér	Óbuda

159 BT 1942: 765 (Nagy L).
160 Detailed explanation, see at the cemetery - 4.II.b. cf. fig. 8/24. NAGY L 1938: 62-64; BT 1942: 765 (NAGY L); PÓCZY 1964: 59, 74, No. 51; SZILÁGYI 1968: 121.
161 BT 1942: 765 (NAGY L).
162 PÓCZY 1964
163 KUZSINSZKY 1934 Beilage I.

deepened as symmetrically and deeply as the original Greek cross.

4.I.b. Marble fragment, fig.12.

Location: BTM 33.781(VI). **Literature**: PÓCZY 1971a: 28; Severin, Catalogue 8.37

White marble, 9,2 x 7,5 cm. It was found in Zsigmond Street. Inscription:

[....] NORA ....There is a chi-rho monogram under the inscription . It is dated to the end of the 4th century.

4.I.c. Clay lamp, fig. 13.

Location: KoJM 71.192.1. **Literature**: NAGY L 1931b: 24 Fig. 16/a; NAGY L 1934: 304; IVÁNYI 1935 Kat. Nr. 984. Taf. XL, 3; NAGY L. 1938: 66 Fig. 31; THOMAS 1977: 42.

On the discus there is a chi-rho monogram that is seen despite slight crock. Monogram is flanked by two filling holes. Rim is decorated with leaves. L: 14.4 cm, H: 3,4 cm, Diam of the disc: 8,1 cm. Lajos Nagy writes the followings about the circumstances of the discovery: 'it is an early Christian lamp from Óbuda. This find was discovered in the Museum in Szentes in the summer of 1931, during redaction of his book.'[164]

4.I.d. Baptistery on the Fürdő sziget (Spa Island), fig. 14a-b.

Literature: BÉL 1737: 12; KIRÁLY 1979: 24. After completing this topography, an article by M. Német about the buildings on the Fürdő sziget was published: A rákospataki híd és az aquincumi Dunaszakasz védelmének kérdései - Die Brücke beim Rakosbach und Fragen der Verteidigung des Donauabschnittes im Bereich von Aquincum. Emlékkonferencia 141-149. Margit Németh declares that buildings found on the island were baths. The article does not deal with the buildings, but with the bridge.

There is only one uncertain note of Mátyás Bél concerning the fact that there is a 'baptisterium' on the Fürdő sziget.[165] This uncertainty is marked by "fama est." There is no further exact data referring to it. By misunderstanding the location of Bél Mátyás, the baptistery was sought on Margit sziget (Margaret Island) and not on the Fürdő sziget. Nagy Lajos wrote the following concerning the Fürdő sziget: "Between the big fortress of Aquincum and the watch-tower along Rákospatak passing the Fürdő sziget opposite, a permanent wooden bridge provided a link. The remains of that bridge were discovered during dredging of the Danube bed in the 1870's, when engineer Gusztáv Zsigmondy drew a map of the remains of the piers."[166] This island was completely dredged during the regulation of the Danube.[167] I can show the Fürdő sziget on an old map. At that time, the present Aquincum was uncovered. However, according to the maps of Marsigli[168] the Roman sites were already known. On the map, the line of the Roman aqueduct shows the direction, which on today's map is the Szentendrei Road. The legionary fortress was situated on the right bank of the Danube opposite Contra Aquincum[169]. The small island next to the Radványi sziget is Hajógyár sziget, where the Governor's Palace was discovered.

4.I.e. Glass vessel, fig. 15.[170] fig. 8/22

Provenience: 170-172 Bécsi Road, grave 122. **Location**. BTM 86.7.270. **Literature**: IIL (TOPÁL) Kat.-Nr. 143.

Fragments of a moulded yellow glass vessel. Under its edge there is an inscription: CH...DALM. There are palm leaves and crosses above the bottom. Diam: 15.5 cm, H: 6.2 cm. It dates back to the first part of the 4th century. ☞ Dunaújváros, 12.I.j.

4.I.f. Burial-chamber, (Cf. 32.I.b.)

Literature: TOPÁL *RégFűz* Ser. 1. No. 39. 1985 68/2.

Excavated in the present Bécsi Road. The explorer completed the manuscript of the early Christian burial-chamber, which has not yet been published.[171] This is one of the very few undoubtedly early Christian finds coming from an authentic excavation.

4.I.g. Building with apses, figs. 16, 17, 18, 19, 20, 21. Fig 8/20.

Provenience: While earthwork carried out in Kiscelli Street ruins of building were discovered. In the section of Kiscelli Street between Fényes Adolf and Kálvin Street. Rescuing was done by Györgyi Parragi. **Literature:** PARRAGI 1976: 177-181; TÓTH E 1990a: 21 Fig. 9; TÓTH E 1994b: Abb.9

In the course of the rescue excavation, an apsidal building was uncovered, fig. 16. The inner diameters of the excavated apses are 380 cm, the max. width is 500 cm.[172] Fig.18a-b. There were many remains of wall paintings and stuccoes found in the level of ashes at the joining of the 'a' and 'b' walls.[173] The backgrounds of the wall paintings are white, and the surface is divided by yellow and black stripes, fig 20a-f. On other pieces, green floral patterns were found. However pink, red and black also occurred. On the black pieces, sedge can be seen as a pattern. Frescos could not be picked up completely. Stuccoes were painted. A fragment of a column painted red was also found.[174] The floor of the building was covered by gravel, where the stone cover survived on some places, fig. 19. After removing it, 5 graves came to light. Graves 1-2 had no goods; there was an iron fragment in Grave 3. There were two bone-combs, two golden rings, both decorated with

[164] Nagy L 1931b: 24.

[165] "Fama est, thermas heic olim fuisse: cuius rei fidem faciunt lacus, et haud obscura baptisterium vestigia. .... Sunt, qui partem insulae Leporariae fuisse aiunt." BÉL 1737: 12. Hungarian translation: Bél Mátyás Pest megyéről. (translated by Béla Szabó) Szentendre 1977, p. 19: the baptisterium was translated as a bathing pool.

[166] BT 1942: 760 (NAGY L).

[167] BT 1942: 358 (NAGY L).

[168] MARSIGLI 1726 Tom I. Tab. II, VII.

[169] This territory was called Transaquincum. However, the latest research has shown that this name is incorrect. Contra Aquincum is here, nevertheless the fort used to be called Contra Aquincum is actually Castellum contra montem Teutani (Castellum Contra Teutanim) - MRÁV 1996: 11-19. Hence according to the latest research Transaquincum = Contra Aquincum = fortress at Rákospatak and Contra Aquincum = Castellum contra montem Teutani = Late Roman fort on Eskü tér (Budapest V).

[170] I thank Dr Judit Topál for allowing me to publish this drawing.

[171] I wish to express my gratitude Dr Judit Topál for shoring with me the knowledge of the excavation.

[172] PARRAGI 1976: 177.

[173] PARRAGI 1976 Abb. 183 (in the publication, signs a and b cannot be seen on the walls.)

[174] PARRAGI 1976. Abb. 184-191.

stones and a pearl in Grave 4. fig. 21a-c. There was nothing in Grave 5. The gender of the dead has not been identified.[175]

The author mentions the excavation of a villa at 10 Kiscelli Street, located close to the apsidal building. The villa had several periods.[176] The first period can be dated to the second half of the 1st century, which ended in a destruction between the years 169 and 171. It was rebuilt after the Marcomannian wars. Certain rooms of the villa were richly decorated with frescos. Among others, there are also green floral patterns in the white background observed also by Györgyi Parragi in the apsidal building. The frescos in the villa depict a hunting scene with animals, for which a parallel was drawn from the Dura-Europos frescos. Analysing the decorations of that villa, Klára Póczy points out that there was a manufacture in Aquincum, whose frescos were observed in six different buildings.[177] I suppose there was a seventh building which preceded the apsidal building, or rather, it assigns an early period of the apsidal building, as we will see it in the description of Györgyi Parragi below. In the third period of the villa, the place used as a bath was painted for the third time and some rooms were repainted. These frescos are in some details similar to those found in the painted tomb II from Pécs.[178] There was also a black painted inscription on the wall, but it could not be read because of the faintness. A total destruction of the buildings is dated to 332 and is connected with the Sarmatian attack.[179] After this destruction, the building was not renovated any more but it remained inhabited. New population that came made plastered ovens, which were not only found in the vicinity, but in other provinces as well.[180] The villa at 10 Kiscelli Street supplies data for the analysis the apsidal building; there were graves inside the villa as well.[181] Furthermore, the analogy of Dura-Europos refers to the presence of Oriental people, which might explain the apses constructed later.

The inner walls of the apsidal buildings were described, as main walls whose width wS 120 cm, and the width of the apse-walls was 80 cm. The walls were built of stones bound by mortar. One can only suspect where the frescos and stuccoes were exactly placed on the wall. They might have been either in the apses or inside the hexagonal part of the building. The latter seems more probable. As for the building, the author suggested several periods. She dates the periods to the time between the 3rd and the 5th centuries without dividing them. The graves were also divided into two different times, I cite: "Part of the tombs are from the second half of the 4th century, others are from the end of the 4th century and from the 5th century."[182] As the graves with goods may be dated to the 5th century, one has to suppose that the author dates the graves without goods to the second half of the 4th, because she presupposes that the building was a burial chapel as a cemetery was discovered nearby. However, it is uncertain whether the graves date back to the 4th century.

These presuppositions might become facts as soon as the relative or absolute chronology of the finds and monuments is clearly made. The fact that there is a cemetery in the neighbourhood does not necessarily mean that the building was a burial chapel. This supposition would be acceptable if the cemetery and the building were from the same period and their purpose can be proved. However, it is rather difficult to answer either this question or the other questions because the place has not been completely excavated[183], and the author herself mentions several times that the right answer can be expected only after another excavation.

Despite the above-mentioned facts, I raise questions and then I shall attempt to answer them hypothetically. I emphasise that these answers are only hypothetical. It is very true that it is impossible to give convincing answers according to our present knowledge.

Questions are:

- What are the periods of the apsidal buildings?
- What conclusion can be drawn from the cemetery uncovered between the building and the villa?
- What conclusion can be drawn from the dated remains of the villa at 10 Kiscelli Street?

a.The dating of building periods of the apsidal building. The author of excavations supposed several phases of building. It is almost certain that the frescos and stuccoes originated from the first period. Traces of burning observed on the frescos, suggest that the building was destroyed at the end of the first period. The destruction may be dated between the years 169 and 171, when the villa in Kiscelli Street was also destroyed. The two buildings were so close to each other that it is possible to imagine the fire spreading from one to the other. Frescos with white background decorated with green floral patterns were destroyed at this time, as is evident from the remains in the villa. Another question is which parts of the buildings were destroyed in the fire and needed rebuilding, which included frescos with different types of patterns. The restored building was further destroyed in the 3rd century. Györgyi Parragi herself dates the destruction to the 3rd century, but

[175] PARRAGI 1976. The ring from the above-mentioned finds could not be seen in the publication, but, instead of the single pearl mentioned in the text, there are two pearls in the photograph.

[176] PÓCZY 1955: 41-87. I explain the periods and data of the villa according to this publication.

[177] PÓCZY 1955: 52: She lists the buildings: villa at 10 Kiscelli Street, building in Laktanya Street, thermae maiores, building at 22 Vihar Street, building at 11 Hunor Street, Governor's Palace on Hajógyár Island cf. WELLNER 1970: 382-383.

[178] PÓCZY 1955: 56.

[179] PÓCZY 1955: 58.

[180] PÓCZY 1955: 60.

[181] PÓCZY 1955: 61. I would like to remark here that cemetery chapels are often referred to in excavation documentation. Obviously, this happened because many graves were found inside the building. On the other hand, it would be a question whether we interpret these buildings as cemetery chapels just because a new population buried their dead in them. I do not think so.

[182] PARRAGI 1976: 179.

[183] We must keep in mind that this was a rescue excavation at a time when construction was being carried out in town; therefore objective factors impeded the complete discovery.

she does not think there was an earlier destruction. The hypothetical periods are as follows:

- A part of the building whose ground plan is not exactly known except for some parts of the walls, which can be seen under the apsidal building. Frescos with floral patterns belong to this.
- A hexaapsidal building decorated with frescos and stuccoes. The thickness of the wall is 120 cm. This was destroyed in the 3rd century.
- After the destruction, apses were attached to the hexaapsidal building. The part of the building with wall 40 cm wide under the second apse disappeared at that time.
- The time of the first burial: Those who buried their dead here did not seem to intend to mark the graves, since the floor was restored with gravel after the burial. According to the finds, this burial dates back to the 5th century and is not connected with the burials around the villa. Klára Póczy points out that finds of the graves around the villa do not differ from the grave goods from the other burials, dated by coins between the reigns of Constantinus II and Gratianus.[184]
- The second burial period: The two graves were marked by cairn "inside the building". This would imply that the building was not in its original use anymore. That is why there are heaps above the graves. The author is right to suppose that there were not only several building periods present, but also several ethnic groups of which the first inhabitants were the Romanised population of Pannonia.

b. Hypothetical chronology of the apsidal buildings. The time when the hexaapsidal building was built cannot be exactly defined. The only certain fact is that the last period was in the 3rd century.[185] At that time, a part of a former building might have stood near the hexaapsidal building. The remains can be seen under the second apse. As another nearby building was discovered as well, the question arises whether the hexaapsidal building was independent or a part of a larger one.

- The apses might have been built in the second half or at the end of the 3rd century. They were probably built after the destruction and during the restoration. The building was still in use in the 5th century, in the time of the first burial.
- The time of the second burial can be dated to the Migration Period when the apsidal building was not used for its original purpose.

c. Hypothetical function of the apsidal building.

Graves found in the building cannot be connected with the building itself, so they cannot be used for the dating of the construction. The cemetery is not connected to the building either, because that would imply the connection with the villa. From the 4th century on the settlement was shrinking and its place was taken by the cemeteries of Aquincum in the whole territory of the canabae. The same process may be detected here, as well. Hence, the building can be connected with burials neither of the first nor of the second period. It was not built as a cemetery chapel. Only with the help of analogies can we find out more about its function during the second period. T. Marasović knows of eight hexaapsidal buildings from Dalmatia[186]. I will present their ground plans published by him, fig. 22. The hexaapsidal building no.4 may be considered as a close analogy, since Parragi writes about the "the largest one", so we have to suppose that one apse is bigger than the others. It is probably the one opposite the entrance as it is seen on the plan of building 4, fig. 22. I have to remark that I do not mean a pentaapsidal building but that of a hexaapsidal type. Therefore, I disagree with the reconstruction published formerly in the viewpoints below:

1. According to the description of the excavation there is a apse that is larger than the rest. Presumably it is the one opposite the entrance; otherwise, the symmetry is not possible. This larger apse cannot be seen on the reconstruction.

2. The estimated size and shape of the entrance are not proportional with the apses. As there was not enough support to complete the entrance, we must use analogues. These analogues contradict the pentaapsidal solution.

According to the plan of the apsidal building, it may be considered Christian in some way, but only on the basis of Dalmatian analogies. As all of my statements are hypothetical, I have to adhere to opinion that a better result would be given by further excavations. The graves uncovered inside do not seem to be connected with the building, or not in the way as publications suggest. The period of the hexaapsidal type may be considered Christian. The analogues are churches whose patron saints are either the Virgin Mary or St Michael.

Finally, I would like to point out to the following remarks made by Margit Németh while discussing an excavation which took place to the north-east of the apsidal building: "Chronologically it is absolutely necessary to observe that late Roman glazed pottery was found in a waste pit under the remains of a monumental building from the last Roman architectural period with a wall-width of 1m. During the reconstruction of this building in the second half of the 4th century, the already existing parts from the 4th century were pulled down. Some of the late Roman buildings were used in the early Middle Ages. Other parts were pulled down and new ones were built above them."[187] Careful mapping and elaboration of the

[184] PÓCZY 1955: 66.

[185] If it was destroyed in the 3rd century; it might be connected with the Roxolan attack of 260, which brought about the destruction of other parts of the province as well.

[186] MARASOVIĆ 1984:158, pl. V. 'Until now there are 8 churches of hexaapsidal type in the pre-Romanesque architecture in Dalmatia: 3 in the vicinity of Split and 5 near Zadar. They consist of a round-domed central nave and 6 apses arranged around it. Many such churches bear the name of the same saint (St Michael, St Mary) and their cult might be the result of the victory of orthodox doctrines in the region that earlier was under the influence of heretic Arianism The most significant structure of a central type from the late Roman period in Dalmatia is Diocletian's Mausoleum...'

[187] NÉMETH 1991b: 93, 99 Abb. 4.

whole region would be necessary instead of different excavation reports.

4.I.h. 'Double church' or old Christian basilica near the eastern wall of the civil town, fig. 54 layout B, fig. 23.

Literature: NAGY L 1940: 250; BT 1942: 370, 766-768 Fig. 37 (NAGY L); NAGY T 1941: 146; SZILÁGYI 1956: 75; MÓCSY 1962: 756; PÓCZY 1964: 69; GAMBER 1968: 26-28; KÁDÁR 1969: 181; KIRÁLY 1979: 23; PÓCZY 1983: [23] fig. 16; RB 1986: 128 (PÓCZY); TÓTH E 1990a: 21 fig. 2 MÁRITY 1993: 139; TÓTH E 1994b: Abb. 2.

The circumstances of the excavation[188] are quoted from the report by Lajos Nagy:" ... the evenly rounded polygonal temple (fig. 54 A) is situated in the far east of the town. The believers could freely move in an empty area next to the temple. This Celtic type of shrine was destroyed around the middle of the 3rd century A.D. The place was levelled as we could observe when a [modern] building was built in 1927. South of it [i.e. the Celtic shrine] there is a new group of buildings [this is the double church] on the ground plan. It [double church] starts a little bit to the southwest, 72 meters away from the southwestern corner of the fence of the pagan shrine. Today [1942], it is situated next to the canteen of the gas-works in a park along the road leading north-south. Some parts of the southern building reach to the street running west and eastwards next to the sports ground of the gas-works. The ground plan shows clearly that there are three-ailed constructions divided by two rows of columns. Their orientation [direction] is remarkable: the walls run nearly north south...The room no. 1 is 19,20m long, 12m wide, divided into two and there is also an entrance-hall (narthex II) in front of the big room. A small passage leads from the entrance-hall to the big room. On the northern side of it, the traces of two columns are seen which mark the base of the destroyed columns; hence, the room was three-ailed. In the southern part of the hall III, a semi-circle made of stone can be seen, where the clergy and bishop had their seats (*subsellium*); and the stone foundation in front of them shows the place of the altar. A small capital carved of sandstone was one of the decorations in this big hall. ... The corridor IV leads to the room no. IX divided by rows of columns. Rooms nos. V and VI are parts of another three-ailed hall. The excavators marked the traces of four columns here. Room VI was separated from the big hall by a barrier/screen built upon the blocks of stone. The two big halls were surrounded by wide corridor VII, whose foundation preserved the traces of a door. In the northwestern part of the corridor the engineer marked a rounded walling well (VIII), ... around which the corridor was roughly flagged. ... It is no wonder that because of the fast building-process in the gasworks, the management of the museum could not rescue any small finds, not even the rough late Roman pottery, which has proved to be very valuable. So now, thirty years later, the ground-plan is the only way of objective stating and dating."[189] The only surviving finds are a capital and the fragment of a marble slab, figs. 24, 25, 26.[190] I do not date the two capitals in the photo to the same time. Their small dimensions suggest that they did not belong to the columns of a building, but only to a marble slab. For their evaluation ☞ Szentendre, a capital from the cella memoriae, 48.I.b.

It follows from the foregoing that the minor finds cannot be any help, only the ground plan can be used for the identification of the building. However, it is obvious from the published ground plan that the building was not excavated completely. There are no stratigraphic data, hence the phases of the building are not known. The levelling dated to the middle of the 3rd century suggests that the whole building was built at this time, but it is not certain at all. And we have absolutely no information about how long the building was in use. The description mentions the 4th century. Its identification as an early Christian basilica seems to be a rash statement; it was rather a supposition than evidence with some archaeological material.[191] Hardly any early Christian church was built in the middle of the 3rd century. The dating for our building to the 4th century is not based on facts, but rather on the theory that it was an early Christian church. I am sure it is not a Christian church, so it is disturbing to use the terminology for Christian churches in the description. The reference to the 'rough late Roman pottery' suggests that there was not only a church here. It is situated near the eastern town-wall.[192] As we have to start from the ground plan, it is necessary to find an analogy for it. The synagogue at Ostia has the most common features with it and G. Calza and B. Becatti briefly describe them in the book:

I quote: "... the Synagogue ... has been excavated at the extreme south-eastern end of the Ancient City, should take the road leading from the Piazzale of Porta Marina to the Rome-Ostia motorway. The Synagogue group of buildings is to be found about 450 yards away. ... They are all built in a mixture of *tufelli* and brick, typical of the 3rd and 4th centuries A.D. There are three entrances on the eastern side giving onto a vestibule paved in white mosaic with several symbols in black *tesserae*. The central part of the Synagogue is slightly elevated and has four large marble columns that were originally part of some other building and a floor of coloured marble tiles. The *Aula Maxima*, to the west, has a slightly apse-shaped brick shrine, which contained the Ark of the Scrolls. It is fronted by two small columns supporting marble brackets carved with the typical seven-branched candelabra, once gilded,

188 Ground plans BT 1942 fig. 37 (NAGY L).

189 NAGY L 1940: 250-252.

190 NAGY L 1940: pl. XLI/1: a marble slab, 2-3: a capital each. According to the photo. caption, the capital has the number 2, and was found in the so-called basilica. However, according to the reference in the text, the same capital has the number 3. – 250 and 252; KIS Á 1987 XV. 73. 74. 75. The dimensions of the capitals are: H: 16.5cm, Diam: 10cm. H: 11cm, Diam: 10.5cm. The marble slab: 30 x 30.5 cm; it is 4.5cm wide at the edges, at the middle it is 3.5cm.

191 In János Szilágyi's opinion, the building proves the existence of an ecclesiastical seat in Aquincum. I think this point of view is incorrect (1956: 75).

192 BT 1942: 768 (NAGY L); RB 1986: 128 (PÓCZY); PÓCZY 1994: 49.

and other Jewish symbols. Turning to the left from the vestibule, you pass into other rooms where there are an oven and a marble pastry-board which were probably used for the baking of unleavened bread. Another hall flanks the one with an apse. A flagstone road runs along the north side of the Synagogue and there is a well near the entrance. Trial excavations below the Synagogue pavement have revealed an earlier construction in *opus reticulatum* dating back to the first century and benches along the walls. A Greek inscription has also been found which mentions a *kibotós*, i.e. the Ark of the Scrolls. Everything thus seems to indicate that the Synagogue already existed in the first century A.D. and was reconstructed during the Late Empire."[193] On the evidence of the description we can say that the group of buildings in question is very similar to the building in Aquincum. L. Michael White - fig. 27 - demonstrates the ground plan of the synagogue according to the article. J. E. Stambaugh explains the location and the features of the building. According to Stambaugh, the word *synagoga* indicates in Rome not the building, but the community. In order to pray they gathered in private places "which, walled off from the surrounding neighbourhoods, served as a prayer hall, a school, and a community centre,"[194] some parts of which were also used as accommodation. This is also supported by the kitchen together with the oven in the southwestern part and the rooms in the wings. A certain rebuilding was carried out at the end of the 2nd or at the beginning of the 3rd century. The inscriptions draw the attention to the storey, which must have existed in the building in Ostia, too. Despite of all the differences some significant similarities can be observed comparing the ground plan of the synagogue in Ostia and that of the building in Aquincum. Room D in Ostia corresponds to Room I in Aquincum. Here the semi-circle identified as 'subsellium' might have been the foundation of a niche. The part called narthex is the same as $C_{1, 2, 3}$. The sizes of the parts D and C are 18.61x12.50m and those of the corresponding room in Aquincum are 19.20x 12m. Thus, their dimensions also have some common features. The orientations of the buildings are different; the one in Ostia is almost east west, while the one in Aquincum is almost north south. Part A is the vestibulum; part G is the kitchen. In Aquincum, number VII corresponds to the vestibulum in Ostia. It cannot be decided whether VIII was actually a well or not. The difference is that there is no passage connecting E and D, as there is one connecting I-II and V-VI. However, one could enter the west wing of the group of buildings from E, while one could not enter the place behind V-VI, therefore a corridor was necessary. In both places the possibility of approaching the rooms within the closed building unit was given.

Because of the similarities, we cannot talk about Christian basilica but a building belonging to the Jews, "which, walled off from the surrounding neighbourhoods, served as a prayer hall, school and a community centre." This way we can understand the existence of rough pottery. It is the gathering place of the Jewish community i.e. the *synagog* which served as a place for prayer and accommodation.

What I doubt is the date, period of the construction and the place of the cemetery belonging to the Jewish community.

4.I.i. Tombstone, figs. 28, 29.

Location: MNM inv. no 62.70.1 A copy in Zsidó Vallási és Történeti Gyűjtemény, inv. no. 64.19.64. **Literature:** CIL III 10611; FRÖHLICH 1891: 233; KRAUSZ 1904: 171; KUZSINSZKY 1908: 98; SZILÁGYI 1956: 92; SCHEIBER 1960: 16-22, Nr. 1; KANAEL 1961: 70, Fig. 57; SZILÁGYI 1968: 114; RADAN 1973: 270-272 Figs. 3-5; THOMAS 1977a: 23; BILKEI 1979: 27 no. 9; THOMAS 1980: 193; FÜLEP 1980: 42; THOMAS 1982b: 258-260; SCHEIBER 1983: 19-25 no.2; SOLIN 1983: 717 n. 287; Onomasticon 1994: 336.

The tombstone came into the possession of Count Szapáry's family, who took it to their castle on their estate in Albertirsa. They bought the tombstone at an auction in Pest in about 1830. Because of this, its find-spot is uncertain. The provenance is most probably Aquincum. Size: 181 x 76 x 28 cm. Above the relief sculpture there is a tympanum with the head of the Medusa in it, and with an acroterion at both corners each. The relief sculpture represents the busts of a couple and a child. A funerary banquet is depicted on the interlace. The inscription is framed by foliation. Inscription: *bon(a)e memoriae / Cl Maximillae / q(ui) vixit ann(os) XXV / et Dom Domnio/ni qui d(e)f(unctus) est in / R(a)etia conigui ei/us q(ui) v(ixit) ann(os) XXXVIII / Aur Urbana et In/genua sorori bene / merenti.* Translation: For a Good Memory. To Claudia Maximilla who lived 25 years and to Dom Domnio who was buried in Raetia, to his wife who lived 38 years. Aurelia Urbana and Ingenua to their sister of blessed memory.

There is a further inscription in Greek on the tombstone. The habit of this latter writing differs very from that of the Latin inscription. The Greek inscription was engraved among the representations, where there was enough place for it. Three menorahs also can be seen here, which were roughly carved into the stone later as the Greek inscription. The Greek inscription: **MHMOPIA ANECTACIO ET ΔHKOYCANI / ET BHNEIAMI ET ΦEΛEIΩ 'NOCTPΩ' EIC ΘEΩC / EIC ΘEΩC EIC ΘEΩC** (sic).

The tombstone was interpreted as a Jewish one because of the menorahs and the Greek inscription. Sándor Scheiber translated the Greek text as follows: In memory of Anastasius and Decusanes and Benjamin, our son. God is one, God is one, God is one.[195]

Kuzsinszky solved the abbreviation CL to Claudiae and mentioned an analogy to Domnio, from

[193] CALZA/BECATTI 1977: 48-49. For a detailed report on the excavation, see SQUARCIAPINO 1965: 299-315. As the first manuscript was finished in 1994, I kept the quotations referring to the synagogue and I did not adopt them from the article by White. WHITE 1997: 23-58.

[194] STAMBAUGH 1978: 599. About Synagogue, meaning and services see DEIMANN 1964 (further scholarly works is enclosed).

[195] SCHEIBER 1983: 23. His comment on the spelling of names should be pointed out here.

Salona.[196] Sámuel Krausz pointed out the wrong spelling in the word θεος. It should be omega instead of omicron.[197] George Radan was of different opinion in terms of some details. According to one opinion the two inscriptions should not be differentiated, and the busts are Jewish. According to the other opinion, and this is what George Radan thinks too, the stone was secondarily used as epitaphium, since the inscription mentions four persons while there are only three depicted.[198] George Radan thinks the fourth person is a baby who has not been given a name. The four persons are a couple, a child and a baby.[199] He considers the depiction of the menorahs apotropaic.[200] Kanael wrote the followings: "Es ist dies die einzige jüdische Bildnisgruppe auf einem Grabstein der römischen Periode, die bisher bekannt ist."

In connection with the explanations and translation, I can present the following. Menorahs and the Greek inscription were engraved later, and this was the secondary use of the stone, as George Radan wrote. First, it was not an epithapium but a real tombstone used by either Jews or others. The inscription among the busts is unusual. The inscription in Greek contains three names written in regular Greek letters, the designation our son without a name, and the set form "God is one" three times. However, the one who wrote it was thinking in Latin because even the conjunction is in Latin: instead of *και* the word *et* is used all the time. At the end of the text, 'our son' is also in Latin, but is written in Greek letters. The scholars must have been influenced by the representation - the couple and the child - on the stone because the names are regarded as the names of a couple. Some scholars agree that there is no connection between the depiction and the Greek inscription. I agree with them. The names are three male ones and the fourth name is unknown; instead of it, only "our son" is there. It is not clear from the inscription, whose son he is. He is rather the son of those who remember, but it is not clear from the inscription either who the remembering ones are. The three depictions of the menorah must have been motivated by the same idea as the repetition of 'God is one'. This gives special significance to the inscription. I would like to point out here that the set form of εiς Θεος and the menorahs might have been used not only by Jews. I think that it was most probably the tombstone of Jews who later became Christians.[201]

4.I.j. Tombstone, fig. 30.

Provenance: The cemetery of the military town. Location: AM 63.10.11. **Literature:** KUZSINSZKY 1934: 176, nr.285; BT 1942: 316 (ALFÖLDI); R.-ALFÖLDI 1956: 89-95; BIRLEY E: Promotion and Transfers in the Roman Army II. The Centurionate. CarJb 1963: 21-33; NÉMETH 1971: 24, Nr. 47. SCHEIBER 1983: 65, 67-68; FITZ 1993a: 928 Nr. 626; FITZ 1995: 1410.

It is an undecorated tombstone. The inscription is framed. *D(is) M(anibus)* is carved in the stone above the frame. I present here the inscription in the transcription by Margit Németh. *D(is) M(anibus) / Ael(io) Silvano (centurioni) leg(ionis) II Adi(utricis) / domo Syriae Palestina / colonia Capitolina stip(endiorum) LXI qui vixit / an(nos) LXXXVI. Ael(ia) Silvana / filia et heres patri / [pient]issim(mo) per Ael(ium) Filoqu[irium...]*

In English: To the Spirits of the Departed. For Aelius Silvanus, the centurio of secunda Adiutrix, from Colonia Capitolina in Palaestina of Syria, who served for 61 years and lived 86 years. Aelia Silvana, daughter and heiress of her father the very honourable together with Aelius Filoquirius.

Aelius Silvanus must have been a Jew by birth because the inscription says that he came from the town of Colonia Aelia Capitolina.[202] Aelius Silvanus served in the Roman army for 61 years, for an extremely long time. He died at the age of 86 at about the beginning of the 190s.[203] Bálint Kuzsinszky thought that Filoquirius was the son of Aelius Silvanus. Jenő Fitz agrees with András Alföldi who regards him as slave of Silvanus he brought with him from the Orient. This slave, later a *libertus*, outlived his master. Sándor Scheiber also considers him a slave, but he explains that he made the tombstone.

Some more finds were also found, but they have not been published yet. These finds give the evidence of the Jewish population in Aquincum. We known only the following: "Schließlich erwähne ich, daß im Budapest Történeti Múzeum sich ein längst in Óbuda zum Vorschein gekommener Sarkophag mit lateinischer und griechischer Inschrift befindet, den Tibor Nagy publizieren wird."[204]

4.II.a. So-called cella trichora, cf. fig. 8/12.

Provenance: Raktár Street. **Literature:** NAGY L 1931b; NAGY L 1938: 62-63; BT 1942: 470, 584-586 pl. LXXXIV. (NAGY L); SZILÁGYI 1956: 75; PÓCZY 1964: 59, 74, Nr. 53 (Nr.53 is in the wrong place on the map by Póczy); KABA 1964: 337-343; SZILÁGYI 1968: 84, 121; KÁDÁR 1969: 184; MÓCSY 1974 333; PARRAGI 1976: 178; KIRÁLY 1979: 24; FÜLEP 1980: 37; HAJNÓCZY 1987: 129-130, fig. 171; TÓTH E 1991a: 123.

Size: 9 x 9 m. There were no burials in it, only in its environs. The graves found here are partly poor and partly without grave goods. Referring to this fact Lajos Nagy declared "the existence of a Christian cemetery for the lower classes."[205] In a sarcophagus "a Roman iron sword and a bronze coin from the 4th century were found."[206] The reports on the excavation and the walls, which can still be seen, suggest that the different buildings and the cella trichora itself should be evaluated together. Lajos Nagy himself classified the walls in five periods. The trichora belongs to the last one and Barnabás Lőrincz dates it to the reign of Valentinian I. by the brick-stamp *figlina Sabiniana.*[207]

---

[196] CIL III 9575

[197] 1904: 171. Scheiber has a similar note –1983: 25.

[198] RADAN 1973: 270.

[199] RADAN 1973: 270.

[200] RADAN 1973: 272.

[201] Cf. SOLIN 1983: 704: Greek use by Jews. Cf. KRAELING 1967: 95.

[202] I would like to thank Dr Barnabás Lőrincz for consultation on the inscription and the time of service.

[203] FITZ, 1993b No. 626 928-929.

[204] BILKEI 1979: 42. Dr Tibor Nagy once mentioned that he found some Jewish finds. I suppose that he also included this sarcophagus. I would like to thank the late Tibor Nagy for his information and consultation.

[205] NAGY L 1938: 62.

[206] NAGY L 1931b: 8.

[207] LŐRINCZ 1980: 50-52.

The question is whether it is actually a cella trichora because the walls uncovered were considerably modified during the reconstruction. The photos taken in the course of the excavation bear the evidence of this modification. Today it is almost impossible to define the obvious traces and the periods of the uncovered wall. The old photos published by Lajos Nagy[208] do not demonstrate clearly the regular trichora. The apses cannot be constructed on the basis of his drawing. Fig.31. The layout to the excavation fig. 32: I quote his report on the excavation: "The uncovered walls belonged to Roman buildings of different ages. ... Walls I. and II are the earliest ones. One of the walls IV runs across above wall I; and the burnt level emphasized the temporal difference. Wall II was 0,70 m deeper than wall III and there was a burnt level above it. Beside this wall two very early South Gaulish terra sigillata pieces came to light that dated the both walls to the end of the 1st century A.D. The structure of the walls is uncommon. Wall I. was 0,3-0,35 m wide; it was built of stones with mixture of loam and earth without outer plaster. The wall IV below the point A crossed the wall I. The latter was 0,42 m deeper than the upper wall [IV]. Wall I disappeared below the terrazzo floor of the western room, which are surrounded by walls IV; and it [wall I] disappeared on the corner of wall III. Thus the room that this wall belonged to was under Hunor Street. The structure and material of wall II were similar. Walls III were more coherent. The structure of the various walls is also different. In the section marked a-b the wall was made of rough stones laid with clay, in section b-c-d-e good mortar was used, it is a so-called opus incertum; in section e-f walls are only made of loam and earth again. The walls b-c-d-e must have gone on to the south-east where there was a building whose courtyard spread on the area of the excavation. A smaller room (k.) extending to the southeast had a brick-floor. The wall e-f closed the courtyard on the northeastern side. The drain ran along the same wall also, but the other side of the drain the wall was 0,20 m high, narrow of earth and stones. The drain had a cistern west of the wall d-e and a branching towards g-h. The continuation of this wall ended at walls VI at west. The bottom of the drain was paved with flanged bricks. Stamps were found on the bricks in several cases and they showed the abbreviated name of legio II adiutrix. These walls are from the second building-period. On the cross section of the supplement I. the cultural level (marked I) was well identified; a burned layer, which showed that the building was destroyed by fire, marked the closing. Period II can exactly be dated. Brickstamps of the drain signal the building period, besides two more stamps, which were among the ruins in the drained room. The abbreviations of legio II adiutrix were in a crescent-shaped frame, and together with the stamped brick of legio X gemina. These stamps are dated to the beginning of the 2nd century, but a precise time-limit is given by X gemina stamps, since this legion was stationed in Aquincum between 102 and 107. The building did not exist for a long time. The Pannonian grey pottery, terra sigillata pieces imported from the west, and small bone and bronze objects found below the burnt layer inform us about the date of the fire: it was between 150 and 170. In the drain-boarded room at point j a fragment of an altar-stone was found. Cl. Severus erected it in honour of Silvanus. The fragment of another altar-stone was found at point i, but only its base has survived. In upper part of the burned layer we found many sherds of flanged bricks therefore we may draw the conclusion that the courtyard was partly covered. The finds of period II occur to the north of the drain walls e-f. These finds are South Gaulish terra sigillata, and that from Lezoux. A waste pit found under the western apse of the cella was also connected with this layer. Besides the sherd of the Verecundus' terra sigillata from Ittenveiler, Raetian pottery, remains of rough kitchenware, and animal bones were discovered in the pit. Walls IV belonged to a larger building. ... The room to the south had a terrazzo floor. The walls were of different width, between 0.3 and 0.60 m. The walls were made of stones placed diagonally, the so-called *opus spicatum*. The walls were once plastered. This was usually rough, but red and yellow painted pieces and marble-inlaid imitations were also discovered in the ruins. In other parts of the building, which was divided into four small rooms later built-in walls (V.) occur, whose function is not known. ... This building belongs to period III. It was once destroyed by fire because the traces of a big fire could be observed in the upper part of the level. ... Its age is estimated between the last decades of the 2nd century and the beginning of the 4th century. There were no finds, apart from a fragment of a model watchtower, and a bronze military-diploma issued by Severus Alexander in 224. ... A small clay object in the shape of a human head is also worth mentioning. ... Walls VI were the best built ones. ... We were able to excavate only two smaller rooms, and the terrazzo-floor of a bigger one. Their walls were 0.55- 0.60 m wide, in opus incertum with good mortar. Their outer part was plastered and the inside parts of the walls were painted red and in imitation of marble. The two smaller rooms were once paved with small floor tiles of sponge-cake shape. These walls are in connection with level IV and they were built in the 4th century. The stamped bricks and the scattered coins found here in 1927 also prove this dating. ... This building was destroyed around the middle of the 4th century. Then the whole area was levelled, the space between the walls was filled in with debris and stones and a new part was built above it. This new building, walls VII, survived only in its traces. ... That part of walls VI, which was not under the walls of the new building, was pulled down in a straight line, which was suitable for the lower layer of the new foundation. From the filling between walls VI, ... two sherds of glass, ... several sherds of green glazed mortar came to light. They date to the 4th century.' Some coins from the reign of Constantius II were also found in the filling. 'On layout I, close to the walls g-h a pavement can be seen *in situ*. Moreover, above this pavement, only quite late Roman objects occurred. ... Some sherds of rough vessels are ... of German taste. The latest coin is from the

[208] NAGY L 1931b figs. 3, 4, 6, 7, 8.

time of Valentinian I."[209]

With the careful reading of the report we are convinced that the excavation was not completed, it actually ended at the very beginning without making clear the function of the walls of different ages and positions.

The evaluation of the report on the excavation. Phases of construction.

Periods 1 and 2

There is a burt level dated to between 150 and 170 by Lajos Nagy. This date corresponds with the destruction, which can be observed all over the province. This destruction can be connected with the Marcomannian wars. A Silvanus altar and the base of another altar are the finds from this era; furthermore, sherds of bricks and sigillata from South Gaul and Lezoux. Lajos Nagy thinks that the waste pit found inside the trichora is in connection with this early level. There were bones and rough kitchenware in it. The walls I II III IV belong to this level, which date to the 1st and 2nd centuries. The building burnt down between 150 and 170.

Period 3

The building and wall V belong to Period 3, which lasted from the last decade of the 2nd to the beginning of the 4th century. 'Once burnt down' means that it burnt down some time in this interval. The finds are as follows: a military-diploma issued by Severus Alexander in 224, sherds of a model of a pharos[210] (fig. 33a), and a clay object shaped as a human head[211] (fig. 33b). The pharos and the head belong functionally together.[212] They have been found together in the course of a modern excavation. The fact that both objects were found together corresponds to the discovery made at the villa in Szentendre, which has been completely excavated; its phases and everything that could be acquired is known about it.[213] This analogy could imply that the building in Raktár Street was also a villa, but an unexcavated one. 'Hercules-villa' (fig. 8/11.) is not far from this place, which proves that the area was suitable for villas in the early Roman times. In later periods burials took place close to 'Hercules-villa'; these graves had no connection with the building.

Period 4, wall VI.

It dates back to the 4th century. The finds were two sherds of glass, a green glazed mortar, and coins of Constantius II. Lajos Nagy places the foundation at wall III and the section g-h in this period. On that surface, coins of Valentinian I and rough pottery sherd of "German taste" were uncovered.

Period 5.

The so-called *cella trichora* together with the wall VII belongs to this period. Based on the brick-stamps, it is dated to the reign of Valentinian I. Finds: "These few sherds of pottery prove that German tribes were settled here towards the end of the Roman Empire to strengthen the limes. Their find spots are the highest level (V) around the cella which corresponds with the age of the cella."[214] Nevertheless, we can state, that the wall of the trichora is above all the listed walls and construction periods so far, since its wall cuts every other wall. To be more precise, not the wall of the trichora but the stones put on the walls cover every other wall. Looking at fig. 34 carefully, the opus spicatum under the stones can be seen. It is also obvious that the substructure of the apses is built together with wall III and only the stone covering gives the impression that it would have been built on top of wall III. If Lajos Nagy had uncovered this apse, we might know more about the part of the building in question. Wall VI was also covered with stones, fig. 35. This wall does not end within the so-called trichora, but goes further the apse. Lajos Nagy wrote that the wall was taken out, and it is well seen. But when did it happen? The wall of the north apse was also taken out at the same time. This is also seen in fig. 35. The stone overlay must have belonged to the last period of the building. The difficulty is that too many stones were removed during the excavation. I am sure (heap of stones will be mentioned below) that the heap of stones in fig. 35 was actually piled up from the removed stones. At the first sight, the outline of a quasi trichora was satisfying enough for everybody to conclude that an early Christian cella trichora was found. Thus, the apparent contradiction in the descriptions can also be understood, i.e. according to the description, the footing close to g-h and the trichora would be from the same time. In fact, footing g-h and the wall under the stone cover are from the same period. But the level of the stone cover is from another period, namely from the fifth period. "German taste" finds, which were also found in Kiscelli Street date back to the 5th century. Hence, I myself would date the so-called trichora or, better to say, the stone cover, which covers also the wall of the trichora, to the 5th century. However, there are some unanswered questions left: 1. What area was covered with stones? What shape did this surface have? Moreover, what was it for?

My hypothesis is the following: The building itself was a part of a villa. This part has a period from the early 5th century when the walls were pulled down to a certain level, even parts of the wall were taken out, and this place was completely covered in stones leaving the existing drain. The drain was not covered, but its gap was narrowed. We can conclude that the drain remained in use and that the buildings were joined. The place covered in stone might have been made into an enclosed or almost enclosed courtyard.

Only one question is left, and that is whether the building was ever a trichora.

[209] NAGY L 1931b: 9-17.

[210] A very close analogy was found in Szentendre: TOPÁL 1994: 322. Abb. 11a.

[211] In the villa of Szentendre a similar object was found. Its function is the same as that of the object from Óbuda; it is also similar in quality. I quote the archaeologist's description of the object from Szentendre (the cited lines immediately follow the description of how the pharos was found): "Auch die auf dem kiesebestreuten Gehniveau etwas weiter gefundene Taubenfigur aus Terrakotta und die kleine, *in lacerna* gekleidete Männerbüste mit Bart dürfen Zierelemente eines solchen Türmchens gewesen sein." TOPÁL 1994: 322 Abb.13.

[212] TOPÁL 1985.

[213] TOPÁL 1994: 321-335.

[214] NAGY L 1931b: 70.

There is indeed a curved wall-section, but it is rather uncertain that it would have formed three apses. Only at the south apse, one can observe the outer edge of the wall where, however, the width of the wall and its inner curving cannot be seen because of the drain and pavement. Although the thickness of the wall cannot be measured anywhere since the wall has no edges on the inner side (and sometimes on the outer side as well). Therefore, we cannot determine the direction of the wall. In any case, it is obvious that there is no evidence that the three apses were constructed here. It is the removed stones that give the impression of a trichora (fig. 36.). It is quite possible that the building had a central arrangement, maybe a round one. In the photos published by Lajos Nagy the following can be observed: a/ There was a drain inside the only real apse to the south, which supposedly flew into the canal that runs south from the apse, joining the wall, cf. fig. 34. b/ From the above cited description we are convinced that the area covered in slabs was not restricted to this inner part of the southern apse with the drain, but that the whole area was covered in slabs which were removed and heaped up, fig. 35.[215] c/ Before laying down the slabs the floor was levelled. d/ The finds do not support the theory of an early Christian cella trichora. e/ We may consider the visible and standing reconstruction to be not only rash but also unfounded, and it has been misleading and almost paralysing further research.
For the sake of verification, I enclose all the other photos of the excavation: Figs. 37, 38, 39.

Summing up the questions and uncertainties I raise the following hypothesis: As this building is an above all wall, it was erected either at the very end of the 4th century or at the beginning of the 5th century. Germans must have used it due to the German ceramic found in it. The early Christian character of the building is uncertain.

4.II.b. Cemetery in Vihar Street, fig. 8/24.

Literature: NAGY L 1938: 62-63; NAGY L 1940: 246 pls XXXIX/1. XL/1-2; BT 1942: 470, 772 (NAGY L); PÓCZY 1964: 59, 74 nr.51; SZILÁGYI 1968: 121.

Lajos Nagy writes the following about the discoveries taken here: "We discovered sepulchres which were small and rectangular or which ended in an apse. Next to them, some graves of various orientation and made of very different materials were uncovered. Most of them were brick-graves with a flat covering, and gabled. There were also some simple earth-graves as well as traces of wooden coffins and lime which was poured over the corpse. Some graves were constructed of tombstones from the previous century. Only one sarcophagus was found, but without any decorations. ... The graves in this cemetery had very poor in finds; we found a ring with a poor glass with the chi-rho monogram in it, but unfortunately the ring could not be saved. Lamps were also found. ...'[216]
We have some excavation photos of the building thought to be a Christian chapel (Figs. 40-41). These photos show the uncovering of a villa, which had a cellar, fig. 42.[217] By no means should this building be considered Christian. All finds that we know from this building are the few pieces mentioned above. Either no other were found, or they have not been published. The contributions only repeat that the reconstruction into Christian chapel took place. It may be possible that during the excavation the reconstruction of the building was determined as this can be observed in several other cases, but these reconstructions do not necessarily point to Christianity. They only mark the periods of the buildings. I disagree with its listing among the Christian buildings because there are some clerical errors in the publications. These errors question whether it is about the same building in the different contributions. Fig.8/28 marks the place of the building in question.[218] Let us see the errors! Lajos Nagy writes: "At the intersection of Vihar Street and Szél Street a large group of buildings was uncovered. It was rebuilt several times and finally it was used as a cemetery basilica."[219] I would like to point out that Vihar Street and Szél Street are two parallel streets, so they cannot intersect each other. What is this building and where is it? According to Klára Póczy "In der Vihar-Gasse 11-13, 12-14 sind längst des römerzeitlichen Strassenkörpers mit wertvollen Wandegemälden verzierte Gebäudespuren bekannt. Von einer dieser wurde festgestellt, daß es die Reste eines Gebäudes sind, das im 4. Jahrhundert in eine christliche Kapelle umgebaut wurde."[220] According to Póczy there was an Attis statue among the finds that were discovered in Vihar Street. Bálint Kuzsinszky reported on the statue earlier: "It was found in late Roman cemetery in Vihar Street in Óbuda in 1933."[221] There is no proof of this chapel except the cited description. This description is not enough to prove the existence of the chapel. Its location (fig. 8/24) is assumed on the basis of street numbers 11-13, 12-14. Lajos Nagy mentions a lamp with chi-rho monogram on its discus.[222] Such a lamp (4.I.c.) from Óbuda really does exist, and he came across it in the museum of Szentes. However, nobody knows when, how, or why it was taken there (cf. fig.13). The lamp's exact provenance is unknown; therefore, it does not prove the existence of an early Christian basilica in Vihar Street. The question is raised in connection with the finger ring as well and it is whether the depiction was explained correctly, as the ring was found in extremely bad condition. Furthermore, I have to point out that recent excavations in Vihar Street have not proved the existence of early Christian graves and finds.[223]

---

[215] This heap is seen in the photo. by NAGY L 1931b fig. 8.

[216] NAGY L 1938: 62-63.

[217] In Balácapuszta, a villa was excavated where a cellar was found in the main building. K. PALÁGYI S: A balácai villagazdaság alaprajza az újabb megfigyelések tükrében - Grundriss der Balácaer Villenwirtschaft aufgrund neuster Beobachtungen. Balácai Közlemények I 1989: 11-34.

[218] In order to avoid misunderstanding, I mention that the building No .41 by Póczy (1964 fig. 9) is not the same as the building regarded as Christian The building 41. in fig .9 is considered to be a market-hall. PÓCZY 1964: 58, 74, No. 41.

[219] NAGY L 1938: 63-64.

[220] PÓCZY 1964: 74, No. 51

[221] KUZSINSZKY 1937: 81.

[222] BT 1942: 772 (NAGY L)

[223] SZIRMAI 1975.

BUT: Lajos Nagy has a sentence "The cemetery in Vihar Street is rich in tombs which refers to the existence of an early Christian cemetery above the ground, ...'[224] I add another record of Lajos Nagy which informs us about Syrian graveyards from Bécsi Road and Leipziger-factory. Early Christians also used these graveyards. Gusztáv Zsigmondy took photos of these graveyards in the 1860s and 1870s.[225] The clarification of the mess is obvious: the archaeologists were aware of the existence of a large early Christian cemetery above the ground, which was destroyed for unknown reasons. For leaving something from the old Christian period, they replaced the graves with buildings, which were either under the ground or perfectly fitted the town planning. We can declare that Christians were living in Aquincum and there were quite a lot of them.

Finally, it should also be pointed out that the terminology used for this particular building is very mixed: burial chapel, basilica, and chapel. These words are not synonyms. This further shows the uncertainty.

4.II.c. Early Christian church - "cemetery chapel", fig.43.

Literature: NAGY L 1940: 253-255; BT 1942: 469 769-770 (NAGY L); SZILÁGYI 1956: 75; SZILÁGYI 1968: 121; GAMBER 1968: 57; MÓCSY 1974: 314 Fig. 52; BARKÓCZI/SALAMON 1984: 174; RB 1986: 128 (PÓCZY); TÓTH E 1990a: 21, fig. 7.

'During the process of a construction a large cemetery was uncovered in the area of the Gasworks of Óbuda, near the Danube, from 1909 to 1913. Although some tombstones were continuously found here from the 1870s, this time a large, contiguous part of a cemetery was discovered. It was a late Roman cemetery. ... Most of the graves in this cemetery on the banks of the Danube were made of stone slabs. There were only few brick-graves and there are not any reports on graves found without grave-furniture. The flagged graves were constructed of earlier tombstones. ... Many sarcophagi were excavated and the ones with inscriptions or decorations were taken to the Aquincumi Múzeum. The ones without inscriptions were partly destroyed and several of them are still in the area of the Gasworks. Most of the graves have a north-south position; very few of them have an east-west position. ... The remains of a building were uncovered in the eastern part of the cemetery near the Danube. It is a chapel closing in a semi-circle and it is 11m long without, and 13 m with the apse, while it is 6.60 m wide. ... The apse faces to the east and in the middle; ... there was a square block of stone, the support for the altar. Two pillars stand out on the other side, next to the straight entrance about 2 m away from each other. ... 2 m away there is a parallel-wall next to the west side of the building that extends for 8 m. ... There is nothing left of the chapel-furniture, at least old reports do not mention any. The graves uncovered around the cemetery chapel are poor. None of the finds were taken into a museum except for a crossbow and some sherds of pottery. ... There are not any known inscriptions of Christian character or guides originating from this cemetery.'[226] Bálint Kuzsinszky points out three tombstones in this area.[227] These finds are not Christian either.

Problems arising in connection with the cemetery

We do not know its periods. According to the description, the graves do not belong to the period although the secondary use of earlier tombstones indicates a different period than the burial in sarcophagi when sarcophagi must have been primary-used. The burial in sarcophagi should be regarded as earlier in time than the graves constructed of tombstones of secondary-use.

There are only a few finds.

There are some graves without furniture, which were studied less carefully.

The facts listed so far do not have any Christian references. The author also admits that nothing Christian was found, not even an inscription. The cemetery might have been simply the cemetery of the town. It has been recorded in other towns in Pannonia that although the cemetery was outside the town-walls, it was also quite close to them and the dead were continuously buried there from the earliest to the latest times. Even if there were some Christian graves among them, they cannot be differentiated from the others because they have not been published properly.

Problems arising in connection with the 'cemetery chapel'

It is uncertain whether the chapel is in connection with the cemetery at all. And if the answer is yes which graves is it in connection with? It is on the edge of the cemetery with only some scattered graves nearby whose position is different from that of the graves in a unit together.

The building has not been excavated completely; there is a wall-extension without a function.

Nothing is known about the materials used for building (stones, bricks, both altered or something different).

There are no essential facts for the dating, which would indicate whether it originates from antiquity.

The church of Sárisáp ☞ Sárisáp, 47.II.a and the church of Ugod-Diópuszta ☞ Ugod-Diópuszta, 59.II.a are mentioned as analogies. In case of the latter one, a verified excavation showed that it was not a church, but rather an agricultural building.

The questions could be answered only by results of further excavations. As long as these questions remain, I list it among the problematic buildings.

4.II.d. Inscribed roof tile, fig. 44.

Provenance: Aquincum, fresco Governor's Palace. **Location:** BTM inv. no R2044. **Literature:** SZILÁGYI 1954: 305-310; MOELLER 1973 pl IV; HOFMANN H 1978: 481-482 with further literature. BARTON 1975: 20-22.

It was found while the Governor's Palace was being excavated. It is stamped by legio II adiutrix. János Szilágyi

[224] NAGY L 1938: 128.

[225] NAGY L 1939: 118 No 11: He says that the unpulished photos. of Gusztáv Zsigmondy are kept in Aquincuimi Múzeum. KAISER / PÓCZY 1985 comment on fig.7: Also at Országos Műemléki Felügyelőség there are some unpublished records made by Gusztáv Zsidmondy.

[226] NAGY L 1940: 253-255.

[227] KUZSINSZKY 1934 No. 59. (p.182), No. 61 (p. 188), No. 344 (p. 201).

reads the inscriptions as follows ROMA TIBI SUBITA, and the magic square ROTAS / OPERA / TENET / AREPO / SATOR. This kind of inscription is very rare in the whole Empire. It has already been connected with Mithras' cult, other mystery-religions and Christianity because the same letters in the same way can be interpreted as the text of PATER NOSTER.[228]

The other side of the brick has no decoration or inscription. Hofmann writes the following about the specimen from Aquincum: 'It contains not only the square SATOR, but also another palindroma, Sid. Apoll.'s epist. IX 14,4 poem: *Roma tibi subito motibus ibit amor*. The habit of the two inscriptions are different, hence they were written by two different people.

Barton thinks that this kind of magic square was used as a Christian cryptogramm only from the 5th or 6th century.[229] According to Guarducci both Christians and pagans used it. We suppose the Christian use if the wording does not run in the regular, but in reverse order. Probably, Christians began thinking that the word *Sator* meant 'Creator' in the magic square *Rotas Opera Tenet Arepo Sator*. This might have been the reason why *Sator* was put to the first place just as it can be observed in other Christian examples - Sator Arepo Tenet Rotas.[230] The inscription starts with *Sator* on a mosaic floor.[231]

The square could be interpreted in a Christian way ,[232] however, as the inscription on this brick starts with *Rotas*, the Christian use is uncertain.

4.II.e. Tombstone, fig. 45.

Provenance: The cemetery south of the military town (the area of the Goldberg factory). **Location:** MNM R-D 162. **Literature:** CIL III 3551; ILCV 3476; PAUR 1859: 221-222. IXa; GYÁRFÁS 1889: 87; NAGY L 1931b: 26; NAGY L 1938: 64 fig. 28; BT 1942: 470 (NAGY L); NAGY T 1944-1945: 268 note 14; SZILÁGYI 1956: 92; MÓCSY 1962: 757; SZILÁGYI 1968: 120.

Size: H. 50 cm, L. 180 cm. Lajos Nagy reads the inscription as follows *Hic quescunt duas matres, duas filias - numero tres facunt (sic!) - et duenas II parvolas qui suscitabit cuius condicio est. Iul. Herculanus vet f(aciendum) c(uravit) in re sua.* Diehl thinks that it is early Christian because of *suscitabit cuius condicio*. A. Mócsy agrees with him. Tibor Nagy suggests that 'advenas' should be read instead of 'duenas', which is possible because of the ligature A-D. Moreover, he writes that this piece is a simple prismatic shaped stone, however it was known as a fragment of a sarcophagus before. Scholars agree that *quescunt* and *facunt* were written instead of *quiescunt* and *faciunt*.

Commentary: The interpretation quescunt = quiescunt = 'here they lay' is correct, but it is difficult to say who the subject is because *duas* and *filias* can only be accusative; although *matres* can also be a nominative, the word *duas* next to it demands an accusative interpretation. So, we have to regard these nouns - in spite of all the accusative forms - as subjects, i.e. we face a linguistic simplification.[233] Therefore, the translation is: two mothers and two daughters lay here ...

This phrase "*numero tres facunt* (*faciunt*)"= "they are three in number" shows that it is not a common funerary inscription, which must refer to *'duas matres, duas filias'*. I think it is a riddle saying that although they are three in number, nevertheless there are two mothers and two daughters at the same time. I shall attempt to solve the riddle saying that a grandmother, a mother and a daughter are in question. The mother is the one who is a mother as well as the daughter of the grandmother, so it is true that although there are two mothers and two daughters in question, they are three in number. Two little ones are also mentioned. The word *quescunt* refers to them, as well.[234] If we accept the suggestion of Tibor Nagy and interpret *duenas* as *advenas,* the attribute of the two little ones is a newcomer, not yet born. But, this orthography as quescunt or facunt is archaism; and *duenas* is also an archaism. I think so. Among the inscriptions of the republic period there is an inscription *duenos.* [235] With the aid of this word (*duenos),* we can interpret the *duenas* as *bona,* i.e. "two good little ones".

The text *qui suscitabit cuius condicio est:* The pronoun *'qui'* might refer to *numerus* because this is the only masculine word in the whole text. So the word *numerus* = number is the subject of the verb *suscitabit.* The translation of this part: the number will reveal the different condition of each person.

In English the whole: Two mothers and two daughters lay here. They are three in number; and two good little ones. The number will reveal the different condition of each person. Iulius Herculanus had it made at his own expense.

If we want to interpret the inscription as Christian, then the pronoun *qui* refers to a person who is not mentioned in the inscription at all. In addition, the orthography *conditio* would be more desirable instead of *condicio.*[236] In this cases the translation: [The One] who [created them] will resurrect [them]. So its Christian interpretation is doubtful.

4.II.f. Tombstone, Figs. 46-47.

Location: It was located in the enormous park of the Károlyi Castle in Fót in 1944. In spite of this fact, I could not find the stone there; though nobody knows about its removal to another place. This garden is the Károlyi István Gyermekközpont (István Károlyi Children's Center) today. I wish to express my gratitude to the Director, László Csáki, who helped my search. **Literature:** CIL III 13382; KUZSINSZKY 1908: 105-106 fig. 44; NAGY T 1944-1945: 266-282 pl. XCI. 2; AnnÉp 1947: 38;

[228] MOELLER 1973: 1-4; HOFMANN H 1978: 513-517; MERRIFIELD 1987: 142-143.
[229] BARTON 1975: 21.
[230] GURADUCCI 1978: 1765.
[231] MOELLER 1973 pl. V. It is the church of Pieva Terzagni near Cremona dating from the 11th century.
While constructing the square, certain requirements had to be observed, BARTA 1989.
[232] While constructing the square, certain requirements had to be observed, BARTA 1989.

[233] Cf. n. 1393.
[234] If *numero tres facunt* refers also to the 'little ones', then they should be embryos (unborn).
[235] CIL [2]I P 2, No.4. p.374: According to the comment on the inscription *duenos = bonus*. I would like to thank László Pintér for calling my attention to this inscription.
[236] Cf. BLAISE 1954: 192 condicio, 193 conditio.

SZILÁGYI 1968: 120; EDH-Nr.: HD019373

Limestone, H: 168 cm W: 54 cm Th: 20 cm. Reading of Tibor Nagy: Domo securitati[s] / Fl(avia) Cal[v]ena nata / Ulmo vix(it) an(nos) XXII / lanxit d(ies) XXVII posuit / memor(iam) Aur(elius) Propincus [5]/ mil(es) leg(ionis) II ad(iutricis) p(iae) f(idelis) coniug(i) / carissim(a)e Fl(avia) Cara / mater dulcissimae / fil(iae) pietatis eius (causa) / f(aciendum) c(uraverunt) [10]/ Φ(ως) I X ligatured Z(ωή). The inscription was translated by Tibor Nagy and in English it is as follows: For Flavia Calvena, who was born in Ulmus and died at the age of 22 after suffering for 27 days, (the tombstone was) put up by her husband Aurelius Propincus, a soldier of the legio II adiutrix and by her mother Flavia Clara by this fulfilling the pietas." Tibor Nagy identifies the place of birth with Ulmus situated on the way between Sirmium and Cibalae. Concerning the shape of the stone, he states that it is a cippus, which was used again at the end of the 3rd and the beginning of the 4th centuries.[237] As opposed to earlier opinions he considers the tombstone to be Christian because of the letters Φ and Z. He has completed the letters as φως and ζωη. The ligature of X and I (i-chi) can be seen between the two letters. According to Tibor Nagy the cross or i-chi flanked by Φ and Z cannot date back to an earlier time than the 6th century in the West.[238] Hence, its occurrence on this stone is earliest with us. This sign is apotropaic as are also some others.[239] Although the expression *domo securitatis* = "the condition without troubles" is used by pagans, the idea is not foreign to the Christian way of thinking because it refers to the salvation of the dead person. Considering all these arguments Tibor Nagy regards this find as early Christian.

The tombstone was listed among Christian finds only because of the φως ζωή and the i-chi flanked by them. The funerary inscription itself is pagan. As the tombstone dates back to the 3rd century we have to think that the φως ζωή dates back to the 3rd century, as well. The habit of the inscription is the same as that of the φως ζωή, hence the possibility of a later carving is out of question. Margarita Garducci dates the earliest use of this mark to the 3rd century and she writes that pagans used it in the same way as Christians in later times.[240] The worship of LIGHT and LIFE was much known in the pagan Rome. I do not think that the stone is Christian.

4.II.g. Fibula and grave goods, Figs. 48, 49, 50.

Provenance: Budapest II, in some publications the find-spot is Budaújlak, but this is wrong. It was found in District II near Pasaréti road. The report of Flóris Rómer: a double grave on area of Henrik Drasche's brick factory; its layout is published on page 103.[241] Aladár Radnóti identified the find-spot on ground of the MNM Records.[242] **Location:** MNM inv. no. 10.1951.227. **Literature:** RÓMER 1864: 101-124; HAMPEL 1886: I. 65; HAMPEL 1894: 24-25; HEKLER 1910; NAGY L 1931b: 29 fig. 19 the set of belts; PEIRCE/TYLER 1926 pl. 64; NAGY L 1938: 64-65 fig. 29: the photo of the grave, fig. 30: the set of belts; BT 1942: 765 (NAGY L); RADNÓTI 1955: 26-27, fig. 12; SZILÁGYI 1956: 92; SZILÁGYI 1968: 120; TÓTH E 1980a: 154 Abb. 14: Here the find-spot unknown, and belongs to the Ráth collection. BURGER 1984; TÓTH E 1986a; FISCHER 1988: 176-177 Abb. 4.

"Among lots of grave goods of a late Roman double-grave a silver belt came to light in 1863 on the area of a brick-factory. Filled in niello, three parts of the belt are gilded. Their decorations are: in the corners, in small medallions there are St Andrews' crosses, in the centre a rosette with eight petals. These motives are very common in the early Christian art. The last of the 42 coins found in the grave dates to the reign of Maximianus"[243], fig. 48. Lajos Nagy published these finds partly in 1942, but the whole of the grave goods in 1938. We can see the fibula here, in the latter publication. The fibula was separated somehow from the grave goods and inventoried later with note "unknown find-spot", fig. 49. Endre Tóth declared this object as an early Christian fibula. Lajos Nagy has published the drawings of the finds, and the layout of the grave in 1938, fig. 50. Although the fibula is not easily seen on the layout of the grave, it can be discerned on the shoulder of the skeleton. Alice Burger, who also supplied the plan of the find-spot, later published the same layout.[244] Consequently, we can state without doubt, that the fibula was found together with the belt in the same grave. Thomas Fischer dated the finds and reconstructed their use: "Alte besprochene Beschlagsätze von silbernen Offiziersgürteln gehören der Mitte bis zweiten Hälfte des 3. Jh an."[245]

The decoration on the pieces consists of motives that are so common that on the basis of crosses we cannot consider them to be Christian objects; and I do not regard them as Christian. The function as well as the dating exclude Christian characters.

4.II.h. So-called altar slab, fig. 51a, b.

Location: AM R917. **Literature:** THOMAS 1978b: 573-587 Abb. 17-18; TÓTH E 1980b: 99; Severin, Katalog 8.10-13 (the pieces from Gorsium and Intercisa, too); THOMAS 1988: 288 Abb.6; NAGY Mi 1988 Abb. 7; UBL 1994: 130 n. 4.

There is a number of stone finds from Pannonia that Edit Thomas regarded as early Christian altar slabs.[246] This point of view was widespread for a long time, although the archaeologists had the impression that this identification was not right. Some scholars agreed with her opinion.[247] Therefore, reconstructions were made and exhibited in Tác and in the MNM. As this opinion was widespread, a neglected refutation by Mihály Nagy must be repeated here.[248] 'The monopodiums or console tables have been identified scientifically in different ways; These finds have a cut of a semi- or a three-quarter-circle shape in the middle of the opposite edge to side on the wall, and resting on a hypo basis or on decorated legs; they were identified as early Christian relic altars dating back to the time between the 4th and 6th centuries. On the basis of parallels,

[237] NAGY T 1944-1945: 267-268.
[238] There were no earlier examples before the article was published.
[239] NAGY T 1944-1945: 271-272.
[240] GURADUCCI 1978: 1749.
[241] RÓMER 1864.
[242] RADNÓTI 1955: 34 note 164.
[243] BT 1942: 765 (NAGY L).
[244] BURGER 1984: 86.
[245] FISCHER 1988: 180.
[246] THOMAS 1978b: 573-587.
[247] TÓTH E 1980b.
[248] NAGY Mi 1988: 135-157.

their function and dating should be corrected. ... We date the Pannonian monopodium slabs to the period between the middle of the 2nd and the middle of the 3rd centuries on the basis of the datable parallels and their decorations"[249] The citation is just a short extraction of the long arguments which have proved that this kind of finds are monopodiums and not early Christian altar slabs.

Similar monopodiums were uncovered not only in Aquincum, but also in Dunaújváros and Tác, but I am not going to list them here. It is enough to present the above one in order to understand the non-Christian use of this kind of objects.

HISTORICAL EVALUATION - AQUINCUM

Hypotheses about Christianity in Aquincum. It might seem strange that I dare put forward some hypotheses about the early Christianity of any town, but hypotheses are always supported by research. The hypotheses are based on the one hand by the archaeological observations, on the other hand by the general legal, constitutional, and economic aspects common in the coexistence of paganism and Christianity.

Until 313, Christianity was *religio illicita* and its practice might have been punished. This fact is shown by the "persecutions". As a defence, a good solution was the *interpretatio Christiana*, which allowed a formal use of the depictions of "pagan" symbols, buildings and objects. This is the first aspect. The silence and the secrecy were obligatory as regards the teaching, and the existence of a Christian community. Moreover, the objects used by the community (the objects used in liturgy, some sacred books, etc.) could not be delivered. This is the second aspect. Of course, the Christians also used "pagan" things, e.g. vessels, potteries, furniture etc. However, the secrecy resulted in occasional use of paraphrases, or of misleading speach, and the representation of their modest pictures and symbols took ambiguous forms. This behaviour proves to be particularly true in "He that hath ears to hear let him hear" (Mt 11, 15). This is the third aspect. Now, when the task is to find evidence for Christianity in Pannonia, I am not doing anything else, but trying to "see" and "hear". That is why I dare to put forward some hypotheses.

Finds, as 4.I.e. and 4.I.f. (fig. 8/18) prove the existence of Christians; but they lived not in large numbers either before or after the change brought by Constantine. After the change of Constantine, the Christians might have lived in the area of the former canabae or even in the area of the former fortress. Evidence for the existence of the Christians are the burials; some burials have been found, some burials are known, and so a church or at least a building serving liturgical purposes should be found. This could be a house-church. Such buildings have not been discovered yet; or we could not recognize them. In such cases, it becomes obvious that the "pagan" surface does not allow the inner Christian character. At the same time, it can be assumed that Christians lived in the civil town and in the environs of the fortress even before the change of Constantine. The presence of Jews and Syrians can be proved; so, this fact makes my assumption possible. Romans regarded Christianity only as a Jewish sect; hence, we have to look for the Christians near the Jewish communities. This kind of coexistence was on the one hand useful and on the other hand compulsory. It was useful because the Jewish religion was *religio licita*.[250] It was compulsory because Jews were ordered to settle with the believers of eastern religions in the same place in Rome. Augustus divided the followers of different religions into different regions[251] and this example was followed in the towns of the provinces. In Rome, the area of the Jews was district XIV.[252] Fig. 8 might help us to "see" the circumstances in Aquincum. Bécsi Road could be the easternmost border of Christian burials.[253] The area surrounded by Vörösvári Road and Szentendrei Road is the site of buildings, streets and workshops, which earlier belonged to the military town. In the later period, as the area of the town became smaller the dead were buried here as well. However, these graves cannot be dated earlier than to the end of the 3rd century. None of the finds, which were found here, refers to Christian burials. So, this area can be considered as a neutral "pagan" zone either as a dwelling place or as a building space. North of this area, starting at Bogdáni Road (fig. 8/10), an excavated but as yet unpublished cemetery was found, that might surprise us all. I base this statement on a publication of a villa (fig. 8/25) by Paula Zsidi.[254] The author writes: "Zuletzt bleibt eine der interessantesten Fragen - wem gehörten die Villenanlagen, bzw. wer hat das Gut bewirtschaftet."[255] The author's answer is based on the inscriptions found. She points out that the municipal aristocracy owned the villa.[256] The villa was built during the last decade of the 2nd century; and at that time, the ordo decurionum in the municipium of Aquincum consisted mainly of the local population with some members, who originated from the East.[257] As the existence of a Jewish community can be supposed in the town, but there are not any data of their burials (at least none have been found so far) their graves will certainly be found sooner or later. Jews must have lived in Aquincum.

The synagogue[258] was situated on the southeastern outskirts of the town. This means that in the municipium of Aquincum this district was the appointed place for worship and dwelling of eastern people. However, this part of the town has not been excavated at a rate that would offer any

[249] NAGY Mi 1988: 158.

[250] BERWIG 1970: 4-24. Tertullianus also reports that Christians could take shelter under the shadow of Jews. - TERTULL. apol 21.

[251] SUET. Aug 21; DIO Cass. 55, 8.

[252] SAUER J: Rom. LThK VIII 1936 350. Former district XIV is in the Vatican territory.

[253] The fact that the pottery fragment 4.I.a. (fig. 8/3) was not found in this area does not contradict this assumption; sherds are easily transportable.

[254] ZSIDI 1994: 292-295.

[255] ZSIDI 1994: 295.

[256] ZSIDI 1994: 295.

[257] SOPRONI 1994: 313.

[258] It is true that only a building period in the 3rd century is known. The antecedent may be assumed on the basis that Jews, together with Orientals, also came to Aquincum in the second half of the 2nd century.

further information on the houses. Considering this situation, I suppose that some Jewish owners lived outside the walls, on the *territorium municipii.* When they passed through the town in the south-easterly direction, they left the town at the west gate, and soon reached the villa-settlement. They buried their dead in the cemetery to the south (i.e. at the cemetery in Bogdáni-Road, cf. fig. 8/10). Of course, I do not suggest that the whole cemetery was Jewish, but rather that Jews buried their dead here as well, and that maybe a part of the cemetery belonged to them.

A group of finds should be mentioned here, which came to light at 23. Szél Street (site fig. 8/21.). While digging for the foundation of a modern building, a substantional find was revealed "by diggers at a depth of two metres, on the terrazzo-floor of a Roman building."[259] It was a closed iron find of 42 pieces. Lajos Nagy published them as a carpenter's tools, fig. 52. Among them were some tools used also by gravediggers. To make a comparison I enclose a drawing by P. Testini where the tools of fossors are demonstrated, fig. 53. Although the find does not allow considerable conclusions, we can still suppose that they were used for digging graves as well. In addition, the tools were found in the block labelled neutrally as "pagan". The owner was either *fossor* or *custos coemeterii.* It is remarkable that the place of discovery is quite close to the edge of the cemetery where Jewish graves should be found, and, at the same time, to the site of the Christian burials as well. Thus, it is possible to imagine that the owner's profession was *fossor* or *artifex*[260], or even *custos coemeterii.*

My other task was to find the traces of Christians' life. If Christian burials were found, at least some kind of liturgical place must accompany them. There must be one, however modest; not only in the military town, but in the municipium as well.

Excavations elsewhere showed and proved an existence of a house-church or a church situated near a mithraeum, e.g. in Poetovio[261] and Rome[262]. Therefore, traces of Christianity should be looked for and found in the neighbourhood of the mithraeums in Aquincum, as well. Regarding the number of mithraeums found, Poetovio and Aquincum were similar because there were several mithraeums in both towns: three in Poetovio and five in the municipium of Aquincum and one in the fortress. This parallel assures us that Christians should be looked for near the mithraeums and maybe near the synagogue either in the municipium (fig. 54) or in the military town of Aquincum.

Let us consider the mithraeums described by Tibor Nagy.[263] The first one flourished on the outskirts of the municipium during the reign of Antoninus Pius and was destroyed at the time of Marcus Aurelius. The next one was built in the centre of the municipium at the beginning of the 3rd century. This was the M. Ant. Victorinus mithraeum. This "ceased serving aims of the cult in the second half of the century." At the same time as the M. Ant. Victorinus mithraeum, another mithraeum was erected on the outskirts of the town, whose the well-known worshippers were among the freed slaves of C. Iulius Victorinus. ... Considering the circumstances of the discovery it is more possible that it ... was set up in a private house." We do not know how long it was in use. The fourth mithraeum is the Symphorus mithraeum, which was built during the time of the Severi and existed until the end of the 4th century. Of the fifth mithraeum, only altars and relief-pieces remain.[264] Examining the mithraeums enables us to gain useful information, such as that some citizens of Aquincum could not only afford to free their slaves, but they did so as well. Further, it was possible for the shrine to be inside a private house, just as the case of the mithraeum inside the legionary fortress proves.[265] In order to give a more precise picture of the mithraeums I present the results of the latest research on M. Ant. Victorinus. Based on the interpretation of several inscriptions, Sándor Soproni writes that Victorianus, the founder of the mithraeum, must have been of eastern origin. He is the same person, as M. Antonius Victorinus, who was the duumvir of colonia Aquincum. His wife, Ulpia Spectata was a worshipper of Terra Mater. M. Antonius Victorinus can be said to have been the owner of the villa, which has been excavated.[266] A coin hoard hidden near the villa was also uncovered and it can also be connected with M. Antonius Victorinus or with his descendents. On the basis of the inscriptions Sándor Soproni could identify the owners of several villas in the Pest County; all of these owners had an official status and they were also Mithras worshippers.[267] M. Antonius Victorinus is particularly important for us because he was probably of eastern origin, has built villas for his family, founded mithraeums and was a member of the *ordo decurionum.* András Mócsy studied the Syrian settlers whose career is represented by M. Antonius Victorinus very well.[268] It is almost certain that M. Ant. Victorinus was of Syrian origin and his example can demonstrate the situation of the Jews as well.[269] His example suggests that Jews had villas in the territory of Aquincum and they buried their dead near the villa.☞Kővágószőlős, ☞ Pécs.

259 NAGY L 1937a: 155.

260 ☞ Rusovce

261 See the general plan of Ptuj. The Christian church was not far from the mithraeum. MIKL-CURK 1978.

262 VERMASEREN M. J. /ESSEN C. C: The Excavations in the Mithraeum of the church of Santa Prisca in Rome. London 1965.

263 BT 1942: 433 (NAGY T) the two quotations are also from this page.

264 BT 1942: 434 (NAGY T).

265 KOCSIS 1990: 709-714; KOCSIS 1991b: 141-153.

266 Fig. 9b / 12: Kamaraerdő.

267 SOPRONI 1994: 312-420.

268 Detailed explanation, see ☞ Dunaújváros, casket-mount, 12.I.c.

269 Earlier research also studied those Jews dwelling in Aquincum. - MÓCSY 1974: 228.

The Christianity of the municipium and the military town through the looking glass of mithraeums - still about hypotheses. Cf. Fig. 54.

The two most important mithraeums are those of M. Ant. Victorinus and Symphorus.[270] I present my suppositions about Christians only on the basis of the ground plan of the town. Most of the small finds remain unpublished, so even if there were any Christian references among them, they are unknown. On the plan of the town (fig. 54), the buildings excavated in the municipium are shown without chronological differentiation. Only a short paper informs us about the chronology of the buildings.[271] The buildings were rebuilt at the beginning of the 4th century, which was connected with the Sarmatian wars during the 330s. Then in the 370s, the town was demolished.[272] The paper points out that the plans of the municipium cannot be used to determine the function of the buildings, and does not show the architectural circumstances during the whole of the Roman period.

The first settlements are dated to the end of the 1st century.[273] In the 2nd century, the town-wall was built surrounding the area today known as the territory of the town.[274] András Mócsy thinks that except for the *macellum* and two public baths there is not enough evidence to suggest any other public buildings.[275] For details of the excavated buildings, maps should be consulted[276]: the first large public bath is situated at the crossing of streets C-D. The *macellum* was situated to the south. The shrine of Fortuna Augusta was discovered between the two. Eastwards, following street E, at the meeting point of the streets E-F to the south of street E and to the east of street F, the other public bath was placed. Further eastwards, along street E a dwelling-house and next to it to the north the M. Ant. Victorinus mithraeum was found. On the south side of street E a group of buildings which was probably a part of the so-called large house[277] with a peristylium was found. The "bathroom" of the so-called large house remained intact. The mosaic of its apodyterium depicts wrestling. The house built on a slope was storied, decorated with mosaics and stuccoes. It was probably constructed at the end of the 2nd century.[278] The so-called large house bordered with street G in the south. South of street G the Symphorus mithraeum, together with the dwelling-house, can be found. The area surrounded by streets D-E-F was populated by craftsmen, where workshops were built and rebuilt several times. Finally, some adobe houses were built on top of the potters' settlements at the end of the 4th and at the beginning of the 5th centuries.[279] At the crossing of streets D-E, in a house, an altar dedicated to Epona was found.[280] In connection with the houses, Lajos Nagy stated that they did not have any regularity in their layout.[281] The area bordered by streets A-C was identified as Forum with a shrine. The *basilica* stands east of street A and northeast of the *basilica* were the shops.[282] Aquincum was granted the rank of municipium by Hadrian and it was a colonia from 194. When Trajan divided Pannonia into Inferior and Superior, Aquincum became the capital of Pannonia Inferior.

Some problems arose in connection with the identification of the *basilica* and the neighbouring buildings, so Tibor Nagy re-excavated the spot.[283] He could state that the so-called basilica had four periods. Earlier, there were workshops and tabernae in the place; it was "an industrial and commercial centre of the municipium."[284] The structural development at the turn of the 2nd and 3rd centuries was in connection with obtaining the rank of colonia. Now the newly erected building matched "basilical aims".[285] With it, the space called palaestra formed a unity, which became the centre of the official religion from the 3rd century. The Forum should be looked for here.[286]

Two statements from the above should be emphasized. One is by Lajos Nagy, i.e. that the houses did not have any kind of common ground plan and the other by András Mócsy who cannot support public features of the buildings except for the three of them. On the contrary, I myself would ask whether a private building has been found at all in the excavated area. If the lack of the common organization is due to the fact that the map shows buildings of all periods together, the obviously organized workshops inside the walls do not make sense. In addition, Lajos Nagy writes the following: "... A burnt, life-sized Aesculapius statue carved of fine marble was found under the modern road almost in the direction of the macellum. It might originate from a shrine of the deity. This is more than likely because of the large bath nearby."[287] In the *insula* system[288] by Erzsébet Márity there is a north-south orientated "religious" *insula* (fig.55) including the unexcavated surface right to the east of the present building

[270] PÓCZY 1991: 45-56. She describes the objects found in the mithraeum in details. Abb. 7. gives the plan and the periods of the mithraeum. The author of the article points out that the shrine could be approached only through the house in which the mithraeum itself was built. (The article assumes the existence of a mint here. I do not want to comment on this question.)
[271] MÁRITY 1993: 137-145.
[272] MÁRITY 1993: 140.
[273] MÓCSY 1974: 161.
[274] MÓCSY 1974: 161.
[275] "... Apart from shrines, there is no definite evidence of other public buildings." MÓCSY 1974: 161.
[276] The detailed map drawing by Katalin Zborovszky was published in a guide, "Aquincum".
[277] Latest research called the rooms of the "large house" by different names. The place of exhibition is called the gathering room; the peristylium became the courtyard of the centre of collegium iuventutis; the bath remained a bath.
[278] BT 1942: 375 (NAGY L).
[279] MÁRITY 1993: 139.
[280] PÓCZY 1994: 16.
[281] BT 1942: 375 (NAGY L).
[282] Latest excavations have confirmed the existence of shops.
[283] NAGY T 1964: 9-54.
[284] NAGY T 1964: 50.
[285] NAGY T 1964: 50. (To avoid misunderstandings it should be pointed out here that in this case, the word basilica is not used in the Christian sense, but it denotes the place of jurisdiction in town.)
[286] NAGY T 1964: 51. The archaeologist observed that the buildings in this territory extended under the present road and might extend further under the line of HÉV.
[287] BT 1942: 377. (NAGY L)
[288] MÁRITY 1993: 148, fig. 3: North of the east-west main road the *insula* a-b-c form a unity, and south of the main road the *insula* j-k-l-m does so, too.

of the museum. Furthermore, a certain religious variety can also be observed. These circumstances would have enabled Christians to live here and to remain hidden at the same time. Consequently, my hypothesis is: the Christians did live here, but not in large numbers. Therefore, we cannot speak about a well-achieved Kirchenorganisation, but we should expect a Hausgemeinde or Hausgemeinden. In spite of their presence, we cannot expect to find any archaeological bequest of the community from a time earlier than the end of the 3rd century or rather the beginning of the 4th century. Hence, we cannot find any archaeological evidence for their presence. At the same time we cannot deny their presence. The Apostles were charged with a commission to teach. We have to consider it because they discharged and handed down this commission. Those who became Christians by one means or another also taught.[289] A stranger who came here intervened in the life in Pannonia, and if he were a Christian, he would be certain to leave at least one Christian behind. The Hausgemeinde was formed of them.

Their dwelling on the "religious" *insula* is shown in fig.56. The particular place where I suppose they lived is marked by ✠.

a/. The environs of the M. Antonius Victorinus mithraeum (Figs. 57, 58), where I think of the close neighbourhood of the mithraeum and the unexcavated area to the east of the museum. The excavation of this area might prove or refute my hypothesis. I suppose there were Christians living in this part from the 2nd century on. I also suppose that Christians appeared around the Symphorus mithraeum during the 2nd century. From the end of the 1st century, Christians might have lived north of Street G on the edge of the excavated area and also east of it. For my arguments see ☞ Dunaújváros, 12.I.c. To decide the question it would be essential to uncover the spot east of the excavated area.

b/. I suppose that the bath in the "large house" was used for baptism after rebuilding. ☞ Szombathely. This place is situated almost in the middle between the two mithraeums.

c/. Christianity can also be presumed in the area where the basilica and the palaestra were located as revealed by earlier research, or where the forum was placed. Christians settled in this area simultaneously with the rebuilding or almost at the same time, rather during the reign of Severus Alexander than in the time of Septimus Severus. Alexander Severus was known for his tolerance and respect for Christians.[290] The lead disc found in this area "is the relic of the cult of Danubian Rider-god"[291]. The disc proves that not only the official religion took place here, but also the unofficial ones. In addition, the place became a religious centre after 313; hence, Christians were likely to appear here. They might have lived around the "*basilica*" and the shops from the 3rd century on, but near the forum only from the 4th century.

The military town and the fortress. The new rescue excavations have brought to light another part of the inscription, fragment of which was found at 167 Lajos Street, in 1937. According to the interpretation of this inscription, the construction of the Danube frontier began at the time of Vespasian.[292] We can read about the explanation of the whole inscription (both the former and new fragments) in a paper by Margit Németh.[293] She deals with the fort of *ala I Tungrorum Frontoniana* stationed in Aquincum. This *ala* is mentioned in the inscription. Németh identifies the fort of *ala* with the uncovered castellum. The general plan intended to demonstrate the conditions of the 1st century; nevertheless, it can be used to show the space where later Christians lived. Fig. 59 demonstrates the condition of the legionary fortress in the 2nd and 3rd centuries[294]. The excavation lasting several years changed the ideas about the fortress significantly; it proved that the *principia* were "arranged like a forum".[295] The "*signa*-shrine"[296] was found in the west wing of the *principia*. In the northeastern wing of the fortress, a building was found with walls decoated by frescos and with mosaic pavement. In the first period, the dado was painted imitation of marble. In the second period, the dado was divided into rectangular panels alternatively with garlands and birds.[297] "Zum Gebäude gehörten noch mehrere Höfe und ein kürzerer Korridor mit einem mit rot-weißen Rhomben verzierten Mosaikfußboden. Der Badetrakt ist jünger. Östlich des mit Swastikamusters, Seerosen und einem schwarzweißen Peltamotiv ausgestatteten Raumes kam ein kleiner Raum mit einem Mosaikfußboden zum Vorschein, der mit einer Teppichimitation verziert war."[298] As it is not my task, I am not going to deal in detail with all the works and articles about the stuccoes, frescos and carvings from the fortress and its environs. Although some of them were mentioned,[299] it is obvious that their evaluation has not yet been completed. This urges me to be more careful while devising my hypotheses. It seems contradiction intending to propound hypotheses of Christian dwelling-places and churches, when the civilian

[289] Just how seriously the conscientious duty to teach must be taken, may be experienced (even by ourselves) when meeting an aggressive and impatient "missionary". There are historical examples for this as well.

[290] LONOVICS 1865: 6: "Cum Christiani quendam locum, qui publicus fuerat, occupasset, contra popinarii dicerent, sibi eum deberi, rescripsit, melius esse, ut quomodocunque illic Deus colatur, quam popinariis dedatur." Lonovics quotes Lampradius, Alex. Sev. c. 49. But the Emperor's attitude is also known from other examples,.

[291] NAGY T 1964: 51.

[292] TÓTH E/VÉKONY G: Beiträge zu Pannoniens Geschichte im Zeitalter des Vespasianus. ActaArchHung 22 1970: 133-161.

[293] NÉMETH/KÉRDŐ 1986: 384-388.

[294] NÉMETH 1991b Abb. 2; NÉMETH 1991a. I have to point out here that there is another interpretation of the ground plan of the fortress, i.e. that the tetrapylon cannot belong to the fortress of the 2nd and 3rd centuries. As a conclusion of the arguments, István Wellner wrote the following: So sind also die im 4/5. Jh. entstandenen Bauten im Zentrum des Legionslagers aus dem 2/3. Jh. wahrscheinlich mit der im Nibelungenlied erwähnten Etzelburg zu identifizieren." WELLNER 1990: 719.

[295] SZIRMAI 1991b: 107.

[296] SZIRMAI 1991b: 108.

[297] SZIRMAI 1991c.

[298] SZIRMAI 1991c: 205.

[299] NAGY T 1967; NAGY T 1976; SZIRMAI 1984b; SZIRMAI 1985b; SZIRMAI 1986a; SZIRMAI 1986b; SZIRMAI 1988a; SZIRMAI 1989; SZIRMAI 1991b.

life generally, and other religious relations have not been evaluated at all. Hence, it is more difficult to do so here, in the military town than in the *municipium*. Looking at the wealthy finds, the question arises whether the *ara provinciae Pannoniae Inferioris* resided not in Tác, but in Aquincum as was earlier suggested by András Mócsy.[300] This is only one example among the uncertainties and unresolved questions. If there were fewer problems, I could present my hypotheses with more certain. In brief, one can say that representative finds, which were uncovered in the fortress and in the *thermae maiores*, have not been properly published and evaluated. Furthermore, also the well-achieved life (mosaics, frescos, stuccos, etc.) in the military town shoul be regarded when examining the fortress. The finds in the fortress ought to be even more representative, for instance, in the bath, even gold mosaics or similar decorations might be imagined. Of course, the elaboration of all the finds will answer these questions.

***Tribuni laticlavi* house in the fortress.** Inside the legionary fortress, in this house, a mithraeum was found.[301] The frescos, amphorae, sigillata and animal bones have been published.[302] László Kocsis made a comprehensive evaluation, whose important points are as follows:

- The late Roman fortress was built in the 4th century.
- The bath of this house can be seen on the north side of the ground plan, fig 60.
- After the Marcomannian wars the house was reconstructed.
- The mithraeum was built at the beginning of the 3rd century.
- The mithraeum was used for 50 years, and then it was walled up "which was never pulled down again in Roman times."[303]
- The house was altered. A bath with several rooms was built in the north of the courtyard.[304]
- Some rooms of this house were used throughout the 4th century.
- The population moved into the territory of the late Roman fortress and graves appear in the other areas.

Let me return to one part of the introduction, namely, to the connection or disconnection of antiquity and Migration Period. The excavation reports generally give evidence of burials among buildings, the shrinking of the town; and the population moved in the fortress. The reason for this must be looked for in the restlessness of the 4th century caused not only by the migration of strange tribes, but also by the repeating attacks.[305] Although the era can be called antiquity, the symptoms of the Migration Period can be observed; this fact is a very important factor from the point of view of Christianity, because this factor shows the different onset of Christianization between Pannonia and other provinces. In order to make more precise what I mean I refer to the "axioma" mentioned after the change of Constantine, when Christianity became *religio licita*. After the change, churches were built and the expansion of the Christianity can be observed in several ways. This seems to have been slightly modified in Pannonia. On the one hand, the changes brought freedom for Christians in Pannonia as well; on the other hand, the repeated Barbarian attacks meant constant danger. Hence, the people were more careful than during the persecutions by the Romans. Therefore, churches were not built, instead old buildings were rebuilt and the already existing ones were used. Despite of the migration of the provincial population, some of them went on staying. Pannonia never become absolutely deserted. Consequently, we cannot expect many Christian finds from this era, i.e. from the 4th century.

Hypotheses for the fortress and the military town. Just like in the case of the municipium, I suggest that the Christian centre, ie the house-church was near the mithraeum. Most probably the part of the house of the *tribuni laticlavii* which was lived in through the whole of the fourth century was rebuilt; in fig.60 it is marked by ✠. I think this building was very modest; it consisted of two rooms. In the larger one, the Eucharistic Offering was celebrated; the other one served for different purposes, as agape, for teaching of catechumens, and for vigilance. On the edge of the military town along the present Bécsi Road, the Christian burial (4.I.f.) was uncovered. This assures us that Christians lived here and thus they must have had a church, at least a house-church; and they were baptized in the bath of the tribunus house. ☞ Szombathely ☞ the municipium of Aquincum.

Summary – Aquincum. The fact that Christians lived in Aquincum can be proved only on the basis of the graves. Their burials have been proved only in the cemetery belonging to the canabae. Christians buried their dead here so they had to live here; the places marked by cross on figs. 8, 56 and 60 show where church and dwelling-places might have been; the analogies and other data helped to gain this result, which is still only a hypothesis. Christians appeared in Aquincum from the 2nd century on. They were not present in large numbers; that is why they formed only "Hausgemeinden". In the municipium, the house of assembly might have been situated in the space between the two mithraeums, and in the unexcavated area to the east of the museum. The other house of assembly might have been around the forum and *basilica*. Here, Christians could live only from the reign of Severus Alexander on; and from the 4th century, their numbers were increasing. In the military town and the fortress, they could live from the 3rd century on. Their house-church could have been inside the house of tribuni laticlavii, in the neighbourhood of the former mithraeum. Christian dwelling-places cannot be identified on the basis of the published material. The building called cella trichora could be regarded as a part of a large building (maybe a villa), which was rebuilt at the end of the 4th century. Even its connection with Christians is doubtful. The "cemetery basilica" mentioned in many places is no more than a hypothesis at a working-stage.

---

[300] Further information, see Tác.
[301] RB 1986: 215-219; KOCSIS 1989 with previous scholarly works; KOCSIS 1991a; KOCSIS 1991b.
[302] MADARASSY 1991a; MADARASSY 1991b; VÖRÖS 1991; BEZECZKY 1991; GABLER 1991.
[303] KOCSIS 1991b: 118.
[304] KOCSIS 1991b: 142.
[305] MÓCSY 1974: 266-296.

This term was used without any ground. In a similar way a cemetery basilica was mentioned in Vihar Street. However, the uncovered building turned out not to be a cemetery basilica; what is more there was no basilica at all on this position. The other "cemetery basilica" was suggested on the border of the municipium, which was excavated but was not totally uncovered; and so its dating and its function cannot be determined. Christians might have used the hexaapsidal building, not as a cemetery basilica, but as a church. The question is which population used it and when. To solve these problems Dalmatian analogies are not enough; further excavations are necessary.

There might have been a synagogue on the outskirts of the municipium, and the Jews might have lived in the municipium. They were not Jewish Christians.[306] The tombstone with the secondary use was an epitaph, 4.I.j. The Jewish centurio from the Colonia Aelia Capitolina can be connected with the fortress. His relations who were also Jewish must have lived in the canabae. So, Jews might have been living here and they probably buried [307] their dead not in a special Jewish cemetery, but in a part of the common cemetery used by everyone. It is difficult to give a precise chronology of the synagogue because no phases were observed. Other data suggest that the Jews started arriving at the beginning of the reign of Marcus Aurelius; and they built a synagogue during the reign of Septimus Severus or in the middle of the 3rd century the latest. At the stage of the present knowledge, it is impossible to tell whether the building, which was uncovered, differs from the previous synagogue or not. We do not know either how long it was used as a synagogue. One of the capitals found here allows us to regard it as a later period. Thus, the building was used for a long period of time. The other question is whether it remained a synagogue or was altered into a Christian church.

5. CSÁKVÁR (FLORIANA)

Jenő Fitz wrote a detailed history of excavation and its archaeological results.[308] I will give an abstract of it. Arnold Marosi started the excavations in Csákvár in 1929 in Templom Street. Some brick graves were found. In 1930, further graves were uncovered. The excavation was carried out in Kiss, Magas and Cigány Streets. Further excavations took place along Petőfi and Ostrom Streets and in Cigány Street in 1932 and 1937. In 1960, Jenő Fitz started the excavation in the area where Marosi's excavation was in 1930. In 1962, there was an excavation in Szent Mihály Square, where only graves were found,[309] fig. 61. Jenő Fitz wrote recently an evaluation of the history of Csákvár.[310] He observed that most of the coins found in the cemetery are from the 4th and 5th centuries. Nádorfi Gabriella is the leader of present excavations and she was so kind as to show me both the finds and the site.[311] Based on these I can state that Christians might have lived in the area, though neither the objects published so far or the objects considered being Christian support this.

5.II.a. Fibula, fig.62.

Provenance: grave 69. **Location:** IKM. **Literature:** SALAMON/BARKÓCZI 1970: 56 Abb.11; THOMAS 1980: 200; TÓTH E 1994a: 150. B4a/6.

A crossbow fibula with three knobs decorated with portraits in niello technique. Edit Thomas thought the fibula was Christian. Certainly, this type of fibula usually has a chi-rho monogram. However, this one lacks it. One cannot suppose the Christian nature of the fibula barely from its type. Hence, I list it among the non-Christian objects.

It is dated to the second half of the 4th century.

6. CSÁSZÁR

6.I.a. Casket-mount, figs. 63-64.

Provenance: grave 1. **Location:** MNM 20.1902.2-3. **Literature:** DINKLER-VON SCHUBERT 1980: 145; GÁSPÁR 1986 Kat.-Nr. 349.

The bronze sheet decorated with Christian and mythological scenes. The reconstructed dimensions of the face side of the casket are 28 x 26cm. Christian scenes framed with medallions were located vertically on the casket. The mythological scenes in aedicula were placed horizontally on the casket. The Christian scenes are as follows: the Good Shepherd, Daniel among the lions, *maiestas Domini*, and Abraham's sacrifice. The spaces beside the medallions are decorated with dolphins and stars. In a small elliptic frame, two human beings facing each other were depicted and in two other, similar frames the *dextrarum iunctio.* Saturnus, Sol, Luna, Mars, Mercurius, Jupiter and Venus, the gods of planets or gods of days were also depicted on the casket-mount. Each deity was set in an aedicula terminating in an arch. Further depictions on the casket-mount are some Bacchic scenes with each figure in an aedicula. Finally, also a frieze of animals belongs to it. For Christian and mythological scenes on the same object see Erika Dinkler-von Schubert's opinion and for a detailed explanation see Dunaújváros, a casket-mount, 12.I.c-d.

7. CSOPAK

7.II.a. Marble fragments of sigma-shaped table.

Location: MNM 54.20.1. **Literature:** THOMAS 1954: 74-76 Taf. XIX; THOMAS 1955: 261-276; KITZINGER 1960: 22 note 9; THOMAS 1964: 28-32. Abb. 11; MRT 2 17/4; KÁDÁR 1969: 189; MÓCSY 1974: 336; Severin, Katalog 8.9 - with a wrong Inv. No.

'In the ground-plot of J. Németh, a clergyman, pottery sherds and marble pieces were found in 1936.'[312] For the general plan of the site, see fig. 65. 'A larger piece of the

[306] Apart from this, we cannot deny that there were Jewish-Christians among the Christians.

[307] This hypothesis includes the cemetery of Pécs, where Jews buried their dead in a separate part of the cemetery. ☞ Pécs, Historical Evaluation

[308] FITZ 1961-1962. For the former literature see p.153 note 16.

[309] Ágnes Salamon and László Barkóczi wrote a detailed report on the excavations. (1970: 35-37).

[310] FITZ 1991-1992.

[311] I would like to thank Gabriella Nádorfi for her kindness and helpfulness.

[312] MRT 2, p.81.

same table was taken to MNM in 1851. According to the documentary archive inventory, this piece was found by J. Kósa among the ruins of caldarium, on the bank of Lake Balaton near Savanyúvíz forrás.'[313] Two of the fragments are straight, the third one is bent. The only one dimension does not refer to the fragments; that is the width of the rim: 11cm. The sunken central area is without decoration, but the rim is decorated in relief; a man is shooting a lion with an arrow. He has only a gown. In the swing of the movement, his gown cannot cover his nakedness. On the two other pieces, some running ungulate (maybe an antelope) can be seen. It is obvious that the lion is hunting. This scene can be repeated in different variations. The rim is framed by mouldings, inside by a single undecorated moulding outside by two – one undecorated and one astragal, fig. 66.

Pannonia research published these marble slabs as fragments of a marble table and as a relic of Arianism. It is a table without doubt; but its being early Christian or even Arian is highly questionable. The circumstances of discovery or the topic of depiction could give clues for answering this question. Edit Thomas lists the known marble sigma-shaped tables, some of which clearly depict Christian scenes.[314] Considering the depiction of the Last Supper in Codex Rossano from the 6th century or the Lord's Supper in the mosaic of Sant'Apollinare Nuovo of Ravenna[315] where the sigma-shaped table can be seen, we are sure that Christians used some of these tables. However, on the one hand it does not mean that all Christians used such tables.[316] It is not always easy to decide whether the fragments of such tables are in connection with Christianity. The researchers working in Hama encountered the same problem when they found some fragments of a sigma-shaped table, fig. 67a-b.[317] I would like to study the marble table of Hama in detail because it is the closest analogy to the ones found in Csopak. I quote its description and the problems with it from the book of Gundhild Ploug: "Fragmented, sigma-shaped table top. ... Measurements: 1.08m (total length), 1.055m (total width), 0.055m (height), 0.11m (width of rim). The top has a sunken central area framed by a flat rim. The central area, which is without decoration, is entirely broken into small pieces, some now missing. ... The underside of the top is chiselled smooth. The rim is decorated in relief and framed by narrow mouldings, inside by a single undecorated moulding outside by two - one undecorated and one astragal. The reliefs depict animal sequences between which are interpolated two anthropomorphic busts - one in each of the two right - angled corners. The animal friezes are not one unbroken sequence but a collection of small groups representing attack and chase scenes, including one antithetic group. ... Of the two busts seen in profile that to the left is male, that to the right female. The male bust, which turns right, is shorthaired, beardless and wears a diadem divided into panels with openwork ornamentation. He wears a chiton and himation with fibula. The female figure, Dea Fortuna, turns left, has a 'puffed' coiffure, wears a mural crown and is clad in a mantle with a string of beads round the neck".[318] The reasoning about the analogies of the marble table in Hama is as follows: Similar slabs were made from the 2nd century onwards and they were still made in the 5th century. The chronological distinction is primarily based on styles. The fragments of Dumbarton Oaks can be dated to 390. A craftsman who worked for the palace of the Emperor Theodosius made it. Moreover, since this table is in connection with the one from Hama, both friezes might have originated from the east. "Although the frieze on our piece (on Hama) has no obvious Christian content, its late date at the end of the 4th cent. AD, as well as the circumstances that a member of the Theodosian Imperial House might be represented on its frieze, indicate that its context was not exclusively pagan."[319]

The comparison of the table of Hama and Csopak. We can observe similarities in style and in dimensions. The width of the decorated rim is 11cm in both cases. Neither of them illustrates explicitly Christian motives. However, both of them illustrate a lion hunt, and human figures; one illustrates some portraits and the other a hunter. The rim is framed by an undecorated moulding inside, while outside an undecorated moulding and an astragal of the same outline. These identical features indicate that the provenance of table in Csopak was also in the East. This possibility is supported by the fact that several Pannonian towns had some connections with the East. These are Sremska Mitrovica ☞ Budapest III. ☞Dunaújváros ☞Pécs ☞Szombathely ☞Szőny ☞ Tokod to mention just the most important of them. Considering the moulding of the astragal with the same outline on both tables, they might have been made in the same workshop. This suggests that they must be dated to the same era, i.e. to the very end or the last decade of the 4th century.

We have difficulties with the historical evaluation of the fragments found in Csopak because no excavations were carried out in the area.[320] Therefore, the building, in which this table was used, is not known. Until excavations reveal the true context of the find, I consider the question open. I would not make the slight concession in case of Csopak that Gundhild Ploug does in case of Hama when she writes "its context was not exclusively pagan", because this is not true in our case. Nevertheless, we can say that in an excavation a convenient representative building may be found. Confirming this viewpoint I quote the Topográfia: "Kőkoporsósdomb was a significant dwelling-place in Roman times for three or four centuries, the early and late cemeteries give evidence of it."[321] The latest research by

[313] MRT 2, p.81.
[314] THOMAS 1955.
[315] Atlas 1959 nos 426-427.
[316] For the use of this type of marble slab, see KITZINGER 1960, especially p. 31 note 58.
[317] PLOUG 1985: 199 fig. 47 a-b.

[318] PLOUG 1985: 199. The reconstruction fig 48.
[319] PLOUG 1985: 201.
[320] Except for the rescue excavation, which found nothing connected with the fragments of this marble table. For short reports on the rescue excavation, see Régészeti Füzetek.
[321] MRT 2 p. 82.

Sylvia Palágyi supports the same. She studied the stone carvings and among others also those from Csopak.[322]

The sigma-shaped tables or the round tables belong to secular life. But their use in the house-church gradually resulted in their functioning as an altar.[323] There is an example for this in Pannonia, in Donnerskirchen,[324] fig.68a. Depictions of sigma-tables can be found in the art of the Middle Ages, as well, fig 69.

8. DOMBÓVÁR

The locations of the finds connected with Dombóvár are not without problems. I adumbrate the difficulties after a summary by István Torma.[325] According to an entry in the documentary archive inventory the site of the brick is Újdombóvár. Aladár Kovács led an excavation here. Újdombóvár is a so-called administrative large village, which was set up in 1886 and consisted of 28 pusztas, 6 farmsteads and 4 farmhouses. After a division of the theoretical large village, the pusztas belonged to different villages. In the territory of the theoretical large village, Vasutastelep was developing from 1918 on in the southeast of present Dombóvár. Vasutastelep has always been called Újdombóvár unofficially. Hence, the same name refers to two different territories in different eras. So the place excavated by Aladár Kovács from 1902 to 1930 cannot be precisely identified. However, the spot of the early excavation and the site of the amulet do not seem to be the same.

8.I.a. Roof tile, fig. 70.

Location: WMM R49.933.1. **Literature:** NAGY L 1938: 48. Archaeology 1980: 200 pl. CXLIX; Severin Katalog 8.52; IIL (PÉTERFI) Kat.-Nr. 316.

Roof tile with chi-rho monogram used to cover a grave. H: 43 cm, W: 27 cm, Th: 6 cm.

[322] PALÁGYI 1996: 108-114, 118-119, Abb.5, 7.

[323] GAMBER 1968 86-93. Here Gamber also deals with the depictions of the sigma-table.

[324] GROLLER 1911 67; NAGY L 1938 97; BARB 1953a; NOLL 1954 73-75; THOMAS 1964 130-137. fig.67: general ground plan, fig.71: a marble slab; BARB 1964 18 Abb. 1a; UBL 1982 303; Severin, Katalog 8.8. Groller excavated the area in the 1910s. Several buildings surrounded by walls were found. In the course of the excavations some marble slabs were discovered. Lajos Nagy regarded building 3, excavated by Groller, as an early Christian church. It is 17.8 m long and 9.5 m wide. It is a rectangular building divided into two. L. Nagy thought that one area was the narthex. The marble slabs were found in this building. A. Barb writes in connection with Donnerskirchen: '...ich möchte betonen: 1. Wenn das fragliche Gebäude in Donnerskirchen keine Kirche war - welche andere Deutung für dasselbe ist denkbar? Ich weiß keine. 2. Wenn die Donnerskirchner Mensa keine geweihte Altarplatte, sondern ein profaner Tisch war - wie ist es dann zu erklären, daß die Fragmente sorgfältig 'bestattet' wurden? So scheint es mir, fast eher gerechtfertigt zu behaupten, daß, während für die Marmor-Tischplatten mit umgebenden Bogennischen ... eine Verwendung als Altarplatten zumindest in einzelnen Fällen feststeht, ein profaner Gebrauch in keinem Fall eindeutig erwiesen ist.' (1964: 18). Because of the marble slabs, Ubl identifies the room as a 'christlicher Kultraum' and dates it to the first half of the 4th century. (1982: 303).

[325] I would like to thank Dr István Torma for helping me with the evaluation of the topographical data of the site.

8.I.b Amulet, fig. 71a-b

Provenance: The cemetery consisted of 7 graves. The third grave was a child grave, from which the amulet came to light. Other grave goods: rings of bronze wire, 2 pearls: one is of black paste; the other is a light blue glass bead shaped as a melon seed. **Location:** WMM 77.65.4. **Literature:** ALBEKER 1978 66 figs. 8/1 and 11/1; THOMAS 1977a 24; SCHWEITZER J. / SZILÁGYI M: A Tolna megyei zsidók története 1867-ig. Budapest 1982: 9; SCHEIBER 1983: 44, 47; SOLIN 1983: 762; SOLIN 1989: 233-236.

Lead. It is in bad condition, broken. Diam: 3.1 cm. Both sides of the amulet are decorated; on the one side, there is a menorah flanked with a horn and a palm leaf, and the whole is framed by pearl border. On its other side, the depiction was also framed by pearl border but its surface is in such bad condition that only some palm leaves can be discerned.

The latest of the coins found in the cemetery date back to the reign of Gratianus. I quote the final evaluation of the cemetery by Mária Albeker: "On the basis of the date of the graves, we cannot determine the exact intervallum for the time the whole cemetery. The earliest burials are datable to the 340s, the latest ones to the time after 375. The year 375 of the coins issued by Valens and Gratianus marks the beginning of a new era in the history of Pannonia, but at the same time, it does not prove that this year was the end of the cemetery."[326]

It cannot be doubted that the object can be regarded as Jewish as well, but it would be too rash to draw considerable conclusions from it. H. Solin[327] warns against it. Sporadic finds show that Jews occurred here sporadically. Alternatively, it is also known that after a time Christians used menorahs as lamps or even as symbols.

9. DUNABOGDÁNY (CIRPI)

The place is known only from geographical writings.[328] András Mócsy wrote a summary about its ancient significance.[329] Its military role is generally known because it is situated in the Danube-bend. An auxiliary fort belonged to it, but the canabae that developed around did not become a town.[330] The fort was excavated, but it was not opened completely. Apart from the finds excavated here, we know that some stones were found sporadically.[331]

Cirpi is listed as a bishopric see.[332] Early Christian finds from the area are not known because of its insufficient investigation. Thus, the lack of Christian finds is not decisive, so the *argumentum ex silentio* cannot be applied here. To confirm this an article by Hermann Vetters should be mentioned here, i.e. his point that the organization of bishopric sees can be proved in former forts.[333]

[326] ALBEKER 1978: 77.

[327] SOLIN 1989: 233.

[328] GRÁF 1936: 101; TIR L 34.

[329] MÓCSY 1962: 651.

[330] TÓTH E 1978: 183.

[331] MRT 7 8/9, p.85.

[332] See the chapter History of Research.

[333] VETTERS 1969.

10. DUNAPATAJ

This site is in the barbaricum. Elvira H. Tóth reported on the archaeological research in the environs. Only prehistoric finds could be recorded. No Roman finds were found.[334] According to the report, the ring was taken to the museum from Dunapataj. It did not come from excavations. It is possible that the donator did not know the actual name of the site.

10.I.a. Finger ring, fig. 72.

Location: MNM 10.1951.236. **Literature:** NAGY L 1938 112: fig. 76. (second row 2).

Bronze, with chi-rho monogram. Its Roman origin cannot be doubted. The size of the head is 2.7 x 2.5 cm.

11. DUNASZEKCSŐ (LUGIO / FLORENTIA)

In 1938, Lajos Nagy wrote the following about Dunaszekcső: "It is situated in the territory of the Roman Lugio later known as Florentia. There was a late Roman cemetery in the fields called 'Püspökhegy'. The cemetery has not been scientifically excavated yet. A glass was found among sporadic grave goods." Although a rescue excavation was led on Várhegy in Dunaszekcső, [335] it has not been published yet. Only one find was published, a bronze head of Marcus Aurelius which was gilt.[336] Lugio is known as a fort along the limes. There were no large-scale excavations in the area.

11.I.a. Pendant, fig. 73.

Location: It was probably lost. **Literature:** HAUGH 1902: 365 fig. 3; NAGY L 1929: 161.

There is a record, which says that 155 finds were taken to the museum of Szekszárd. Among the finds, there were some also from Dunaszekcső. One of them is a disc made of gold sheet with a hole in its middle. A chi-rho monogram is represented on the disc. Punching little points made the picture. Two little loops were applied to the disc. These loops were decorated with mouldings and scotias in turns. Diam: 1.8 cm. According to a description by Lajos Nagy, the object was made of gold. There is no detailed information about its site. Its particular dating is uncertain. Béla Haugh remarks that the loops are too big for the disc.

11.II.a. Fondo d'oro fig.74.

Location: MNM 3.1934 purchase, collection Fleissig. **Literature:** NAGY L 1929: 158-163; PAULOVICS 1938b: 95 fig. 132; NAGY L 1938: 42-47 fig. 11; FÜLEP 1967a: 239-248; FÜLEP 1968: 404-406 fig. 2. pl. I.2; KÁDÁR 1969: 198; FÜLEP 1980: 42; THOMAS 1980: 199 pl. CXLII; THOMAS 1982b: 275 Abb. 20; THOMAS 1986a: 531-536; BARKÓCZI 1988: 550.

On the gold leaf sealed between layers of glass, half-length figures of a couple are represented. The legend is: SEMPER GAVDEATIS IN NOMINE DEI. Diam: 8.7 cm. The glass is very cracked. Both portraits are depicted en face. Big, wide-open eyes, semicircular eyebrows and a short forehead characterize the faces. The woman's hair is parted in the middle. From here on the right and left, it surrounds her face in wavy section. The hair is covered with a headdress. The necklace of ball-shaped pearls is around her neck. A jewelled collar lies over her tunic. She is wearing a stole on both shoulders. Only her left hand can be seen. She is stretching out her thumb and forefinger. The man's hair is short. His tunic is long-sleeved, above which he is wearing a *chlamys*. The latter is held together with a fibula on the right shoulder. The type of the fibula cannot be discerned, it might be a disc-fibula. On the right shoulder one can see that that his garment is ornamented.

Opinions about the fondo d'oro. F. Fülöp wrote that the ones in Pannonia were made at the time of Constantine or later and that they were all cups and birthday presents for children.[337] Edit Thomas thinks that they were presents for baptism and confirmation.[338] According to László Barkóczi the figures on the gold leaf are imperial officials of higher rank.[339]

The ideas about its use are just ideas that might be true in this particular case. The question that remains is why it is not the scene of baptism that is depicted if these objects were meant to be presents for baptism. Among biblical scenes one can find representations of baptism and those of baptism of Jesus.[340] However, there are fondo d'oros that were meant as presents at marriages.[341] A fondo d'oro itself is not early Christian. It has some antecedents listed and classified by Vopel. He points out in his book that a fondo d'oro should not be regarded as Christian unless the depiction on it suggests so.[342]

According to our present knowledge, the site does not justify its Christian classification and the depiction on our fondo does not support this either. It was listed among early Christian objects because of its legend. The inscriptions on the beakers usually express best wishes. Most of them do not have a particular expression on the basis of which it could be clearly regarded as Christian. The expression *in nomine Dei* is characteristic of the period, it is almost a style of the era. However, the use of this expression does not prove or deny whether the object is Christian. It could be called the 'diplomacy' of the 4th century, ☞ Szabadszállás. It might be said that its date does not contradict its Christian use. The depicted people, as Barkóczi wrote, were officials of high rank. This aspect is worth pointing out because the bronze head of Marcus Aurelius was found here in excellent condition whose analogy in quality, size and manufacture is the portrait of

[334] H.-TÓTH E 1990 items 20-21, p.87.

[335] In the 1970s.

[336] KOVÁTS 1983: 181-184; KOVÁTS 1984: 89-92.

[337] FÜLEP 1968: 412.

[338] THOMAS 1986a: 536.

[339] BARKÓCZI 1988: 216.

[340] There are similar depictions, e.g. one of the frescos in the Catacomb Callixtus (Sakramentskapelle A 3 WILPERT 1897: 27,3). Wilpert and Pierre du Bourguet think that it is the baptism of Jesus Christ. (1965 No. 9), Orazio Marucchi interprets it as a baptism with symbolic meaning (1912: 309-312).

[341] E.g. Dumbarton Oaks Collection - KANTOROVICZ 1960 figs 8: Amor pronobus, 9: Hercules pronobus. Legend: ORFITVS CONSTATIA IN NOMINE HERCVLIS, 10: Christus pronobus pp.6-7.

[342] VOPEL 1899: 2-20.

the rider Marcus Aurelius in Rome.[343]

This fondo d'oro can be dated on the basis of its style. It is usually dated to the first half of the 4th century and László Barkóczi thinks so, too. A similar half-length portrait of a married couple can be seen on the cover of the Proiecta casket.[344] The woman wears an ornamented collar and the man's pose is similar, too. The depiction of the face with big eyes, semicircular eyebrows and a short forehead satisfies the face type of the 4th century, especially in the eastern part of the Empire.[345] Taking the above said into consideration the dating can be narrowed to the 40s of the 4th century.

Summary. On this fondo d'oro some officials of high rank are depicted but there is no reason to list it among Christian objects.[346] ☞ Dunaújváros 12.II.b. I think its historical evaluation is possible when the other finds are published. ☞ Szabadszállás 42.II.a.

12. DUNAÚJVÁROS (INTERCISA) fig. 75.[347]

This is one of the most excavated Roman sites in Pannonia. Eravisci lived here before the Romans. The earliest Roman settlement dating back to the 1st and 2nd centuries was found on Papsziget. The fort was built later. Intercisa became important because of the military station. The canabae, a civil settlement was established. Before the excavations between 1963 and 1971, the north south road of the fort was thought to have run in a direction later proved to be wrong. The road ran contiguously across the canabae, past the cemetery in the south where it bent a little to the west.[348] The cemeteries were located south, southeast and west of the canabae. According to the latest research, there was a cemetery north of the fort, too. The excavation of the cemeteries began at the beginning of the 20th century and was led by Ede Mahler. By fits and starts, the excavations have been carried out continuously since then. The grave goods of early excavations were discovered on the hillside of Öreg-hegy. Since Intercisa was situated along the limes in Pannonia, its military significance was highlighted by research; the excavation of the fort and the canabae provoked debate on the questions connected with them.[349] Some of the cemeteries have been published.[350] Besides the official excavations, there were many sporadic finds whose evaluation is difficult even if they were taken to a museum. Eszter Vágó uncovered 1528 graves and different buildings in the canabae. One of the buildings excavated earlier was regarded as an early Christian basilica.[351]

There were probably many immigrants from the east. Of the soldiers stationed here, Syrians are significant. Because of the presence of eastern population, and because of an inscription, it is supposed that Jews lived here in diaspora and a synagogue was built for them. It could not be determined which of the uncovered buildings the synagogue was.

No mithraeums have been found yet, but three cult pictures with tauroctonia were discovered.

12.I.a. Casket-mount, fig. 76.a, b, c

Specialities: The most part of the casket-mount was found in 1903. The sheet with Orpheus broken horizontally in two. The lower part was found in 1903, but Eszter B. Vágó in grave 1023 found the upper one in the 1960s. The two Orpheus-sheets fit together, as it is shown in fig. 76a. Apart from this sheet, there was a glass vessel in grave 1023. The leader of the excavation found the grave destroyed. **Location:** MNM 64.1903.19-24 and IM 70.1023.2. **Literature:** GÁSPÁR 1986 Kat.-Nr. 382.

Bronze. The dimension of the sheet with Orpheus is 6.5 x 7.3 cm. The sizes reconstructed of the face of the casket are 27.5 x 30.9 cm. Depictions: The deities were depicted in panels: Jupiter twice, Mars three times, Mercurius twice, Minerva once. The male deities are standing while Minerva is sitting. The panels are situated on both sides of the face of the casket under each other. The sheet with Christian scenes is in the middle. The space between the metopes and the sheet with Christian scenes is decorated with lattice of bronze stripes. The Christian scenes can be seen in pairs under each other. They are as follows: the raising of Lazarus, the Moses striking the rock, the Good Shepherd, Moses unlooses his shoes,[352] Jonas taking his ease in the shade, the healing of a blind person. In two big medallions, the same scene was depicted: it is Orpheus enchanting animals with his lyre. Only a small piece of the first medallion survived. The other Orpheus scene is complete. On the sheet, there are further representations: in small medallions Sol, Luna,[353] chi-rho monogram and an eagle can be seen. In an oval-shape field, *dextrarum iunctio* was

---

[343] PARISI PRESICCE 1990: 48-60.

[344] BUSCHHAUSEN 1971: B7, B Taf. 26; SHELTON 1981: 72 pl. 4. "The portraits are those of a richly dressed man and woman. The woman wears a large jewelled collar over what appears to be a long-sleeved tunic girdled below the breasts; ... the man wears a long-sleeved tunic beneath a chlamys, fastened at his right shoulder with a bow or crossbow fibula. His left arm is covered, and his right gestures in front of his body, with thumb raised and two fingers extended indicating speech. The woman's hair is parted in the middle, brushed back in wavy section, and brought up in a braid circling the crown of the head. ... Both have heads and hands relatively large for their bodily proportions.'

[345] BIANCHI BANDINELLI 1971: 29-38.

[346] I would like to allude again to the 'diplomatic' ambiguity characteristic of the first half of the 4th century. Because of this ambiguity, it is difficult to decide whether something is Christian or not.

[347] BARKÓCZI 1954 fig.2. Each of the sites of early excavation gives the name of a landowner. Jenő FITZ demonstrates the neighbourhood of Intercisa, and its situation along the limes on the maps: Őrtornyok Intercisa és Annamatia között. (Burgi zwischen Intercisa and Annamatia). ArchÉrt LXXXII, 1955: 69-79 figs. 2-3.

[348] VÁGÓ/BÓNA 1976: 123. The topographical situation resulting from the latest excavations is shown in VÁGÓ/BÓNA 1976 Abb. 156.

[349] Intercisa Volume I. and Volume II. by several authors, ArchHung 33 and 36; FITZ 1959; FITZ 1962b: 102-103; VÁGÓ 1970; FITZ 1972a; MÓCSY 1973: 385; SALAMON/BARKÓCZI 1973; VISY 1978: 242-247; VISY 1980; VISY 1981: 44-46; VISY 1989: 96-101; each of them with further literature.

[350] VÁGÓ/BÓNA 1976.

[351] FÜLEP 1980: 41-42.

[352] The scene has been interpreted differently: Christ's baptism - SUPKA 1913: 182; the sacrifice of Isaac - NAGY L 1938: 56 fig. 19; BUSCHHAUSEN 1971 A65; Moses unlooses his shoes - GÁSPÁR 1971 No.12; GÁSPÁR 1986 Kat.-Nr. 382.

[353] BUSCHHAUSEN 1971 p.134: he does not mention Luna. He writes about a woman's portrait 'mit antoninischer Lockenfrisur, deren Haare im Nacken zu einem kleinen Knoten zusammengefaßt wereden.' Cf. Iulia Soaemias or Iulia Paula: WESSEL 1946: 62-75.

depicted.

12.I.b. Lid of casket, fig.77.

Location: MNM 29.1906.4. **Literature:** GÁSPÁR 1986 Kat.-Nr. 401.

It is a bronze rectangle plate with hinge. Size: 9.5 x 7.3 cm. It was cast and decorated with embossed bronze sheets, which are destroyed. However, from the published literature, images can be reconstructed. An ovolo decorated the edge of all the four sides of the lid. The ovolo is framed inside by a moulding, which was decorated with points. In addition, a disc with the portraits of Peter and Paul could be seen on the lid. Traces of fastening of five further sheets can be discerned. But we have no data of their pattern. I think four of the traces were covered with embossed sheets and the fifth marks the place of the handle.

12.I.c. Casket-mount, fig. 78.

Provenance: It was discovered at Antal Hekler's excavation from grave 109 of the field XXII. It was a disturbed stone grave. Other finds were an iron ring and a coin of Valentinian I. **Location:** MNM 100.1912.196. **Literature:** GÁSPÁR 1986 Kat.-Nr. 550.

Dimension of escutcheon: 6.8 x 3.2 cm. The face of the wooden lid is 27.5 x 6 cm. The width of a bronze band is 6 cm. These bands are decorated. On one of the bands, heads of Medusa and portraits in medallions can be seen: Medusa, a man's portrait and portraits of Peter and Paul. On the other band, there are biblical scenes in panels: Daniel among the lions, Moses striking the rock, miracles at Cana. Among the medallions, some 'Zwickelmotive', a bird, a dolphin, a bee and a jug, are depicted. The scenes are repeated. The portraits of Peter and Paul opposite each other is in the same medallion. Above and between their heads a chi-rho monogram can be seen. There is another mount, which is the same as this one. It was found in Intercisa, but is preserved in Mainz. The latter one is more intact = 12.I.d.

The man in the portrait looks left. There are several opinions concerning its identification and summed up by Erika Dinkler von Schubert,[354] who thinks that the antecedent of this mount was the portrait of Constantine on a coin dating back to about 330.[355] It should be mentioned here that the depiction of an earlier coin-picture could occur on the mount.[356]

Concerning the casket-mount preserved in Mainz, Erika Dinkler-von Schubert writes the following: "Derartige kleine Zwickelmotive finden sich auf pannonischen Kästchen auch sonst ..., meist als «apotropäische Zeichen» gedeutet. Supka hielt sie auf unserem Beschlag für christliche Symbole. Stattdessen glauben wir - mit gebotener Vorsicht - für die Zwickelmotive unseres Beschlages Beziehung zum Mithraskult in Erwägung ziehen zu dürfen. Von dorther könnten sie - mit bereits abgeschliffener Bedeutung - in den Motivschatz der pannonischen Werkstätten gelangt sein. Aus dem Vorstellungkreis dieses in Pannonien stark verbreiteten und noch Anfang 4. Jh. nachweislich geübten Kultes lassen sich manche Hauptdarstellungen der Baschläge erklären, so Tagesgötter und Jahreszeiten. In dem Bildschmuck von Mz 04651 käme damit ein synkretistischer Zug, wie er für die Beschläge dieser Gegend weithin charakteristisch ist. Es entspricht dies der religionsgeschichtlichen Situation in dieser Donauprovinz, die mitgeprägt ist durch das starke orientalische, speziell syrische Einwohnerpotential."[357]

Developing this train of thought, I would like to refer to István Tóth. After elaborating the Mithras-finds in Pannonia he pointed out that a special Mithras religion was practised in Pannonia and it cannot be compared with the earlier Persian cult.[358] In the course of our conversation, we concluded that the name *Mithras Pannonicus* would be suitable to denote this particular cult. Still developing this train of thought, I would like to refer to the connection that can be observed between mithraeums and Christian churches based on excavations. Because of this observation - 'mit gebotener Vorsicht' - I put forward the hypothesis that there was not just a Christian mission in Pannonia, but the baptism of Mithras worshippers, or rather baptized Mithras worshippers.[359] Therefore, the first Christian churches were not the buildings closing in an apse,[360] but houses or parts of houses,[361] which having rectangular rooms, could resemble the ground plan of mithraeums. The rebuilding of the house was not necessary for the celebration of the *fractio panis*. "Während man bei den Juden früher allgemein bei Tisch saß, wurde es damals üblich, ... wie die Griechen und Römer, bei Tisch zu liegen und zwar auf Ruhebetten, die mit Polstern und Teppichen belegt waren. Auch Jesus hat sich beim letzten Abendmahl an diese Sitte gehalten. Bei den antiken Mahlzeiten wurde entweder das Triclinium verwendet, eine Gruppe von drei Ruhebetten, auf denen jeweils bis zu drei Personen Platz finden konnten. ... Oder man benützte das sigmaförmige (halbrunde) Polster (stibadium oder accubitum. Hinsichtlich der Reihenfolge der Plätze herrschte eine strenge Regel."[362] However, early Christian communities were not lying but sitting while having their meals. "So spricht auch der Jakobs-Brief (2,2-4) nicht von *accubita*, sondern von Sitzen."[363] The seating plan was strictly determined. Fragments of such a sigma-shaped table were found in ☞ Csopak.

Another fact should be considered here. In the East, in the

[354] DINKLER-VON SCHUBERT 1980: 144.

[355] DINKLER-VON SCHUBERT 1980: 146.

[356] BUSCHHAUSEN 1971 A16; GÁSPÁR 1971 No. 51. The reverse of an Antoninus Pius' coin is on the mount.

[357] DINKLER-VON SCHUBERT 1980: 145. ☞ Császár 6.I.a.

[358] TÓTH I 1994: 1: 'it is not the adaptation of a Persian cult, but a mystery from the period of the Roman Empire, which is the result of a deliberate foundation of a religion. Hence, this religion can be regarded as an eastern cult only indirectly; and beside the eastern connections of the cult, its origin in the region of the Danube and in Pannonia is emphasized.'

[359] It is possible that they were already baptized when they arrived.

[360] They did not become churches, not even in the 4th century. Constantine also brought changes in connection with the buildings. Earlier the house-church, or part of the house, was a secular building where the eulogion was celebrated. Cf. DEICHMANN 1983: 68-88. In the 4th century, early Christian basilicas in the present sense gradually became a sacred space, a Heiligtum.

[361] Like e.g. in Dura-Europos - GAMBER 1968: 15 Abb.1.

[362] GAMBER 1968: 70.

[363] GAMBER 1968: 71. I have to point out that in the letter the word community is expressed by συναγωγη.

province of Syria, early spreading of Christianity has been proved. For example, Antiochia in Syria could be mentioned where "wurden die Glieder der neuen Gemeinschaft als Χριστιανοί bezeichnet; hier spielte sich der Streit zwischen Paulus und Petrus"[364] Let me mention even the Abgar legend.[365] When talking of the arrival of Syrians to Pannonia, the 150 years between their converting and the arrival is never taken into consideration, regarding the intervallum between Paulus apostolus and Marcus Aurelius. Syrians regularly came to Pannonia since the reign of Marcus Aurelius. Thus, we may well say that even if they were soldiers there were some Christians among the Syrians arriving. That is why I think that there were Christians in Aquincum as well as in Dunaújváros probably from the 2nd century on and in Aquicum maybe from the end of the 1st century on.[366] It is obvious that in this early era the forms of *modus vivendi* guaranteeing the undisturbed process of liturgy had to be found. Syrians lived isolated from local population. András Mócsy proved this by the inscriptions.[367] A trace of Syrian population was discovered also in ☞ Pécs, and in ☞ Kővágószőlős.

12.I.d. Casket-mount, fig. 79.

Location: Römisch-Germanisches Zentralmuseum, Mainz O 4651. **Literature:** GÁSPÁR 1986 Kat.-Nr. 686.

Dimension: 24.5 x 30.5 cm. The depiction on it is the same as that of the former casket-mount. Both of them might have been made with the same die. They are two identical pieces. This one is in better condition than the other.[368]

12.I.e. Casket-mount, fig. 80.

Location: Römisch-Germanisches Zentralmuseum, Mainz O 4653. **Literature:** GÁSPÁR 1986 Kat.-Nr. 688.

The male figures framed with panels are considered being Apostles according to analogies ☞ Ságvár, 36.I.a.

12.I.f. Casket-mount, fig.81.

Provenance: It was found in a brick grave in an excavation led by Antal Hekler in 1912. The names Frigeridus and Terentius were on the brick stamps. **Location:** MNM 100.1912.201. **Literature:** GÁSPÁR 1986 Kat.-Nr. 551.

The scenes are depicted in panels. The width of panels is about 4.8 cm. The casket is so broken that it cannot be measured exactly. The depicted biblical scenes are the miracles at Cana, the Good Shepherd, and the raising of Lazarus.

12.I.g. Casket-mount, fig. 82.

Provenance: "Im Jahre 1949 auf dem Gebiet des Lagers, aus der obersten Schicht des Grabens K1 hervorgekommen."[369] **Location:** MNM 5.1950.69. **Literature:** GÁSPÁR 1986 Kat.-Nr. 576.

Two fragments with inscription; ΣESV, and O. They are written out as follows: Zesu[s] = Jesus, [M]o[ses] = Moses. They probably depicted the raising of Lazarus and Moses striking the rock. A fondo d'oro in the Museum of Vatican (fig. 83[370]) gives a good analogy for the habit of the letters.

12.I.h. Casket-mount, fig. 84.

Location: Staatliches Museum für Völkerkunde (Inv. No.: Berlin IV d 3756c). It was lost during the war. **Literature:** GÁSPÁR 1986 Kat.-Nr. 685.

Bronze. Two horizontal sheets with the same depictions; they are framed with medallions. The depictions are: a/ A riding Emperor has Victoria handing over a wreath. At both sides of the main scene, a soldier stands holding a *signum* in his hands. In the segment under the scene, two prisoners are turning towards each other. b/ Constantinopolis and Roma. In the segment under the scene, two prisoners are depicted as well. c/ The Good Shepherd. There is an ornamented band between the two sheets. I took over the description as well as the photo from a work by Lajos Nagy.[371] The iconographical interpretation of the depiction is the *profectio*.[372]

12.I.i. Mirror, fig. 85

Location: MNM 27.1906.1 **Literature:** Jelnetés a Magyar Nemzeti Múzeum 1906. évi állapotáról 37. Fig. 5; NAGY L 1930b: 245; NAGY L 1938 Fig. 24; RADNÓTI 1957 Kat.-Nr. 17B; BUSCHHAUSEN 1971 A85, Fig. 6.

The bronze mirror soldered with white metal and decorated with concentric circles. In the middle of the mirror and between the two inner concentric circles there are five Greek crosses. Lajos Nagy thinks that it depicts the 'Universe'. Diam: 14 cm. I agree with Lajos Nagy, and every scholar according to whom paganism and Christianity could use the same symbols. First of all the cross is such a symbol. The cross symbolizes the CENTRE. The Centre is the point where heaven and earth are

---

364 HAMMERSCHMIDT 1986: 648. It is true that Syria is large and it is not certain that Orientals came to Pannonia from Antiochia, but it is provable that Orientals came to Aquincum from Dura because the uncovered frescos show the influence of Dura. ☞ Aquincum, 4.I.g. and the villa of Kiscelli Street. The same influence can be observed in the frescos of Kővágószőlős ☞ Kővágószőlős, 26.II.a. There were Christians living in Dura, see baptistery in insula M8. It does not mean, of course, that Syrians and Orientals came here only from Dura. In respect of Christianity, it is important that liturgy was not homogeneous in the Syrian territory. Western and Eastern types can be distinguished - BOTTE 1963: 17. It is not sure whether such a difference can be proved based on archaeological finds.

365 H. RAHNER: Abgar, König. LThK 1 1986: 43. (It should be pointed out that Edessa remained an independent kingdom until about 200 - BOTTE 1963: 18.)

366 In short, I do not refute an early mission in Sirmium. Thus it is also possible that Christians appeared early in Aquincum as a result of the migrations going on inside the province.

367 MÓCSY 1974: 228. 'The Syrians in Intercisa married only among themselves, their sole outside links being with Syrian Families in Aquincum; the reason for this is mainly that some soldiers' sons from the circle of the cohors Hemesenorum were enlisted in legio II Adiutrix. The Syrians in Ulcisia Castra, too, had connections only with the Syrians of Aquincum.' Furthermore, he writes about their becoming members of *ordo decurionum* and their being the richest in Aquincum as well as in Brigetio. Cf. M. Ant. Victorinus ☞ Hypotheses about the Christianity of Aquincum.

368 Erika Dinkler-von Schubert expounded her opinions concerning this item, not the former one.

369 This information comes from Aladár Radnóti. RADNÓTI 1957: 346. The sign K1 for ditch is probably wrong. Although László Barkóczi also mentions ditch K1 (BARKÓCZI 1954: 21), in the description of the finds ditch M is mentioned (BARKÓCZI 1954: 22: 'Ditch M is at a depth of 40 cm, the third layer of a stone fort') and ditch M may also be seen in the general plan, so I regard the latter denotation to be more probable.

370 TESTINI 1966 Tav.II.

371 NAGY L 1938: 56-58 figs. 21-22.

372 Richard BRILLIANT: Simboli e Attributi, 2. Arte Romana. In: Enciclopedia dell'arte antica. Roma, VII 1966: 308. fig.391.

connected, where cosmic layers are penetrable. The symbolism of the centre is expressed in the road-system of Roman towns,[373] in the height where a temple is built and in every sign denoting *axis mundi* for a community; the axis of the world, the axis of the centre. The Centre is the place of happiness where 'Lost Paradise' is. Christ's cross is the Centre where fallen people can find the way to 'Lost Paradise'. The cross symbolizes 'the way to absolutum'[374] for pagans as well as Christians.

12.I.j. Glass vessel, fig. 86.

Location: IM 77.112.1. **Literature:** VISY 1977 37 Abb. 40.

A sherd of a glass vessel with engraved stars and the chi-rho monogram. The chi-rho monogram was engraved later. Originally, only the stars were engraved. Even a man's head remained on the sherd. Diam of mouth 20 cm. Dimension of the sherd: 7.8 x 4.3 cm. Because of the depictions, the sherd represents one of the objects which point to a particular religiosity in Pannonia, which was discussed by Erika Dinkler-von Schubert. ☞ A casket-mount 12.I.c. and 12.I.d. ☞ Aquincum 4.I.e.

12.I.k. Standard lamp, fig. 87.

Provenance: sporadic find from the building C4 of the vicus. **Location:** IM 79.73.5. **Literature:** THOMAS 1988: 294. Abb. 17.

Edit Thomas published it as a censer. The object consists of three parts. One is the base Diam: 9.1 cm, the second is the column L: 11.1 cm, the third is the plate, whose side is decorated with openwork Diam: 11 cm. Originally only the column and the base belonged together. The upper part was fastened to the column later. However, the column and the plate could not be used as a censer because censers have only their lid open worked. Originally, the 'present base' was not a base; its rim that it is standing on now is rather unfinished, or rather, it seems as if a higher rim had been cut. The present upper part is not fastened to the column. It can be removed easily and it was made of a different material. Although the whole object was made of bronze, the quality of the present upper part differs from other examples.

The following conclusions can be drawn from the observations above: a/ The present upper part did not belong to this object; but it was the lid of a censer. However, all the other parts of this censer are missing. b/ The column and the present base belonged originally together, but applied to each other in another way. c/ The present base was the upper part, i.e. a plate, on which a bronze lamp could be fastened. d/ The true reconstruction of the original form, is, in my mind, that presented on fig. 88.[375] László Török dates these kinds of objects to the late 4th century.[376] Based on this, the object from Intercisa can also be dated to the late 4th century.

The question remains why was the standard lamp altered. It could not function as a censer, not even in the form it is shown in fig. 87. However, it might have been used as a candlestick. A stone located in Museo Laterano shows stylistic parallel to such a candlestick,[377] fig. 89.

12.I.1. Pendant ☞ Dunaszekcső, pendant 12.I.1=11.I.a.

I quote the record: "We have only one early Christian find from Dunapentele: a golden pendant with chi-rho monogram. Location, the museum of Székesfehérvár. (A kind information by Arnold Marosi)."[378]

On the basis of this short record, the object cannot be identified. There is not a golden pendant with chi-rho monogram preserved at the museum Székesfehérvár. There is no data proving that a similar pendant has ever been preserved there. That is why, we may well say that the information refers to a golden pendant of Dunaszekcső taken to Szekszárd and published by Pál Haug.[379] Arnold Marosi might have seen this pendant, but he could not remember the site exactly. Albin Balogh knew about it only from a verbal message. I think it is most likely that there were not two such pendants, but only one.

12.II.a. So-called church in the parcel no. 2023.[380] fig. 90.

Literature: OROSZLÁN 1920-1922: 93-96; NAGY L 1927: 124 note 37, 129. fig. 56; PAULOVICS 1927: 20; BARKÓCZI 1954: 41; SÁGI 1954: 84; THOMAS 1964: 240; PARRAGI 1976: 178; VISY 1977 fig. XII; FÜLEP 1980: 41.

During the 1922 excavations, Zoltán Oroszlán uncovered a large stone building of which he wrote the following: "From the middle of the site, the graves occurred less frequently, and the apse of a bigger building was found here. Only 5-6 m long, 80 cm wide and about 60 cm high part of the wall was found, the rest had been carried away." A pebbled layer was observed around the building, which was the remains of the road leading to the building. "The remains of a lead water-pipe were discovered close to the house," and fragments of frescos and stuccoes.[381] Above the fragments with figures there were many pieces painted red and blue. "The stucco was divided into several stripes, each of them decorated with different patterns: leaves in medallions, panthers with a cantharos among them, a bull, harpy-eagles; all these pictures are interrupted here and there by kyma, which is strong provincial in its style."[382] Paulovics thought that the building was an early Christian basilica and Lajos Nagy regarded it as a Roman building converted into a church. Just as Károly Sági wrote, the finds suggest a villa.[383] Sági analysed the building, which limited the south border of the cremation cemetery. Some skeleton-graves were unearthed close to the building. Stucco fragments were found in the fill from these graves.

---

[373] It is the intersection of *cardo* and *decumanus*, for which there are lots of examples in a book by FINLEY (1979).

[374] ELIADE 1997: 64.

[375] As analogies, the objects published by László Török were used for the reconstruction. TÖRÖK 1988 Table 113/12, 66 and Table 186/3. The reconstruction was drawn by Sándor Ősi.

[376] TÖRÖK L 1988 § 28.

[377] TESTINI 1966 Fig.96.

[378] BALOGH 1927: 199.

[379] HAUG 1902 365. Fig.3.

[380] According to the ownership in 1920, it was in the plot of János Szórád.

[381] NAGY L 1926: 109. For a detailed descripton of the stuccos found all over Pannonia, see NAGY L 1927. The decoration on a stucco of Intercisa is similar to the horn of Crete, but NAGY L. gives another interpretation; NAGY L 1927: 123 fig. 55.

[382] OROSZLÁN 1920-1922: 96.

[383] SÁGI 1954: 84.

Thus, the building can be dated to the early Roman times. There are different opinions concerning its function. Because of the apse, it was regarded as a Christian church, although no finds or architectural elements were found, which could show the Christian character. The finds are the following: a lead water pipe, a wall painting, stucco remains, sigillata sherds and moulded imitation of Samian ware. Zsolt Visy wrote the following: "On the other side of the same way a late Roman building was discovered. The entrance opened to the way, but this part of the building was collapsed when the way was destroyed. It was a nice building with an apse, on the basis of its ground plan it was most probably an early Christian church."[384]

There is no doubt that the building did have an apse, but an apse does not belong exclusively to the features of Christian buildings. In addition, the finds are not Christian at all; on the contrary, the depictions on the frescos and stuccos deny the Christian character.[385] Hence, on the basis of present knowledge I do not regard this building as Christian.

12.II.b. Fondo d'oro, fig.91.

Location: MNM 63.5.1, purchased from Károly Majsi. **Literature:** FÜLEP 1967a; FÜLEP 1968: 401-412 fig. 1 pl. I/1; KÁDÁR 1969: 198; VISY 1977 fig. 41; FÜLEP 1980: 42; THOMAS 1980: 199 pl CXLII; THOMAS 1982b: 275; THOMAS 1986a: 531, 534; BARKÓCZI 1988: 551.

Diam: 9.2 cm. The half-length figures of a man and a woman are depicted. There is a child in front of the man who is putting his fore- and middle fingers on top of the child's head. The figures are not in proportion to each other. The man occupies two thirds of the space. The woman is forced into the background. The woman is wearing a necklace and an ornamented stola. The man is wearing a clamys on the long-sleeved tunica. A crossbow fibula with three knobs holds the clamys together on his right shoulder. The shoulder of the tunica and the edge of the sleeves are ornamented. The child's garment is similar to the one of the man, but he wears a disc-fibula. The woman has a headdress. There is a four-leaved clover between the two portraits. The faces are of eastern features. This is the reason why Ferenc Fülep thought that the depicted persons were Syrians or Jews.[386] An analogy, which is dated about 100 years earlier, demonstrates this eastern type very well.[387] These two objects are different in every aspect except for the eastern features. The legend is VIV(as) INNOCENTI CVM TVIS IN DEO. In English: Innocentius, live together with your relations in God.[388]

There are not any Christian scenes and motives in its depiction. Because of the inscription, it is usually listed among Christian objects. For an evaluation of the inscription see ☞ Dunaszekcső, 11.II.a, ☞ Szabadszállás, 44.II.a.

The object dates back to the second half of the 4th century. I consider its being Christian questionable.

12.II.c. Votive plaque, Figs.92, 93, 94.

Provenance: "Rómer saw it built in the wall of a building behind the post in Dunapentele."[389] **Location:** MNM 154.1874.2. **Literature:** CIL III 3327=10301; ILS 3981; CIJud 1 676; RÓMER 1864a: 60, no. XXVIII; PETSCH 1893: 194; HAMPEL 1906: 230, pl. III/7; SUPKA 1913: 166; DOBÓ 1940: 165, 176 nr. 58; ERDÉLYI G/FÜLEP 1954, no. 329; FÜLEP 1954: 216, 219; SCHEIBER 1955: 191-195; ALFÖLDY 1958: 177-180; SCHEIBER 1960: 29-37, Nr.4; FITZ 1962b: 102; MÓCSY 1962: 595; FÜLEP 1966: 93-98; DOBÓ 1968, no. 287; SANIE 1968: 1101; FITZ 1972a: 175-177; RADAN 1973: 268-269. fig. 2; MÓCSY 1974: 228; THOMAS 1977a: 22; VISY 1977: 28; SCHEIBER 1983: 28-31, 33, no. 3; BARKÓCZI 1986: 184; SOLIN 1989; RIU, 1051; AnnÉp 1990: 823; TÓTH I 1995, no 1; EDH-Nr: HD024864; GÁSPÁR a paper in manuscript.[390]

Dimension: 58 x 82 x 22.5 cm. The inscription is framed by a tabula ansata. However, there are some letters outside the frame of tabula ansata, between the ansa and the moulding of the tabula and under the lower moulding of the tabula. The interpretations so far have considered only some letters outside the tabula, namely letters at the ansa. The height the letters reduces downwards. Ligatures are often used. The engraver used half-line large letters placed either in the upper part of the line or inside another letter, e.g. Votum: the letter O is in the letter V. The generally accepted reading of the inscription is: Deo aeter/no pro sal(ute) do(mini) / n(ostri) Sev(eri) A[lexand / dri] p(ii) f(elicis) Aug(usti) [et Iuliae / Mame]ae Aug(ustae) mat(ris) Aug(usti) vot(um) / red(dit) l(ibens) Cosmius pr(aepositus) / sta(tionis) spondilla synag(ogae) / Iud(a)eor(um).[391]

The last word is thought to be the meaning of the letters outside the frame, on the right. This word and the last line are the subjects of the argumentation and discussion. In one opinion, Cosmius is the president of the synagogue and at

[384] VISY 1977: 35-36.

[385] NAGY L 1927: 124 note 37. He writes: 'This stucco decorated a building with an apse which stood in the middle of a cemetery, but it was not a cemetery chapel. The pieces of the stucco found in graves show that the building was ruined at that time.'

[386] FÜLEP 1968: 411.

[387] A fondo d'oro depicting the family of Vunnerius Keramus. It dates back to about 250 and was framed in a cross of the 7th century - GROENEWEGEN-FRANKFORT / ASHMOLE 497 fig. 700.

[388] An analogy to the inscription on a casket-mount of Carnuntum. BUSCHHAUSEN 1971 A6, A Taf.7 - 'Das 3.6 x 7.5 cm große Bronzeblech zeigt zwei nebeneinander gestellte, perlbandgerandete Medaillons. Im rechten Kreis sehen wir den Bildniskopf eines Herrschers im Profil, nach links ausgerichtet, angetan mit Diadem und abflatternden Nackenbändern; im linken Kreis die frontale Büste einer weiteren männlichen Gestalt, deren Chlamys auf der rechten Schulter durch eine Fibel zusemmengehalten wird. Zu Seiten dieses Kopfes stehen die Buchstaben IN DEO / VIVAS. ... In das 4. Jahrhundert datieren. ... Die Ausführung der Arbeit ist sehr präzise und höher zu bewerten als die der übrigen pannonischen Arbeiten.' Research regarded it as Christian because of the inscription. The frontal portrait depicts the younger son of Constantine (DINKLER-VON SCHUBERT 1980: 151-152), so the era may defined as Christian. The Christian connections of the inscription next to the portrait of the Emperor can hardly be evaluated as the Christian mission in Carnuntum or in the province. For greetings of best wishes of this type, see DINKLER-VON SCHUBERT 1980: 145. The inscription VIVAS VINCAS on a casket-mount in Dunaújváros should be mentioned here as a variation of these inscriptions. (BUSCHHAUSEN 1971 A97 A Taf.99)

[389] ERDÉLYI G/FÜLEP 1954, No. 329.

[390] Many have written about the inscription and there are many opinions about its reading and historical evaluation. I explain in detail these different opinions in the article; I point out the important elements about antecedents and I summarize my findings briefly. I therefore do not propose to remark, or discuss the evidence here in detail; only the results are given.

[391] ERDÉLYI G/FÜLEP 1954 No. 329.

the same time, he is the praepositus of the custom station in Spondil.[392] Otherwise, the synagogue is not in Intercisa. According to the other opinion, Cosmius is not the principal of the synagogue. However, he is the principal of the custom station in Spondil.[393] József Hampel points out that Cosmos[394] was named after his native town and an adjective was formed of it.[395]

Commentary:

1. There are many engravings on the stone, but not all of them belong to the inscription. Such can be seen inside as well as outside the tabula. It is obvious that there is a palimpsestus on the stone because of these engravings. Therefore, the stone was re-carved. Thus, the letters outside the tabula cannot be connected to the inscription inside the tabula. In addition, the word Iudaeorum cannot be read on the stone.

2. Earlier readings did not notice some ligatures in names.[396] The name Cosmius is wrong. The letter-order: C and S can be seen; between these two, there is a small O, then a ligature. The following letters are ligatured: M I A N. This ligature is followed by a half-line large V in the upper part of the line and finally a letter S. The name COSMIANVS [397] or COSIMANVS can be put together; although, I regard the second form as less possible. The stem of the word is Cosm and the adjective derivative -ianus is added to it.[398] Hence, it is really a place name, just as Hampel wrote about it.

An unnoticed ligature occurs elsewhere in the inscription, but there, it modifies only the transcription of the word.

After *Spondill a*...there is interpunction that has been noticed only by H. Solin so far. Therefore, the reading *archisynag* is not possible just as H. Solin has already pointed out. Only SYNA can be seen. The G is uncertain.

The *Deo Aeterno* does not necessarily suggest monotheism. This epitheton occurs in other contexts as well, e.g. SILVANO DEO AETERNO.[399]

My **reading** of the text is: Deo Aeter/no pro sal(ute) d(omini)/ n(ostri) Sev(eri) A[lexan]/[dri] P(ii) F(elicis) Aug(usti) e[t Iuliae] / [Mame]ae Aug(ustae) mat(ris) Aug(usti) vot(um) /red(dit) Ti. Cosmianus pr(aeposi)t(us) sta(tionis) Spondill (...) a syna(goga).

Its **interpretation** is: It is known that Syrians lived in a block in Intercisa, they formed a real community usually called *synagoge / synagoga*. Thus, the supposition is historically established that it is not about the building, but about the community where the man, who set the inscription, belonged. The preposition "a" does not mark the distance from the building, but it refers to the community-origin. Thus the opinion of Hampel, who was the first to suggest this, can be proved.

In English: To the eternal god, for the welfare of our lord Severus Alexander, the pious felicitous emperor and of Iulia Mamea empress, the mother of emperor, Ti. Cosmianus, the chief of the station of Spondill ..., from the (Syrian) community fulfilled his vow.

The stone was set during the reign of Severus Alexander (222-235).

Conclusions: The inscription contains no information about the Jewish population of Intercisa, if any Jews were living in Intercisa. The excavations so far have not proved the existence of a synagogue as a building. Sándor Scheiber did not think that there was a synagogue. Thus, the interpretations of the inscription have not been in harmony with the results of the excavations. However, the presence of Syrians could be proved. H. Solin wrote a summary of this.[400]

12.II.d. Tombstone, fig. 95.

Location: "It was discovered in the Szitányi-estate. Then, Péter Dubecz gave it to the secondary school in district I.of Budapest. Later, it dilapidated in the school." **Literature:** CIL III 10.315; FRÖHLICH 1890: 231-232 nr. 32; AnnÉp 1891 59; MAHLER 1905: 232; KUZSINSZKY 1908: 86 fig. 15; RITTERLING RE XII 1448; ERDÉLYI G/FÜLEP 1954, no. 17; NAGY T 1955: 242; SCHEIBER 1955: 195-196 no II/1; SCHEIBER 1960: 37-39; FÜLEP 1966: 93-98; FITZ 1972a: 175; RADAN 1973: 272; SCHEIBER 1983: 33-35 no. 4; SOLIN 1989: 233; RIU 1189; AnnÉp 1992 1456.

Dimensions: 40 x 60 x 14 cm. It is a fragment of a tombstone. The panel of inscription is damaged. The frame of the inscription is intact on the left, but it is missing at the right hand side. Thus, the readings supposed lines of different length. Some scholars think that only the frame had been broken, but the others think that as many as three letters were broken above the frame. The different readings are the following:

1. M. Aur. Malc[h]ia[s mil.] / leg. II adi(utricis) strat(or) off(icii) [cos] / Mocur signifer c[oh(ortis) (mill)] / Hemes(enorum) q. v. ann. LV [Aur?] / Pulchra uxo[r] ip[/sius] fuentissim[a (sic) posuit.][401]

2. M(arcus) Aur(elius) Malc[h]ia[s mil(es)] / leg(ionis) II adi(utricis) strat(or) off(icii) [co(n)s(ularis) et] / Mocur signifer c[oh(ortis) (milliariae)] / Hemes(enorum) q(ui) v(ixit) ann(os) LV / Pulchra uxo[r] ip[sius] / fuentissim[is(!) posuit][402]

3. M(arco) Aur(elio) Malci[h]ia[no] mil[iti] / leg(ionis) II Adi(utricis) strat(ori) off(icii) [co(n)s(ularis) sum]/mo cur(atori) signifer[o] coh(ortis) |(milliariae)] /Hemes(enorum) q(ui) v(ixit) ann(os) LV[3] / Pulchra uxo[r] ip[sius 3] / faventissim[o fac(iendum) curavit][403]

[392] Fitz, Fülep, Mócsy, Radan, Scheiber, Tóth I, and Visy maintain this opinion.

[393] Solin and Hampel maintain this opinion. The latter also assumes that the word synagogue denotes, not a building, but a community. Solin says 'daß die Inschrift nicht jüdisch ist. Andererseits zeigt sie, daß in Intercisa oder in der Umbebung sich eine jüdische Synagoge befand. ... Fast überraschender ist, saß sich sonst keine Spuren von Juden im syrischen Lager von Intercisa finden lassen.' SOLIN 1989: 336.

[394] In an inscription Cosmus, ILS 1778: C. Iulius divi Aug. l./ Cozmus (sic) speclaria. / Agrippianus, hic situs est. / C. Iulius Cosmi l. / Sabinus.

[395] HAMPEL 1906: 230.

[396] Revision was possible after the stone was cleaned.

[397] The name, as an imperial libertinus, also occurs in other places: STEIN: Kosmos RE XI, 2 1498-1499.

[398] A similar formation may be read in another inscription. ILS 5205: scaenici / Asiaticia/ni et / qui in eo/dem cor/pore sunt, / vivi sibi fe/cerunt.

[399] AnnÉp 1903 No. 68. Cf. APPEL 1909: 98. Cf. 52.II.c.

[400] SOLIN 1983: 763-764.

[401] ERDÉLYI G/FÜLEP 1954, No. 17.

[402] AnnÉp 1891 59.

[403] AnnÉp 1992 1456.

Opinions:

- Róbert Fröhlich supposes that M. Aur. Malchias was a consularis, or at least centurio because otherwise he could not have occupied the post of the strator officii.
- Bálint Kuzsinszky mentions two persons; one of them is M. Aur. Malchias and the other one is Mocur. Both of them are considered to be of Syrian origin. Pulchra was Mocur's wife who was called Aurelia Pulchra.
- On the basis of the names, the tombstone is regarded as a Jewish one by Sándor Scheiber. He thinks that M. Aur. Malchias began the cursus honorum as a soldier of the legio II adiutrix and finished it as the signifer of cohors milliaria Hemesenorum at the age of 55. His wife Pulchra set the tombstone. Scheiber's reading shows that he misunderstood the inscription.
- According to János Szilágyi Mocur was Malcihia's father. Mocur was a signifer, Malcihia was the groom of the procurator.[404]
- Solin points out two things in connection with the stone and the readings. One is that Sándor Scheiber misunderstood the inscription. The other is that a Syrian soldier and his family are in question; none of Jewish soldiers and families.

My own point of view. The reading of the cognomen is very questionable because the letters can hardly be seen. Another tombstone can be taken as an analogy where the name on the stone is not damaged. Here it says M. Aur. Malcihian[us], who was a veteranus of legio II adiutrix.[405] Another cognomen from Intercisa is the brick stamp of *Malcianus*.[406] It is difficult to doubt the reading because the stone was lost, so the readings cannot be checked. However, I should point out here that the M of miles could clearly be seen in the photo of the stone. In addition, I agree with H. Solin that the soldiers mentioned here were not Jewish ones.
The stone dates to the 3rd century.

12.II.e. Sarcophagus of M. Aur. Sallumas, fig. 96.

Location: MNM 154.1874.4. **Literature:** CIL III 10.318; RÓMER 1875 45-46; ORTVAY 1875 123. no. 20; HAMPEL 1906 223-224; KERÉNYI 1941 187, no. 2168; ERDÉLYI G/FÜLEP 1954, no. 133; SCHEIBER 1955 197, no. II/3; SCHEIBER 1960: 45-47; FITZ 1972a: 175; RADAN 1973: 272; SCHEIBER 1983: 41-42. no. 6; SOLIN 1983: 761; SOLIN 1989: 233-236; RIU, 1202.

It was found in the vineyard of Gáspár Farkas in 1874.[407] Dimension: 40 x 150 x 12 cm. Inscription: D(is) [M(anibus)] / M Aur(elius) Sallumas vet(eranus) / ex tess(e)r(ario) coh(ortis) (milliariae) Hemes(enorum) domo / Hemesa annorum LXXXII vivo / sibi et Aureliae Mat[---] / quondam coni[ugi --- / --- / --- /[408]

Sándor Scheiber points out "the Jewish origin of the veteran official is indicated by his biblical name."[409] The name is in connection with Sallum, so Scheiber thinks that it is wrong to identify it with the name Salmas.[410] H. Solin thinks that he is not a Jewish, but a Syrian soldier. In the Syrian cohors mostly Syrians served. On the basis of her arguments I agree with her and I do not list the stone among Jewish finds.

In English: To Spirits of the Departed. M. Aurelius Sallumas, the veteran clerk of the cohors milliaria Hemesenorum, originated from Hemesa, lived 82 years. Erected it in his lifetime to himself and Aurelia Mat..., who once was his wife.
It dates back to the 3rd century.

12.II.f. Tombstone of Aurelia Baracha, figs. 97, 98.

Location: MNM 22.1905.12. **Literature:** MAHLER 1905: 225-226, no. 4; AnnÉp 1906 no. 107; HAMPEL 1906: 223, 270 fig. 49; HAMPEL 1907: 314; SCHOBER 1923 Nr. 198, Abb. 103; ERDÉLYI G/FÜLEP 1954, no. 19; SCHEIBER 1955: 196-197, no. II/2; SCHEIBER 1960: 40-44, no. 6; FITZ 1962b: 96; FITZ 1972a: 175; RADAN 1973: 272-273 Fig. 6; BARKÓCZI 1982: 36, 40, 41, 43 Abb. 12/2; SCHEIBER 1983: 36-39, no. 5; BARKÓCZI 1983: 137-140, 142 Taf. 22 /28; SOLIN 1983: 761; SOLIN 1989: 233-236; RIU 1161; EDH-Nr: HD021905.

The stone was found in the vineyard of Károly Tiszuczky in 1905.[411] Dimensions: 193 x 90 x 17 cm. According to RIU the inscription reads: D M / Aureliae Barachae [v]ixit / ann(os) XXXV et Aur<e>l(iae) Ger/manilae vixit / ann(os) IIII et altera / filia Aurelia Ger/manilla vi[xi]t ann(os) / II et Immostae matri su(a)/e vixit ann(os) LX German/ius Valens mil(es) coh(ortis) (milliariae) Hem/es(enorum) uxori et matri et fi/liis posuit et sibi {v}vi(v)us fe/cit.

In English: To the Spirits of Departed. For Aurelia Baracha who lived 35 years and for Aurelia Germanila who lived 4 years and (for) the other daughter (for) Aurelia Germanilla who lived 2 years and for Immosta, her mother, who lived 60 years. Germanius Valens, a soldier of the cohors milliaria Hemenesorum. He erected it for his wife, mother and daughters and made it for himself while he was still alive.

Sándor Scheiber thinks that the people mentioned on the stone are Jews on the basis of their names. However, the fact that the soldier served in a Syrian cohors refers to his Syrian origin. In connection with the study of the names, Sándor Scheiber writes that the name Baracha is a male name both in its biblical and Greek form Βαραχος. However, it is explicitly a woman's name here.

I have to point out that neither Ede Mahler nor H. Solin in modern times listed the stone among Jewish finds.[412] We have to do so because style and habit of the tombstone refer not only to the province Syria, but also to Palmyra in Syria, just as László Barkóczi stated this in connection with the workshop of the tombstones.[413] Other groups of finds demonstrate the same characteristic, i.e. the Syrian

[404] János SZILÁGYI, Intercisa helyőrsége. In: Magyar Múzeum II. 1947: 11.
[405] RIU 1190; AnnÉp 1971 341.
[406] LŐRINCZ 1978: 108 No. 9/37. Evaluation: FITZ 1994: 1270-1271 No. 909.
[407] The topographical lot-number of the plot owned by the Farkas inheritors is 1794 - BARKÓCZI 1954: 4. There is no further accurate data in the scholarly works about the plot of Gáspár Farkas.
[408] RIU 1202.
[409] SCHEIBER 1960: 47.
[410] SCHEIBER 1983: 42.
[411] The topographical lot number could not be identified according to literature.
[412] MAHLER 1905: 226; SOLIN 1983: 762; SOLIN 1989: 233.
[413] BARKÓCZI 1982: 18-50.

connection was strong in Pannonia at the age of Severus.[414] On the basis of the above arguments the tombstone cannot be listed among Jewish finds.
It dates to the years between 25 and 50 in the 3rd century.[415]

HISTORICAL EVALUATION - DUNAÚJVÁROS
Dunaújváros is in a particular situation because relatively many finds of Christian features were found in graves. While a settlement was being excavated, only a standard lamp became known. Therefore, there are no more finds indicating Christians in the settlement. In this respect, Dunaújváros is similar to Aquincum where there are also Christian burials by the canabae. However, the settlement does not provide enough evidence (at least under the present stage of publication because the finds are mostly unpublished). In Dunaújváros, the only building regarded as a basilica is not a basilica. It is situated not in the settlement, but on the southern border of the early cemetery.
It is certain that Christians lived here. Hence, theoretically some finds referring to Christianity might be found in the settlement, as well. Christians were probably arriving here from Syria from the 2nd century on. It is almost certain that the Christians in Intercisa were of eastern origin; the depictions on casket-mounts and the standard lamp of eastern type demonstrate this.
Hypothetically, the following can be suggested:

- Intercisa tends to be the place where Christians were mostly among soldiers. I would like to remind to the fact that the casket-mount with the inscription ΣESV was found inside the fort. Therefore, if it is true that soldiers contributed to the spreading of Christianity, this case confirms it even stronger.
- In the army of Intercisa, Syrians, in a wider sense people of eastern origin, were the most numerous. The presence of Jews among them was also supposed, but this was based on the misinterpretation of inscriptions.
- Since Syrians married only Syrians, the relations of the soldiers were also Syrians/Orientals; hence, their finds might be found also in the settlement. I would like to point to the standard lamp whose construction has eastern features. If a soldier was Christian or became one, his relations were also Christians or became Christians.
- The dwelling-places of Christians and the buildings for liturgy should be found around mithraeums here just like in Aquincum. Nevertheless, the places of mithraeums are not known here, so this theory is more hypothetical than the above ones.

The building considered as Christian, but actually not of Christian use. The building was listed among early Christian basilicas because of its apse. Today, buildings with an apse are automatically regarded as Christian churches.[416] In the ancient time, it was not so. In antiquity, not all the buildings with an apse were churches and it is also true that not all the churches had an apse. A church like any other object had countless symbols which had the aim to remind either Christian or pagan people of God, and thus distracting their attention from secular things.[417] Mankind professing God's perfection attempted to look for the perfect material form in the secular world. It was the sphere.[418] Thus, the Pantheon became the perfect temple of Rome.[419] A Christian church is not only the place where God's statue is preserved, but also a space where worshippers have enough room and take part in the liturgy. Christians themselves did not call these spaces churches. *Domus ecclesiae* was a frequent name[420] where the word *ecclesia* denotes the community itself similarly to the word *synagoga.*[421] A church is not man's creation, but that of God.[422] "Diese Vorstellung vom Kultbau als ein dem Gottesdienst der Gemeinde geweihtes Haus ließ die Auffassung der frühchristlichen Gemeinde klar zutage treten, daß die eigentliche Ecclesia, der eigentliche Gottesbau nicht die Kirche aus Steinen, sondern der aus lebendigen Steinen zusammengesetzte Tempel Gottes, der Leib Christi sei."[423] The continuation of this idea is the *Ecclesia*, the church, they "sind die Berufenen, die Erwählten, die aus aller Welt herausgerufen werden."[424] The building created by man is a *domus*. When a basilica or a church is built, the building (house) should represent perfection.[425] This perfection was expressed in different ways. One was the octagon.[426] At the same time, unity

[414] BARKÓCZI 1981: 65-66; BARKÓCZI 1986: 183-184.
[415] BARKÓCZI 1982: 41; BARKÓCZI 1983: 142; BARKÓCZI 1992: 31.

[416] In ancient time, the word 'templum' did not mean the same as today. See: St WEINSTOCK: Templum RE V/A 1 (1934) 480-485. P. CATALANO: Aspetti spaziali del sistema giurdico-religioso romano. *Mundus, templum urbs, ager.* Latium Italia. ANRW II 16 452-491. Cf. DEICHMANN 1964. I use the word according to its present sense.
[417] A 'pagan' did not always behaviour in a 'pagan' way, as we read at Livius; LIVIUS I, 21. P. 58. (Auctores Ltini III. Budapest 1967): *Ad haec consultanda procurandaque multitudine omni a vi et armis conversa, et animi aliquid agendo occupati errant et deorum adsidua insidens cura, cum interesse rebus huémanis caeleste numen videretur, ea pietate omnium pectora imbuerat, ut fides ad ius iurandum pro summo legum ac poenarum metu civitatem regerent.*
[418] BRENDEL 1977. Cf. note 427.
[419] Although I quoted Doris and Gottfried GRUBEN's work, nevertheless, the quotation is H. Drerup's: "Griechische Bauten sind gestalthafte und gestaltete Dinge dieser Welt – römische Architektur führt hinaus in die Transzendenz." (GRUBEN 1977: 3).
[420] GAMBER 1968: 14. Gamber translated the term *domus ecclesiae* "Haus der Gemeinde (Gottes)" – GAMBER 1968: 13.
[421] As for the rest of the inscription 15.II.a, the word *domus* was misunderstood. ☞ County Fejér. Derivation of the word, see SAUSER 1966: 466: "Das Wort, das etymologisch aus 9k kaleĺn abzuleiten ist, bezeichnet normalerweise die Vollversammlung der Heraufgerufenen, … Im hellenistischen und christlichen Wortgebrauch versteht man darunter das versammelte Volk, … "
[422] Vulg. Is. 66, 1. Klaus GAMBER examines the terms used for church (1968: 13-17): " … die Kirche als ein *tabernaculum,* ... auch *parocia* genannt." (p. 14). The church is built not by man's hand – MARTIMORT 1963: 186-187.
[423] SAUSER 1966: 464-465. He deals with this subject in detail on 466-495.
[424] SAUSER 1966: 468.
[425] The picture of the celestial Jerusalem: SAUSER 1966: 472. He writes on page 473 that Eusebius was the first to call this building a church. (hist. eccl. X, 4).
[426] GAMBER 1987 Fig. 20, p. 65: "Es findet sich als Grundriss von

might have been expressed architecturally in the pagan or in the Christian antiquity: the sign is the circle. The circle starts from itself and returns to itself.[427] The centre of this circle is also important - Apoc. 4,2. "Mittelpunkt (Thron) ist das τὸ ἕν Platos, der anfang- und endlose Kreis das ἄπειρον (Unendliche). Auf diesem Kreis haben unendlich viele Punkte Platz. Das Wesen einer Zahl macht die Beziehung ihrer 'Punkte' zum Mittelpunkt aus."[428] The A and Ω belong in this range of symbols. I would like to refer back to the mirror, which is the symbol of this endless circle whose centre is marked. Considering this interpretation of the symbols, it is obvious why the ancient people in the East of the Empire preferred central space. In the west of the Empire, central space was less frequently used than the axial structure with an apse. The apse as a quarter sphere or a half-sphere represents the *axis mundi.*[429] It also represents the throne, the centre, and the perfect. Thus identifying the occurrence of an apse as a Christian church seems to be logical. However, since these symbols are primordial and they occur on other Roman, "pagan" objects, they cannot be solely connected with Christianity. Neither chronologically nor in its idea does the establishment of a domus require a spherical house, a central space where quiet life is possible. To fulfil the aim of the perfection it has to satisfy the celestial patterns or ideas, together with its perfection. The depiction of the seven planets is the special way of this.[430] Here I have to refer back to the above opinion of Erika Dinkler-von Schubert (see casket-mount, 12.I.c.) and also to the casket-mount in ☞ Császár (6.I.a.) where the seven planets and the days of the week are depicted. Moreover, it should be pointed out that the cult pictures of Mithras also depict the seven planets. Considering the symbols 'taken over' by Christians we concluded that the apse was the latest to be aqcuired. Christians did not use the apse before the time of Constantine I. Even more, the present sense of the church, as the house of God, is identical with the pagan sense of the templum; and this sense appeared also late with Christians. The first such church was built in 313.[431] Here, in Pannonia Prima and Valeria, there were probably no Christian churches with an apse before the dynasty of Valentinian.[432] Hence, we may well say that especially in Intercisa, where most of the population was of eastern origin, *interpretatio Christiana Pannonica* was adopted. This meant that the place for Christian liturgy was rectangular, divided into two rooms with a wall built parallel to the shorter side. Let us regard the space developed in this way as a unit of a church; this unit is the *sine qua non.* The domus may have further rooms whose function is not easy to determine. Now I would like to repeat and propose the hypothesis – of course - "mit gebotener Vorsicht", that in Pannonia not always a Christian mission was in question. The Mithras worshippers were baptized following the example of Syrian Christians; that is why, the first Christian churches here were not buildings closing in an apse, but rectangular, mithraeum-like spaces, which developed according to the requirements of the Christian liturgy. The transformation was possible when shrines were rebuilt into spaces of Christian use, into *domus ecclesiae* that became secular Christian spaces. I know well I have to handle this hypothesis with care, although it seems more than a hypothesis here in Intercisa. For a correct evaluation of the unearthed buildings, it is important to know and define the new liturgical requirements in space and time. To fulfil this aim I used the scholarly works listed below.[433] As regards the archaeology, it is necessary to date the buildings in question as precisely as possible. I quote Gamber about this: "Dabei wäre jedoch eine genaue zeitliche Festlagung von großer Bedeutung; um zu wissen, ob der getreffende Kirchenbau vor oder nach der änderung der Liturgie liegt. Im allgemeinen bildet der Beginn des 5. Jh. für uns (sc. Alpen- und Donaugebiet) die Grenze."[434] In addition, the Syrian influence must be taken into consideration in Pannonia and Intercisa. The liturgical changes happened earlier in the province of Syria and local customs (buildings and equipments of the meals) were also different from those in the west.

Summarizing, I list the important issues:

- Among the finds of the cemetery, there are several Christian ones.
- There was one Christian find discovered in the fort.
- The standard lamp is known from the canabae.
- The finds from the cemetery are not proportionate with the finds of the settlement, so further Christian finds are expected from the settlement.
- The features of the finds suggest a particular eastern Christianity in Intercisa. There might have been soldiers among them.
- Because of the special features, symbols had to be studied: the sphere, the circle, the centre, the apse, the seven planets, the Zodiac, the A Ω, and the cross.
- On the basis of this study it can be supposed that the scenes of Christian liturgy were not the apse buildings, but the rectangular spaces divided into two.

13. EGYHÁZASHOLLÓS

13.II.a. Marble screen, fig. 99.

Provenance: It was built in the wall of the church. **Literature:** THOMAS 1975-1976: 105-160 Abb. 36; THOMAS 1977b: 79.

Baptisterien (wegen der Auferstehung), aber auch von Kirchen und wird hier als Durchdringung der irdischen mit der himmlischen Welt (2x4!) gedeutet."

427 GAMBER 1987: 38 – He quotes Ambrosius. Cf. n.418.

428 GAMBER 1987: 38.

429 VANYÓ 1995: 35-44.

430 GAMBER 1987 deals with the Christian interpretation of the seven planets. HÜBNER (1983: 37-43, 120-128, 144-147) deals with the same subject and the signs of the Zodiac. BULLINGER rep. 1974: the firmament with star-pictures reveals the Heavenly Volume of Divine Revelation.

431 DEICHMANN 1964: 58.

432 ☞ Aquincum, 4.I.g. and the Historical evaluation.

433 JUNGMANN 1948; LIETZMANN 1958; GAMBER 1968; GAMBER 1971; GAMBER 1971; GAMBER 1976; JUNGMANN 1977; GAMBER 1981.

434 GAMBER 1968: 18. It occurs, in connection with the Pannonian archaeological finds, that liturgical practice dates the finds, instead of the finds dating liturgical practice.

A grating carved in a frame. It seems to be open work, but it is not. The ligature of a cross and a Greek X designed the motive. Edit Thomas identified it as chi-rho monogram. ☞ Szombathely, 51.II.a. Since several such barrier pieces survived, I am going to evaluate them together with the piece from Szombathely.

14. ESZTERGOM-BÁNOMI-DŰLŐ

14.I.a. Tombstone of Kassia, fig.100a-b.

Location: Zsidó Vallási és Történeti Gyűjtemény, previous Inv. No. 2500, present Inv. No. 104.1963. Inventar number published in Budapesti Zsidó múzeum. It was taken from the primatial collection to the museum above in 1934. **Literature:** CIL III 10599; FRÖHLICH 1891: 236 no. 26; KRAUSZ 1904: 172-173; SCHEIBER 1960: 25-29 No. 3; RADAN 1973: 272-273 Fig.7; THOMAS 1977a: 23; SCHEIBER 1983: 15-19 No. 1; SOLIN 1983: 76; A Budapesti Zsidó Múzeum. Ed. Ilona BENOSCHOFSKY / Sándor SCHEIBER. Budapest 1987: 8; RIU 787.

The stone was found near the primatial brick maker in 1890 and was taken to the primatial collection. Dimensions: 148 x 76 x 11,5 cm. It depicts a menorah. According to statements, the inscription was written both in Greek and Latin. Fröhlich's reading as follows: Memoria(e) Iudati pat[ris] / memoria(e) Kasete / Eul ... There is an R instead of P in the word "memoria" and Fröhlich marked this. Krausz did not accept this reading. He modified it as follows: Memoria Iudati pat[ris] / memoria Kasete matr[is] / Eva. Krausz could discern further letters following the woman's name and he mentioned them. Following the example of the first line, he interpreted these uncertain letters as *matris*. Krausz thought that the woman's name in the last line is the person who erected the stone. Consequently, a name must be there. He had no doubt that the Jews set it. Both the depiction and the names would prove his opinion. He thought that the name Iudat is a form of Iudas. Sándor Scheiber published the following reading: Memoria Iudati patiri et memoria Cassi(a)e ευλ(ογια). The translation by Sándor Scheiber: Memory for father Juda and memory for Cassia. Blessing. According to Scheiber, the word *patiri* stands instead of *patri*. Radan's reading: MEMOPIA IVDATI PATIR/MEMORIA KACCIE/EYΛ. He does not interpret the reading.

My own point of view: Both Greek and Latin letters were applied in the word "memoria" because there is a capital letter P instead of R. Therefore, it should be supposed that the same may occur in the inscription again. First line: The last letter is kappa and not a capital R as it has been interpreted so far. The letters, which were regarded to be a T, are dubious. Hence, the readings Iudati or patiri or patri are certainly not correct. It is possible that it is not only one word, but also the abbreviations of several words put down in a continuous way.[435] One of the abbreviations is the person's name that the memoria was erected for. Just like in the second line where there is a name after the word *memoria*. The reading of the name Kassie seems to be wrong. The correct reading is Kacete, which can also contain abbreviations. I think the Greek word in the third line should be completed as eul(ogion) because in a case of a tombstone it is more appropriate to mention a funeral sermon.[436] Thus the inscription is: MEMOPIA IVBATIPAFIK / MEMOPIA KACETE / EYL.

[435] An inscription like this was found in Syria, not far from Antiochia, in Dar-Ali-el-Hochami. DACL 11/1 sp 307: MEMOPIA BATI IVLI / LNI / ECOMBICEMALOS DI / ACONVS FECI = memoria b(e)ati Iuli/ani/ ego Bicemalos di/aconus feci.

15. COUNTY FEJÉR

15.II.a. Building inscription

Provenance: Rómer and Desjardins think that the stone was discovered in Göböljárás in Fejér megye in 1845.[437] Later it was kept in the castle of Alcsut. It was handed over to Magyar Nemzeti Múzeum as a bequest of the Palatine József January 13th 1847. Some think that it was found in Temesvár, but I do not think so. **Location:** MNM R-D 129. **Literature:** R-D pl. LII; CIL III 3370; NAGY L 1938: 66; ILCV 518; BT 1942: 766 n. 9 (NAGY L); MÓCSY 1962: 756; HOFFMANN D 1969: 74, IV 191.

There is no depiction on it; it is a simple stone tablet. The reading by Mommsen as in CIL: Fl Iovinus ex pp / militum Histricrum (sic) / et Fl Paulus biarcus / pater et filius domum / a fundamentis / iusserunt fabricari. Th. Mommsen thought that the word *domus* denotes a private house,[438] however, L Nagy and A Mócsy regard the building to be a Christian church. Thus, the inscription is considered an early Christian one which would point to the Christianity existing here.

It is obvious that the idea of the house of God, the *domus ecclesiae* suggests the Christian interpretation of the inscription. The expression could not be used arbitrarily; J. E. Stambaugh explained this in his work.[439] In early times, Christians gathered in private houses to celebrate the remembrance. The Last Supper[440] took place in a private house, so they gathered in private houses to follow the example. This gathering and celebration was called κοινωνια. Apostle Paul uses the expression *caena dominica* denoting the meal, which was held together with eucharist, but this was never called an agape.[441] The different activities became better defined by the end of the 2nd century. Agape meant treating the poor from the beginning of the 3rd century; and it was not connected with the celebration of eucharist.[442] The meal mentioned together with the Eucharist had its own historical process. It was increasingly separated from Eucharist and finally the Eucharist could be celebrated without a meal. This process would finish at different times in different areas, but

[436] Eulogium = funeral sermon. Eulogia: see STUIBER, A: Eulogia RAC VI, 900-928.

[437] It was found in 1841, according to the record in CIL.

[438] It would be in use in at the time, theoretically, but we cannot prove it based on the text.

[439] STAMBAUGH 1978: 602-605; also the *domus Dei* was mentioned instead of the *domus ecclesiae;* see this subject in detail: GAMBER 1968: 13-18.

[440] The Jews held a banquet among friends, therefore the liturgists have asked the question whether the Last Supper was such a friendly banquet (chaburah), or perhaps a banquet celebrated in advance of Easter – JUNGMANN 1948 I 11-14; GAMBER 1968: 64-71; JUNGMANN 1977: 9-10, 20-21. GAMBER's opinion: "Doch scheint dies mehr ein Streit um Worte zu sein, ..." 1968: 64.

[441] GAMBER 1968: 66; JUNGMANN 1977: 21-23; Jungmann mentions a source, which would be an exception, but it cannot be interpreted clearly.

[442] JUNGMANN 1948 I 18; JUNGMANN 1977; 22.

generally, it was developed by the 2nd century.[443] These feasts took place in houses whose original aim was to dwell in. Although it is true that houses were rebuilt on bases of older buildings, this inscription tells us different because the house in question was built from its foundations, and there are no words about rebuilding.
The interpretation: Flavius Iovinus is the father, Flavius Paulus is the son. Both of them were soldiers as it was common in the late Empire. To secure the new generation of soldiers military service was inherited. As a result of this, boys grew up with their fathers in the army. The eldest son inherited the title *biarchus*.[444] This was indicated by the title following their names. EX PP = ex p(rae)p(ositis). HISTRICRVM = Histric(o)rum, of the superiors of the soldiers of Histria. László Várady identifies the title BIARCVS = biarc(h)us with the title sergeant.[445] Th. Mommsen comments on the inscription as follows: "biarchus et ipsi milites pertinent ad scholam agentium in rebus ..." In English: Flavius Iovinus of the chief of the soldiers in Histria and Flavius Paulus, a superior of provisions, had the house built from its foundations.
The inscription tells only about the fact of having the house built. It is datable to the middle of the 4th century. The inscription cannot be regard as Christian. Another inscription found in Szentendre[446] should be mentioned here in connection with the title biarchus occurring together with the name Flavius.

16. FELSŐDÖRGICSE

16.II.a. Carved stone plaque, fig.101.

Provenance: It was secondarily used in the course of building a medieval church. **Location:** MNM 51.1882.4. **Literature:** KUZSINSZKY 1920: 153; MRT 2. 18/2; TÓTH E 1974a; FÜLEP/TÖRÖK GY 1977: 313 Abb. 393c; TÓTH E 1980b: 98 Abb. 2; KISS Á 1987: 21 Nr. 24; Ch. ERTEL: Römische Architektur in Nordwestungarn, In: CSIR (Ungarn) VIII, Die Skulpturen des Stadtgebietes von Salla Mogentiana sowie des Balaton-(Plattensee) Oberlandes in der Kommitaten Zala und Veszprém. Bearb. Ch. ERTEL / S. PALÁGYI / F. REDŐ. Budapest 1999: 150. Kat.-Nr. 136.

An excavation was led on the site in 1959. The following finds were made: "Two stone blocks, sandstone balk, a red-sandstone capital decorated with leaves, a hypobasis of a sandstone statue and finally the fragment of a lid of a sandstone sarcophagus. The blocks were used as a stone fence of a burial garden. Tihamér Szentléleky dates them to the last decades of the 2nd century. ... He thinks that one of the sarcophagus fragments date back to the 4th century, which he regards as an evidence for the continuous Roman settlement after the 3rd century below Kőhegy. ... He suggests that a Roman tomb building was rebuilt during the Middle Ages."[447] One of them is decorated with an animal frieze. This was considered by Endre Tóth to be an early Byzantine stone carving. On the basis of the animal frieze which resembles that of an ambo from Ravenna, he stated that this stone carving was part of an ambo or of a screen.[448] Applying as analogy the ambo of Ravenna, he dated ours to the third quarter of the 6th century; although, there are some differences both in quality and depiction of animals, e.g. animals on the ambo in Ravenna look upwards and towards the middle, on the Pannonian fragment the animals look in different directions. The carving cannot be dated to the 6th century, but rather to the time indicated in the reports on the excavations. Ch. Ertel identified it as a Kassettenplatte and decided that it and the other fragment belonged together and formed one object.
Because of the wrong analogy, it could be considered as early Christian, however, it is not that.

17. GYŐR (ARRABONA) fig.102.

17.I.a. Finger ring

Location: Bencés Gimnázium (Benedictine secondary school), Győr. **Literature:** BÖRZSÖNYI 1904: 254; NAGY L 1938: 90.

It was found in a grave in the former Honvéd liget together with another ring. Both were made of bronze. One has a depiction of a hand, and the other has a Greek cross. Among other finds, there were coins of Constantine I, Constans, and Constantius II.

17.I.b. Clay lamp, fig.103.

Location: XJM 53.237.78[449] **Literature:** ALFÖLDI 1931-1932: 1 fig. 1; IVÁNYI 1935 No 989 pl. XL, 8 and LXXVI, 7 and LXXVIII 47; NAGY L 1938: 91.

L: 10cm, Diam: 6,4 cm, H: 4,5 cm. Discus: chi-rho monogram flanked with filling holes. Rim: series of circles alternating with double lines. On the bottom PE are ligatured, which is interpreted as p(alma) e(merita) by András Alföldi and as Petrus by Margherita Guarducci.[450] Type Abbiani 1-a.

18. KAJÁR (SZENTLÁSZLÓ)

18.II.a. Roof tile fig.104.

Location: FM 78.274.1. **Literature:** CIL III 11467; NAGY T 1978; THOMAS 1982b: 265 Abb. 7 – Szentendre is the wrong provenance; IIL (MARÓTI) Kat.-Nr. 318; GÁSPÁR 1997.

Tegula with an inscription. The inscription is framed with finger-drawn mouldings. Dimensions: 45 x 34 x 3 cm. The reading according to CIL: NV S / SVRVS QVI / OFFICIVM DE/DICATVM HABET / VIVAT PER MVLTA/ SECVLA SEMPER. Edit Thomas considered it

---

[443] JUNGMANN 1977: 23. Everybody who had just been baptized, immediately took part in a thanksgiving and received their first Eucharist. On this occasion they would be given one calyx with milk and honey and another with water. The milk and honey is interpreted as a drink for children; the water symbolized the spiritual *castitas*. – JUNGMANN 1948 I 21.

[444] VÁRADY 1961 23.

[445] VÁRADY 1961 23: If the father died, the eldest son, the *bioarchus*, was obliged to take care of his younger brothers. That is why the building (*domus*) may be interpreted as a forsight.

[446] RIU 906: D(is) M(anibus) / Fl(avius) Superianus biar/co (!) qui militavit Co/nstantia vixsit (sic) ann(os) / ... Constantia = Castra Constantia = Szentendre; BARKÓCZI 1973: 77 Abb. 7; BARKÓCZI 1982: 28 Abb. 5; BARKÓCZI 1983: 128 Taf. 8.11

[447] MRT 2. 18/2.

[448] TÓTH E 1974a.

[449] Presently it is recorded as an unknown site. Nevertheless, Dóra IVÁNYI clearly wrote of Győr as a definite location. Regarding the inventorizing campaign in Hungary during the 50 year period, it is more than probable that the initial data was correct.

[450] GUARDUCCI 1978: 1753.

early Christian because of *secula semper*. She defined letter R as being upside down above the inscription.

The inscription on the tegula has two separate parts which were not written by the same person. One can imagine that two workmen were standing opposite each other and they were writing on the brick each on his own side before the brick was burnt. Hence, the two inscriptions are opposite each other. The reading of the other inscription is ROM[A]. Let me leave out dealing with is in details because I have done it elsewhere. I do not consider the piece to be Christian.

19. KAPOSPULA-ALSÓHETÉNYPUSZTA / Alsóhetény, (Iovia?) fig. 105.

Literature: SOPRONI 1974a; SOPRONI 1978: 149-150; FITZ 1980a: 56; BARKÓCZI/SALAMON 1984: 168-169; SOPRONI 1985: 17-21; TÓTH E 1986b: 6-13; TÓTH E 1988a; TÓTH E 1988b; TÓTH E 1989b; TÓTH E 1991a: 123-127; TÓTH E 1994b Abb.11.

RESEARCH HISTORY

Sándor Soproni unearthened parts of late Roman Binnenfestigung in this area. He started the excavations in 1969 and Endre Tóth continued them. Ditches to the west and to the south surrounded the fortification. Sándor Soproni distinguished three architectural periods next to the south gate. In the first period it had a horse-shoe shaped tower, in the second a round one, and in the third period "covered postern was built in defence of the gate inside of the fortification."[451] Some fan-shaped towers from the first period were found at the corner. The first period of the fortification could be determined. South of the fort "dwelling and refuse pits from the 4th century were found."[452]

According to the scholarly works, the following is known about the fort. Fig. 106.[453] Dimensions: 500 x 475 m. The wall of the fortification is 150 cm wide. The wall surrounds the north south oriented valley with the river, which provided the fortification with water. Next to the north gate, parallel to the wall there is a storehouse measuring 60 x 21 m. It is divided into three parts; the part in the middle is deeper, which is divided into three by eight pairs of columns. A row of pillars runs outside the wall. There is a rectangular room in front of this. The building was undecorated. It had no floor either. South of this building, the walls of a workshop were found and an oven used for baking bread was also discovered. Next to the east gate barracks were uncovered. According to Endre Tóth the fortification has strong military features. It was built during the reign of Constantine I. or Constantius II. And it was rebuilt between 375 and 380. Civil population lived south and west of the fortification and in later periods took shelter in it.[454] A chapel (with graves) was uncovered between 1984 and 1986. This is grave building I. At the same time, 41 graves were also uncovered; among them were cremated graves. All of them dated back to the 4th century. E. Tóth thinks that the cremation indicates a foreign ethnic group, probably the Carps who were settled in this region in 295 by Galerius. The traces of further 5 or 6 chapels could be observed. The graves of the chapel had no grave goods; the only find were the remains of the textile of the mortuary cloth, which was woven with golden fibres. The burials inside the chapel were different from the burials in the cemetery. The other graves were earthen ones. E. Tóth thinks that the commander of the fortification was not of such a high rank that he could build the chapel. Only the Christian community could do so. That is why he regards the place as the see of the bishop Amantius. He thinks that the object described below consideredto be a fibula, supports this idea.

The annual reports on the excavation show that many Roman stone carvings were built secondarily in the fort wall, as well as marble fragments, stone slabs, tombstones, lion statues and fragments with inscriptions, all of them originating from former Roman cemeteries.[455] Further reports report on the investigation of the chapel and cemetery.[456] Sándor Soproni thinks that the fortress was partly built over them.[457]

According to Sándor Soproni, comitatenses were stationed here; E. Tóth also thinks that the fortress served military aims. Building 4 would prove this opinion because E. Tóth holds it as a barrack.[458] András Mócsy considers it a town.[459]

The largest building was regarded a horreum with a west foreground. The bath was also excavated.[460] Endre Tóth dates the first period of the fortification to the second third of the fourth century and he dates the second period to the time after 375 based on the second period of the fortification in Ságvár.[461] ☞ Ságvár.

The cemetery is situated about 600 m southeast of the fortification. In the cemetery, some sherds of glazed tegulas were collected.[462] In the cemetery, there were cremation graves, cremation burial with scattered ashes, inhumation graves and graves with gabled roof. Grave 26

[451] SOPRONI *RégFüz* Ser I. No. 24 1970: §44.

[452] SOPRONI *RégFüz* Ser I. No. 25 1971: 71: according to Soproni, the measurements of the fortification are 500 x 450 / 475m (width of wall 260 cm). It had more than 50 towers: SOPRONI 1974a: 182.

[453] After Tóth E.

[454] Balga ALEKSOVA (1982-1983: 52) writes that Theodosius issued a decree, which allowed the inhabitants to move into the fortifications and palaces of the imperial offices. I quote this decree; CodTheod. IX, 14, 2. (1. July 391.) IMPPP. VALENTINIANVS, THEODOSIVS ET ARCADIVS AAA. AD PROVINCIALES. Liberam resistendi cunctis tribuimus facultatem, ut quicumque militum vel privatorum ad agros nocturnus intraverit aut itinera frequentata insidiis adgressionis obsederit, permissa cuicumque licentia dignus ilico supplicio subiugetur ac mortem quam minabatur excipiat et id quod intendabat incurrat. Melius est enim occurrere in tempore, quam post exitum vindicari. Vestram igitur vobis permittimus ultionem et, quod serum est punire iudicio, subiugamus edicto: nullus parcat militi, cui obviari telo oporteat ut latroni. DAT. KAL. IVL. TATIANO ET SYMMACHO CONSS. (App. KRVGER P, Edd. MOMMSEN TH et MEYER P M. Berolini ²1954 Tom II p.457.)

[455] TÓTH E RégFüz Ser. 1. No. 37 1983 §79; No. 38 1984 §72.

[456] TÓTH E RégFüz Ser. 1. No. 39 1985 §70; No. 40 1986 §76; No. 41 1987 §64.

[457] SOPRONI 1974a: 182-184.

[458] TÓTH E 1988b: 29.

[459] MÓCSY 1973: 394; 1977: 391.

[460] TÓTH E 1988b: 29.

[461] TÓTH E 1988b: 32.

[462] TÓTH E 1988b: 36.

was a painted double grave. Fig. 107. There were no skeletons in the grave, but a Probus coin and a glass jar with two handles. "The sides of the grave were painted sepia. There is a fence and flower pattern on the long sides, and life-tree on the shorter ones. The trees are decorated with lines between two concentric circles."[463] The largest building in the cemetery is mausoleum no. I, picture of which I am reproducing after the drawing by Endre Tóth. Figs. 108a b, 109. Endre Tóth holds the building to be a mausoleum; however, he mentions baptisteries as analogies. He dates its construction to the reign of Constantius II or to the time after him.[464]

Among the scholarly works, the longest paper appeared in Antike Welt where E. Tóth deals with the fortifications of Környe and Ságvár. Furthermore, he introduces some finds, which cannot be seen in other publications: two of the tombstones built in the wall of the horreum.[465] He found a piece of a textile woven with golden fibres in a stone sarcophagus[466] and he mentions an analogy with the textile from Viminacium.

In the small building next to the mausoleum, a bronze hanging lamp was discovered. He supposed that this rectangle building was used for burial ceremonies and for keeping of burial equipment.[467]

He identified the site with Iovia(?) on the basis of the historical conclusions of the excavations; thus, Alsóheténypuszta might have been the see of bishop Amantius.[468] However, we cannot leave Ludbreg out of consideration.[469] I myself consider any kind of identification to be too early. ☞Ságvár.

19.I.a. Fibula? (A pallium-pin?) fig. 110.

Location: MNM 83.3.1 purchased from István Ács. **Literature:** TÓTH E 1988b: 59 Abb. 24-25; TÓTH E 1989; TÓTH E 1991a: 127-130.

A gilt bronze object cut out of a sheet. The craftsmen did not care about accuracy and beauty of the outline. It consists of a medallion and a shaft. A chi-rho monogram is framed with pearl border on the medallion, on which a trapezoid-shaped sheet is fastened. The shaft is simpler. The complete length is 10.3 cm. Diam of the medallion is 1.8 cm. According to Endre Tóth, it was a pin used for fastening a pallium. A similar object is preserved in the museum of Zagreb.[470] The sheet the object is made of is so thin that it could not have been used as a fibula. It would have broken right away. It was probably fastened to another metal object. It had no other function, but to ornament something and to be a symbol. The shape of the object is strange; it might have been an ornament on a helmet, more precisely on the nasal or above the forehead.

19.I.b. Hanging lamp, fig. 111, 112.

Provenance: "Aus dem Schutt des Gebäudes kamen eine Bronzelampe und das Kettenfragment einer Hängelampe zum Vorschein."[471] **Location:** MNM. **Literature:** TÓTH E 1990b: 267 Textabb.1 Taf.28/1-2.

Bronze. The lamp is 12.5 cm long, 8.3 cm wide and 4 cm high. The length of the chain is 39 cm, the length of the decorative elements is 7 cm.

I think that regarding the proportions it is also possible that the lamp did not belonge to the chain. Therefore, I reconstructed it in a different way using a paper by Catherine Metzger.[472] Fig. 113.

SUMMARY

The investigation is significant, nevertheless the most part of the finds are unpublished. More is known about the position of the buildings. But there is less information about the connection between the small finds and the buildings. Only the excavator himself could give an evaluation. I put down my opinion based on the publications. I start with a sentence quoted by Plough: "the context was not exclusively pagan." Except for the so-called pallium-pin there are no finds that could be regarded as Christian. The other finds do not exclude the presence of Christians, but they do not prove it either.

The so-called pallium-pin was identified as a pin belonging to the pallium because the excavator regarded the territory as the see Iovia(?). But:

- It has not been proved yet whether Kapospula is the same as Iovia(?), thus it has not been proved whether Kapospula was a see of a bishop, either.
- The so-called pallium-pin is not a pin. A thin bronze sheet can match the function of a pin.
- Technically, it must have been fastened on a metal object. Hypothetically, I suggested the helmet because the chi-rho monogram was depicted on the Roman armours.
- The chi-rho monogram on a helmet might refer to someone being Christian or to a Christian soldier. The dating of the fortification does not exclude the presence of Christianity or that of a Christian army. However, let us ask what was the task of the military forces: a/ Were the military troops stationed here? b/ Did they defend the population? c/ Were they bodyguards?

20. KECSKEMÉT

20.II.a. Finger ring - Roman (?) fig. 114.

Location: MNM 73.1905 purchased from Salamon Weiszberger. **Literature:** NAGY L 1938: 112 Fig. 76. I line 1.

[463] TÓTH E 1988b: 39.
[464] TÓTH E 1988b: 39-52.
[465] TÓTH E 1989 Abb. 4-5.
[466] TÓTH E 1989 Abb. 14.
[467] TÓTH E 1989b: 38.
[468] TÓTH E 1988b: 52-60. Sándor Soproni was the first to identify Heténypuszta with Iovia, based on the Itinerarium Antonini (264,8); the map shows Iovia along the way between Sopianae and Brigetio – SOPRONI 1974a: 185.
[469] The previous research identified Ludbreg as Iovia, whose bishop was Amantius, (EGGER 1962a: 57-67). 'Iovia, probably known originally as Botivo, is mentioned in the Itinerarium Burdigalense ... as a civitas' (MÓCSY 1974: 222). About Ludbreg see MIGOTTI 1997 I.a.4. Cf. JARAK 1994: 175.
[470] TÓTH E 1988b: 59 Abb. 26; MIGOTTI 1997 V.a.7. Chi-rho monogram framed pearl band. The object is broken. The Department of Archaeology of the Eötvös Lóránd Tudományegyetem (Universitas de Nominate Eötvös Lóránd) preserved a collection of photo-negatives assembled by András Alföldi and his students over 30 years. These negatives show this object together with lead votives.
[471] TÓTH 1990b: 267.
[472] METZGER 1975. Sándor Ősi has drawn the reconstruction.

A golden ring made of a thin sheet. Its head is rectangular with chi-rho monogram on it. Diam: 2.2 cm. It weighs 8 gramm. Lajos Nagy wrote that it was brought to the MNM from Kecskemét. Hence, Kecskemét may be a secondary provenance, and not the original. We can also doubt whether the object is Roman. Arguments are followers:

- The use of the chi-rho monogram was not restricted to Roman times.
- The feature of the head is not in the Roman style.
- The whole construction of the ring is not in a Roman style.

21. KÉKKÚT

Árpád Csák and János Sági led an excavation in Maktányi dűlő in Kékkút in 1902.[473] Bálint Kuzsinszky published the two uncovered Roman buildings.[474] Fig. 115.

The two buildings are known as basilica I. (fig. 116.) and basilica II. (fig. 117) of Kékkút in the scholarly works. Since that time, the basilica II. was investigated again The new excavation was led by Károly Sági, who, in his publication on the results, mentions only some small finds.[475] We know nothing about the phases of the building.[476] On the other hand, its ground plan became more precise, fig. 118.[477] Károly Sági identified Room 6 as a peristylium whose passage was protected by screens. The columns were made of Perm limestone. They supported the roof.[478] Room 4 and room 5 could not be investigated. In room 5, a millstone was found. Based on this, Károly Sági thinks that it was the kitchen of the villa. Therefore, Károly Sági regards that the pipe found here did not serve the heating system, but that it was a water drain.[479] In rooms 2 and 3 a rebuilding could be observed, fig. 119.[480] After the rebuilding, the room 3 became smaller. Károly Sági supposes that there were corner towers next to room 1 both to the right and to the left. Sporadically, he found fragments of openwork brick; but he could not decide whether they belong to the known earlier brick or to a third one.[481] Both buildings are dated to the 4th century based on the bricks decorated with a chi-rho monogram. Károly Sági states that the villa was built in the 4th century and it had no precedent.[482] Its ground plans is uncommon in Pannonia. The stones from Kékkút are located in the Balatoni Múzeum. These stones were used secondarily except for the columns.[483] Referring to András Alföldi, Lajos Nagy wrote that there frescos were also found.[484] András Alföldi had found some fragments of violet-coloured mortar. Sylvia Palágyi[485] studied the stone carvings, the columns, their pattern, workshops and chronology. She pointed out that there were two types of ornamentation in Kékkút: "geteilte Vollblattkapitelle und Kapitelle mit Eckblatt-Volutenverzierung".[486]

Commentary in connection with the reports on the excavations.

- There is a contradiction between the two statements of Károly Sági: On the one hand, he did not consider the excavation sufficiently to establish the phases, on the other hand, he declares that there are no phases because the villa from the 4th century had no antecedents.
- If the villa had no antecedents why and when was room 3 rebuilt? Thus, it should be supposed that either there was an antecedent, but it is not known what it was like and what it was. Alternatively, it should be supposed that the reconstruction took place in the 5th century, and the antecedent was the villa built in the 4th century.
- About the excavation in 1903, Károly Sági writes "the walls were much higher at that time. ... So two doorways were observed in 1903, which could not been identified any more.[487] Consequently, Károly Sági could unearth the foundations; but at the same time, he uncovered a wall-section in room 1. that could not be seen in 1903. In addition, this wall-section suggests that reconstruction was possible; hence, the villa had an antecedent.
- No conclusions can be drawn from the drain in room 5 because it was not excavated. A drain must be linked, but this cannot be observed in room 5.
- He states that the building was not a church, but only a villa. Considering this, the openwork brick with the chi-rho monogram raised questions, which should be answered. ☞ Dunaújváros, Historical evaluation.

21.I.a. Openwork clay brick, fig. 120.

Location: BM 55.24.1. **Literature:** KUZSINSZKY 1920: 135. Abb. 175. NAGY L 1931a Abb. 21; NAGY L 1934 300; NAGY L 1938: 119. Abb. 46; KÁDÁR 1969: 180; SÁGI 1972 126 Abb. 9-10; THOMAS 1964: 52; MRT 1/22, 3; THOMAS 1980: 196 Taf. 8/2.; THOMA 1982b: 269 Abb. 9; THOMAS 1988: 285.

Two openwork bricks with chi-rho monogram were found at the entrance of the larger building. One of them was collapsed; the other was taken to the museum. This was damaged during World War II, but reconstruction based on a photograph was possible. Generally it is supposed, they were parts of the windows in a building.[488] Size: 45 x 45 x 5 cm. The fragments mentioned by Károly Sági belong to these pieces or it is the fragment of a third brick.

21.I.b. Basilica I. fig.116.

Literature: KUZSINSZKY 1920: 133 fig. 172, columns in fig. 173;

[473] MRT 1, 22/3.
[474] KUZSINSZKY 1920: 131-141; NAGY L 1931a: 31; NAGY L 1938: 79-81.
[475] SÁGI 1972: 121-137. According to Sági, further excavations are needed to indentify the periods clearly.
[476] KUZSINSZKY 1920: 132-133 gives relative chronology.
[477] SÁGI 1972 Fig. 3.
[478] SÁGI 1972: 123-124.
[479] SÁGI 1972: 126.
[480] SÁGI 1972 Fig. 7.
[481] SÁGI 1972: 127. Fig. 10 shows the fragments of the brick.
[482] SÁGI 1972: 134.
[483] BILKEI 1990 Nos 5, 8, 18, 31, 32, 35. Géza Alföldy commented on the No 5 of Bilkei. ALFÖLDY 1990 No. 8.
[484] NAGY L1926: 127.
[485] PALÁGYI 1996 111-119.
[486] PALÁGYI 1996 Abb. 6, 9.
[487] SÁGI 1872: 123. Károly Sági led the excavation in 1965.
[488] BELOŠEVIĆ J: Il complesso dell' architettura paleocristiana a Crkvina di Galovac nei pressi di Zadar. Atti 1998 III 69-104 deals with openwork building fragments. They were located in various parts of a building and they were not only used in antiquity, but in later periods as well.

ALFÖLDI 1920-1922: 128; NAGY L 1931a: 32 fig. 20; NAGY L 1934: 298-299; NAGY L 1938: 80, 119 fig. 44, reconstruction fig. 79. NAGY T 1939: 212; NAGY T 1941: 146; MIRABELLA ROBERTI 1953: 235; THOMAS 1964: 52-60; KÁDÁR 1969: 181; TÓTH E 1974a: 174; MANO-ZISI 1978: 16; PALÁGYI 1996: 134.

The scholars defined the smaller building as a hall-church without an apse. L: 37.5 m W: 16.5 m. It had two rooms. The larger one occupied three quarters of the whole building. A doorway was found in the south wall of the larger room (this is the longitudinal side of the building). According to Lajos Nagy, columns divided the larger room because plinths were found on the narrower sides. Bálint Kuzsinszky reconstructed this building with the aid of analogies,[489] and unearthed a plan. Everyone refers to this reconstruction. He dated it to the second half of the 4th century and to the beginning of the 5th century. Not only bases but also capitals were found.[490] Figs. 121, 122a-c, 123. Bálint Kuzsinszky recorded two more columns, on which the pulpit of the church in Zánka presently stands,[491] fig. 124. Kuzsinszky wrote that the two capitals, which serve as bases today, are probably from Kékkút. The capitals, the shafts and the bases do not belong together. The bases are from the Roman times, H: 21cm. He does not date other parts of the two columns.

Small finds from this building are not known, only the plan is known. The plan indicates the definition of a church without an apse. However, a large-scale excavation would help to solve the uncertainties. The modern investigations can modify former theories.[492] Thus, hypothetically I would propound the possibility that the ground plan is not complete, so the definition and dating are not right. As L. Nagy regarded the building as an early Christian basilica, he naturally dated it to the end of the 4th century without having any data supporting this. In addition, it must be remembered that the brick with the chi-rho monogram was found in the other building! Its connection with the other building should also be clarified. The possibility that they belonged together in their function cannot be excluded.

21.I.c. Basilica II. fig.117.

Literature: KUZSINSZKY 1920: 135 Fig. 175; ALFÖLDI 1920-1922: 128; NAGY L 1931a: 36-38; NAGY L 1934: 298-299; NAGY L 1938: 80-81, 119 fig. 45; NAGY T 1939: 212; NAGY T 1941: 146 fig. 4; MIRABELLA ROBERTI 1953: 235; THOMAS 1964: 52-60; KÁDÁR 1969: 181-182; MRT I/22, 3; SÁGI 1972 121-137, reconstruction fig. 19; TESTINI ^{2}1980: 774 fig. 427; THOMAS 1980: 196 ground plan: fig. 38, reconstruction: fig. 39.

85 m away from Basilica I, another building was uncovered and assumed to be Basilica II. Tibor Nagy considers it a church developed from a private house. Since neither the *subsellium* nor traces of it was found, it might have been made of wood, as it was mentioned at ☞Sümeg. Károly Sági led the verifying excavation of this building; at last, he doubted whether this building was a basilica. He reported on the poor finds of earlier excavations, namely some pieces of wall paintings, clay spindle whorl, iron knives, bronze fibula and bronze buckle.[493] Iron nails, lead discs, spearhead, scissors and door hinge were found in the course of the verifying excavation. However, it must be pointed out, that the minor finds were not published completely in the reports. These finds suggest that there was actually no basilica here. In spite of this, there would have been a room inside the building, which would match a *domus.* Nevertheless, it is probable that some parts of the building were lived in. The heating system in rooms 5, 7, 9, 11, 12, 13 shows signs of habitation. On the other hand, this fact does not exclude the possibility that one or two rooms were used for liturgy. According to the scholarly works, the openwork bricks were found at the entrance of room 1, which was probably a hall. Thus, the bricks indicated that the gathering place of *ecclesia* was in this building and that this house was the *domus ecclesiae.* Although Károly Sági thinks that rooms 2-3 and rooms 4-5 had a different function, I suppose the possibility that these separable rooms fulfil the requirements of the places for liturgy in Pannonia, as regards their ground plans.[494] The rooms 2-3 were reconstructed on the basis of the verifying excavation. This space could have the same function after the Constantine's turn. I would like to point out the possibility that Christians did not give up their dwelling houses and other buildings just because they were not built as basilicas. Furthermore, if they had a church they could use them as before. It was not necessary to build a new one immediately.

Edit Thomas thinks that there was a bishopric see in Kékkút. Although the presence of a Christian community suggests the presence of a bishop, this bishop was most of all the pastor of the community and not a leader of territorial organization.[495] According to the 6th canon in which the Council of Serdica defined the office of bishops, I do not think that Kékkút was an ecclesiastical see. The finds and other known sources do not prove the existence of an ecclesiastical see.

SUMMARY

In spite of many uncertainties the following can be stated:

- Christians must have been present in Kékkút. There were not many of them, maybe just a Hausgemeinde, hence their church should be called a house-church.
- The connection of the two buildings should be supposed, but they should not be called Basilicas I and II. It seems wrong to call them basilicas at all.
- The verifying excavation was restricted to the larger

[489] Basilica of Parenzo, or Pola: GNIRS 1919.

[490] KUZSINSZKY 1920 fig. 173; PALÁGYI 1996: 105-119 – the new conception is connected with the capital and base.

[491] KUZSINSZKY 1920 150, fig 188. Cf. KISS Á 1987 VI,58 Raf. 41,7.

[492] e.g. Balácapuszta – Balácai Közlemények II 1992: Sylvia Palágyi wrote about the main building (7-15, 123-124, 219-231, 272-273); Márta H.-Kelemen about building No. II; Judit Regenye about building X. The situation is the same at ☞ Ugod-Dióspuszta 59.II.a.

[493] MRT 1, 22.

[494] I am referring to the places divided into two, as was mentioned at ☞ Aquincum, Historical Evaluation and ☞ Dunaújváros, Historical Evaluation.

[495] The definition of bishops and their sees was altered just in the period here examined; therefore the chronology is very important. The Council of Serdica (342-343) altered the former conception and customs. See the canon VI. EnchF No. 503, p. 308. Cf. POLASCHEK 1937: 1028; Cf. note 1429.

building, but its area and connection with the other building were not explained. The larger building itself is by no means complete. Considering other villa-settlements in Pannonia a bath should also be expected here, but at least more buildings should be expected here. Nothing is known about them, so the territory can be regarded as unexcavated. It is only the bricks on the basis of which the building can be listed in Group I.

22. KESZTHELY-FENÉKPUSZTA, (Valcum) fig. 125.

Although the area has been populated since early Roman times, or rather, there was a fort constructed of earth and wood, nevertheless, the main subject is the later period of Fenékpuszta. The most interesting is the fortified settlement called Binnenfestigung.[496] Some of the questions are connected with the buildings inside the fortification. There is a third group of questions concerning the cemeteries. The questions are the following:

- When was the fortification built and what do we know of its periodization?
- What was the function of the fortification?
- Who and when dwelled in the fortification?
- When was the basilica built? How many basilicas were built at all?
- Who built the basilica?
- When was the "horreum" built?
- What is the connection between the horreum and burials around it?
- Who buried their dead here?
- To whom does the "Keszthely culture" belong to?
- To which phases of the buildings are which historical events connected?
- What was the continuity like?

Although only the questions concerning the basilicas are important for this catalogue, the above list of questions cannot be ignored because basilicas have to be studied within the wider context. Before the verifying excavation, two basilicas were recorded, namely I and II, early Christian basilicas. Both "basilicas" were uncovered inside the fortification. Their periods and dating cause the most problems and disagreement.

RESEARCH HISTORY

The history of the investigation until 1966, were summed up in MRT I/21, 33. Vilmos Lipp led the first excavation in 1883. Árpád Csák continued this in 1899 and buildings A and B were found. Building C was found also by Árpád Csák. He was excavating the area again from 2nd to the 31st December 1907, and uncovered one of the corner towers in the wall surrounding the settlement. The whole diameter of the tower was 17 m, the wall was 2.6m wide. The finds date back to Roman times and to the Migration Period and most of them were destroyed in 1945. We know many of these finds, which were published in Kuzsinszky's work.[497] In 1908, Csák uncovered another building; later, Lajos Nagy identified it as an early Christian basilica. This is Basilica II. He wrote about the finds as well.[498] Aladár Radnóti started the verifying excavation in 1947; among others, Károly Sági continued the excavation. It was he who published the results of the excavation without the small finds.[499] It is known that the antecedent of the early Christian basilica was a Roman house, which was rebuilt several times. At last, it had the plan and outline, which was uncovered. Layers of destruction could be observed; and with the aid of these, one could determine the periodization of the building. On the other hand, dating the periods caused disagreement, and is still causing disagreement.

Árpád Csák uncovered further buildings – among others Building 4. Kuzsinszky published the results of this excavation.[500] Lajos Nagy identified Building 4 as an early Christian basilica and he called it early Christian Basilica I.[501] The finds of this building were destroyed during the second word war. The different periods of the building cannot be determined on the basis of the published ground plan. Between 1949 and 1952, Aladár Radnóti led another excavation and László Barkóczi, Ferenc Fülep, Ilona Kovrig, András Mócsy, Tamás Pekáry and Edit Thomas worked together with him. This excavation is unpublished. In the excavation led by Radnóti the southwest and the northwest corner towers were uncovered. The southwest corner tower proved to be smaller, its diam only 13 m wide; and in the uppermost layer of this tower, Avar ceramics were found.[502] He unearthed the two south gate-towers and gates and determined that the south wall of the fortification was 348 m, the west wall 392 m long. In 1959, László Barkóczi uncovered a horreum south of basilica II, and Roman walls east of the horreum, a coin-hoard from the 4th century and "an unusually rich cemetery of 31 graves from the end of the 6th century" east of the horreum,[503] fig. 126. Károly Sági uncovered further graves outside the walls in 1963. The excavations have not ended yet. We may well say that excavations have been continuous with shorter or longer interruption and short reports inform us about the results in RégFüz.[504] Apart from minor reports, the late Roman fortification has not been published in details. This might be the reason why "research must rely on hypotheses when answering lots of questions about the date of building, its function, its periods, the ethnic groups dwelling in it and its history until the 9th century."[505]

[496] Such fortification was built in Ságvár, Alsóheténypuszta, Környe, and Mórichida-Kisárpás.

[497] KUZSINSZKY 1920: 55-58.
[498] NAGY L 1938: 78-79.
[499] SÁGI 1961: 398-456.
[500] KUZSINSZKY 1920: 58-67.
[501] NAGY L 1938: 78-79.
[502] MRT I, 21, 33 p. 85.
[503] MRT I, 21, 33 p. 87.
[504] SÁGI /HORVÁTH L RégRüz Ser 1 No. 24 1970 §32; MÜLLER R RégFüz Ser 1. No. 26 1973 §42; ERDÉLYI I/SÁGI 1978-1979; ERDÉLYI I 1982-1983; ERDÉLYI I /SÁGI RégFüz Ser 1. No. 37 1984 §51; MÜLLER R 1978; 1979; 1987; 1988; TÓTH E 1987; SÁGI 1989; 1991.
[505] MÜLLER R 1987: 105.

Data and dimensions: width of the walls: 2.5 m. The area, which is enclosed, is rectangular and measures about 377 x 358 m. The towers are round, measuring 15 m in the diameter, and 17 m at the corners. North and south gates were fortified with rectangular towers. A building that would be especially suitable for military accommodation has not been discovered inside the walls. The data above show the results of the excavation in 1988.[506] Fig. 127.

Problems

The antecedent of the fortification. Károly Sági pointed out that there would have been an earlier settlement,[507] which could be proved by the investigation.[508] The fortification was built in the space of the former settlement. Róbert Müller dedicated interest in the early earthen-built fort which can be observed in the air-photos and whose importance was lost when the limes was built. The fortification was built in the place of the civil settlement belonging to the early fort. Its dimensions ranged between 9500 and 11500 m^2.[509] Cf. Fig. 125. Consequently, there was a settlement before the fortification would have been built. This is the fact, which should be emphasized, because the existence of this settlement can be proved, so it is not a hypothesis.

The building of the fortification. The main point of the discussion is when the round towers, which belonged to the second period, were built. Comparing to the fortification of Ságvár and that of Alsóheténypuszta, we can observe that the earlier types of towers can also be found under the round tower,[510] but these early periods are missing in Fenékpuszta. The building has been dated to different times from Tetrarchia to the times after Valentinian I[511] on the basis of coins and coin circulation. However, as Jenő Fitz pointed out, dating only based on money circulation and hoards of coins is not enough and is ambiguous.[512] The earlier buildings and the identified graves together can help define the interval for the reconstruction more precisely, i.e. as a period between 350 and 383.[513] The question is still undecided.

The function of the fort. This is also questionable. András Mócsy thought that it was a center of a latifundium contributing to the provision of the army, but that it did not have a defensive function.[514] Károly Sági pointed out the civil population of the fortification.[515] Sándor Soproni stated that the fortification was built for the accommodation of elite military forces.[516] Endre Tóth and Róbert Müller have the opinion that "fortified settlements inside Pannonia had two functions, on the one hand they had a military function, on the other hand they had an important role in the provision of the limes."[517]

My hypotheses.

a. Groups of buildings called the inner fortified settlements of Pannonia – in spite of their similarities – have different features. The late Roman forts had irregular outline because they adapted to the plateau; but in spite of this, they have common features.[518] Thus, the research which compared the different archaeological estates, even those, which had no round towers on the walls, is informative.[519] Every such estate gives evidence of civil life based on the buildings, which belonged only to civilians. At this point the change of names of the fort in Szentendre should be mentioned, i.e. *Ulcisia Castra* became *Castra Constantia*. The word *Castra* may be important because of the later usage of the word *castrenses;* this word denoted the civil population of the fortified settlements in the Middle Ages. In the early Roman times, words *castrum, castra* naturally denote military buildings. Thus, we generally suppose that *castrum, castra* refer only to military estates. However, in the time of the transition of Roman times to the Middle Ages we cannot stick to former Roman meanings; the Latin language itself also changed as compared with the era of Caesar and Cicero. Based on this fact, I propose an idea in connection with the change of names. This change might have been the result of the change of the lifestyle, i.e. the settlement fortified by walls began growing into a town until it became a medieval one. However, we also can say that the late Roman fortified settlements had potential to develop into a medieval town. Alternatively, even an established bishopric see resulted in the change of the name.[520] Furthermore, I suppose that the late Roman fortified settlement of Keszthely-Fenékpuszta actually turned out to be an early medieval town. Keszthely-Fenékpuszta is in this respect different from the other inner fortified settlements.

b. Let us suppose that military units were stationed in every fortified settlement, as the researchers of Pannonia think. Sometimes, the presence of soldiers can indeed be proved. Nevertheless, we should ask whether there is some other reason for the presence of a military unit. Let us think of the Emperor's decree! He made the gate of fortifications open before the civil population who took shelter here, the former fort became a town, and the soldiers became the defenders of the town. On the other hand, these places were not forts at all; and the task of the soldiers from the beginning was to defend the settlement.

c. Furthermore, if the army was garrisoned here at

---

506 MÜLLER 1988: 270.

507 SÁGI 1968a: 26.

508 MÜLLER 1987: 105-107.

509 MÜLLER 1987 106.

510 The fan-shaped and horseshoe-shaped towers. MÓCSY 1974: 104-111, Fig 18, for the chronology of building forts and towers. TÓTH E 1985, for the military buildings in the 4th century.

511 RADNÓTI 1954: 495; SÁGI 1961: 403-405; SÁGI 1970: 151; MÓCSY 1974: 140-141; SOPRONI 1974a: 184-185; SOPRONI 1975: 179; SÁGI 1978: 47; ERDÉLYI I/SÁGI 1978-1979: 153; FITZ 1980a: 53-59; BARKÓCZI/SALAMON 1984: 168-169; SOPRONI 1985: 100, he mentions it as a fortified town. TÓTH E 1985: 125-126.

512 FITZ 1980: 40.

513 MÜLLER 1987:108.

514 MÓCSY 1974: 304-305; later, he modified his point of view and summarized his opinion connected with the inner fortified settlements; Pannonia 1990: 234-236 (MÓCSY).

515 SÁGI 1970: 150-151.

516 SOPRONI 1974a; 1975; 1978;

517 MÜLLER 1987: 106 quoting TÓTH E1975a: 188.

518 MÓCSY 1974: 282; SOPRONI 1978; 1985; THOMAS 1964.

519 BARKÓCZI/SALAMON 1984: 147-187.

520 Cf. VETTERS 1969. He writes that the former forts (*castra*) became bishopric sees. I suggest here that, in fact, places where bishopric sees were located changed their name, and were later called *castrum, castra.*

times, then the suitable number of civilians[521] was necessary for the provision of troops. This is also a tendency towards the Middle Ages, a period when certain habits were taken over from earlier experiences, in this case from the late Antiquity. At the beginning of the Middle Ages it was natural that the army had to be provided for where it was garrisoned.

d. Different viewpoints connected with the fortified settlements do not contradict each other. All of them have archaeological proof. Therefore, we have no difficulties with the viewpoints, but rather with the fact that we want to think according to the traditional 'well arranged order' of the Roman Empire. Although this 'order' was not upset yet, barbarian tribes, especially German ones, did everything to destroy this 'order' and to form a new way of life. This was one effort. The other, the opposite effort, was to maintain the 'order' of the Roman Empire. These two tendencies towards the new order resulted in changes. If the different viewpoints of scholars took account of these changes, we could hope to make conclusive statements. Róbert Müller and Endre Tóth started this method with the above opinions.[522]

The destructions, the periods of the fortification summarized by Róbert Müller. The fortification fell victim to a barbarian attack in 374. This was followed by an almost immediate reconstruction. The population lived undisturbed inside of the walls of the fortification during the reign of the Huns. In the middle of the 5th century, after the death of Attila, the whole fortification was destroyed and this is indicated by the burnt layer. Eastern Goths destroyed the fortification, but also rebuilt it.[523] The destruction at the end of the 5th century is the single fact, which is accepted by every scholar. The time following this era is already the time of Migration Period.

My hypothesis is in connection with German tribes and I propound it very carefully. The fortification was destroyed and rebuilt by them, although it does not seem to be a logical deed. Considering the relations, disagreements, battles, contracts and the settling of the *foederati* of Rome or Eastern Rome and the German tribes in its historical process, certain conclusions can be drawn, e.g. not even the contracts lead to a unity with the German tribes.[524] There was some fundamental disagreement remaining.[525] Germans did everything to disturb the Roman order. The most obvious example for the disturbances was the occupation of territories. They wanted to reach territories where some unity could be perturbed; for example, their settlements around Pavia and Aquitana.[526] To abstract my opinion I dare say that Germans took every chance to annoy the Romans and Eastern Rome with the purpose to upset the order and to take advantage of the increasing weakness of the Roman Empire. The archaeological evidence of this 'disturbance' was the barbarian destruction.

"Keszthely culture" Research refers as "Keszthely culture" to the defined group of archaeological material found in 17 graves of Christian population from the time of early Migration Period. This material is characterized by basket earrings, long ornamented hairpins and snake-shaped bracelets. Several hypotheses have been put forward about the owners of the rich finds. Károly Sági's publication made the different opinions known;[527] István Bóna - They were prisoners of war from Illyricum and Thracia taken captive by the Avars. Róbert Müller - Prisoners of war cannot wear gold jewels. His argument refers to the richness of the cemetery. Károly Sági: "It can be regarded as the cemetery of a Christian family and their relations who were wealthy and Romanized Christians. These people coming from the South might have brought the custom preserved in Keszthely culture."[528]

It is not my task to analyse the questions, but I cannot ignore it completely. Considering this culture in the context of historical events, abolishing the praefectura from Sirmium to Thessaloniki in the 5th century can be regarded as the historical reason for a group with a certain number of people coming to Pannonia from the South and whose descendants were buried in the 17 graves. Similar jewellery came to light in Greece, on the Aegean Islands, namely on Astypalea where jewellery of this type have been worn up to now and this is supported by Greek ethnographers.[529] This ornamented jewellery belongs to bride's jewels.[530]

FINDS

22.I.a-b. Basilica[531] **and *horreum*,** figs. 128, 129

Literature: MRT 1 21/33; L. NAGY 1938: 79; SÁGI 1951; SÁGI 1960; SÁGI 1961: 398-408; KÁDÁR 1965: 416-417 Tav. 2; KÁDÁR 1969: 181; BARKÓCZI 1968: 276-277; SÁGI 1970: 195, reconstruction Abb. 12; BARKÓCZI/LENGYEL 1971; FÜLEP 1980: 46; BARKÓCZI/SALAMON 1984;. TÓTH E. 1987: 257-264; MÜLLER 1987: 109-114; MÜLLER 1988: 270-272; SÁGI 1989: 271-285; TÓTH E 1994b Abb. 6.

Károly Sági[532] and everybody afterwards applied a 'periodic order' that was a help to find the ultimate shape of the basilica. I have already mentioned that Károly Sági received the right to publish the objects excavated within an appointed time though he in fact did not lead the excavation. This refers to the basilica as well. From his writings, we can get to know that this 'periodic order' is not accurate considering the early Roman period. In the case of the basilica, on the other hand, the later periods are obscure; however, we cannot keep their question distinct

[521] Remembering SÁGI 1970: 150-151; he proved the influx of peasant settlers.

[522] Same as note 517.

[523] MÜLLER 1988: 270.

[524] VÁRADY 1984: 37-61, 68.

[525] For its legal and economical references, see VÁRADY 1984.

[526] FERLUGA J: L'Istria tra Giustiniano e Carlo Magno. AV 43 1992: 175-190, fig. 2: divisione politica dell'Italia ca. 590-751. One sees well that Regnum Longobardorum was inserted into the regional unity.

[527] SÁGI 1991: 135-137.

[528] SÁGI 1991: 137.

[529] MNENAKH A. E: Εκδιδομεναι Ελληνικαι ενθνικαι ενδυμασιαι Αθηναι 1954: 66, pl. 80.

[530] DELIVORIAS A: Guide to the Benaki Museum. Athens 1980: 153.

[531] Originally, the basilica was numbered II, as the research also considered a further building a basilica (given the number I). Nowadays, however, we are aware that the first building was not a basilica, therefore the numerical distinction has become unnecessary. Henceforth simply 'basilica' will be used.

[532] SÁGI 1961: Abb. 3.

from that of the dating of its building.

The date of the building of the basilica is a much-debated question. According to Tibor Nagy, it was built around the 5th and 6th centuries,[533] Károly Sági traces it back to the 6th century, Endre Tóth puts it between the 6th and 7th centuries, while Zoltán Kádár speaks about the 7th century. László Barkóczi and Ágnes Salamon do not separate the basilica and the *horreum* which was excavated near it, but they give a relative chronology that draws a comparison between them: 'Innerhalb der Baukomplexe wurden jederzeit die *horrea* früher errichtet als die Basiliken.'[534] What they also take into account in their chronology is that the cemetery close to the *horreum* is of the 6th century; moreover, it supposedly belonged to the basilica as well as to the *horreum.*[535] In this state of the research, the fact mentioned by Miklós Béla Szőke is not well known: the burial place was around a 'small wooden church' in front of the *horreum.*[536] This fact is essential from two viewpoints: on the one hand, we always have the question where, by whom, and when were wooden churches used; on the other hand, the second viewpoint is connected to the function of the *horreum*, and it confirms that it is probably not a *horreum* though we had considered it as such.

A rebuilding that was worth attention and consideration was discovered during the excavation of Diocletian's palace in Split. Under the Benedictine church and cloister, which can be found at the northern wall, finds from the 6th century were noted: tombstones, fragments of a screen, graves and walls that are indicative of an early Christian church.[537] The ground plan of the church is shown on fig.130.[538] This exemplifies the possibility that the basilica of Fenékpuszta - even if we do not know whether it existed in the early Roman period - was rebuilt later. It might have been rebuilt in the 6th and 7th centuries and, what is more, a medieval addition is possible. The reason why I dare suppose this is the fact, that the examples, both western and eastern show that, in most cases, a medieval church is built above the Roman. This possibility has been actually left out of account,[539] and a future decision can be made only with the help of excavations. It is also worth indicating that the forming of the Benedictine order is put to the year 529, thus the Benedictine church noted by Migotti could have been built, at the earliest, either in the middle or at the end of the 6th century, or at the beginning of the 7th century. This can be so, for 5th- and 6th-century finds were noted related to the layers under the church. We need not reject this analogy, for we can reckon with the possibility that this area was a concern of Emperor Iustinianus[540] or even of the Benedictines.[541] My hypothesis about the periods of the Keszthely basilica, based on the foregoing are the following: 1. At the very end of the 5th century, a single-aisled church in basilican style was built. Its early antecedent might have been a house-church, whose periods are rather uncertain. 2. The single-aisled building might have been rebuilt into a three-aisled church at the end of the 6th century. What is also worth mentioning here is that the footing level of the middle apse is higher than that of the two lateral apses. This is illustrated properly in a photograph taken during the excavations and afterwards published by Károly Sági[542] (fig. 131). 3. A further addition to this three-aisled building in basilican style was a small room in the apse. Due to the Ω pendant, which was found in the basilica, we can assume that the building itself must have been extant, and was used when thc settlers of the Migration Period and the descendants of the local Roman population were already living next to each other (or, better to say, at a time when the latter were controlled by the migrants).[543] Its further periods are much beyond the time limit of my present writing.

Close to this basilica lies the building named *horreum*, next to which László Barkóczi found the above-mentioned graves.[544] Barkóczi divided the graves of the cemetery[545] into two groups, based on the results of Imre Lengyel's chemical examination of the bones.[546] This examination determined the blood types and the Personalindex; the latter one was the result of all the tests carried out on the bones in different ways. If we project the result of this examination onto the map of the cemetery, we can see the division of the two groups. The determination of the Personalindex made it possible to compare this cemetery with others, among which the following are interesting:☞ Pécs, Geisler Eta Street; Környe[547] and ☞ Hegykő. The presence of these graves makes the consideration of the '*horreum*' as *horreum* dubious.[548] Moreover, in Barkóczi's opinion, it is also curious that the cemetery consisting of 31 graves spread only along the *horreum.*[549] He also finds it strange that where *horrea* are located, basilicas are present as well. Therefore, it is presumable that, similarly to the cases of ☞ Gorsium,[550] ☞

[533] NAGY T 1941: 145.

[534] BARKÓCZI/SALAMON 1984: 174.

[535] BARKÓCZI/SALAMON 1984: 180-181.

[536] SZŐKE 1998: 259.

[537] MIGOTTI 1990: 33-34.

[538] MIGOTTI 1990 Taf. XVI/3.

[539] We did not make allowance for it as there were also doubts over its continuity. However, some continuity exists, cf. KAHL 1980: 43-46.

[540] JARAK 1998: 782, 784: M. Jarak analyses the source *Epitome chronicorum Cassinensium*, where those cities listed in the *donatio Iustiniani* are enumerated. Although the author considers the donatio unacceptable, at the same time she thinks that the enumerated cities had something in common with Iustinianus. She stresses the importance of Valcum and Sopianae.

[541] Donatio Iustiniani. My own set of hypothesis in connection with the basilica, its building and rebuilding, and dating.

[542] SÁGI 1961 Taf. XVI.

[543] MÜLLER www.huniufr.hukeszt/1keszt.htm

[544] BARKÓCZI 1968, the map of the cemetery is in fig. 2; the description of the graves is on pp. 278-286. In the fifth grave there was a gilded ring whose inscription was BONOSA. Cf. AnnÉp 1968: 426 and notes 110, 1406 here.

[545] This is a cemetery of 31 graves within the inner fortified settlement - BARKÓCZI 1968: 275-311. During recent excavations, seven further graves were brought to light: SÁGI/ HORVÁTH L. RégFüz Ser I No 24 1970 45.

[546] BARKÓCZI/LENGYEL 1971: 191-199.

[547] SALAMON/ERDÉLYI I 1971.

[548] BARKÓCZI/LENGYEL 1971: 181.

[549] BARKÓCZI 1984: 172.

[550] Barkóczi refers to the building considered as an early Christian

Sirmium and Keszthely-Fenékpuszta, there could have been basilicas next to the *horrea* of ☞ Pilismarót and ☞ Tokod as well.[551] These doubts warn us that many of the buildings so far considered to be *horrea* had in fact different functions, therefore we have to revise our views on these buildings. H. Leclercq describes this type of ground plan as those of the synagogues of the Diaspora. I insert the picture published by him in fig. 132.[552] H. Leclercq, and later, Noël Duval, both bring the ground plans of synagogues, because they agree that these were influential in the building of Christian churches.[553] Here one can see a possibility of examining the effect of synagogues as buildings on the constructions of Christian churches. For, I assume that the *horreum* functioned as a church or dwelling place or teaching place and not as a *horreum* after some time past, though it may have been used as a *horreum* previously. The settlers (not the Germanic tribes but the Christians coming from the eastern part of the Imperium) recognised[554] the synagogue-like building and they used it as a church until the completion of the building work on the basilica.[555] As for the ground plan of the *'horreum'* of Fenékpuszta cf. fig. 129. After having compared the settlement of Fenékpuszta with the provenances considered as inner fortified settlements excavated quite well like ☞ Ságvár, ☞ Kapospula-Alsóheténypuszta, we have to conclude that there is some similarity between them. Not only because of the fortifying walls and the identity of the towers; but, as we can read in the scholarly literature, the features of the inner buildings resemble each other as well: namely, the existence of an imposing palace-like greater building, a building (or buildings) called *'horreum'*, and smaller dwelling-houses. The scholars have recorded such settlements at ☞ Kisárpás and Környe,[556] however, the latter one has not been excavated enough to verify or disprove the fact that it belonged to the settlements of the previous kind. There were no Christian finds recorded at Környe indeed. In recent years, Csákvár has been dealt with in this respect.[557] Csákvár is being excavated, and, according to the signs, Christian finds might be expected, ☞ Csákvár. Comparisons with forts and other provincial parallels, and also between each other, have been made and analysed in detail by László Barkóczi and Ágnes Salamon.[558]

basilica.

[551] BARKÓCZI 1984: 173.

[552] LECLERCQ 1907 fig. 104. His argument may be found on pp. 340-353.

[553] LECLERCQ 1907: 348-349; DUVAL 1978.

[554] I would like to note that the missionary of the pagans, the Apostle Paul, while travelling, usually preached in synagogues; the *Actus apostolorum* can prove this. Therefore, synagogues or synagogue-like buildings were, even in later times, suitable for the practice of religion for the Christians converted from paganism. No wonder that Leclercq and Duval have raised the question.

[555] I do not think that these *horrea* were originally synagogues.

[556] RADNÓTI 1941; RADNÓTI 1954.

[557] FITZ 1961-1962; SALAMON/BARKÓCZI 1970; SALAMON/BARKÓCZI 1982.

[558] BARKÓCZI/SALAMON 1984.

22.I.c. Omega, suspended by chain, fig. 133.

Provenance: It was found in the basilica. **Location:** BM 59.11.1. **Literature:** SÁGI 1961: 454 Taf. 14,4; KÁDÁR 1965: 416; Severin, Katalog 8.22 Taf. 57.

Made of bronze and suspended by chain. L: 16.3 cm. The figure is taken from Károly Sági's paper. The letters α and ω are hung on the arms of a cross (crux gemmata) with the help of a chain. There are more illustrations of this kind.[559]

22.I.d. Silver spoon, fig. 134

Location: BM 75.172.1. **Literature:** THOMAS 1980: 199; THOMAS 1982b: 275, Abb. 19 a, b, c (from different views); Severin, Katalog 8.56 Taf. 62 is identical with Abb. 19 a.

Silver spoon decorated with a chi-rho monogram. L: 14.5 cm.

22.I.e. Hairpin

Location: BM 63.67.1. **Literature:** Severin, Katalog 8.59 (Róbert Müller). The picture of the object is not published.

I cite the description: 'Bruchstück einer silbernen Haarnadel mit Inschrift Alpha und Omega. Als Streufund aus dem Plattensee geborgen. Länge 3.7 cm.'

22.I.f. Casket mount, fig. 135

Provenance: 'Im westlichen Turm des nördlichen Tores der Festung von Keszthely-Fenékpuszta waren zwei Kornspeichergruben eingetieft, die nach der Zerstörung der Festung mit Abfall aufgeschüttet wurden. Die Kornspeichegrube Nr. 2 besteht aus zwei Räumen (2-2a). Die Kästchenbeschlage lagen im Raum 2a zusammengelegtem Zustand und gehörten zu den Überresten einer Bronzewerkstätte.' Fig. 136. Not only casket-mounts with biblical scenes have been found, but also casket-mounts with other decorations have been recorded. **Location:** BM 72.181.1-6. **Literature:** GÁSPÁR 1986 Kat.-No. 792.

Made of bronze, the length of its restorable part is approximately 39 cm. The remaining maximum height is 7.5 cm. The illustrations in the metope with a pearl border are, from left to right, the following: the raising of Lazarus, Daniel among the lions, the Good Shepherd and Moses striking the rock.

22.II.a. So-called basilica I. fig. 137.

Literature: MRT 1 21/33 enumerating and expounding the previous literature, therefore I will leave the earlier literature out – except for a few special articles; NAGY L 1938: 124; NAGY T 1941: 145; GOSZTONYI 1944-1945; E. TÓTH 1972a: 45-46; FÜLEP 1980: 46; THOMAS 1980: 196 fig. 40 – ground plan, 41 - reconstruction; THOMAS 1988: 285; MÜLLER 1988: 270; SÁGI 1989: 289 fig. 33.

The so-called basilica I seemed less significant to the scholars. They did not find it as problematic as basilica II; almost everybody accepted the reconstruction carried out by Edit Thomas. She supposed a portico with a peristyle in front of the church and drew the ground plan according to this. She gave the reasons for her solution by mentioning building III found at Gyulafirátót-Pogánytelek.[560] This analogy is, however, not so evident, since its shape is related to buildings of another kind.[561] If it is supposed that the building was a basilica, a better

[559] E. g. A. LIPINSKY: Ein vergessenes Christus-Symbol: der Karfunkel-Stein oder Alamandin. Atti 1978 II 353, Abb. 4. The illustration shown here is put to the 5th and 6th centuries.

[560] THOMAS 1964: 44-45.

[561] BIRÓ 1974: 32.

parallel would be the church found near Bitola.[562] According to further research, however, our building was not a basilica but a reception room. In Róbert Müller's article, it is summarised like this: 'Außer Wohnhäusern sind bis jetzt ein 102 m langer Palast, ein fünfschiffiges Empfangsgebäude, eine altchristliche Basilika und ein großes Horreum ausgegraben worden.'[563] A general plan of the site is also included in this paper so as to present the fortified settlement and the buildings that have been excavated so far. This general map shows how big the building, regarded as basilica I, was. cf. fig. 127. Károly Sági, fig. 138, first carried out a modern excavation.

22. KESZTHELY-FENÉKPUSZTA, HALÁSZRÉT, fig. 139

22.II.b.Tomb, Figs. 140, 141

Literature: SÁGI 1960: 188-196; FÜLEP 1962a: 31; KÁDÁR 1965: 414-416 Tav.1; SÁGI/FÜZES 1967: 80 Abb. 2; MÜLLER 1987: 108.

The burial chamber was excavated and published by Károly Sági. According to him, in the 5th century a cella memoriae was built above a grave from the 4th century. This grave was made of brick, and was plundered. He stated that there had been more than one periods of its building. The last period was when a 'cella memoriae' was built above the tomb.[564] He imagined the vault of the tomb based on analogues. From one of his remarks it turns out that the burial chamber had been painted, but the war damaged the frescos totally. He does not give any description of the frescos.[565] There were several burials arranged in the chamber, and Sági points to the use of wooden coffins, based on the iron nails that were found. The burial chamber was built in a cemetery. The grave-goods of this cemetery do not show any Christian features.

The tomb's dimensions were 3 x 2.5 m. Materials of a former construction were used for building the chamber. State (a) indicates the grave of brick above which, in Sági's opinion, the burial chamber was built. State (b) presents the chamber already built. In period (c), a grave in the ground was located beside and north of the one made of brick. In period (d), the superstructure was built.[566] Zoltán Kádár, having compared the burial chamber with the memoria to Synerotas in Sirmium, suggests that the chamber might be an early Christian find as well. However, I keep on saying that the finds recorded and published here do not prove that there were Christian funerals in the chamber.

Róbert Müller regards this tomb as a significant link in view of the periods of the fortified settlement, since its presence verifies an ethnic continuity up to the 5th century.[567] Besides, from the cemeteries between Halászrét and Dobogódomb, he concludes that there has been a dense population in this area.[568]

[562] CAMBI 1978: 154 fig. 11: Ramenska Tumba near Bitola, Macedonia.
[563] MÜLLER 1988: 270, Abb. 2.
[564] SÁGI 1960: 192-196. The periods are presented by Abb. 9. one by one.
[565] SÁGI 1960: 192 note 26.
[566] SÁGI 1960: 194-195.
[567] MÜLLER 1987: 108.
[568] MÜLLER www, huniufr.hukeszt/1keszth1.htm

Commentary on the illustrations and descriptions of the excavation:

- It is difficult to imagine a burial chamber whose wall cuts the brick grave above which it is built. This can occur on an occasion when it is of indifference to the burial chamber that there has already been a grave there.
- It is curious that, next to an important grave, another is located virtually occupying the 'main place'.
- The superstructure was started, but also interrupted, to be finished only in a later period.
- On the ground plan, the leader of the excavation did not indicate the places of other graves. He paid attention to them only in the text.
- I do not regard this tomb as one related to Christianity, although its dating and provenance do not run contrary to such a consideration.
- It is analogous with Pécs, ☞ 32.II.g. Having learnt from the lesson of tomb IX, we can conclude that this tomb should be put to an earlier date than Károly Sági supposes. The burial chamber itself might have already been built in the 2nd century, but its presumably continuous further additions and graves are from later times. According to its recent documentation, it is impossible to produce a relative chronology and choose a date, which would represent early Christianity.

23. KESZTHELY-DOBOGÓ

On the site, Károly Sági has found 132 graves out of which seven were early Roman. Two out of seven were urn graves. In the cemetery, there were graves from the 9th century as well.

23.I.a. Spindle whorl

Location: BM 62.22.1. **Literature:** SÁGI 1981 Grab 32/2; Abb. 5/27; Severin, Katalog 8.67.

Decorated with a chi-rho monogram. Diam: 3.5 cm.

23.II.a. Pottery vessel with incised pattern, fig. 142

Provenance: Cremation grave. **Location:** BM 65.67.4. **Literature:** SÁGI 1968b: 394 Abb. 3.

Károly Sági was inclined to think that there was a reference to Christ's cross on the base stamp, and he connected it with the engraving on the roof tile from Becsehely (3.II.a). He dated the vessel to the middle of the 4th century. Even supposing that the pattern on the bottom of the vessel is a stamp, the idea that it is cross-shaped is rather dubious. I do not hold it to be related to Christianity.

24. KISDOROG

24.II.a. Engraving on a roof tile, fig. 143

Provenance: Recently, there has been a rescue excavation on the site, namely, in the fields of Kisdorog-Hegyiszántók, where brick graves from the Roman age were found during soil cultivation. The machines destroyed the richest graves. The rescued graves can be dated back to the second half of the 4th century according to the grave goods.[569] István Bóna's article quoted below is proving the actual provenance of the roof

[569] PÉTERFI RégFüz Ser. 1. No. 39 1985 71.

tile; the article is far too important, because Edit Thomas has given wrong provenance and excavation conditions in her articles. In the fields of Hegyiszántók, a partly devastated, partly rescued place in a graveyard, in István Bóna's views, belongs to that part of the cemetery where the roof tile was also found. **Location:** WMM R.54.933.2. **Literature:** THOMAS 1973-1974; THOMAS 1977b: 45-46; THOMAS 1980: 200-204 Fig. 5. Taf. CL-CLI; Severin, Katalog 8.51; THOMAS 1986a: 520-526; IIL (PÉTERFI) Kat.-No. 317; LAKATOS 1993: 10: Mentioned with provenance Tevel. I BÓNA: A kisdorogi római tégla lelőhelyéről és előkerülési idejéről - Über Ort und Zeit des Fundes des römischen Ziegelzeichens aus Kisdorog. Emlékkonferencia 11-20; D. GÁSPÁR: Pannoniai téglakarcolatok. Emlékkonferencia 37-44 (See 4.I.d. for a comment on the last two writings).

The size of the roof tile: 42.5 x 28.5 x 7 cm. In Edit Thomas's opinion the human figure on the roof tile is Arius, since she read the letters above as ARIO. She thought the engraving under its left sleeve was a PE ligature, with a cross, below. She identified the engraving next to the sleeve of the clothes with a harpoon. The figure, as she thinks, is holding a short shepherd's crook in its right hand. The meaning of the ligature is *palma emerita*. The figure is 16 cm tall. This interpretation of the engraving made it possible to suggest that the roof tile had Christian relations.

Let me add the following comments on the interpretation of the illustration: there is nothing in the right hand of the figure; its palm is as much opened as the left. The engraving interpreted as a PE ligature is not a letter ligature. If we intend to recognize a letter in it at all, then it is a reverse letter 'S' whose line happens to be a bit angular. Under this, the lines interpreted as a cross, are in fact a vertical line, which is, in the middle, divided into two by a very short horizontal line. The image interpreted as a harpoon is a rather forced definition, the drawn lines much more resemble snakes; two snakes crawling away from each other. The explanation of the inscription is again problematic. In the above mentioned article I found the following readings: ARSQ(ue), which is connected to the illustrated person, who is an *artifex*. Its interpretation: master and his craft. I raised another solution, namely that the inscription is ARCO, and this is the name of the illustrated person, who is standing on an arch, i. e. an *arcus*. Its interpretation: arch to the Arch.[570] Consequently, I do not consider the definition of the figure as Arius, especially that very Arius, probable. Thus, I deem the argumentation attached to the roof tile in connection with the spread of Arianism in Pannonia unfounded.

25. KISMÁKFA fig. 144

25.II.a. 'Paten', 'chalice', buckle ?, Figs. 145, 146

Location: MNM 74.3.1, purchase from Imre Bürki. **Literature:** THOMAS 1975-1976: 105-160 Abb. 41-50; THOMAS 1977b: 83-87; THOMAS 1980: 198 CXXXIX; THOMAS 1982b: 273; THOMAS 1986b; THOMAS 1988: 290 Abb. 9; E. TÓTH 1994-1995.

All the three objects are silver. The bowl is original. The 'chalice' and the buckle have been lost. The seller himself drew chalice and the buckle as a child. This drawing and his account gave a basis to create the object. The drawing of the objects called a chalice and a buckle (brooch) are in fig. 146.

The diameter of the bowl is 9.8 cm, its height is 1.3 cm, and its weight is 42 gr. According to Edit Thomas's definition, the object is a paten whose rim is ornamented with 13 stamped birds and 13 stylised human figures. Its middle is decorated with a seven-pointed star, between the arms, there are fish, and, among the fish, there are concentric circles. Edit Thomas interprets the symbols as follows: the star in the middle is a sun with seven beams, the Sol, the symbol of Redemption. The circles symbolise bread. The fish may symbolise Christ, but they may refer to the miraculous multiplication of loaves and fishes. The stylised human figures depict the twelve apostles and Christ.

She dates the object to the 5th century.[571] The two objects were not found in a grave, they had been in a hoard.

On the site, Edit Thomas carried out newly an excavation, where she found some pottery sherds from the period of the early empire, but nothing that would explain the presence of the silver finds was found.

Recently, Endre Tóth has dealt with the objects.[572] The results of his study are summarised briefly as follows. His statements can be grouped in three themes; one is the technique of the elaborating, the other is the iconography, the third is the question of usage. He touched upon the dating of the objects in all the three themes.

Technique: it was executed with a stamp, which is unknown in 'late antique silverwork'.

Iconography: he enumerates a rich store of analogies from different periods from the 4th to the 7th centuries; among these, the most convincing is a stamped vessel from Nissedal on which the stylised human figure is similar to the one on the bowl.[573] Endre Tóth thinks that the stylised human figures allegorise Christ and the twelve apostles. The birds refer to the apostles too. As for the fish, he mentions the meaning *tria genera animantium* although the representation of the mammals is missing from the paten. Related to both the birds and the fish, he lists numerous interpretations that are mentioned by other authors. His conclusion is that it belongs to Christianity based on its iconographic interpretation.

Its function: because of its small size, it was impossible to use it as a paten. Therefore, he thinks that it belonged to a priest as a portable object, who might have carried it on his missions. His argument is as follows: '... some bowls that must have been used at ceremonies have remained from the 6th century, and they, because of their small size, are comparable with that from Kismákfa: such are the bowls in the finding of Canoscio (diameters: 13.5, 15, 32 cm ... the diameter of the Galognano paten is only 20.3 cm.)'[574] Let me remind you that the diameter of the

[570] In the article I touched upon the problems of the 2nd and 4th declinations.

[571] Quotation: 'Der in Kismákfa aufgefundene Kelch und die Patene sind der älteste eucharistische Fundkomplex aus dem 5. Jahrhundert, ...' THOMAS 1982b: 273.

[572] TÓTH E 1994-1995.

[573] TÓTH E 1994-1995: 128. fig. 1. The dating of the vessel is 450-500/550. It was found in a grave with a skeleton.

[574] TÓTH E 1994-1995: 141.

Kismákfa bowl is 9.8 cm.

It's dating: I quote: 'On the basis of the same or almost the same dating of the different impacts, I put the Kismákfa bowl to the second half of the 5th century'.[575]

Commentary. Independently from Endre Tóth, I found the small size of the bowl impossible to be a paten, since the loaves that appear in the scene where Jesus breaks the bread on the antique illustrations are of a size for which would be enough space on the bowl. The patens listed by Endre Tóth are all bigger in diameter and their list could be lengthened with a piece from Hispania whose diameter is 22 and its height 3.5 cm.[576] I do not even find the solution of the portable paten convincible since it is lacking an acceptable basis in the history of liturgy. This fact alone raises the suspicion that we are not inevitably dealing with a paten.

My arguments are presented in the following points:

- the paten as a liturgical object independently from the object of Kismákfa
- illustrations on patens and on this object
- formal analogues of the Kismákfa object
- its possible functions and dating.

The paten. It was not significant in the antiquity. The practice was different. The bread was put on the table, as it is shown in the illustrations, as we can see it in Ravenna on the mosaic of the San Vitale: Abraham entertains the three angels, the sacrifices of Abel and Melchisedec. We should not forget that all began from everyday life; from the circumstances of the friendly supper and so from its utensils. Therefore, originally, they did not take any 'liturgical' tools, since wherever they went; the necessary things for everyday usage were present. It was required by everyday life - simpler or more elegant, poorer or richer - but there was a bowl to hold the bread. Thus, the little serving tables[577] and bowls should be taken into account.[578] The diameters of the tables are nearly the same as the ones of the bowls.

The table. On this question, Alphons A. Barb writes in more detail, we can say, exhaustively, under the title Mensa und Patene. Several scholars have dealt with the question what purposes were served by those marble finds executed in a circular or a semicircular shape with a rim.[579] Its function of the table was taken over by the paten in the 6th century; at least this becomes transparent in the fact that we can really talk about patens from the 6th century on.[580] The Riha paten has remained from the 6th century, another one from the 7th century.[581] Barb presents the picture of such a mensa in connection with which he writes the following: 'Das ist also ganz die Form der Altarplatten von Besançon oder vom Lechaion in Korinth und entspricht auch der von mir abgebildeten Patene im Louvre'[582] fig. 147. As for the making of the paten, he cites the work of Theopilus: '... fac in unoquoque spatio fac circulum dimidium, ut sint quasi octo arcus, quos cum rotundo malleo percuties donec cavi fiant, et inferius ductili opere percuties angulos inter ipsos arcus et limbum circa eos ..., qui superemineat aequalitatem totius patenae ...'[583]

- *Illustrations.* The illustration of the object is so schematic and general that different interpretation is possible. Its surroundings and usage will determine its interpretation. However, if we examine the above-mentioned patens, then we can state that their illustrations are concrete and easily connectable to Christianity. Therefore, based on its illustration, I abstain from classifying the Kismákfa object as Christian.
- *Formal characteristics.* As far as the form is concerned (i.e. the working out of the rim, the deepening of the bowl part, its slanting and grooving), the patens are not analogues of the Kismákfa object. Nevertheless, this does not mean that our object would lack analogues in this respect. From merely formal aspects, this bowl matches the ones used by shamans in the 19th century at sacrificial rites, fig. 148.[584] In the collection of the Néprajzi Múzeum (Ethnographical Museum), under inventory numbers 3759, 3760 and 3761, we can find such little bowls, their diameter is 9.1 cm, and they are from 1831 and 1888. By this, I do not wish to say that the Kismákfa object necessarily belonged to shamans, and that it was made in the 19th century, I would simply like to emphasise that we have to be very careful in interpreting it.[585]
- *Function.* I doubt the usage of the bowl as a paten in any case. I hold its belonging to shamans thought-provoking, which cannot be neglected in the course of further research. I do not want either to answer or to close the question. Its dating is uncertain, it may stretch between the 5th and the 19th centuries. It is surely not Roman. Here I

[575] TÓTH E 1994-1995: 137.

[576] PALOL 1964: 312, 316.

[577] Those frescos where meals are painted. The wall paintings of Pompeji may exemplify this, but catacomb paintings might serve as examples where these illustrations can also be found.

[578] I am thinking of those large bowls that form part of a dinner-set, e.g. the Seuso treasure, or the bowl of the Mildenhall treasure. Age of spirituality No. 130 - Diam: 60,5 cm. Furthermore, let me allude to those treasure troves that included dinner-sets even from sacred districts. Annemarie KAUFMANN-HEINEMANN: Götter und Lararien aus Augusta Rauruca. Augst 1998.

[579] Etienne MICHON, Revue Biblique (Paris) N.S. XIII 1916: 121-170; BARB 1964. As a table, it could have been the table for breaking bread; as a marble slab, the cover of a grave, or an altar in a church. Barb elaborates on the fact that these functional areas are not contradictory but rather in correlation.

[580] ELBERN 1979. The form and size of the patens listed here differ considerably from those of the Kismákfa object.

[581] Age of Spirituality, The Riha Paten, diam: 35 cm. Paten with monogram, diam: 44 cm - Nos. 547, 548. Plate of Paternus, diam: 61 cm - No. 546.

[582] BARB 1964: 22.

[583] The quotation was taken from BARB 1964: 22. He used C. R. Dodwell's publication 3, 44 on page 94: '... make in the middle ... a circle, divide the circle into eight equal parts and draw a semicircle into each so that they produce eight arches. Beat them with a round-headed hammer until they become deeper, and at the bottom, beat the angle back inwardly on the deepened parts, then produce ... a rim on it so as to make it stand out from the whole surface of the paten ...' translation by Vilmos TAKÁCS - Teophilus presbyter: A különféle mesterségekről - On different crafts. Budapest 1986: 100.

[584] Schamanen 44. Abb. 44.

[585] I would like to thank Dr Terézia Horváth for her assistance and advice in ethnography. The ethnographers would have accepted the Kismákfa object as a shaman's bowl, regarding both its size and form.

agree with Endre Tóth, who is of the same opinion. I have doubts about its Christian use, especially considering the illustration of the lost chalice. Its ornament cannot be found in the Christian world. It can, however, be found in the 'liturgy' of shamanism. The object is a debated one.

26. KŐVÁGÓSZŐLŐS (Kővágószöllős), fig. 149

THE HISTORY OF RESEARCH

The first excavation on the site was carried out in 1867.[586] A vaulted and painted Roman burial chamber was found, which collapsed at the opening. Due to its similarity to the tombs of Pécs, L. Nagy mentions it among the Christian relics; although he adds that the Christianity of this relic is dubious.[587] Its pavement used to be covered by a mosaic depicting human and animal figures, which has been destroyed, according to Lajos Nagy.[588] Interpreting the fresco, he mentions that, while here the motifs were flowers and birds, at Zalalövő, they were ornamental foliage with red flowers.[589] On the other hand, Flóris Rómer wrote about the inner walls of the underground sepulchral vault that 'they had been painted with human and animal figures of different colour. Inside, there was an undecorated sarcophagus'. The latter one had been broken.[590] János Dombay published an account of the traces of a Roman villa, where he mentions that some traces suggesting a painted burial chamber were also found next to the Alsóheténypuszta villa.[591] In E. Thomas's opinion, the burial chamber is either Christian or pagan.[592] In 1975, Gábor Kárpáti and Borbála Maráz conducted a rescue excavation in the area. They unearthed the pits of a late Bronze Age settlement, and the pits of a Roman settlement, which existed later than the villa, in the 4th and the 5th centuries.[593] A newer excavation was carried out by Alice Burger, who, as a result, defined the burial chamber evidently early Christian, and the territory the residence of Arian Bishop Paulus.[594] Perhaps, the restoration of the object as a burial chamber took place too quickly.[595] However, during the excavation, a dwelling house was also found,[596] from which the latest coin dates back to the reign of Gratianus, fig. 150. The patio - peristylium - was paved. 'Rather nice iron finds, remnants of frescos and window-glass, fibulas from the turn of the first and second centuries and pottery sherds came to light.'[597] In the report of the excavation, we can read about both the villa,[598] and its finds.[599] The size of the villa was 40 x 30 m, the bath joined to its southeast corner was 15 x 20 m, and the length of the entrance wing at the southwest corner was 25 m. According to Burger, the bath is not complete in its present form, because its other parts have been destroyed; as she found such traces during her excavation. The remnants of the buildings in the southwest part were defined as an ornamented entrance, which used to belong to the plumbing-system. The latter suggestion could not have been proved completely. According to the report of the excavation, there were so many finds, that their usual publication would be impossible; therefore, she chose the solution 'Catalogues I-XVI'. Here, she enumerates the finds without pictures and according to their types. There were 2800 animal bones found. In the peristylium of the villa, there were three times as many bones as in other rooms. There were skeletons of dogs and cats among them, too. There were fewer bones at the bath, and in the southwest part there were more. North and east of the villa, in the 3rd- and 4th-century layer of the settlement, there were many bones.[600] There was an incomplete skeleton of a dog in the corridor. The number of remnants of cattle was growing on the settlement from the 4th and 5th centuries. There were bones of pigs, sheep, goats, horses and donkeys as well. The assessment of the piece of wood found in the villa is the work of Ernő Horváth, who assessed it as being an elder, whose grooved stalk was used for wine-growing purposes.[601] The villa was built in the middle of the 2nd century and it was still standing at the end of the 4th century. Burger identified it with the central building of a latifundium.

We can read the short description of the excavation in the *Itinerarium*.[602]

26.II.a-b. Tomb with burial chamber and mausoleum, fig. 151 is the ground plan.

The building defined as an early Christian mausoleum was found 16 metres south of the villa. Its description based on Burger is summarised as follows. The length of the building was 16 m. The upper part was called a chapel. The lower part was the painted burial chamber. An unpainted, narrow corridor led to both. The floor of the burial chamber was made of pink terrazzo.[603] In the burial chamber, there were two burial sites: a sarcophagus and a grave made of brick. These two were located next to each other. The styles of the paintings found in the antechamber and in the burial chamber are different. The walls of the antechamber (fig. 152 A) were covered by geometric decoration, while the burial chamber was ornamented with representational painting. The pink wall was divided into fields with 18 cm-thick black vertical stripes. Within these

586 RÓMER 1868: 189; Without name, Századok II 1868: 69.
587 NAGY L 1938: 42.
588 NAGY L 1938: 42.
589 NAGY L 1926: 124.
590 RÓMER 1868: 189.
591 DOMBAY 1957: 308.
592 THOMAS 1964: 279.
593 VÖRÖS 1985-1986: 243.
594 BURGER 1985-1986: 165-179. Tibor NAGY thinks that Bishop Paulus' residence was in Pécs – 1939: 210.
595 HAJNÓCZI 1985-1986: 229-236.
596 Based on BURGER 1985-1986.
597 BURGER RégFüz Ser. I. No. 32 1978 71.
598 BURGER 1985-1986: 66-164.
599 BURGER 1985-1986: 180-210; money circulation 211-218.
600 VÖRÖS 1985-1986: 243-247. This article represents the total processing of the bone material. I have taken the description of the bones from here.
601 HORVÁTH E. 1985-1986: 237-241.
602 103-105.
603 Did the above-mentioned mosaic cover the terrazzo floor, or else, where was there any room for the destroyed mosaic floor?

dividing stripes, "light blue embossed floral patterns were painted. Each black stripe was fluted in the middle line, and these flutings were bordered by regular circles ..."[604] In each field, standing figures in white clothes and with feet taking steps were illustrated. A 20 cm-high frieze uniformly bordered the wall painting with an identical motif both in the antechamber and in the burial chamber. Its pattern was a yellow-black-red wavy line. The friezes in both rooms were made at the same time.

Afterwards, the author states that close relation to the mausoleum of Pécs as was been expected could not be established. She explains this by showing the chronological difference between them. She dates the building of Pécs to the end of the 4th century, while, on the basis of a coin from the age of Constantine II, she puts the Kővágószőlős building to a period before Julian.

In the mausoleum, some coins from the 3rd century have also been found. The latest coin unearthed here was that of Valens (No. 129 in the catalogue of the finds). She points out that the pottery fragments and remnants of animal skeletons found in the corridor refer to a fireplace, which was left there by a later population.

The most important photos made during the excavation based on Alice Burger are appended:[605] Fig. 153: the two apses of the mausoleum and the entrance with the corridor (cf. fig. 151 D, E). We can observe here on the western wall of the entrance an addition. Moreover, in this photo, it seems as though the masonry of the entrance wall and the apses would differ, which poses the question whether the burial chamber and the upper part were built at the same time. Fig. 154: a photo taken from the entrance showing the western wall of the corridor, where frescos with geometric decoration can be seen. Fig. 155: a photo taken from the entrance showing the western wall of the burial chamber. The wall surface with the fields divided by the stripes is visible. Right beside the wall of the burial chamber, the wall of the superstructure begins. Fig. 156: the northern wall with the fresco, and, in the foreground, walled-up grave 2. Cf. Pécs 32.II.f. Fig. 157: the southern wall of the burial chamber with the entrance and the place of the built grave and the sarcophagus. Fig. 158: the northern apse of the superstructure with the buttress, which is marked C3 on the ground plan.

Commentary. None of the small finds is indicative of Christianity. The fact that the building belongs to early Christianity can be asserted only on basis of frescos; however frescos mentioned here could have equally been the decorations of a synagogue as of a Christian building. However, I consider this fact insufficient to think of episcopacy in this case. Anyhow, it is highly regrettable that the building was so quickly reconstructed and renovated. The difference between the 'mausoleum' found in Pécs and this one confirms the suspicion that we cannot reckon with a significant number of early Christians at Kővágószőlős. I have to note that Lajos Nagy actually did not regard the painted burial chamber of Kővágószőlős as one related to Christianity.[606] Nevertheless, we have to acknowledge that, as we can find it in one of the excavation reports of Margit Cs.-Dax, a burial chamber can be of other kind different from Christian too.[607] During a rescue excavation at Balatonfűzfő, she found 'two late Roman burial chambers made of brick, ... out of which the vaulted one is two-parted.' A fragment of a bronze jug and an undisturbed skeleton were unearthed here. From the other one, fragments of relief were unearthed: "heads of a god and a horse of the eastern type'.

It is a particularity that it is located near a villa. The villa can be considered as a luxurious private house that was fitted with all conveniences and to which outbuildings belonged as well. The excavation report informs us about the reconstruction and the periods of the building,[608] but we do not know which period of the dwelling house was contemporary to that of the building of the 'mausoleum'. She is of the opinion that the burial chamber and the part above it were both constructed in the 4th century, for a coin from the age of Constantine II was found there. A new examination of this dating cannot be avoided. I am going to carry it out according to the frescos. The coin can at most indicate that the building was still used then, but it does not show that it was built at that time.

In Pannonia, it seems that burials took place near a dwelling house. In fact, this phenomenon occurs in other places too: '... die Gräber häufig in gemischt genutztem Vorstadtgelände zwischen Gärten, Villen, Gutsbetrieben, aber auch Tavernen und Gaststätten lagen.'[609] At present, there is one excavated building in the villa settlement whose dimensions can only be roughly estimated. The mineworkers could estimate some data concerning the dimensions of the villa. Northwards and westwards, only its continuation can be presumed, but its dimensions cannot. At the same time, a huge dimension can be taken into account towards the south and the southeast. The same is probable eastwards too. The southern and southeastern part was the place of the workshops and the outbuildings. Therefore, though such burials are strange for us, they were usual and acceptable for ancient people. Only one question remains open: can the burial be proved in fact?

The evaluation and the dating of the frescos. Those frescos whose remnants can be observed in the burial chamber betray an earlier time than the 4th century. With the help of analogues, the fragments can be defined more easily. The most suitable analogues either of the imitation of marble incrustation or of the figural illustrations can be found in Dura-Europos.

Geometric ornament in the corridor: fig.159a, b. Another one resembling this very much is in the torah

[604] Quotation from BURGER – 1985-1986: 171.

[605] BURGER 1985-1986 figs. 157, 158, 162, 163, 164, 172, 179.

[606] NAGY L 1938: 42.

[607] DAX RégFüz Ser. I No. 24 1970, 29.

[608] A general ground plan can be seen in BURGER 1985-1986 Figs. 155-156.

[609] HESBERG/ZANKER 1987: 10. The town of Pompeji is as an example of this; at the Herculaneum gate, along the street of the tombs, there is a row of shops, workshops and also Cicero's villa right beside the sepulchres. PURCELL (1987: 27-29), among other things, analyses the *auctors*' data connected with this phenomenon as well.

shrine of the synagogue in Dura-Europos,[610] fig. 160. We can also observe that our fresco is slightly similar to its analogue regarding that the rhombus framed by a rectangle is irregular and cannot be accurately placed in its rectangle, as a result of which the two angles of the rhombus are missing from both the Dura fresco and the one of Kővágószőlős.

Figural illustrations. Fig. 161a, b. One of the fresco-details in the same synagogue illustrates scenes from the Exodus,[611] fig. 162a, b. Such position of the human figures' legs can be seen on the wall of the Christian house-church, too, where the remains of the frescos illustrate the scene of 'the women at the tomb'.[612]

Bordering frieze. Fig. 163a, b. We can see a similar ornament in Dura-Europos, in the baptistery, fig. 164.[613]

Dating. The frescos of the Dura synagogue can be dated accurately. 'On the basis of dated paintings, others with stylistic or iconographical affinities may be assigned to approximately the same period. This analogical method is of course not without dangers, since the subjective element plays so large a part, and scholars who have previously discussed the Dura paintings do not always agree as to the affinities and differences among them. No surviving paintings can be dated prior to the second half of the first Christian century, and the majority come from the last century of the city's history.'[614] Above the history of the city is described by Perkins, whose end is described in his book as follows: 'In fact, the latest dated document from Dura, a coin of A.D. 256, was found on one of the bodies and gives us the approximate date for the fall of the city.'[615]

The baptistery: In Block M8, a dwelling house was built in 232/233, which was a private house. It was rebuilt into a Christian house-church inside the baptistery. The date of the rebuilding is obscure, somewhere between 232 and 256. Logically, Christians used it for about 10-15 years.[616] The synagogue is in Block L7. The fall of Dura was caused by the third campaign of Shapur (254-259/260), which was conducted against the Romans.[617]

However incomplete the Kővágószőlős frescos may be, the analogies cannot be regarded as accidental according to the listed discernible drawings. However, if the analogous frescos were destroyed in the middle of the 3rd century, then those of Kővágószőlős cannot be put to the 4th century, but rather to the turn of the 2nd and 3rd centuries, or, at the latest, to the first part of the 3rd century. This, of course, would indicate the date of their form, which would not exclude that the building existed later on. The analogy also shows that the painters were Syrians, Hellenised Jews, maybe Hellenised Jewish Christians,[618] who had brought the motifs from there. This refers to the burial chamber and its frescos too. The burial chamber - for its paintings - is often mentioned in the special literature as a catacomb, a definition that in my opinion is wrong. I would rather suggest the term hypogeum for this type of tomb, if we can at all talk about a burial here. I have to note that the presence of Syrians in this territory has not at all been brought up by the research.[619] We have already mentioned the building of villas by the Syrians,[620] as well as the Dura-Europos frescos as analogues.[621] On the richness of tombs ☞ Pécs, historical summary. This means that we do not have to doubt the presence of Syrians. The example of villas around Aquincum suggests that the villas of Syrians arriving at Pécs should be searched in the neighbourhood of Pécs; one of them is at Kővágószőlős.

The superstructure. If one accepts Alice Burger's statement i.e. that the burial chamber and the upper building were built at the same time, then the part named as mausoleum had to be built in the first half of the 3rd century, which is referred to by the 3rd-century coin.

The single remarkable characteristic of the superstructure are its thick walls compared to its size. The walls of the entrance apses are 80 cm thick; the thickness of the middle buttress of the northern apse is 100 cm. The northern apse has five buttresses; the two southern ones have one each. There is also one at both walls between the apses. If we draw a parallel between this and the walls of the villa on the general map, it is even more apparent that the small building has thicker walls. I have to note that this characteristic is shared by the tombs of Pécs (☞ 32.I.c., 32.I.e.). Consequently, classifying these buildings, we have to start out from these features, which is in correlation with their function.

Its function. Alice Burger defines it as a mausoleum in which there used to be a sarcophagus and a brick grave joined to it; the latter one's size on the outside was 240 x 125, on the inside 210 x 65 cm. 'The grave has yielded no articles or other anthropological material.'[622] This grave is not mentioned in the publication of 1867. We have no data as for the findings of the sarcophagus. Thus, we have no basis for deciding whether there was a Christian or a pagan burial here. The other matter at issue is the connection between the burial chamber and the superstructure in time and function.

In the literature, functional terms are often used for buildings. In cases similar to the building at

[610] PERKINS 1973. Fig. 20.
[611] PERKINS 1973. Fig. 21; below.
[612] KRAELING 1967: 38-39 pl. XLII, 1-2.
[613] ROSTOWTZEFF 1938: 100-101 pl XXVIII; KRAELING 1967 pl XXXIX, 2; PERKINS 1973 Fig. 18.
[614] PERKINS 1973: 35.
[615] PERKINS 1973: 7.
[616] ROSTOWTZEFF 1938: 101; KRAELING 1967: 38-39. There is another opinion that says that the dwelling house was built in the 1st century and was reconstructed into a Christian building between 165-167. KRAELING 1967: 34.
[617] KRAELING 1956: 336.
[618] On the settlement of Jews in Syrian territories and establishing diasporas cf. ROSTOWTZEFF 1938 100. On the local population of Dura and artistic influences PERKINS 1973: 34-35.
[619] A complete examination of the question will require the small finds as well. Lacking these, I comment on the question based on the interrelations registered in the province, to the extent that the phenomena allow.
[620] ☞ Budapest III, historical evaluation, hypotheses on Aquincum's Christianity, M. Ant. Victorinus.
[621] ☞ Budapest III 4.I.g.
[622] BURGER 1985-1986: 171.

Kővágószőlős, the most frequently used terms are mausoleum, cella memoriae, martyrium or chapel. The term chapel does not stand for a small church now where masses are celebrated, but a type of building where a sarcophagus is kept. This building right is called an εὐκτήριον. The mausoleum is also a modern term.[623] The *cella memoriae* is an antique one.[624] The *martyrium* is such a *memoria,* which was built over a martyr's tomb. In connection with these, there is a widespread misconception, namely, that the martyr's grave or sarcophagus serves as an altar, and masses used to be celebrated in such a place. The excavations opening such finds do not support this widespread view.[625] The εὐκτήριον is 'eine besondere Kapelle mit dem Sarkophag der Märtyrerin war an die eigentliche Gedenkkirche (ἐκκλησία = μαρτύριον) angebaut, in welcher der Altar für die Eucharistie stand.'[626] This statement is valid for both eastern and western parts of the empire. A change in the cult of martyrs was made by Pope Gregory I (590-604), who 'ließ für den Altar von St Peter in Rom eine erhöhte Estrade in das Sanktuarium und in die Apsis einbauen, so daß das bisher sich frei im Raum erhebene Grabmonument von einer Krypta unter dem Altar umschlossen wurde.'[627] Before this time, we can hardly use this terminology for the opened objects. Here, at Kővágószőlős, by no means.

An analogy with the thick walls. Above all, I would like to mention Anastasius's mausoleum unearthed in Marusinac,[628] fig. 165. Marusinac is a place near Salona where, on the occasion of the excavation, a mausoleum was found. This was the built burial place of Asklepia and her husband. Into this tomb, Asklepia buried Anastasius, who was martyred in Salona. The time of this martyrdom was 304. Asklepia hid the body for two years and buried it in about 306. During the excavations, sarcophagi were found on the lower floor. Anastasius's sarcophagus was placed in an apse separated by a wall, and Asklepia's and her husband's sarcophagi stood in front of the apse. There was no burial upstairs. Here, the anniversaries of the funerals could be celebrated; however, there was room for only a few people. 'Der Märtyrer erhielt also in Marusinac keine eigene Begräbnisstätte und vor allem keinen eigenen Kultbau, sondern er wurde mit anderen Toten zusammen beigesetzt. ... selbst noch zu Beginn des 4. Jh., ... blieb also das Matyrergrab in der privaten Sepulkralsphäre, und es war nicht gegenüber anderem Bestattungen sichtbar hervorgehoben. Es handelt sich in Marusinac zugleich um das erste bekannte Beispiel der Beisetzung eines Märtyres in einem Mausoleum, und damit in einem geschlossenen Gebäude: aber dieses Gebäude gehört seinem architektonischen Typus nach zur antiken Sepulkral-Architektur, ebenso wie die Apsiden der anderen salonitaner Friedhöfe.'[629]

Marusinac and Kővágószőlős. If we make a comparison between the ground plans, it is noticeable that both have thick walls compared to the overall size. In both places, there is a lower and an upper floor. The term private area is applicable to both territories. However, there is at least 100 years between the two buildings. In spite of this fact, the parallel provides certain clues.

In the case of Kővágószőlős, we cannot say that the grave was hiding a body of a martyr. This does not apply to the martyrs under Diocletian's reign, but to those of an earlier persecution, since the frescos were painted before the first half of the 3rd century. Therefore, if someone built this tomb for their own family, then they could only have entombed a martyr of an earlier persecution. This could have happened in the time of Traianus Decius. Hypothetically, this is possible, but we have no evidence. Only the the frescos, on basis of parallels, can justify this suggestion. The private area however can be considered obvious and proved. The function of the upper building was to serve for the celebration of the day of death i.e. for a heavenly birthday. If we assume that we are dealing with Christians, then a prayer, singing or a vigil took place here.

The Marusinac analogue. It might be curious that the frescos follow the pattern of the Dura frescos almost completely; nevertheless, I have brought a Dalmatian analogy for the building. For the solution of this process that can be seen as a contradiction, I can only come up with a hypothesis. It is as follows:

It is a fact that Pannonia and Dalmatia were originally one administrative unit. It is also a fact that the finds from the territory of Savia and, to the north of Savia, along Kapospula and Pécs are different from the finds in other parts of Pannonia. The area of Pécs and its vicinity - in view of its finds - resembles southern territories such as Valeria or Pannonia Prima. I can explain this phenomenon partly[630] by the fact that the original administrative unit helped the interference, which could spread and remain intensively up to this line. I regard this as a basic culture, which would later meet the settling Syrians' culture, and out of this, a special style would be born, which I propose to call a 'Syrian provincial' art, following the pattern of the different provincial arts in Roman provinces. Considering either the finds of Kővágószőlős or ☞ Pécs, the Syrian, or, in a wider sense, the oriental impact can be traced. The local features - limiting the phenomenon to the burials - come from heroon architecture.[631]

SUMMARY

At its construction, the burial chamber has to be looked upon as a hypogeum, whose vault used to be above the surface of the ground. Its building can be dated to the

623 The antique Mausoleion is the name of one special building; it is the tomb of Mausolus, King of Karia (from around 350 B.C.); this is where the name derives. It was also used in the early Christian period, but not earlier than the middle of the 4th century.

624 DEICHMANN 1954: 942. The name has remained in an inscription, ILCV 1583.

625 DEICHMANN 1983: 54-67.

626 DEICHMANN 1983: 61.

627 DEICHMANN 1983: 62.

628 MARIN 1988: 56-59, 111; MIGOTTI 1990 No. 106, pl IX, 1. On the eastern origin of this type of building see EGGER 1962b. Cf. note 694.

629 DEICHMANN 1983: 58.

630 I am using 'partly' because I do not find the presented explanation satisfactory.

631 GRABAR 1946 Vol. I 33.

first half of the 3rd century. Despite all the excavation observations, the question may occur whether the upper floor was built later. The superstructure can be called a mausoleum in the sense that it held the graves, and, in its upper part, it could have been place for the celebration of anniversaries. If we think of Christians, then the community might have shown their respect by praying, singing hymns and by vigils, which could last until dawn.

From Pannonia we do not know of any codified saints prior to the martyrs of the Diocletian's persecutions. This does not exclude the possibility that the local community would respect somebody as a saint above whose sarcophagus they erected a mausoleum. It is even more possible in Pécs. We can presume its Christian relation but we cannot verify it. The circumstances do not rule out the Christian usage which is expressly supported by the analogues of the frescos; although, its a disturbing condition that none of the plentiful small finds shows Christian symbols, illustrations, or any other Christian relations. The coins of Constantinus II and Valens suggest that we could expect such relations. Since this excavation cannot be regarded as complete or finished, we can hope that a further excavation will result in such a material of finds that can diminish this uncertainty. Up to that time, to be careful, I range it among the debatable finds.

In front of the entrance of the present reconstructed tomb, a few metres southwards, there is a door-stone, which seems to be *in situ* and is observable on the surface. This was not recorded on the published ground plans. It might be the entrance of another mausoleum. Without further excavation, however, it cannot be explained.

The rest of the excavated site may have belonged to a villa estate. The habitation house had an atrium floored with tiles. There were further rooms around it. To this part, in the east, a bath was joined which has not been excavated but whose details are - in spite of destruction - visible. Nor does the excavator locate the entrance to the point wherev it would be logically expected, but to the eastern part of the building, in the room marked P3, which was the *vestibulum*. The part joining the building from the southeast has not been uncovered yet, either; A. Burger defines that as an ornamented entrance. At the Y4, the wall runs further. However, the excavation site was not determined by scholarly, but by another objective factor, namely, by the territory of the uranic mine.[632] That is the reason why the part of the building reaching beyond the mine has not been opened yet. At present, further traces of buildings can still be noticed on the surface, located northwards from the unearthed part. Presumably, the same building could be found there continuing with the actual part of the peristylium. The part called an 'ornate entrance' had also some extension northwards. Nevertheless, over these territories, the excavation was not pursued; accordingly, we cannot say anything with certainty or concrete. According to the mineworkers' remembrance, there had been a building (or buildings) built some 10 metres towards the southeast from the excavated area. Evidently, they stand for those outbuildings whose presence can be supposed on the strength of the examination of the wood or the animal bones. The local geographical conditions make a wine growing viniculture possible here as well as animal husbandry indeed. Moreover, let us not leave out of consideration that the examination of the animal bones shows many bones right in the southeastern part. Finally, I would like to make a quite lyrical but here and now potentially informative remark on what has been expounded before. The surroundings, the remains of the building, the vegetation and the hillside really exhale a Mediterranean 'fragrance'. Oriental people can feel at home here.

27. LEÁNYFALU fig. 166.

A flagon was found among the ceramic finds on the excavation of a Roman *burgus*. The decoration of this flagon has Christian character. The excavation was lead by Bálint Kuzsinszky in 1915 and András Alföldi published the finds of this excavation. The finds were taken to the museum in Aquincum and later were given to Ferenczi Múzeum in Szentendre. Modern building-estate was planned on that area. That is why Sándor Soproni made some excavations in 1963 before the construction. The inner size of the *burgus* was 16.28x16.11x16.12m.[633] The thickness of the wall was 1.6m. The *burgus* was located in a yard, which was surrounded wall. The wall was 0.7m thick. We do not know much about its *fossa*. Mostly pottery was found in the *burgus*, which - according to András Alföldi - was considered to be barbaric. Analyses and comprehensive work defined it as Roman ceramics, even those that are smoothed.[634]

27.I.a. Jug fig. 167.[635]

Location: BTM 66.5.12. **Literature:** NAGY L 1938: 77 fig. 42; ALFÖLDI 1932: 85 XXXI t. 4; fig. 20; OTTOMÁNYI 1991: 17-23, 42-52.

A pottery flagon, whose shoulder is decorated with incised patterns. Cross-, pine branch, and three circles in-group. H: 18cm, base Diam: 5.5 cm. The mouth of the flagon is broken, which was later restored. Flagons of these type were smoothed; dated to the end of the 4th and the first half of the 5th centuries.[636]

28. MAGYAREGREGY

28.II.a. Finger ring figs. 168a, b.

Provenance: Tibor Nagy writes about the surroundings of the find-spot: "The ring was found during the road building between Kisvaszar and Magyaregregy in county Baranya close to Magyaregregy." Besides the ring, a bracelet was taken to the museum as well, which was found at the

[632] It is an essential factor that the whole object is in the territory of a uranium mine. Therefore, acknowledgements are due to those who have promoted or carried out the excavation.

[633] MRT 7/11, p.114.

[634] MRT 7/11, p. 116 OTTOMÁNYI 1991.

[635] I am grateful to Dr Margit Nagy for allowing her drawing to be published.

[636] OTTOMÁNYI 1991: 19

same place at the same time. According to Edit Thomas, they came to light from the cemetery which might have belonged to a villa. In addition, Tibor Nagy thinks that they are from a cemetery but not necessarily from the same grave. **Location:** JPM old inv. no. 1922/135 = 1531. **Literature:** NAGY T 1949; THOMAS 1964: 279-280 Abb. 146; Severin, Katalog 8.63 – Here the provenance is Kisvaszar and the inv. no. is incorrect.

Silver ring, Diam: 1.7 cm, head Diam 0.6 cm. The chi-rho monogram can be seen on the head. Letter "P" turns to left. The ring - upon Mashall's typology[637] – can be dated to the 3rd century A.D. This typology differentiates two types of rings. The one consists of only a simple circle and the other one is richly decorated and has a head, which is circular or rectangular. Tibor Nagy disagrees with Marshall's dating due to the chi-rho ligature. He dated the ring back to the period of Constantine,[638] although we have many data, which testify the appearance of chi-rho ligature earlier than Constantine's rein.[639] Therefore, we do not have to change the dating according to the typology but we should not list this ring to the Christian finds. In addition, András Alföldi does not always list objects with the chi-rho monogram to Christian ones. "Nevertheless these finds are from the Christian period, and bear the chi-rho monogram, we do not list this sign strictly to Christian usage."[640] I have also tried to prove the non-Christian usage of this sign.[641]

29. MÓRICHIDA-KISÁRPÁS (MURSELLA I)

One of those places that the Hungarian researchers named inner fortified settlements (Binnenfestung) is found in Kisárpás. Relating to the others, it is less researched. However, the find-spot has been known since the last century, regular excavations were carried out only between 1975-1980, led by Eszter T. Szőnyi, who gave a report of the results.[642] Due to the excavations, an early settlement was proved to be here. As the term *"Mursella municipium"* occurs on the inscriptions, we are convinced it had the municipal rank.[643] The existence of a later inner fortified settlement is not supported by the results of the excavations.

29.I.a. Casket-mounts, fig. 169.

Provenance: from plundering. **Location:** MNM 62.344.1. **Literature:** GÁSPÁR 1986 Kat.-Nr. 843.

It has been found in fragments. The scene Moses striking the rock is still recognised in the pearl-bordered panel. In front of the face of Moses, there is a chi-rho monogram. There are other non-Christian depictions on the casket-mount, such as frieze with animals, head of Medusa and an astragal.

[637] MARSHALL 1968: 88. XV.t.518.

[638] NAGY T 1949: 81.

[639] Ex. A. B. COOK: The Cretan axe-cult outside Crete. Transaction of the Third International Congress for the History of Religions. Oxford 1908, vol. II. 184-194.

[640] ALFÖLDI 1931-1932: 3.

[641] GÁSPÁR 1993a: 207-212.

[642] E. SZŐNYI 1981.

[643] RIU 328.

30. MUCSFA

30.II.a. Roof tile, fig. 170.

Provenance: It is from a brick-tomb. **Location:** lost. **Literature:** THOMAS 1982b: 262 Abb. 5

Depicted is a man, who carries a human head in his left hand. The tile is exactly cut in the line of man's his neck, so his own head could not be seen at all. Edit Thomas sees a beheaded martyr in this engraving,. The tile does not exist any more; therefore, it is impossible to check the drawing and the interpretation. If we could find an accurate analogue, we would be able to solve the secret of the depiction. Now, it would be better to list this object as a doubtful one.

31. NAGYBERKI-SZALACSKA

31.I.a. Lamp, fig. 171.

Location: RRM, Depot. **Literature:** BAKAY 1975: 31 Abb. 22/3; IIL (NÉMETH) Kat.-Nr. 315.

A bronze lamp with chi-rho monogram. L: 8.4 cm, H: 3 cm.☞ Szombathely (51.I.g.) With Southern-Pannonian analogues, ☞ Zalaegerszeg, 60.I.a.

32. PÉCS fig. 172.[644]

RESEARCH HISTORY

According to F. Fülep the Roman city was founded in the 2nd century.[645] He presupposes that the inhabitants were Illyrians and Celts. Based on András Mócsy's nomenclature the immigration of Italians are evident.[646] The research into the town history began in the 18th century. József Koller, canon of Pécs, discovered a burial chamber in 1780.[647] Today it is known as painted tomb I. He also found 13 other graves.[648] István Möller and Ottó Szőnyi discovered a *cella trichora.* Gyula Gosztonyi found the burial chamber II. and the *septichora* in 1939.[649] Gyula Török also undertook excavation in Pécs in the 1940s. Aladár Radnóti and Ferenc Fülep started the more recent excavations. Ferenc Fülep led the excavations in Pécs for more than 20 years. He published the results and his opinion about the city in 1984. He excavated in Geisler Eta (Street) utca. A painted double grave and three burial chambers were found under the house No. 8. He also unearthed a cemetery under the house No. 14. There were

[644] On the basis of FÜLEP 1984, I enclose the changes of the street names. The earlier names are on the map, but utilisation of the new names is frequent.

previous:	new:
Doktor Sándor	Hungaria
Geisler Eta	Apáca
Kossuth Lajos	Király
Kulich Gyula	Papnövelde
Landler Jenő	Klimó György
Sallai	Ferencesek utcája
Székesfehérvár	Vörösmarty

[645] FÜLEP 1974: 31.

[646] MÓCSY 1974: 120. Caesernii.

[647] CZOBOR 1879: 170.

[648] CZOBOR 1879: 170.

[649] FÜLEP 1987: 31-32.

more than 100 graves, three burial chambers and three cemetery chapels, one of which has been reconstructed. He also excavated on Szent István tér (square), where a building he named mausoleum was found. Although several publications excavations exist,[650] complete presentations of the finds and excavations have not been prepared until now.[651] The 18th century excavations which would indicate their geographical locations, and their complete map, which would show the relation and the distance of the different find-spots have not been published. The maps that were published do not contain any marks to show chronologically different buildings and find-spots. The preconception that everything in Pécs is early Christian is overstated. However, it is true that we should not underestimate the Early Christian finds but the exact role of these are not known in the history of the city and our province. We do not know the local separation of the pagan and Christian burial. The evaluations of these places have not given many data to the Christian history of Pécs or Pannonia. First of all we should re-examine whether everything, which was thought to be Christian, is in fact Christian, or not. We should not disregard the fact that excavations are due to the modern constructions and other earth-works in a populated modern city. Therefore, they are incidental concerning either location or time. We should not neglect the fact either that there are also non-Christian finds in Pécs. For example, during the renovation of the library, under the house no. 3 in Székesfehérvár Street, on the area of approximately $400m^2$ 87 late Roman graves were discovered. They were earth-graves, brick-graves, stone-graves and built tombs. These were situated above each other in several levels[652]. Late Roman graves and limestone sarcophagi came to light in Geisler Eta Street, as well under the houses no. 6 and 8.[653] A short study by Imre Lengyel informs us about the dating of the late burials compared with other cemeteries[654] ☞ Keszthely-Fenékpuszta. F. Fülep gave a short summary of changes during the four centuries, and the early population in the town, which became smaller by the later times.[655] In connection with the history of the town, E. Tóth thinks the followings:[656] Pécs was neither municipium nor colony until 200-250. It was rather a settlement at a road-junction, and the money circulation of this settlement increased under the emperor Gallienus. After the destruction of Gorsium in 260, the *concilium provinciae* Pannoniae Inferioris might have moved here.[657] On the other hand, András Mócsy pointed out that the conditions did not match to form an urban life from the *civitas peregrina* in Pécs up to the age of Trajan.[658] He also cautiously mentions this alteration by the later time: "As a town with municipal status, Sopianae (Pécs) on the southern slope of the Mecsek mountains comes into consideration."[659] Let me mention its bishop and the episcopacy. Tibor Nagy thinks that Paulus Arianus had his see here. This bishop played a very special role among Arians since he was very connected with Valens.[660]

In Beška, which waso nce in Pannonia but is located today outside the Hungary, a tomb which must be mentioned was discovered.[661] It is important because it shows the direction of the spread of this burial type, so it makes easier to draw some consequences concerning other territories of Pannonia.

Commentary: The description of the finds and sites of Pécs are not discussed in the usual way, namely separating the Christian and non-Christian objects; but in the territorial units which are suitable because of their find-spots.

---

650 FÜLEP 1959a; 1959b; 1961a; 1961b; 1962a; 1962b; 1963; 1967b; 1969a; 1969b; 1969c; 1973a; 1973b; 1974; 1977a; 1977b; 1984; 1987; FÜLEP/BURGER 1974; FÜLEP/DUMA 1972a; FÜLEP/DUMA 1972b; FÜLEP/FETTER 1969-1970; FÜLEP/BACHMAN/PINTÉR 1988; RIU 4 pp. 12-20.

651 The monograph published in 1984 covers only part of the whole material.

652 KATONA GYŐR Zs RégFüz Ser. 1. No. 35 1981 88/3.

653 KATONA GYŐR Zs RégFüz Ser. 1. No. 37 1983 84/1, 84/3.

654 BARKÓCZI and LENGYEL 1971: 191-199.

655 FÜLEP 1967b: 45 note 283. In connection with the inscribed stones located in Pécs, Endre Tóth is convinced that their provenances are unknown, as most of them stood in Mursa: TÓTH E 1988c: 24.

656 TÓTH E 1988c: 23, 25-26; There is no connection between the name Quinque Ecclesiae of Pécs and the ancient finds, according to the recent research. TÓTH E 1991c: 101-108, with previous scholarly works. So I have omitted those scholarly works, for example the contribution of Dezső Simonyi: Sull' origine del toponimo "Quinque ecclesiae" di Pécs, ActaAntHung VIII 1960: 165-184. From the same author about the origin of the name of Pécs: "Pécs 'Quinque ecclesiae' nevének eredetéről". AntTan 6 1959: 87-103.

657 The viewpoint of Endre Tóth in connection with the Gorsium and location of *concilium provinciae*, see also at Tác.

658 MÓCSY 1974: 137; Jenő Fitz writes that there is no town under the territory of Pécs. (J FITZ: La Pannonie sous Gallien.[Latomus 148] 56).

659 MÓCSY 1974: 222.

660 NAGY T 1939: 210. ☞ Kővágószőlős.

661 The find-spot was a cemetery situated around the village of Beška, 5 km away from the Danube on the hill known as Selište. The excavations took place in 1965, when a painted burial chamber was also discovered. The excavators opened 111 graves (all skeletons apart from 22 cremations). This burial chamber and cemetery are discussed by Mirjana MARIJANSKI-MANOJLOVIĆ (Romska nekropola kod Beške u Sremu. Novi Sad. 1987). The burial chamber is 197 cm long, 130 cm wide and 156 cm high. The walls were decorated with frescos, the floor covered with floor tiles. The narrower western side: the upper part, and lower right corner have been damaged. The depicted scenes are hardly visible. "The central figure.... is the master of the tomb with his wife" (p. 84). Eastern narrower side: upper part has been damaged. According to the author, there are three female figures, which might be the *Parcae*. All three figures hold the stick of *signum*. Only the middle figure's head is intact enough to identify the halo around it, curly hair and big open eyes. The remains of the frescos allow us to assume that all three figures had halos. They are the saints worshipped by the local people. The southern longer side: the wall was divided horizontally into five zones. On the lowest zone, four persons were painted in rectangular panels, whose sizes are different. A female figure stands in the panel outermost to the right. The further three figures are men. According to the author, the three men hold funerary offerings. The next zone is decorated with fretwork; trailing vine with leaves (with birds eating the grapes) follows this zone. The two highest zones are decorated with spiral patterns. The northern longer side: The lowest panel (dado) is divided vertically into three fields, painted with imitations of marble incrustation. Then fret-like ornamentation follows. Above this zone, peacocks and other birds are discernible in the remaining parts of the frescos. The author dates the burial chamber after 330. The burial was disturbed and therefore the grave-goods are poor: golden, T shaped brooch, "hemispherical cup, ...glass jug, ... glazed lamps, coins URBS ROMA". The author gave the dating based on the coins. They were minted in 330.

FINDS AND SITES

32.I.a. 13 graves

Literature: KOLLER 1804: 33-34; SZŐNYI O 1907: 19-20; NAGY L 1938: 36; GOSZTONYI 1939: 105; GOSZTONYI 1941: 42; FÜLEP 1977b: 11; FÜLEP 1984: 42, R/1-13.

The modern research nearly forgot these graves, although they are important as much as the other finds. It seems necessary to cite the witness: "non multo post detectum hoc Sacellum (sc. painted tomb I), dum juxta struendis aedibus (sc. the residence of the parish-priest.) fundamenta jacerentur, tredecim sepulcra e Romanis lateribus facta deprehendimus, omnia eius formae, quam Tab. XIV. Fig. 7. exhibemus, (here fig. 173). In eorum uno, quod est Fig. 7. ad caput, et pedes monogramma Christi pictum in latere fuit. In singulis sceletum hominis, in aliis minus integra, in eo, quod refertur Fig. 7 intcgrum eo, quem repraesentamus, positu. In duobus sepulchris ampullae vitreae; in altero, ut Fig. 8. in altero, ut Fig. 9. ..."(Here fig. 174). After Koller, Szőnyi writes about these graves and glasses as well.[662] The location of the discovered graves is more accurate at Szőnyi: "...The southeastern tower of the Cathedral, 7 meters to the east..." was dug the ditch of the parish residence leading to the N-S direction. The graves came to light from this ditch.

Gyula Gosztony's viewpoint about the cemetery: The burial systems in Pécs significantly correspond to the Syrian ones. The fact is that this burial system, except for a few Gallic examples is completely unknown to the north and west from Italy."[663] ☞ Research history, Beška.

Commentary: Both glasses are undoubtedly Roman. The brick grave was built in Roman way. The deceased was laid on the side in the grave. This grave is different from the other 12 not only because of the intact skeleton but also because of the chi-rho monogram, which was painted onto the inner surface of the brick. The glasses and the bricks without doubt prove that we face a Roman burial in this case. What is more, chi-rho monogram confirms that this grave really is a Christian burial. Further excavations are almost impossible now due to the modern buildings.

It has to be underlined that Gyula Gosztonyi's viewpoint is very worth considering.

32.II.a. Tomb III, fig. 175.

Literature: GOSZTONYI 1939: 109; GOSZTONYI 1943: 28, f, 11 drawing 3 (unpainted tomb 2)[664] NAGY L 1938: 36; FÜLEP 1977b: 12; FÜLEP 1984: 46-49 pl. XXII, 2.

It is situated to the south of the painted tomb I. Szőnyi and Möller discovered it in 1913. Dimensions: 3.88 x 3.12 m. N-S orientation. The entrance is in the southern wall, W: 95 cm, fig 176. In the north wall, there was a 62 cm wide and 42 cm deep niche, under which a sarcophagus was placed. The lid was broken and was decorated with acroterion. There were no grave goods at all. The wall of the tomb remained up to the vault. The latter was destroyed on the construction of a modern corridor leading to the painted tomb I. This tomb was unpainted. There were six graves around the tomb: R/14, R/15, R/16, R/17, R/18, -R/19 child's grave.[665] N-S orientation of these graves. Grave-goods were found only in nos. R/15, R/18, and R/19. 15: a bronze brooch and an iron knife.18: 49 various pottery. 19: 85 glass-past pearls, a bronze ring, two glasses and three coins (Tacitus, Julian, Valentinian I). On the finds, neither the tomb nor the graves indicate Christian character.

32.II.b. Tomb IV, fig. 177.

Literature: GOSZTONYI 1939: 117; GOSZTONYI 1943: 26-27, c, 11 drawing 1 (unpainted tomb 1); FÜLEP 1977b: 14; FÜLEP 1984: 49-50 pl. XXIV.

It was to the west of the painted tomb I. L: 2.47 m, W: 2.07 m in the north, and 1.98 m in the south. It was rectangular and barrel-vaulted. The walls were built of stone and the vault of brick. The thickness of the wall was 45 cm. The semicircular entrance was 125 cm high, 65-68 cm wide and faced to the south fig.178. An antechamber might have existed. Traces of painting could not be seen. The niche was built in the western wall, and was broken through, figs 179, 180.[666] Because of this, Gosztonyi thought of plundering.[667] A 70 x 193 cm sarcophagus stood along the northern wall. Grave goods: sherds of vessels, chicken bones, bronze key, and two glazed roof tiles.[668] Above the tomb, three built graves were found.[669] Szőnyi describes these tombs: their longer sides were built of stone, shorter ones of brick except the third, whose longer sides were also built of brick. Orientations of these graves were W-E. He does not say any word about grave goods. It is quite possible that they did not indicate any Christian character.

32.II.c. Tomb V.

Literature: GOSZTONYI 1939: 117; GOSZTONYI 1941: 42; GOSZTONYI 1943: 30, k, 6 drawing g (tomb 7); FÜLEP 1977b: 14; FÜLEP 1984: 50; Itinerarium 99.

It was found in 1913, but the uncovering was impossible because of technical difficulties. Gyula Gosztonyi excavated here but he did not have the possibility for the complete opening either. It was situated to the east from the painted tomb I. The floor level was 70 cm higher than that of the painted tomb I. Traces of painting were not found. Ferenc Fülep did not lead excavation in this area. The *Itinerarium* published in 1995 gives a layout of the tomb itself, and the whole circumstances, but without commentary.[670] Due to the lack

[662] SZŐNYI O 1907 fig. 2-3.

[663] GOSZTONYI 1941: 44.

[664] The numbering of Fülep and Gosztonyi differ. The concordance to their numbering:
(Roman numerals – Fülep, Arabic numerals – Gosztonyi).

III	2	V	7	VII	-	IX	5
IV	1	VI	3	VIII	6	X	4

[665] FÜLEP 1984: 47-49: with description of finds and graves.

[666] FÜLEP 1984 Pl. XXIV, 1, 3.

[667] GOSZTONYI 1943: 28.

[668] FÜLEP 1984: 50.

[669] FÜLEP 1984: 50. R/20-22.

[670] Itinerarium p. 99: Also the tomb V. is discernible on the layout, labelled Saint Peter and Paul tomb plans.

of sufficient information, one cannot decide on he nature of this tomb.

32.II.d. Tomb VI, fig. 181.

Literature: GOSZTONYI 1939: 117; GOSZTONYI 1941: 42; GOSZTONYI 1943: 28, g, 11 drawing 2 (unpainted tomb 3); FÜLEP 1977: 15; FÜLEP 1984: 50-51, fig. 14.

This was found on a square in front of the Cathedral in 1922. The walls were built of stone and the vault of brick. The remaining inner size of the tomb was 2.40x1.15 m according to Fülep. Gosztonyi gives the complete size: 2.30x2.30 m. The entrance of 5 4 cm wide was in the southern wall. The northern wall was destroyed by a later construction. On both sides of the entrance, two sarcophagi used to stand, (R/VI/1-2). There were no finds in them. The R/VI/2, however, contained a skeleton. This was a sarcophagus on the eastern side of the tomb.

32.II.f-g. Tombs VIII-IX, fig. 182.

Literature: GOSZTONYI 1941: 41; TÖRÖK GY 1942: 207-211 pls XXIV-XXVI. (tomb A); GOSZTONYI 1943: 29-30, i-j (unpainted tombs 5-6) 11 drawing 5-6; FÜLEP 1977b: 16-17; FÜLEP 1984: 59-61.

Gyula Török and Aladár Radnóti discovered two tombs altogether with the graves in the vicinity, in 1940. Tomb VIII has been damaged, figs. 183, 184, 185. It was oriented N-S. The entrance was on the southern side. Dimension: 595 x 430 cm outside, and 465 x 300 cm inside, H: 230 cm. The wall of the vault was 30 cm thick. The sidewall up to 90 cm was made of stone above that of brick. Although the inside was plastered, there were no paintings on the walls. On the western side, wall fragments 15 cm heigh extended above the vault. According to Gyula Török, this would imply the chapel above the tomb. He found two tombs in the chamber; tiles covered one, stone slabs the other (VIII/1-2).

Tomb IX has been found 2.30 m to the north from Tomb VIII. Dimension: 530 x 435 cm outside, and 415 x 310 cm inside. The reconstructed height is 235 cm. There is an entrance in the southern wall, with an antechamber of 130 x 260 cm, fig.186. A grave was attached to the burial chamber from the north, measuring 140 x 435 cm, fig. 187. Gosztonyi remarks that the grave walls can be dated to a later period than that of the tomb. The floor of the grave lied 160 cm higher than the floor of the tomb. Dimension of the niche in the northern wall: 120 cm high, 30 cm deep, and 65 cm wide. The floor was tiled twice. The distance between the two floor levels amounts to 10 cm. The walls were plastered and unpainted. Nobody was buried in the burial chamber itself, however a grave was added to the north wall. The floor of this "chamber" was tamped.[671]

Further four graves (C, D, E, F = R/36, R/37, R/38, R/39) were discovered around Tomb VIII. A coin from the period of Claudius I was found in D=R/37 grave. This child-grave was disturbed. The other three graves were totally plundered and even the skeletons were missing. Some of the grave goods have been found on the floor, such as a bronze buckle and sherds of pottery, fig.188. "The grave was plundered several times as the nearly half meter thick remains of medieval skeletons and medieval pottery fragments showed."[672] A bronze hair ring became known from the antechamber of Tomb IX.

Gy. Török writes that the distance of the levels of the tombs A and B (id. e. VIII and IX) is due to natural features. All over the tombs, Roman painted plaster fragments (fig 189), brick fragments, and glazed tile fragments enfolded the area. The thickness of this level varied between 5 and 80 cm. Two fragments of a console decorated with animal head, Roman glazed jug sherds, and Celtic pottery fragments came to light, as well.[673] Plaster fragments could have been found even at the bottom of Tomb VIII.[674]

Fülep lists more finds such as bronze spur, iron key, iron ring, pottery sherds and bronze bracelets.

Commentary: Beside the wall of Tomb VIII, the wall of another building, named by Gy Török the upper chapel, is well visible on plate XXIV, 2 of Gy Török (cf. fig. 184). An analogue is given for the two walls being tightly close to each other, ☞ Kővágószőlős 26.II.a-b. The burial chamber was painted in Kővágószőlős but there is no information about the paintings of the upper building. In Pécs, the wall painting fragments have been found above Tombs VIII-IX, hence, it is right to assume that they belonged to the "upper chapel". The poor finds do not contradict the possibility that Tomb VIII at least, and the "upper chapel" could be dated to the early Roman time. The published fresco fragments indicate walls decorated with floral patterns. They might have been floral decorations which vertically divided the wall surfaces into smaller panels, as on frescos of Pompeii.

Tomb IX is not a single example of its type in Pannonia. Károly Sági recovered similar tomb in ☞ Keszthely-Fenékpuszta, Halászrét (22.II.b). If the periodization given by Sági is correct then we should presuppose the long usage of the tomb of Fenékpuszta; and accordingly we could date its construction to the early Roman period, as might be the case for Tomb IX in Pécs. Let me indicate briefly the fact that the Romans used to offer offerings, lay presents in the tomb, even serve funerary banquets at the anniversary of the funeral, or on the day of *dii manes* or on particular occasions, which would be wished by the testator. That is why a tomb could have a superstructure, some times a representative one.[675]

Tombs VIII and IX could be dated to the early Roman times. I think so. Tomb VIII was earlier, it might have been finished during the 1st century, whereas Tomb IX was erected in the 2nd century. Gyula Török has found even Celtic pottery sherds, which also prove the early date of the tomb.

32.II.h. Tomb X, figs. 190 a, b, c.

Literature: SZŐNYI O 1906: 44; NAGY L 1938: 39; GOSZTONYI 1939: 124; GOSZTONYI 1943: 29, h (tomb 4); KISS 1973: 31-32;

---

[671] TÖRÖK GY 1942: 208-209.

[672] TÖRÖK GY 1942: 209.

[673] TÖRÖK GY 1942: 209.

[674] TÖRÖK GY 1942: 210.

[675] For example, tombs from the time of Augustus, or street of tombs in Pompeji: ESBERG/ZANKER 1987: 12-13.

FÜLEP 1977b: 17; FÜLEP 1984: 61-63.

A mosaic pavement came to light in 1841 under 6. Káptalan Street during construction works, 285 meters deep under the floor. First, researchers interpreted the find to belong to a bath, but they raised the question of it being a tomb because of a mosaic pavement. A painted tomb with an undecorated limestone sarcophagus (R/X/1) was unearthed on this site in 1927, in course the building of drainage system. The sarcophagus has broken and the lid has disappeared. A plundered skeleton and a Diocletianus follis have been found in it. These finds became known not by an excavation but only by digging a ditch. That is why there were no possibilities for observing any connections. In addition, it was forbidden to deviate from the canal.

Descriptions about this site slightly differ from each other. Gyula Gosztony does not mention the mosaic pavement but writes about the burial chamber, which is very similar to the others. As regards the painting of this tomb, we have a different information: "According to an earlier publication yellow traces were found, so it was one colour; the later publication, however, mentions figured panel. It is highly possible that the painted fragments came to light from an upper filling level as it occurred at *septichora*, ..."[676] Gosztonyi cited even some lines from a contemporary daily paper Dunántúl napilap (Trans-Danubian):[677] "...I have found a semicircular wall painted yellow, and eastern and western walls of a room whose door was 95 cm wide and its columns built of Roman bricks." Fülep notes that one assumed that the mosaic pavement and the tomb belonged together .

Lajos Nagy mentions only the mosaic, because it was thought to be Christian. Namely, the patterns on the mosaic pavement seemed to scholars to have been equal-armed crosses.

Commentary: It is essential to underline that the data about the mosaic, or the bath, or the tomb are occasional since they are chance finds. Only digging of a ditch brought them to light. It is very likely that the walls mentioned above belonged to a tomb, but its connection with the mosaic pavement remains uncertain.

The use of sarcophagi became fashionable during the rein of Diocletian or later. The pattern of the mosaic pavement does not resemble a cross, but rather a knot. So there is no reason to list this site among Christian ones.

32.I.b. Painted tomb I. (Cf. 4.I.f.)

Literature: KOLLER 1804: 25-34 tab XI; HENSZLMANN 1869: 78-84, 123-146; HENSZLMANN 1873a: 155; HENSZLMANN 1876: 37-39; CZOBOR 1880: 327-339; NÉMETHY 1886: 242; KUZSINSZKY 1895 CCLI; SZŐNYI O 1907; SZŐNYI O 1913; SZŐNYI O 1913a; SZŐNYI O 1913b; SZŐNYI O 1913c; H. LECLERQC: Funfkirchen DACL V/2 2715-2716, figs 4760-4763; SZŐNYI O 1926; SZŐNYI O 1929; NAGY L 1934: 299-302; DYGGVE 1935: 64; NAGY L 1938: 99; KÁDÁR 1939: 4 -5, 10-11, 18, 20, 29, 31; KÁDÁR 1940-1941: 62-77; GOSZTONY 1941: 42; GOSZTONYI 1943: 16-22; GERKE 1954: 147-199; FÜLEP 1962a: 23-41; KÁDÁR 1969: 183; FÜLEP 1969a: 151; FÜLEP 1970: 177; MÓCSY 1974: 313, 334-335 Fig. 57; FÜLEP 1977b: 11; PELEKANIDIS 1977: 87; THOMAS 1977b: 42; FÜLEP 1984: 36-41; FÜLEP/BACHMAN/PINTÉR 1988: 22-27; THOMAS 1988: 286 Abb. 3; TÓTH E 1988c: 30-31; TÓTH E 1994b Abb. 3; Itinerarium 96-103; VALEVA 1998: 779 No. 55 fig. 9

Plan and section: Fig. 191a, b. "A" is a modern passageway leading to the tomb. The tomb is divided into parts "B" and "C", fig. 193a, b. Part C has a small niche. Gosztonyi himself has prepared a reconstruction, 192a, b, c.[678] Bricks stamp: IMP N.[679] The description of the tomb follows the plan made by Ottó Szőnyi, but sometimes it is necessary to follow the results of the new excavations.

B: This is the antechamber. A doorway (g) leads to the antechamber from the modern corridor (A). Dimension of the doorway (g): H: 140 cm, W: 89 cm. Dimension of the doorway (h) between the antechamber and the burial chamber: H: 136 cm, W: 56 cm. The antechamber (B) is trapezoid, approximately 2.35 m². The diameter of the airing hole in the antechamber is 67 cm. This hole was made in the last century in order to protect the frescos from the louver. Lengths of walls besides the doorway (g): 75 and 69 cm; besides the other doorway (h): 33.5 and 46.5 cm. The length of the eastern wall is 118 cm and that of the western wall is 122 cm. Sedilia are running along the eastern wall, the height of which is 37 cm, and the width 31-36 cm.[680]

C: This is the burial chamber. Size: 351 x 282.4 x 219.4 cm. Fig. 194: The fresco on the south wall. In the northern dado, there is an opening of 72 x 26 cm; along the northern wall, there is a cornice of 75 cm high and 17-18 cm deep. This cornice is made of brick. Barrel vault starts above the cornice. The diameter of the vault is 282.4 cm, and the height is 144.4 cm. There is a rectangular opening in the northern wall, which leads to the semicircular niche behind the wall. The size of this opening is 44.8 x 44 cm. It is flanked by 5-6 cm thick red border. The full-length figure of Peter to the right and that of Paul to the left are visible. They are 80 cm tall. The length of Peter's head is 14 cm, and that of Paul is 15 cm.[681] They are wearing tunica and pallium. Paul is holding a scroll. A *capsa* with scrolls stands besides his feet. The letter "Z" is depicted on his pallium. According to F. Fülep, Peter is clutching the fold of his clothing. Originally, both would be holding scrolls but that of Peter has rather fainted.[682] Both apostles are pointing to the chi-rho monogram framed in a circle. Among the branches of chi-rho, eight petal rosettes are depicted. Set of garlands follows the vaulting above the apostles.

Szőnyi's photos show the panels as a complex one, therefore it is worth enclosing them: Figs.195, 196, 197. The other pictures are based on Fülep: 198, 199, 200, 201, 202, 203, 204.

Eastern and western walls were plastered only up to 54-60 cm high. The painting starts from that height. Each scene is situated in reddish-purple framed panels on

[676] GOSZTONYI 1943: 29.

[677] 1927 number appeared at Christmas.

[678] GOSZTONYI 1943 fig. 6.E. Dyggve commented on that plan, and noted that the tomb originally had only one entrance (which could have been covered up with a stone).

[679] KOLLER 1804: 25.

[680] FÜLEP 1984: 36.

[681] The sizes quoted from Ottó Szőnyi slightly differ from those of Ferenc Fülep (sometimes by 1 cm); the reader will find the exact sizes on the Henszlmann's plan, see Fig. 191a, b.

[682] KÁDÁR 1939: 4-5.

both sides three-three. Each field is rectangular and measures approximately 60 x 74 cm. The scenes of the western wall are as follows, starting from the north: 1. Adam and Eve (fig. 203.) at the apple tree with the snake, generally so interpreted. Koller disagrees with this concept.[683] The whole scene has not remained, but only the portrait of a woman and a so-called apple, which has a rather strange shape. 2. This scene is in a rather bad condition. One can recognize bent knees depicted with black contour. Ottó Szőnyi explained this – on analogues – as Daniel among lions. Koller disagreed with him.[684] 3. Story of Jonah (fig 204), which is represented in a special way here, in Pécs. Namely, the three scenes of his story are composed in a single panel; Jonah being precipitated from the ship, Jonah is ejected by the whale, Jonah is taking his ease in the shade. The picture of Pécs is: "Jonah is standing on the rostrum, and falling into the sea and simultaneously the ivy tendril appears as shade."[685] Zoltán Kádár considers that the ivy has ichnographical importance, since Jonah is usually depicted taking his ease under the 'booth'. Zoltán Kádár interprets the ivy as the continuation of Bacchus' representation.[686] Eastern wall: 1. From the scene of Noe (fig. 202.), only a 43 cm high figure is seen as he stands in a box. 2. A woman's head 12 cm long is interpreted as the Virgin Mary with the infant Jesus (figs 200a, b). 3. Three figures are represented on this panel; they are the three Magi, according to Szőnyi,[687] fig 205a, b; they are the three young men in the furnace (fig. 201), according to Fülep; and according to Kádár, the three Magi appear as Persian Magi wearing trousers, cape, and Phrygian cap."[688]

The barrel vault: The fresco forms a rectangular panel with a reddish-purple band. Dimension of this panel: 320 x 247.7 cm; it is divided into three fields by a 7.5-8 cm wide reddish-purple band. From the north, the first field is 65 cm wide; the third is 60 cm wide. The middle field is almost rectangular. The first and third fields are decorated with the same pattern, namely red and green wave lines with acanthus leaves. Even a 10-12 cm wide yellow and 6 cm wide brown bands frame them. The yellow band frames the second field only on the east and west sides. In the middle of the second (rectangular) field, there is a medallion of 71 cm diameter. This frames the green wreath, which circles the chi-rho monogram that is in a very bad condition today. The pictures that divide the second field into four parts touch the medallion. Imre Henszlmann calls them cornucopia, Ottó Szőnyi and Ferenc Fülep define them as bunches of flowers. Two-two peacocks, and two-two pigeons flanked two "bunches of flowers". Four smaller medallions are framed with green and red contour in the four corners of the second field. The medallions surrounded four portraits in the blue background. According to Szőnyi, all portraits represent women as ideal-portraits,[689] fig. 206. According to Zoltán Kádár, they depicted local saints that are not ideal-portraits, and they may not all be women.[690] Zoltán Kádár reminds us of Henszlmann's opinion who explained this hair dress as Pannonian fashion. The diameter of the small medallions is 43 cm. A small bunch of flowers has been painted in each corner directed to the medallions. Flowers and leaves fill out the spaces among the representations.

On the south wall of the burial chamber, the decoration consists of different stripes, and wavy line with acanthus leaves. They all boarded the arch of the doorway on the one hand, and the vault on the other hand. The stripes and wavy line run down to the dado, where they form a horizontal frame. The antechamber was also painted but very little of it remains. A candlestick is recognized in the southeastern corner,[691] fig 207.

Walls were found above the tomb in the 19th century. Buttresses also were found on the wall of tomb. On these, Imre Henszlmann presupposed a chapel above the tomb.

Sarcophagi might have stood at the sedilia in the burial chamber, according to Fülep. He supposed this because of the lack of plaster. Later sarcophagi might have been broken; therefore, they do not exist any more.[692]

Dating of the tomb. Fülep dates it to the end of the 4th century, no later than 395. Based on other frescos, which these typologically resemble, Kádár dates it back to the age of Pope Damasus, between 366-384.[693] Dyggve mentions the Anastasius-mausoleum of Marusinac close to Salona as an analogy.[694] On the bases of the analogy, Dyggve thinks this tomb was a mausoleum, namely a storied structure, whose lower part was the tomb and the upstairs was the "Kultsaal". One can go downstairs from the "Kultsaal". He thinks that the opening in the north wall, under the chi-rho monogram is the *fenestella*. He has made the reconstruction of Tomb I matching this theory.[695] He also raises the possibility of worshipping a local saint; consequently, the solution of this question is possible after having got to know the Pannonian martyrs. Pelekanidis discussed the relation between the frescos in Pécs and the ones found in the basilica of Skhira analysed by him. The latter one was dated back to the 4th century.[696]

Commentary:

I. Paintings. It is impossible to agree with the explanations of the scene in every case. Despite the bad condition of the frescos, I shall attempt to reinterpret them.

1. Eve, fig 203. The head and part of the body remain of the scene. The face and the neck is skin colour, but the bust around the neck is greenish brown. 'Eve' seems to have a necklace or a decorated collar. So the depicted person is wearing a necklace, hence this figure

683 KOLLER 1804.
684 KOLLER 1804.
685 KÁDÁR 1939: 29.
686 KÁDÁR 1939: 30.
687 SZŐNYI O 1907 figs 25, 27.
688 KÁDÁR 1939: 31.

689 SZŐNYI O 1907: 33.
690 KÁDÁR 1939: 15.
691 SZŐNYI O 1907 pl. 2.
692 FÜLEP/BACHMANN/PINTÉR 1988: 26-27.
693 KÁDÁR 1939: 8-9.
694 The scholarly works for this mausoleum, see Kővágószőlős; 26.II.b. Fig 166. Cf. note 628, fig. 165.
695 DYGGVE 1935 Fig. 1, 3, 4.
696 PELEKANIDIS 1977: 85-87.

cannot be Eve. The habit of the figure also differs from the usual form as we see on Eve in the mausoleum, 32.I.e. 'Eve' faces to the right and is standing on the right of the panel; so she looks out of the picture instead of watching at Adam and the snake or the tree as usual. Here, 'Eve' is turning away from Adam and the tree if we insist on this being the scene of Adam and Eve. Her face does not show any sign of repentance. The habit of the head and the hair resemble a wall painting fragment, which was unearthed in Balácapuszta, and which showed maenads. Klára Póczy already pointd out to the similarity between these two frescos[697]. The lower part of the body has been preserved at Balácapuszta; the posture of the forearm indicates that the depicted person's left is in front of the body, covering her loin. This single movement would suggest the definition of Adam and Eve was correct. Apart from this, the habit of this woman is completely different from the usual and iconographically suitable depiction of Eve. Her necklace or collar is similar to the necklace of the woman on fondo d'oro, ☞Dunaszekcső 11.II.a. If we assumed that the so-called Eve indeed was a maenad as we see on the wall painting of Balácapuszta, then we should refer to the casket-mount of Császár (6.I.a), on which the Bacchic motifs appear beside biblical scenes. We may speak about a "Pannonian practise", which does not preclude Bacchic motifs in Christian surroundings. Moreover, let me return to Zoltán Kádár's viewpoint that considers the ivy leave in the Jonah scene to be a Bacchic motif. Obviously, this assumption is a hypothesis. Since the other part of the scene is invisible, we cannot receive any help for our definition.

2. The Virgin Mary, fig 200a. The veiled woman was defined as Virgin Mary. There is a partly-preserved painted circle above her right shoulder; and this was explained as the remains of a head of a child. The presence of a child determined that the represented person was the Virgin Mary; however, the circle could be anything else. The old photographs taken when these frescos were better preserved, show that there was no head of a child present,[698] fig. 200b. There is a curved line visible over the chest which could indicate a wreath or a *patera*; both were used in a funerary celebration. The further arguments, which contradict the definition, refer the sadness and fear in woman's glance. These feelings are not suitable in the iconography of the Virgin Mary. Further, Klára Póczy pointed out that this so-called 'Virgin Mary' resembles the veiled female heads on the frescos in Aquincum and Balácapuszta; and they are not considered to be the Virgin Mary.[699] The head of the infant Jesus is not above Mary's shoulder on the catacomb paintings. For example, in the Priscilla catacomb, Virgin Mary embraces the infant Jesus who is so close to his mother that her veil covers the child.[700] In the Domitilla catacomb, Virgin Mary is presented with the worship of the Magi. The child is sitting on her left knee.[701] At last, let me mention the Callixtus catacomb or the Capella Graeca,[702] in the Priscilla catacomb,[703] or an epitaphium,[704] where Mary is depicted in a very different way. On the fresco in Pécs, some leaves are visible near the head. Hence, the milieu of the scene is presumably lifelike. The glance is excited as a glance of a woman who has just deceased and is being accompanied for reckoning to heaven. The wreath besides this lady means her victory. Such a scene is visible in the Domitillan catacomb where the dead is standing among some flowers and plants, fig. 208.[705] Let me suppose that the leading of the dead is depicted on this panel instead of Virgin Mary. This dead might be the person buried in the burial chamber.

3. The Three Magi or three young men in the furnace, fig. 201. The definition Three Magi came from the definition mentioned above, namely "Virgin Mary with the infant Jesus". Since the latter presumably cannot hold any more, consequently the other cannot either. Let us examine the other assumption, the "three young men in the furnace". The three young men in orante are standing in the furnace, in which flames can be seen or directly standing in the flames without a furnace. Here, however, the persons are seating as we can conclude from the position of the legs. Therefore, this definition is not precise, either. Some parts of the fresco are in better condition than others. The feet, the head and the neck of the figure sitting on the left are better preserved. The man faces left. The new catacomb serves here as an analogy;[706] namely, the students are sitting around the master. The students are holding their legs in a posture, which is discernible on the fresco in Pécs as well, fig 209. The other analogy is the depiction on a glass medallion;[707] here the apostles are sitting in the same pose. The scene is interpreted as Jesus healing the paralytic[708]. This interpretation may be acceptable in the case of the fresco in Pécs. What seems to be obvious is that the master is Jesus Christ and the seated disciples surround him.

4. Northern wall, Peter and Paul with the chi-rho monogram,[709] see fig. 195. Both apostles are pointing to the chi-rho monogram framed a circle. Among the branches of chi-rho, eight petal rosettes are depicted. Set of garlands follows the vaulting above the apostles. As regards the habit of the monogram, the best analogy is the bronze monogram of unknown provenance; ☞ 62.I.c. Ribbons are hanging from the garlands. These hanging ribbons are very similar to the ones depicted in the so-called mausoleum ☞ 32.I.e., according to Fülep. Let me mention that the chi-rho monogram of the 32.I.e. is not analoguous to the monogram of this burial chamber; the joining of the circle around the monogram and the garlands is not analogue,

[697] PÓCZY 1984: 54 Abb. 3/1 Balácapuszta: heads of maenads.
[698] BALOGH 1936: 249.
[699] PÓCZY 1984: 53. Abb. 2/1: Aquincum, Vihar Street, 2/3: Balácapuszta.
[700] TESTINI 1966 fig. 136.
[701] KAUFMANN 1922 Abb. 164.
[702] TESTINI 1966 fig 139.
[703] GASSIOT-TALABOT 1966 p. 74.
[704] TESTINI 1966 fig. 63.
[705] TESTINI 1966 fig. 111.
[706] TESTINI 1966 fig. 183: Ipogeo di via Compagni, la cosidetta "lezione di medicione". Cf. DU BOURGUET 1965: 111.
[707] CAMBI 1975: 54 fig. 4.
[708] In connection with the interpretations, see Pölöske, 35.II.a.
[709] FÜLEP 1984 CX.

either. The circle with the monogram seems to be a suspension of a lamp, and it is really hanging on the garlands in the burial chamber of Tomb I. On the other hand, the garlands are not exactly connected with the circle around the monogram in the so-called mausoleum ☞ 32.I.e.

5. The portraits in medallions and the ceiling, fig. 198. A tomb was uncovered in Or ha-Ner.[710] On the vault of the central hall are portraits painted in medallions, seven on each side, fig. 211. These portraits can rightly be used as analogies for our portraits. I cite the most important detail of the description; "The lower frieze a dado contains red-coloured imitations of marble panels. The middle frieze consists of a pattern of pointed leaves in red and green with groups of red dots between them. The upper frieze contains the major paintings, portraits of men and women, ... and in the northern corner appear two more candlesticks or torches. The spaces between the medallions are filled with flowers with eight petals painted alternately in red and green.... The medallions, .58-.59 meter high and .56-.57 meter wide, are almost circular. ... All the portraits are en face. The eyes are large and directed upward or to the side. The brows and lips are thick, giving a stern expression to the faces. ... In the manner of representation, the paintings display some similarity with Coptic art of the fifth-sixth centuries..." The excavator who is also the author, Yoram Tsafrir dates the building to the 4th century,[711] fig. 210a, b. Using this analogy, we can go on to analyse the fresco of Pécs. The flowers with eight petals fill the space among the branches of the chi-rho monogram. The medallions are painted on the both longer sides of the ceiling. Among the medallions are running branches with leaves, and pomegranates. According to the data of Szőnyi the diameter of these medallions are 43 cm;[712] and everybody has presupposed that they are circular. It would be useful to check their measurements if the condition of the frescos allow this at all. All the portraits - male and female - are en face here as well. The faces are similar to the eastern type in the manner of representation. The eyes are large, the brows and the lips are thick, fig. 198. It is impossible to recognize what was represented between the portraits. The first descriptions do not give any help because they are different.[713] Today one could hardly see anything. Cantharos with flowers for dividing the panel are painted on the ceiling of the "O" cubiculum in the new catacomb along the via Latina.[714] That is why Szőnyi's viewpoint that these representations are flowers, is more probable. The dado with marble imitation, and the scenes in framed panel generally characterized the frescos in Pécs.

Portraits circled with medallions occur also in a non-Christian catacomb in Palmyra.[715] This fact is important as regards the different influences to the frescos of Pécs. Some signs point to the oriental population that immigrated to Pécs.

Here, the early Christian tomb of Naissus should be mentioned. The apostles Peter and Paul with the chi-rho monogram in a wreath, and the Paradise are represented.[716] This tomb is considered to be an exact analogy to Tomb I of Pécs. The style of the pictures displays no similarity with each other, or the manner of the funerals. The *loculi* were simply cut into the painted walls in Naissus, but we have not experienced this method in Pécs. In addition, we have had no exact evidence of funerals due to plundering.

Let me deal with another analogue tomb, in Qusayr 'Amra,[717] which is decorated with a rich wavy line. The antechamber is barrel vaulted but the burial chamber cross-vaulted. On the intersection of the vaults, trees or bunches of flowers are painted in the same way as certain floral decoration on the vault of the burials chamber I in Pécs. Looking at either the painting framing the doorway or the decoration of the vault in Qusayr 'Amra, we assume that the culture, which has created and painted the tomb in Qusayr 'Amra, could give ideas for wall painting of Tomb I of Pécs. This culture is oriental.[718] With help of these oriental paintings, the secret of 'Eve' composition can be solved. The female figure[719] is wearing a necklace (similar to Eve in Pécs), her upper body is nude and lower body is covered with clothes. Its definition is *Venus Marina*.[720] The figure in Pécs belongs rather to the type *Venus Pudica*,[721] fig. 212. At last, it may be declared that the female figure is *Venus Pudica* instead of Eve.

Frescos in Pécs. Comparing all the frescos found in Pécs, we can discover similarities, which draw our attention to the fact that, on the one hand the paintings in different tombs cannot be discussed isolated, on the other hand that the artists, belonging very presumably to a group, worked together, and matched each other in character. That is why we can observe similarities. Simultaneously, we can observe also differences; because each craftsman might have his own feature. Their activity would begin after the rein of Constantine I. and would end at the late 4th century. Consequently, the frescos in Pécs were completed during these 40-50 years. Average character of these frescos remembers the viewpoint of Pelekanidis who considered these frescos analogue to the frescos created by a Hellenized culture. Hence, we suppose that population or some groups of the population Hellenized rather than Romanized were living in Pécs and around Pécs. They

[710] Encyclopedia IV 936-938. The quotation is from here.
[711] Encyclopedia IV, 936, two portraits.
[712] SZŐNYI O 1907: 71.
[713] KOLLER 1804 Tab. XI. Koller and Szőnyi first described these representations.
[714] DU BOURGUET 1965 Abb. 113.
[715] LECLERCQ 1907: 516-517 fig. 145.

[716] MIRKOVIĆ 1956: 85-100.
[717] BLAZQUEZ 1983.
[718] BLAVQUEZ 1983: 196: "Weinlaub, Trauben Flanzenornamenten dem künstlerischen Geschmack des Späten Kaiserreiches entsprachen. Als Vergleich dienen auch palästinische Mosaiken, die syrischen Künstlern zugeshcrieben werden. ... Alle Abbildungen zeigen deutlich den Einfluss der hellenistischen Kunst, obwohl der darstellte Frauentyp mehr dem arabischen Geschmack angepasst ist."
[719] BLAZQUEZ fig 12.
[720] BLAZQUEZ 181. The author refers to the mosaics of Djemila, where this Venus type is also found.
[721] Stephanie BUSCHER / Susanne TASSINARI avec la colloboration de P. -M. DUVAL / J. -P. BOUCHER: Bronzes antiques du Museé de la civilisation Gallo-Romaine a Lyon I Inscriptions, Statuaire, Vaissele. Lyon 1976 No. 65.

were who had brought some eastern culture mainly Syrian and Jewish culture. We may talk about Hellenized Jews or Syrians who preserved their cultures as Christians, as well.

II. Buildings: Tomb, the so-called chapel and its reconstruction.

The assumption that there was a chapel above the burial chamber led to the opinion that these tombs had definitely been superstructures where the agape was served and the eucharistia celebrated during the persecutions. This tomb I, however, can be dated back to the period in which there were no persecutions.[722] On the other hand visits to catacombs, or other tombs, to offer a sacrifice, or pray, would have occurred if there had been no fear of persecution; pagan funerary rites, or Christian ceremonies, attracted relatives or the community to the tomb in order to express homage to the deceased. It was an act of veneration to erect a building (*cella memoriae*) above a tomb. It both marked the tomb and provided a location for specific rites ("pietas"). The actual form of the structure was defined by two criteria: function[723], and the necessity for a handsome exterior, the latter being influenced by the landscape.[724] Here in Pécs, the tombs, including tomb I, are situated on the slopes of the hill. Gyula Gosztonyi treated the natural features of the site in 1942.[725] He demonstrated that the angle of the slope of the land explained the difference in height between the sub- and superstructure. Excavation records state that the foundations of the superstructure began below the barrel vault of the substructure. These descriptions, however, are doubtful and we are forced to ask the question whether a superstructure indeed existed. The theory of these superstructures is based on reports by Henszlmann who writes: "The long passage leading underground to the tomb was found after the discovery of the tomb. Originally, the steps, which led down, were in a confined area in front of the tomb. A small chapel seems to have existed above the tomb; the chapel was destroyed, as indicated by the condition of the apse."[726] The A-B section of the plan made by Henszlmann, (figs. 191a, b) shows a wall, higher than the vault, the function of which is unclear. The tomb, with its associated apse, might have been used for depositing relics, as suggested by Dyggve. It must be mentioned that there are no reports of the approach to the antechamber. Dyggve stated that the door of the antechamber was not contemporary and that entry was only possible via the upper storey by a trap door. Gosztonyi and Fülep denied the supposition, stating that the wall of the antechamber continued towards the south.[727] It is certain that there could have been a ramp in front of the antechamber. As the refill is partly from the Middle Ages and partly modern (cf. 32.I.d), the depth of the present footings do not provide many clues.[728] The pavement of the burial chamber is 480 cm deep and the barrel vault's height is between 204 and 220 cm;[729] therefore a wall of 260 cm would need to have been built below ground if the wall of the superstructure had only been rested on the wall of the tomb.[730] The extant wall section that rises above the vault might have static importance. The northern wall could have been supported by soil, while the southern side and major part of the vault could have been of earth, as a result of the hill's slope. It is almost certain that there was no superstructure above the tomb, since it was clearly visible, like a hypogeum. Gosztonyi also shows that the tomb could be approached from the ground.[731] I maintain that there was no chapel above tomb I, and will not comment, therefore, on the proposed reconstruction.

[722] It would be useful to mention that mass was not celebrated in the catacombs.
[723] This subject was treated above.
[724] The superstructures above the tombs are not of Christian tradition. The pagan world has structured several memorials like these. See: Gräberstrassen 1987: 43-60, 125-138, 239-250, 281-290.
[725] GOSZTONYI 1942: 200.
[726] HENSZLMANN 1876: 38.
[727] FÜLEP 1984: 40.

32.I.c. Painted tomb II.

Literature: GOSZTONYI 1940: 56; GOSZTONYI 1941: 42; GOSZTONYI 1943: 22-26; GERKE 1952: 115-137; GERKE 1954: 147; PÓCZY 1955: 56; FÜLEP 1959a: 399; KÁDÁR 1969: 183, 188; FÜLEP/FETTER 1969-1970: 91-103; MÓCSY 1974: 313; FÜLEP 1977b: 11; FÜLEP 1984: 42-46;FÜLEP/BACHMAN/PINTÉR 1988: 27-31; TÓTH E 1988c: 31-32; Itinerarium 96-103.

The tomb was first excavated by Gyula Gosztonyi in 1939 and completely uncovered in 1964. The southeastern corner of the burial chamber could not be approached because of the wall of a new cellar, fig 213a, b, c, d.

The burial chamber is N-S oriented and barrel vaulted; the entrance is in the south wall. Gosztonyi's measurements give H. (inside) 230 cm; W. 245 cm. The thickness of the walls alters between 55-60 cm. There is a niche with rounded top in the northern wall (W. 56.5 cm; H. 65 cm; Diam. 36 cm. The niche is not located along the central axis. There is also a niche in the western wall (W. 37 cm; H. 30 cm; Diam. 30 cm). The size of the entrance (with a rounded top) is: W. 72 cm; H. 102 cm. Gosztonyi continues: "A grave with a double bottom was found along the northern wall. .... The size of the grave is 85 x 218 cm. The spacing of the double bottom is 15 cm. The lower section is made of bricks (30 x 30 x 6.5 cm bricks), as is the upper one (60 x 60 x 6.5 cm), but the bricks are bored. At the western end of the grave there is a walled 'cushion', fig. 214). The grave was covered with carved stone slabs, broken into three. The grave has been completely plundered, and, except for a few scattered bone pieces, no finds were made. The space in front of the grave is 86 x 218 cm, and it was covered with bricks (30 x 30 x 6.5 cm)."[732] A funerary chapel was built above the tomb on N-S axis, however it had no architectural link with the tomb. The foundation of the substructure starts 85-90 cm above the burial chamber. Inside, the width is 303 cm and the length (incomplete) only 320 cm: the total length could be 445 cm. The thickness of the walls is 65 cm. Three pillars on both the eastern and western sides projected out 65-70

[728] This measurement is good only for comparing the footings.
[729] FÜLEP 1984: 36.
[730] This is another reason why a trap door is impossible.
[731] GOSZTONYI 1943: 37-40.
[732] GOSZTONYI 1942: 196.

cm from the wall. The walls of the chapel[733] remained up to 160 cm, of which a section of 35 cm was foundation. In his excavations of 1964, Ferenc Fülep revised the dimensions. His main findings are: "Eine außenordentlich wichtige Beobachtungen der Freilegungen von 1964 war, dass an der Schlussmauer der Kapelle eine Tür mit Gewolbe gefunden wurde. Die Tür liegt hier ebenfalls nicht in der Mittelachse: ihr Rand ist von der SO-Ecke 135 cm, von der SW-Ecke 105 cm entfernt. Ihre Schwelle liegt 500-520 cm tief unter dem heutigen Niveau. Die Breite der Eingangstür beträgt 62 cm, ihre rekonstruierte Höhe 90-92 cm. An ihren oberen Teil waren die schräg angebrachten. Anfangssteine des oben aus Ziegeln gefertigen Gewölbes an beiden Seiten vorgefunden worden. Das Gewölbe brach ein ung sein oberer Teil fehlt zur Zeit in 40 cm-Breite. Der Giebel dieser Türöffnung ist ein unter der in die Kapelle führenden Schwelle in 50-55cm Tiefe vorhandeln."[734] "Wie wir ... angedeutet haben, setzte Gosztonyi bei dieser Grabkammer einen kleinen Vorraum (Narthex und einen über Treppen verfügenden Abgang voraus. Zwischen der neu entdeckten Eingangstürm (Schwellentiefe 615 cm) besteht etwa ein Niveauunterschied von etwa 100 cm."[735]

The dimensions are given on the special drawings, figs. 215, 216, 217, 218, 219, 220.

Wall paintings. Figs. 221a, b, c, d. Some traces of the vault paintings have survived. They show geometrical ornamentation. The northern wall was divided horizontally into two zones with a wide red stripe, and green, yellow and black lines that are interrupted at the niche. A greenish, brownish jug and beaker[736] are represented in the niche on a blue background colour, fig. 222. The niche is framed with red stripes contoured with black lines. The upper zone is decorated with vine tendrils, green leaves, and black grapes. The lower zone is divided into two panels framed with stripes and decorated with marble imitation. The panel west of the niche is reddish, the other is bluish black. There is also a framed panel below the niche, but this is now damaged. In addition, the eastern wall is decorated with yellow and black rectangular panels, framed with double lines, between which is a set of concentric semicircles. The south wall by the entrance, and where the east and west walls join, are decorated with lattice pattern with flower imitations in the centre of the lattices.

Fülep says that the jug and beaker together refer to an Eucharistic motif and that the lattice pattern symbolises Paradise. Gerke deals exhaustively with the iconography of this pattern. He found analogies with Roman catacomb painting[737] and remarked on the blue base colour of the niche ("als überirdisch-himmlisches Symbol gekennzeichnet").[738] He also considered the jug and cup an expression of the Eucharistic offering. The geometrical patterns, and the architectonic motifs, and the schematic representation of gardens, featured with Biblical scenes on the analogies mentioned, referring to happiness after death. The architectonic style and the "park motifs" are characteristic features of Roman art; one of the most charming examples being the painting in Livia's House at Prima Porta. Constantine I went on to revive this style of wall painting. Gerke also expounded the symbolic meaning of the representations in tomb II, "Der Maler hatte den Auftrag, den kleinen tonnengewölbten Raum in zwei Hälften zu zerlegen: in einen durch Inkrustationsimitation angedeuteten Innenraum und eine paradiesische Gartenlandschaft. Indem der Maler hiermit zwei Prinzipien der Malerei des 4. Jahrhunderts zur Geltung bringt, eint er sie Zugleicht in einzigartiger Weise zu neuer Symbolkraft. ... die andere Hälfte des Raumes mit dem vordergründigen Gartenraum belegt wird, weitet er diesen Teil unwirklich aus und täuscht mit primitiven Mitteln die Unendlichkeit und den Türcharakter des Paradiese vor."[739]

Fülep writes that the funerary chapel was used after the Roman period. The fireplace found in the south-western corner relates to this. Fülep's viewpoint is that the upper section served worshippers as they gathered for agape, prayer, and Eucharistic offerings. He dates it to the end of the 4th century and the beginning of the 5th century.[740]

Commentary:

- *Burials.* Fülep disagrees with Gosztonyi concerning the number of burials. Fülep assumes that two deceased were buried here because of the space available. The first grave is dated to the end of the 4th century[741] and was discovered completely plundered.
- It is true that tombs may contain several graves, but it was not obligatory. This was especially so where local inhabitants greatly revered their dead. Hence, there is a local unwritten, tradition that this tomb contained the remains of St Bartholomeus. This is an oral tradition without proof, but perhaps future research may uncover decisive documents. If the tradition were true, then the painted tomb would contain only one grave. Dyggve has already assumed that a Saint of a Christian community was buried in tomb I.[742] This assumption might be truer for the painted tomb II, because the door leading to the burial chamber is smaller than that of any other tomb. This fact may mean that the community did not intend to bury anyone else inside the painted tomb II.
- *Buildings.* There is no structural connection between the tomb and the superstructure. Both Gosztonyi and Fülep state this. The vault of the burial chamber had already collapsed when it was discovered. Among the debris there were fragments of frescos and bricks. In

[733] I use the term "chapel", and not "chamber", for the superstructure to avoid the confusion.
[734] FÜLEP/FETTER 1971: 98; FÜLEP/BACHMAN/PINTÉR 1988: 30.
[735] FÜLEP/FETTER 1971: 99.
[736] FÜLEP 1984, photograph pl. XX, interpretation 45. ☞Pilismarót, jug and cup have been found in the cemetery.
[737] Okeanos-Crypt in Callixtus catacomb, lattice pattern and inlay decoration in the Petrus and Marcellinus catacomb, the main corridor of the catacomb I. of Naples, the catacomb of the family Jordanus.
[738] GERKE 1954: 122.
[739] GERKE 1954: 132.
[740] This tomb is reconstructed and now exhibited.
[741] FÜLEP/ BACHMAN/PINTÉR 1988: 31.
[742] DYGGVE 1935: 75.

addition, Gosztonyi pointed out that no pieces of the superstructure's paving were found in the debris. This statement is important. If the superstructure was indeed immediately above the painted tomb,[743] and if it had already existed before the collapse, pieces of paving should have been found amongst the debris. Therefore, it is more likely that the vault collapsed as the result of something having been built above it. The questions arise, who built it? When and why? If we take into consideration the other tombs, especially 32.I.e. and 32.I.g,[744] then extremely careful consideration can be applied. The plan of the superstructure does not have any similarities with any other chapel types. It might be a *cella memoriae* marking the tomb, as with many other constructions of pagan Rome.[745] The walls are 60-65 cm thick, the pillars 65-70 cm, and the extent is approximately12 m². The door of the funerary chapel is smaller than the entrance of the burial chamber. According to Fülep the door of the funerary chapel is 90 cm high and 62 cm wide. It is rather difficult to imagine how believers would have entered the chapel. Measuring from the surface, the paving of the burial chamber is 665 cm deep and its threshold was uncovered at a depth of 615 cm. Consequently, the footings are deeper than the doorstep, but not by 50 cm, fig. 223. The paving of the superstructure was discovered 426 cm deep, and its threshold 520 cm deep. Based on this data, the paving and thresholds present a problem.[746] The height of the burial chamber is approximately 290 cm. Consequently, if the paving of the burial chamber were 615 cm deep then the top of the barrel vault would have to close at 325 cm. However, Goszonyi states that the foundation of the superstructure started 85-90 cm above the tomb.[747] Therefore, Ferenc Fülep could not have found the floor of the superstructure 426 cm deep and its threshold 520 cm deep. Presumably, they are different constructions one being from the superstructure of painted tomb II, and the other from a building, probably a tomb, located nearby.

- *The function of the superstructure.* The superstructure above painted tomb II - as far as the dimensions of the entrance are accurate - cannot be considered a *cella memoriae*, mausoleum, or *oratorium* at all. We assume it was probably also used for funerary purposes, but in a later period than the painted tomb below it. If the supposition is true, then the grave of this superstructure has also been plundered and destroyed. A section of paving (130 x 90 cm) from the "superstructure" was found in 1964, and hence ash stratum covered the pavement. Fülep thought that this building was inhabited during the Migration Period. I date the superstructure to the end of the 5th century, based on the analogies of Pécs. It is not relevant to comment on the reconstruction[748] as it shows a chapel.
- *Frescos.* A Biblical scene is not delineated in the fresco, only motifs that may be interpreted as Christian motifs, for example the lattice pattern. The jug and beaker, as symbols of the funerary cult, may be found on tombstones in even completely pagan surroundings.[749] They do not refer to the Eucharist. Looking at the shape of the jug, it represents a real type that dates back to the end of the 2nd century and the 3rd century.[750] This date is worth considering because the frescos could well have been painted at the end of the 3rd century. The scenes represented in the paintings cannot be elsewhere than Paradise. According to early Christian thinking, the martyr appears immediately before God.[751] Honouring the martyr would depict Paradise and the newly created saint would enter the lost Paradise. Jesus' words to the thief on his right side - "Hodie mecum eris in Paradiso"[752] - provided grounds enough for depicting Paradise. Continuing, let me assume that saints would never have been buried in chambers where biblical scenes decorated the walls, but only where glimpses of Paradise were frescod. Biblical scenes were painted for the simple believers who lived and died in the hope of gaining Paradise. Let us not forget that the person buried here was honoured as a saint, according to unwritten tradition.

Sedila run around the walls. These could have been made at the funeral itself or after the painting.

Summary. We may probably state that the so-called superstructure was not related to the tomb at all. It was more likely to have been a separate building of some kind (the size of the entrance precluding its usage as a chapel). We have no evidence of its function. Concerning the frescos, the jug in painted tomb II indicates the 3rd century. The jug and cup do not refer to the Eucharist but are connected rather with the funerary cult, better to say the *refrigerium,* obtained by the local saint, Bartholomeus, who may have been buried in the chamber.

32.I.d. Cella septichora, fig. 224. Photographs of the excavation: figs. 225, 226a, b, c.

Literature: GOSZTONYI 1940a; GOSZTONYI 1940b; GOSZTONYI 1943: 11-12, fig. 1 drawing 1. KÁDÁR 1969: 183; FÜLEP 1969a: 156-157; MÓCSY 1974: 335; FÜLEP 1977b; TESTINI ²1980: 655 fig. 328; FÜLEP 1984: 57-59; TÓTH E 1988b: 36; Itinerarium 96-103.

"The *cella septichora* building in the cemetery, in the close vicinity of the Cathedral of Pécs, was discovered in 1938; since then, it has been continually excavated from the Cathedral to the southeast, partially under the parsonage

743 Gosztonyi reconstructed the collapsed vault. According to this reconstruction, the vault reached the pavement of the superstructure.
744 For more detailed explanation cf. mausoleum.
745 HESBERG 1987 Abb. 3-4.
746 Figs. 215-220 help to understand these measurements.
747 These are the real data.

748 FÜLEP/BACHMAN/PINTÉR 1988 XXIV.
749 For example, V. RUSU-BOLINDET: Representations de vaisseaux en céramique sur les monuments sculpturaux de la Dacie romaine. Ephemeris Napocensis IV 1994: 113-148.
750 K SZABÓ: Pot à anse, en bronze, ornée d'un pied humain, provenant de Pannonie. Antiquités Nationales 14-15 1982-83: 86-96.
751 DEICHMANN 1983: 54-67: the summary of the archaeological evidence proving the martyr cult.
752 Lc. 23, 43. (Novum testamentum Graece et Latine. Ed H. Jos. VOGELS, Düsseldorf 1922, pars Prima p. 235)

and partially to the east under the Erzsébet tér (square) and the road to a depth of 6.00 m. The construction consists of seven apses located on the E-W axis; the apses form an extended octagonal space."[753] Dimensions: full exterior L: 22,80 m, W: 17.35 m, interior L: 20.30m, W: 15,15 m. Diam. of the east apse 490 cm, depth 360 cm. Diam. of the other apses: 390 cm, depths 285 cm. The thickness of the walls vary between 110-125 cm.[754] The pavement in the south-eastern apse was found to be a "hard concrete level".[755] The walls were built alternately of stone and brick. The size of the bricks is 44 x 30 x 5 cm.[756] There are no traces of plaster or paintings. Sherds of imbrex and clay are also discernible in the stone walls. The western wall is straight, and a doorway has been observed. Gosztonyi considers that the walls discovered belong to the underground part of a funerary building.[757]According to Gosztonyi, the remaining walls could have been approximately 300 cm high, and that there were two rows of bricks alternating systematically between two rows of stone.[758] The building is under the Middle Age moat, hence it is assumed that the *cella septichora* was destroyed on this occasion. Gosztonyi supposed that the apses were vaulted but could not determine the function of the building. His team reconstructed a Roman level of footings higher than the pavement of this building.[759] They supposed that it might have served some funerary purposes: sarcophagi stood in the apses.[760] Supporting this idea, Ferenc Fülep analyzes the depth of the building and makes comparisons at certain points with the tombs, he finds them similar.[761] Fülep also discussed the function, considering the possibility of a *basilica discoperta*. He compared the St Gereon basilica in Köln with *septichora*.[762] Gosztonyi thought there were analogies to be made after observing the apsidal buildings in Rome, both "pagan" and Christian, as well as some in Armenia and along Adria.[763] Endre Tóth tells us that the glazed roof tiles from the *septichora* are now located in the museum of Pécs.[764] Mirja Jarak assumes that the *septichora,* considered to be from the 6th century, could be one of those sites that Iustinianus gave to the Benedictines.[765]

Commentary:

- If we compare the size of the *septichora* to the tombs in its vicinity, the difference is striking. The *septichora* is even lager than the so-called mausoleum. Consequently, the tombs cannot help in determining the function of the structure and it would be better to concentrate on its size.
- The plan of the *septichora* differs significantly from the plan of rectangular tombs.
- Its E-W orientation also differs from that of the tombs oriented to the north. They are situated rather like a street, in rows and facing the same direction. The *septichora*, however, is oriented to the east and does not relate to the street of tombs. Orientation would be empirical evidence of such a building and would infer Christian usage. Orientation to the east and west occurred equally in the 4th century but in different regions. From the 5th century on, buildings oriented to the east became more frequent.[766]
- Graves or other funerary traces did not come to light at all. This also proves that the *cella septichora* was not used as a tomb.
- The walls were built with the technique known as *opus mixtum*, fig. 225. The rows of bricks alternately follow the rows of stones.[767] A later version of this technique was modified by Byzantine influence and occurred later than the 4th century. Cf. 32.I.e, 32.I.g.
- A few finds were found. There are no features or objects to indicate any Christian characters. There are no objects at all to indicate the function.

Summary. The *cella septichora* is larger than any other building in the neighbourhood, and its orientation differs from that of the tombs. None of the graves cut the walls of the *septichora* and the walls of *septichora* do not cut into any of the graves. No evidence of funerals was observed inside the building. The foundations were carefully built. There were no small finds discovered here. We have no evidence to indicate how long it was in use. Was it destroyed at the same time as the tombs or did it stand until the Middle Ages as Gosztonyi supposed? Moreover, we also have to ask a further question; what was its function? We have no clues other than its ground plan. There are no present analogies for its seven apses and outline.[768] Let me assume that its plan matches an enlarged sigma table (which reveals the function), namely *locus orationis*, *oratorium*, where the believers commemorated the *fractio panis*. The *cella septichora* is not a church. Hence in the 5th century, here in Pannonia, the Christian community was convinced that the believers themselves are the church of God, because any house built by human beings is not a

[753] GOSZTONYI 1940a: 56, fig.2.
[754] GOSZTONYI 1940a: 56.
[755] GOSZTONYI 1940a: 60.
[756] This brick is bigger than the bricks in the 4th century: 38 x 25 x 3.3 cm, at the buildings of the Caesar Galerius. KLEINBAUER paid attention to the bigger bricks (1984 II 244) ☞ 32.I.e. and ☞ 32.I.g.
[757] GOSZTONYI 1940a: 57.
[758] GOSZTONYI 1940b: 210. Fig. 1: drawing of the walls.
[759] GOSZTONYI 1940b: 210.
[760] GOSZTONYI 1939: 100; FÜLEP 1984: 58.
[761] FÜLEP 1984: 56.
[762] The St Gereon consists of three parts; one of them is the apsidal structure, which has nine apses. RISTOW G: Christentum, die neue Kraft. Römer-Illustrierte 2 1975: 196-197.
[763] GOSZTONYI 1940b.
[764] TÓTH E 1988b: 36. Such tiles were also unearthed in Alsóhetény.
[765] JARAK 1998: 785.
[766] GAMBER 1976: 9-14.
[767] Similar walls may also be seen in Pompeji, in the cemetery of Nuceria gate; a similar kind is the *opus mixtum*, in which not only the bricks, but also the stones were laid in order systematically. On the other hand, in Pécs, both excavation documents and the sizes of the bricks suggest a later time; we may assume that the stones laid in order served the foundations, and there is no significant importance concerning the dating of the building. Cf. 32.I.e, 32.I.e.
[768] BRATOŠ 1994: 41. He summarizes the architectural specialities of our region, which is called "illyrisch alpine Architektur". Its characteristics are, "---ungewöhnlich oder sogar Ausnahmenfälle sind,..." His statement is suitable to septichora as well. Pannonia also belongs to the region known as 'illyrisch-alpine'. Cf. former notes.

church.[769] The usage of the sigma table, and consequently its effect, is much greater than we might think. Gamber detailed this influence.[770] The refectories in the later period were equipped with tables that imitated the sigma tables, either circular or horseshoe shaped (cf. fig. 68). These analogies enable us to put forward the above hypothesis. To repeat it, I suggest that the community celebrated the Eucharist in the *cella septichora* and hence I assume its Christian character.

32.I.e. The so-called mausoleum, figs. 227 a, b[771]

Literature: FÜLEP 1977a; Ausstellung 1986; BACHMANN Z: A pécsi mauzóleum helyreállítása, védelme - The reconstruction and protection of the mausoleum of Pécs. Magyar Építőmûvészet LXXVIII 1987 12-13; FÜLEP 1987: 33-44; FÜLEP/BACHMAN/PINTÉR 1988: 15-22, 51-55; TÓTH E 1988c: 33; TÓTH E 1990a Abb. 6; TÓTH E 1991a: 123; TÓTH E 1994b Abb. 12; Itinerarium 96-103.

The excavation was carried out in 1975-1976. The site is Szent István Square, a park south of the Cathedral. Figs. 228a, b are photographs of the site by Ferenc Fülep and the excavation leader has provided the following information:[772]

A two-storey building was found whose upper floor is a single-nave chapel in a west-east direction. The outer length of the building is 18.1 m. and its inner length is 13.8 m; its outer width is 9.5 m, and inner width 4.8 m. The walls are 100-120 cm thick. In the apses, however, there is a recess where the thickness of the wall is only 50 cm. The building ends in an apse to the east. Both in the apse and on three further sides of the construction, three pillars stand out from the wall. Inside there are pillars in the north and south, and there are also two pillars at the beginning of the apse. The excavators found the original footings in the northwest corner. The rest was reconstructed accordingly. The entrance to the superstructure was put to the southern, longer side. Although it was Ferenc Fülep himself who stated that there was no correlation between the burial chamber and the building found over it, he still treats them as a unit. His opinion that the burial chamber is a crypt and that the upper building is a chapel, matches this idea.

Burial chamber. According to the ground plan presented by Fülep and his colleagues,[773] the burials took place in rooms *A* and *B*; and *C* - in Fülep's view – is the antechamber of the burial chamber up from where a staircase used to lead. The combined length of parts *A* and *B* is 6.8 m, with a width of 3.3 - 3.4 m. They were once vaulted. The length of room *A* is 4.25 - 4.3 m. This room was painted. It is separated from room *B* by pillars (*X* and *Y*). These pillars are ornamented with imitation marble frescos.

The description of the frescos in part A *by Ferenc Fülep*: The scenes had rectangular frames. Starting from the *X*-pillar and moving to the east, the first scene is 'The Fall' (fig. 229) followed by a geometric imitation marble decoration. Then follow 'Daniel among the lions' and a floral ornament, fig. 230. There is a little niche on the eastern wall (marked No. 3.), and above it a chi-rho monogram with surrounding garlands, fig. 231. The niche was decorated with wheatears. On the left side of the niche a figure dressed in white is sitting on a throne. The southern wall had no painting apart from geometric imitation marble decoration near the *X*-pillar, fig. 232. The head of a grazing donkey can be seen above the pillar of the southern wall.

Finds from the burial chamber according to Ferenc Fülep's description: in the burial chamber the lower parts of three sarcophagi (numbered 1-2-3 in part *A*) were found, fig. 233. The first sarcophagus was made of marble, whose analogue, as Ferenc Fülep puts it, is the sarcophagus of Iunius Bassus. The sarcophagi were smashed and only their lower sections remained. The marble fragments were left in the burial chamber and restorers tried to reassemble the sarcophagus. By doing so they realized that all the parts of the sarcophagus were not originally carved. Its ornamentation did not seem to refer to early Christianity at all; putti represented the seasons. The second sarcophagus was made of grey limestone and only its base remained. The third sarcophagus was the best preserved, but it belongs rather to part *B*. No human remains were found in the sarcophagi. All the skeletons and finds had been scattered and the remnants of fourteen skeletons were collected. Seventeen coins from the years between 350 and 375 were also found. A few other finds were unearthed, of which only a candlestick is mentioned by Ferenc Fülep.

As for the building, he writes the following. There are no connections between the lower and upper structure. This is an important difference in relation to the analogues as well. First the marble sarcophagus was lowered and burial chamber *A* built above it, with its doorway originally near the pillars. The doorway disappeared when burial chamber *B* was built. Burial chamber *B* was built when they installed the third sarcophagus. A slight break in the surface of the wall verifies this addition. This was also when part *C* was built, serving as a passage leading down (no traces of stairs were found). The upper building must have been constructed when the burial chambers were already there. Burials took place continuously up until the beginning of the 5th century.

Analogues of the wall paintings may be found at Pécs, although their origins stretch back to the catacomb paintings of Rome - as Ferenc Fülep asserts. They are based on pattern books and were carried out by wandering Italian painters.

Fülep dates the destruction of the sarcophagi to the 5th and 6th centuries, the action of some Hunnish, Germanic and Avar raiders to make the burial chamber habitable for themselves. In his opinion, the argument that they settled down here is proved by the fact that pile-dwellings, with a kiln, were attached to the building from outside; 9th-10th-century pottery fragments have been

---

769 Cf. Szombathely, locus orationis and Dunaújváros historical summary.

770 GAMBER 1968: 78-86: about the survival of the sigma-table, 86-93: furniture of the house-church.

771 Based on FÜLEP/BACHMAN/PINTÉR 1988.

772 FÜLEP 1977a; FÜLEP 1987: 33-41.

773 FÜLEP/BACHMAN/PINTÉR 1988: 17, fig. 5.

found. This concludes F. Fülep's account.

Commentary:

1. F. Fülep links the construction of the burial chamber with the installation of the marble sarcophagus. This opinion raises a certain difficulty. There were three sarcophagi in the burial chamber and fourteen burials can be verified. It would be more logical to suppose that an entrance existed through which the sarcophagi could have been brought in and further burials take place. The conclusion may be drawn that the excavations are not yet complete.
2. The painting of the early Christian frescos and the sarcophagi cannot be interconnected in terms of cause and effect in any case. The excavation observations allow only the fact that the burials using these sarcophagi are earlier than the early Christian frescos of the burial chamber. The analogue of the marble sarcophagus itself can be found in Pannonia, which can be put to the 3rd century at the latest; it might also be earlier than.[774] Consequently, we cannot date the burial in the sarcophagus to the second half of the 4th century. In addition we cannot ignore the fact, as he also writes, that there was no symbol or illustration on the sarcophagus that could refer to Christianity. Thus, we cannot relate it to the sarcophagus of Senator Iunius Bassus.[775]
3. If we attribute the systematic destruction of the sarcophagi to the peoples of the Migration Period, then we should expect the 9th and 10th century finds not only outside and upstairs, but also in the crypt. The fact that only huts were found attached to the outside makes it probable that these people did not intend to turn the burial chamber into a dwelling. If it were not so, then we would have to explain the 300-400 year gap between the destruction date of (5th or 6th century) and the dating of the finds (9th or 10th century). If the Ostrogoths could despoil Keszthely-Fenékpuszta in the 5th century, after Attila's death, then we may suppose that the Pécs relics could also have been destroyed around the same time. I would not accuse Attila and his followers of such a barbarous deed. The pile-dwellings were joined to the superstructure, and not to the burial chamber.
4. The previous remarks raise questions about the origins of the burial chamber, both in terms of the constructional work and its burials. It is presumable that the chamber was continuously used, and, if so, we should be able to trace the beginning and the end. The sarcophagus could mark an early period in the 3rd century, signifying both the building of the chamber and the first burial. The vault had collapsed, but Ferenc Fülep found no remains of it, or did not report his findings. We may conclude that as the chamber was used, it was also tidied up. What caused the collapse and why was not it restored? We can only guess. There is, however, a problematic phenomenon. Although the excavation report fails to mention that the excavation was left incomplete because of certain problems, in point of fact the complete opening of the chamber never took place. This can be shown from the traces on the southern side where the custodians' room is located. Using Zoltán Bachman's reconstruction plan, I have made the following observations and measurements (fig. 234).[776] Along the *d* section of wall, which is the extension of the southern wall of room *C*,[777] we can notice the beginning of a barrel vault to the south, i.e. in the direction of the present custodians' room (to the south of room *C*). On my map, the custodians' room is marked *D*. This means that we may suspect a further burial chamber. The present restoration does not allow us to observe the possible continuation of this barrel vault along parts *A* and *B* of the burial chamber, although this would be the only means of detecting the history of its building and the system of these chambers. Therefore I hypothetically state that there used to be a further burial chamber south of the present one, and its building occasioned the damage to part *A*.
5. Part *C*, presently called the narthex, could also have been a burial chamber. It can be seen on the wall-section between rooms *B* and *C*, i.e. on wall-section *a-b*. Here we can observe a brick arch, an arching in its size equal to the barrel vaulting of parts *A* and *B*. It may be the outer wall of the burial chamber whose entrance could have been where we can today approach the corridor of the so-called narthex (the wall of part *D* on my map). The thickness of the wall dividing parts *B* and *C*, (i.e. wall-section *a-b*) is 57.5 cm until the beginning of the barrel vault. From here on it is only 27 cm thick. It makes us conclude that on the other part, on the remaining 30.5 cm (where there is no wall today and only its position can be seen), a barrel vault could have begun. Its beginning is 112 cm from the former floor; from there the height of the wall is approximately a further 59 cm, so we can imagine a wall of a total height of 171 cm. The level of the former floor is known, as a tiny part of it remains in front of wall-section *a-b* in room *C*; it is covered with terrazzo.[778] The upper part of wall *c* is a modern complement. The question is whether or not it was a misconception to build this wall as the outer wall of room *C*.
6. The tiered ground between parts *B* and *C*, interpreted as stairs, is presumably the remnants of tombs. We observe in every burial chamber at Pécs that the tombs

[774] BURGER 1991 Kat.-Nr. 97, 98, 99, 101. This sarcophagus also belongs to the group in connection with which there is good advice: 'dort wäre wohl auch der Herstellungsort zu suchen.' This place refers to Pannonia, more closely the environs of Sopron. - KOCH/SICHTERMANN 1982: 328.

[775] The dating makes it impossible, too. We know that Iunius Bassus died on 25 August, 359, at the age of 42. His inscription says that he was a Christian - 'neofitus iit ad Deum'. At Pécs, no inscription has remained, and we know nothing about the deceased. On the other hand, the sarcophagus appears rather from the 3rd century; its carving cannot be related to that of Iunius Bassus' sarcophagus, e.g. Friedrich GERKE: Der Sarkophag des Iunius Bassus. Berlin 1936.

[776] Freehand drawing based on FÜLEP/BACHMAN/PINTÉR 1988 Abb. 20.

[777] This is a modern wall at Bachmann.

[778] The hatched part on my sketch-plan.

were built on an elevated platform, either they were made of brick or the burial took place in a sarcophagus. These parts may have been altered when the present painted burial chamber was created. As a consequence, originally (i.e. in 'pagan' antiquity) four interdependent burial chambers or crypts were built, approachable one from the other, which were used for several burials. In the Early Christian Age - as we know today - only one was provably used, tomb *A-B* at Fülep, or actually two chambers if we divide parts *A* and *B*.

7. The archaeologist suggests that the southern side of part *A* was not painted. As the mortar remains in patches we may at least suppose that it was mortared.
8. As for the frescos, I find the interpretation of the one on the eastern wall uncertain - it is the white figure sitting on a throne. As the fresco is rather worn, it is very indistinct and thus its interpretation will always be uncertain. However I have doubts about the figure on the throne, specifically that the clothes of a sitting figure would not settle like that, and that the clothes of a white-clothed figure remain white even when covering the legs.[779] The remaining fragment in the Pécs fresco resembles an object, rather, surrounded by some vegetal environment, like a well. We are reminded, perhaps, of the scene 'Jesus meeting the woman of Samaria'. We can undoubtedly put the frescos illustrating biblical scenes to the 4th century, and more specifically to its second half.

The superstructure. When was it built, by whom, and why? We should be able to find answers to these questions. It is already clear that the term *mausoleum* is a false one. In this building there has never been a burial and it has no connection with the crypt either. The director of the excavation also notes this. Even the possibility that the vault fell down when the superstructure was being built can be ruled out. At this point of the excavation, it seems that it was built over the tomb, but in a later period. We can also note that the foundation of this building cannot be seen at the wall of the crypt (see fig. 227), which means that the builders must have known about the tomb below. Although the bricks of the superstructure are Roman, the brick courses alternate with rows of stones in mortar (*opus mixtum*).[780] On the southern wall - where we can see the doorway on the reconstruction - there is no Roman brick wall for a long way beginning from the first pillar at the apse. The stone wall stops between pillars Ny2 and Ny3 on the western side. This missing stone wall was completed at the reconstruction. This observation is echoed by Fülep, at the excavation of the so-called tomb XIII (see 32.I.g), that under the stonewall there is a foundation of a brick course. Thus, if the then community used the *opus mixtum* as a walling technique (including the brickwork with *opus latericium* walling) we may suppose that brickwork and stonework was used together. It is difficult to decide whether the brick wall is merely the foundation, or whether simply very little has remained from the upper part of the wall, explaining why we cannot observe the alternation of brickwork and stonework. The latter version has an analogue, the wall of the Acheiropoietos Basilica in Thessaloniki.[781] Here we can see the alternation of the *opus latericium* and stonework. The bricks are Roman, their size is 45 x 30 x 5 cm.[782] The walling in the case of the Acheiropoietos was dated to the middle of the 5th century by Kleinbauer, which is especially significant in the case of the so-called mausoleum. The above observations suggest that the builders knew about the tomb but that they did not use it as it did not belong to them. Respect for the Christian corpses made them erect a superstructure over the crypt, which should be called an oratory. I date its construction to the end of the 5th century, after the time of the destruction.[783] I also have to note that this structure and the *septichora* are the only buildings at Pécs whose apses are orientated; the walls being parallel to the walls of the medieval Cathedral. The other superstructures at Pécs face north.[784]

Summary. The excavators also agree that there is no relationship between the crypt and the superstructure. Consequently the term *mausoleum* is wrong.

As for the burial chambers, we may state that we have not one but rather a system of them that has been altered. Its dating is uncertain at the present stage of excavation; we may only make presumptions. With a good deal of uncertainty, we can perhaps say that the pagan chamber was altered by Christians, but not to a larger extent than was necessary. We may suppose an extension. They did not intend to hide the marks of the alterations, but to create a worthy resting place for the respected Christian dead.

We have to take notice of several periods. The excavation reports indicate that the broken sarcophagi are not newer than the 3rd century, or at least the marble one is not. The other sarcophagi, although they are not necessarily the same age as the marble one, do not contradict the presumption that they were made in the 3rd century. Therefore, I would not put the first period later than the first half of the 3rd century. For the sarcophagi, a burial chamber was built that would later be altered by the Christians. This alteration - according to the style of the frescos - is not older than the middle of the 4th century. Furthermore, we have not clarified the time of the destruction (mentioned by the excavators) and when the bones and relics were scattered around the floor. We can by no means put this destruction down to those Christians who had their burials here in the earlier burial place. I put

779 For instance the seated Abraham in the Priscilla catacomb. - DU BOURGUET 1965: 72.

780 This building method was noticed and written about by Gosztonyi - GOSZTONYI 1943: 48-52.

781 The dating argument of the article written on the chronological debate about the construction and the alteration, refers exactly to the walling technique. - KLEINBAUER 1984 II 241-234, fig. 3. The Basilica of Pula, may also be cited as a parallel, see MARUŠIĆ 1978.

782 KLEINBAUER 1984 II 244. The author notes that these bricks are bigger than those in the buildings erected during Galerius. Their size is 38 x 25 x 3.3 cm.

783 Cf. MIGOTTI 1992.

784 It is peculiar because the Christian buildings in the 4th and 5th centuries face either west or east, depending on their geographical location. GAMBER 1972; GAMBER 1976: 7-27.

the destruction to a later point in time. I propose that it occurred in the 5th century, after Attila's death, when the Ostrogoths also ransacked Keszthely-Fenékpuszta. At this time every relic at Pécs was ravaged. The Christians supposedly built tombs for themselves, suggested by the fact there were destroyed graves in tombs I and II. We must emphasise that no undisturbed tombs were found by the excavators in any of the burial chambers.

The superstructure was built later. It can be dated to the end of the 5th century, based on the walling technique and other circumstances.[785] The descriptions so far have used the terms 'chapel' or 'church' for the building although they do not supply any explicit evidence. Both the reconstruction and the definition were based on the preconception that the object could only be an early Christian mausoleum, chapel or church. Consequently, certain observations required for revealing the truth were superficial. However, based on analogues and considering the rest of the objects at Pécs generally, we may presume that it was simply a *locus orationis*, built, not at the same time as the burials, by a Hellenised population whose presence can be traced at Pécs. They could as well be descendants of an oriental pagan and an oriental Christian population that lived here earlier and had known about the burial chambers.[786] We also have to assume that life was consolidated during the time leading from antiquity to the Migration Period, and that this oriental population prepared for a longer stay. I do not accept Fülep's idea that the destruction was carried out by the population of the 8th - 9th century. Those who seek a home will not willingly destroy it. A certain damage may occur over the years, but settlers would certainly not deliberately destroy their environment. Their relics (pottery, fireplaces and other smaller finds), found on almost every site, suggest a longer settlement (cf. 32.II.1).

Finally, I have to note that I think new excavations are needed to find evidence for the great number of presumptions.

32.I.f. G/4 tomb, fig. 235.

The object defined as a tomb or burial chamber differs in numerous aspects from other burial chambers. These phenomena warn us that we are dealing with a special relic; therefore, all the observations are highly significant.

First, I will disclose Ferenc Fülep's opinion, whose first summarising statement is that it is likely to have preserved a Jewish custom.[787] The opened tomb is quadrilateral. Its sizes: 'Interior l: 804 cm, w: 400 cm at the northern end, 410 cm at the southern. ... The present H of the walls, northern 219 cm, western 183 cm. ... There were two round holes for ventilation in the northern wall of the burial-chamber, at a height of 118 and 110 cm, measured from the bottom ... The entrance ... on the southern wall, ... the width is 76 cm. ... The height of the threshold measured from the bottom of the burial-chamber was 120 cm. ... the tombs were not found inside the burial-chamber.'[788] The floor has deteriorated, but we can suppose its traces at the northern wall. There are holes on its longer sides, interpreted as vent-holes by Ferenc Fülep. The wall used to be mortared, there are white and pink marks on it, perhaps the marks of painting. Only the beginning of a vault has remained on the western side, where two bricks were found *in situ*. On the part closer to the northern wall, both on the western and the eastern walls, the plastered and the unplastered surfaces regularly alternate. The plastered surfaces are wider. Ferenc Fülep interpreted these parts, stating that the sarcophagi, or walled graves, used to stand here. He prepared the reconstruction of the tomb accordingly, figs 236a, b.

Commentary and my description. I prepared a sketch on the spot (fig. 237). It is not accurate in its proportions, but it records the peculiarities found in this tomb. The entrance (*A*) is on the southern wall, very high compared to the floor of the tomb; therefore stairs would have had to be used. (Ferenc Fülep supposes three or four steps.) The site ascends northwards, rather significantly, but it cannot be perceived in the chamber. However, if we regard the vent-holes mentioned by Ferenc Fülep as real vent-holes, then, contrary to all appearances, the inner room is also ascending, since the height of the vent-holes, according to Ferenc Fülep, is 118 cm. They are 110 cm above the inner floor while the door-stone of the entrance on the southern wall is 120 cm above it. This fact also means that the destroyed barrel vault of the chamber was above the ground in Roman times. Only two bricks remain from the barrel vault on the western wall of the northern side. The drawing and the reconstruction (see fig. 236a, b) made by Ferenc Fülep show that the thickness of the supporting wall is approximately twice as large as that of the barrel vault. Therefore, we have to suppose that the other barrel vault began towards the opposite direction, otherwise, it would have been senseless not to benefit from the thickness of the wall. Furthermore, I would like to remind the reader that we could observe the same phenomenon in the case of the so-called mausoleum.

In the area excavated by Ferenc Fülep, the plastered and unplastered parts show regularity. In my own sketch (fig. 237), the odd numbers indicate the locations of the unplastered stripes, while the even numbers indicate the plastered ones. Unplastered parts cannot be observed up to the highest level, only up to 60 cm. This regularity prompted Ferenc Fülep to create his reconstruction. The reason I disagree with this reconstruction is that the archaeologist himself wrote that he had found no traces referring to burials in the burial chamber. Only three finds

[785] Although the walling technique itself is typical of the middle of the 5th century, considering Pannonia's difficult periods and the following consolidation, we can work with the beginning of the 6th century.

[786] It may be bold to assume their survival as all the sources speak about the escape of the Pannonian population. There is an opinion which exaggerates this migration to the picture of a deserted Pannonia. I maintain my above opinion. Modern history has produced series of migrations, but there were always people staying in their places, continuing the lives they knew. We must not suppose a total migration in antiquity. Cf. MÓCSY 1974: 350-354.

[787] FÜLEP 1984: 79-81, figs. 23, 24.

[788] FÜLEP 1984: 79.

were unearthed: '1. Side fragment of pottery flagon, 2. Gallienus coin, 3. side fragment of glass vessel.'[789] It would therefore appear groundless to reconstruct tombs or sarcophagi here, especially not fourteen as shown in the reconstruction. With no traces indicating burials, we must find another solution to explain the existence of the plastered and the unplastered parts besides saying that there must have been sarcophagi standing by the unplastered parts. Moreover, we have to pay attention to the fact that the whole construction is in a cemetery,[790] see fig. 235. This cemetery, however, is not a cemetery of a single period as there are earlier graves as well as later ones.[791] In addition, excavations are made even more difficult at 8, Geisler Eta Street today, as the library occupies the site. The exposed section of the cemetery is in the library yard, but we do not know the extent hidden by the building. The tomb (to be discussed shortly) is partially covered by an 18th-century building whose foundations can be seen in the chamber. (Fortunately these foundations did not completely destroy the chamber.) The whole building cannot be fully excavated westwards. Eastwards, the library yard allowed more access but the 18th-century building limits the area of the excavation.

Returning to my sketch, there are holes in both the western and eastern walls. In the western wall there is a larger one (*B*) that is equal in size to the bigger hole (*C*) in the eastern wall. I suggest that the smaller hole in the eastern wall (*a*) and the one in the northern wall (*b*) were windows. Contrary to Fülep's opinion, I do not regard the holes in the western and eastern walls, which are facing each other and are of the same size - as vent-holes, but doors. Their proportions are approximately the same, or perhaps a little bigger, than the entrance door in the southern wall (*A*).

The phenomena described above are hypothetically assessed as follows.

The section excavated by Ferenc Fülep is the corridor of the actual tomb and this explains why there were no burials in it. The holes considered as vent-holes (fig. 237 *B-C*) are passages to the actual burial chambers where a vent-hole also opened (fig. 237 a). The alternation of plastered and unplastered sections of walls marks the locations of rows of pillars, as can be seen in the so-called mausoleum. The actual burial chambers have not yet been opened, obviously because they are under a building. As an analogue of this arrangement, I offer the tomb already mentioned that was unearthed in Or ha-Ner, the photograph and isometric reconstruction of which are presented based on the author's drawing,[792] fig. 238. I have made the reconstruction of the G/4 tomb (fig. 239) by using the Or ha-Ner tomb as a model. For the barrel vault leading to the western and eastern sides, a support is indicated at the point where only a part of the thickness of the wall is utilised, see fig. 236.

This hypothesis reinforces Ferenc Fülep's suspicion that the tomb preserved a Jewish custom. I would date it back to the 4th century. It is difficult to decide whether it is Christian or not as no motif, or find, has been found in this respect. If we compare it to the architectural methods observed in the so-called mausoleum, the painted pillars in the chamber, or the painter's solutions in the painted chamber I (i.e. the portraits in tondos), we may suppose that it was used by Christians, moreover, Jewish Christians. Its relation to the graves in the cemetery is indefinable. In any event, there are several notable peculiarities in this part of the cemetery containing the double painted grave (G/1-2) and the G/25 brick grave. The latter are very similar in style to the brick graves in the ☞ Ságvár cemetery. According to Ferenc Fülep the following distribution seems to be logical: the painted burial chambers belong to Jewish Christians, those graves with a special brick covering belong to the Jewish, and the sarcophagi and other graves to the 'pagans'.

32.I.g. Tomb XIII, fig. 240.

Literature: FÜLEP 1984: 99-107, fig. 51 - in the description and interpretation of the building, this picture is referred to and its numbers are used so that the relevant detail can be identified; TÓTH E 1990a fig. 5; Itinerarium 96-103.

F. Fülep excavated at 14, Geisler Eta Street in 1969, and the construction he called tomb XIII was found then.[793] Graves L4, L40, L43 and L45 were found under and inside these walls. The dimensions of the opened building were: L: 760 cm, W: 580-600 cm. Wall 2 on the ground plan on the western side could be followed to a depth of 210 cm. The presence of these graves explains the term 'tomb'. Fülep does not state whether he had the chance to examine whether the walls continued after a certain length, and, if so, in which direction. Cf. excavation photographs, figs 241, 242, 243.

As for the walls, Fülep writes about a *subsellium*, an altar and an altar screen, making the reader believe he has opened a church. He produced a reconstruction of the walls he found.[794] The outstanding peculiarity of this reconstruction is the screen shown attached to the altar; it is a functional nonsense. If we suppose that the semicircular part is an altar, then the screen should stand south of it, but the only possibility is provided by wall-section 15. In Fülep's reconstruction we have a church whose sanctuary occupies two-thirds of the total building, whereas its nave is only one-third of it. The reconstruction represents the object not as a funerary-chapel but as a church: he does not apply his terminology consistently.

In his opinion there is a connection between the building and the graves. Let us examine whether his statements are credible:

Grave L4a,b,c contained three skeletons, two females (a-b) and a male. All three were disturbed by collapsing bricks. The grave goods from 4a are two

[789] FÜLEP 1984: 81.
[790] FÜLEP 1984, fig. 23. Here, in G/4, we have the supposed graves.
[791] FÜLEP 1984: 174.
[792] Encyclopedia IV 1978: 938 (TSAFRIR).

[793] We have to note that Ferenc Fülep used the terms church and grave chapel as synonyms, so when disclosing his opinion, I follow his terminology.
[794] FÜLEP 1984, after fig. 55.

irregular shaped beads of glass paste. 4b had no grave goods. 4c had a bronze bulla with textile(?) remains in it.

Grave L40 contained the remains of a child's skeleton; no grave goods. A brick grave whose top has collapsed.

Grave 43a,b,c contained three skeletons, two females (a-b) and a male. There were no grave goods next to skeletons a and b; three glass vessels and the fragments of a fourth were associated with skeleton c.

Grave L45a,b,c,d,e contained five skeletons, two females (a, d), and three males. Of the five remains, only 45a had grave goods associated with it: a coin (Julian), an iron bracelet, two bone bracelets, two bronze bracelets, a bronze earring and a bronze finger-ring.

Building XIII was found in a cemetery (like tomb XIV). The map of the cemetery is shown in fig. 244. Building XIII raises many unanswered questions, and I propose to give my interpretations with those of Ferenc Fülep.

We can hardly suspect a connection between the graves found under the walls and the building, as the walls cross the graves. Grave L45 is exactly under the 'altar', where there were four further, and later, burials. Grave L43 is crossed by wall 15. There were two further burials in this grave. Only graves L4 and 40 were not under the walls. The later burials make us suppose that the building was only erected when the burials had been discontinued. The connection, therefore, between the building and the graves is quite doubtful. If there were a connection between them, it would not occur that any wall belonging to the superstructure crossed and thus destroyed the graves. The fallen bricks show that the builders did not bother with the graves. We can only state that there is no connection between the superstructure and the burials.

The grave goods do not reveal any Christian characteristics. The coins were made in the second half of the 4th century; the other finds also belong to the 4th century. The coins mentioned by Fülep, found between the walls, very probably belonged not to the building, but to the graves.[795] The excavator himself suspected this. Because the walls cross some graves, we may suppose that some graves were disturbed and that some of their goods might have been transferred to the area between the walls.

The outer side of wall XIII used to be plastered. It raises the question whether the building was plastered outside, or whether the wall now described as an outside wall used to be inside. The report does not refer to walls or sections of walls joined to the building.[796] We may be suspicious of such things as Ferenc Fülep does not refer to this plaster on his map or in his report. The plaster and the painting may still be observed. The inner, smaller semicircle was also painted. The marks of white paint have remained on the outer side of the 'outside' wall. The excavation photographs[797] were taken at different stages of the excavation, and show different phases of the work (cf. figs 241, 242). The localisation of the walls is difficult because of grave 45, which is just under the semicircle defined as an altar by Fülep. In Fülep's opinion, the building has an earlier and a later period. Walls 14, 16 and 17 belong to the later period and wall 16 used to divide the room into a western and an eastern part. It was joined by wall 17, which was plastered. However, the relationship between the two periods is not clear in Fülep's disclosures and we may say the same about the process and the purpose of the later alterations. Moreover, the periodization of wall 13 is also uncertain as he once writes that it was built together with No. 14,[798] but he omits wall 13 from the list of the walls belonging to the later period. The ground plan suggests that they were built together, and wall 14 does not belong to the later period. According to the photographs, wall 14 reached wall 2 on the eastern part (see figs 241, 242, 243). But this wall stops on the western part, where one descends to the lower side (i.e. the area south of wall 14). Here a section of the floor has remained in the southeast corner, next to the entrance. Based on Ferenc Fülep's observations, this floor ascends northwards continuously. This ascent cannot be perceived, but it can be pointed out by precise measurements. However, we may tell without instruments that the floor of the so-called sanctuary is higher than that of the so-called nave.

The explanation of the phenomena observed in the excavation photographs is sometimes uncertain. A north-south wall divided the so-called sanctuary into a western and an eastern part. This wall was above both walls 14 and 13, and also above the so-called altar. During excavations this wall was pulled down, and it cannot be found on the ground plan either. Where does this wall belong; is it the remnant of a modern building, or does it belong to a third period? Another excavation photograph (fig. 243) shows the opening of the basement of the altar at the opening of grave 45. The depth of the basement is not mentioned, and it cannot be observed in the photograph, so we have no real information about it. Fülep defines the area between walls 15 and 16 as a 'burial chamber', but it contradicts a previous statement, namely that wall 16 was said to be newer, belonging to the second period. The definition *burial chamber* is also untenable as both walls 15 and 16 cross the grave. This section must be interpreted in the larger context of the whole building. Another question we face is why Fülep thinks that walls 14, 16 and 17 are newer. Furthermore, we cannot neglect the fact that the graves and the building were presumably made at different points of time. The graves, according to the grave goods, belong to the 4th century, and as the building crossed them it could only have been built later. We may suggest the second half of the 5th century, at the earliest. However this is exactly the period when the outside and the interior of the 'churches' change with the alterations in the liturgy.[799]

[795] FÜLEP 1984 101, between walls 13 and 18, it may well have belonged to grave 45.

[796] It should be noted that the excavation was conducted in an urban site, thus the opening of segments and trail trenches was limited. We cannot really be certain whether the opened building was just a part of a bigger one.

[797] FÜLEP 1984 XLVII. Tables 1-2.

[798] FÜLEP 1984: 100.

[799] ☞ Dunaújváros, historical evaluation.

We should need a very accurate dating to interpret the object. Fülep defined it as a burial chamber although he uses the term *chapel* as a synonym. He thought it belonged to the graves. Consequently, based on the found coins, he puts the date of the building to the second half of the 4th century. However, it must be newer.

Summing up, we may state that it is not a burial chamber, but a church; the 4th-century Roman cemetery with graves (W-E) spread here. According to the grave goods found, these have nothing to do with the Christians and the church has no relationship with the burials. We still have to ascertain when the church was built and its size, and when it was destroyed.

The sanctuary faces north, and it is supposedly only a part of a larger building complex. We observe at early Christian churches that next to the church there used to be rooms attached with different functions. Such a construction resulted in a 'town', which altered even the original system of streets.[800] We assume that we are dealing with such a church, whose ground plan - without new excavations - can only be guessed at. However, even in its present state, we can already say that the building matches the architecture of East Illyricum, and more generally the architecture of the whole Mediterranean area.[801] To be more concrete, we can mention the analogue of the *extra muros* basilica at Philippi, where there are many graves below; there is also a burial chamber.[802] If we want a geographically closer analogue, then there is the Banjica church near Peć.[803] Sections of the analogues, they may be earlier or later, belong to the 6th century. Figuratively, parallel events seem to have met in the 6th century, making it very probable that the superstructure (building XIII) also belongs to this century (but no earlier than the end of the 5th), presuming that the opened part was the room of a building complex belonging to a basilica facing east or west. I will try to define its function according to this. The semicircular part interpreted as an altar serves as a starting point. If we do not interpret it as an altar, but as a table executed similar to sigma tables, then the room may have been the place of love feasts.[804] This also explains why the room faces north. The changes in the liturgy also support this idea. Thus, we have an *extra muros* basilica built at the end of the 5th (or beginning of the 6th) century that could have been the culmination of the building activity that erected oratories and other buildings,[805] establishing a real 'Städte außerhalb Städte' town.[806] It seems very likely that we can say the same of buildings XIV and XV.

Building XV, called a burial chamber by Fülep, raises a new question. Is it, or could it be, a part of building XIII, and what function can we attribute to it? Cf. fig. 240.[807]

Pécs's large cemetery and scattered graves also raise the question where the town was that might have had such an *extra muros* basilica. Where was the town whose supposed territory supplied numerous graves opened by Ferenc Fülep?[808] Cf. fig. 172b. Was Pécs a town? More exactly, do we have to seek Sopianae here? Or else, do we have to reckon with burials inside the town? The cemetery was required to be outside the town. To bury in a church in town was an imperial privilege.[809] In 381, the law that prohibited burials within towns was still confirmed. It was Leo I (457-474) who cancelled it.[810] Despite all this, there were examples of burials within towns: 'The Romans believed that their early custom had been to bury all the dead in the precincts of their town houses (Serv.ad. Aen. 6, 152). The law that prohibited such practices was never absolute: exceptions could be made *honoris causa*, and the tombs of some of the greatest families of the patriciate were to be seen near the centre of the city, for example the Valerii and the Cincii had tombs between Palatine and Esquiline, (Dionysius of Halicarnassus 5, 48; Cic. leg. 2, 58; Varro ap. Fest. 262). Other ancient families placed their tombs, naturally enough, just outside the line of the walls.'[811] Thus, by the time this basilica (marked as burial chamber XIII and XV by Fülep) was constructed, the law prohibiting burial activities in towns had been changed. When assessing excavation results we must consider this date and this fact.

There are further observations to be made concerning building XIII. For example, there is the comment relating to a course of bricks under the stone wall ('The stone layer built in a makeshift fashion on top of the row of bricks forming wall 14 ...'[812]). This observation is highly significant as the same feature is present at the so-called mausoleum, where the stone-wall also rested on a brick wall. I must note, however, that the same phenomenon is in evidence at tomb G/25.[813] Another significant feature of this burial chamber is that it resembles the double grave unearthed in the previously mentioned *extra muros* basilica at Philippi.[814] The latter also had a double bottom similar to the grave in painted tomb II at Pécs. This correlation suggests that there was a 4th-century cemetery here that was built upon at the end of the 5th or at the beginning of the 6th century. Pelekanidis dates the cemetery excavated by him between 331 and 361,[815] which also reinforces the above assumption.

Summarising the analysis of object XIII, we may conclude that it is not a burial chamber but a small chapel, as shown by the reconstruction. The reconstruction itself,

[800] G. LAVAS: Die Städte der 'christlichen Basiliken': ein Beitrag zum Städtbau des Illyricum Oriental. Atti 1984 Vol I. 581-623.
[801] Atti 1984 Vols I-II; D. PALLAS: les monuments paléochrétiens de Grèce découverts de 1958 à 1973. Città del Vaticano 1977.
[802] PELEKANIDIS 1977: 333-379. On the construction of Christian buildings outside the town see MIGOTTI 1992.
[803] NIKOLAJEVIĆ 1978: 680 fig. 3.
[804] BARB 1964; GAMBER 1968: 86-93; ☞ County Fejér, 15.II.a.
[805] ☞ the so-called mausoleum, 32.I.e. ☞ septichora, 32.I.d.
[806] ELIADE 1983: 59-61.
[807] FÜLEP 1984: 105 writes about tomb XV, whereas Suppl. VII. shows the relationship between XIII and XV, as well as the situation of the cemetery.
[808] FÜLEP 1984 Suppl. VI.
[809] EGGER 1962c: 15.
[810] CIC Constitutio 53. (E. OSENBRÜGGEN 1875 III. p. 796-797).
[811] PURCELL 1987: 27.
[812] FÜLEP 1984: 100.
[813] FÜLEP 1984: 82, 85. Figs. 23, 26. pl. XLIII.
[814] PELEKANIDIS 1977: 372-375, eik. 34.
[815] ibid. 372.

however, is inaccurate. The building is not related to the graves under it; the graves and the building are of different periods. The graves - according to their material finds - are of the 4th century, or to be more precise, of the second half of the 4th century. The excavated building is only a part of a much larger edifice, a building complex in which we have to suppose a Christian basilica as well. The latter was probably built at the beginning of the 6th century. This is also witnessed at other similar sites. This basilica is obviously the church of the community to which the descendants of the former oriental population of the 4th century, and the survivors of Pannonia's stormy years, belonged. Cf. 32.I.c, 32.I.e.

32.II.i. Double grave at 8, Geisler Eta Street, figs. 245a, b.

Literature: FÜLEP 1969c; FÜLEP 1984: 76, G/1, G/2; VALEVA 1998 779, No. 56 fig. 10. At 8, Geisler Eta Street, Kornél Seitl directed an excavation in 1958. He found Roman graves at the same level as the library yard. This is why one side of G/1 was discovered when the library was built.

Grave G/1. Oriented west east, L: 250-262 cm, W: 97 cm. The thickness of its western and eastern walls is 60 cm; the thickness of the northern and southern walls is 45 cm. Pieces of roof tiles were found in the grave. Finds: glass beaker, shoulder of a flagon, fragment of an iron nail, fragment of a fresco, brick rim. Grave G/2: built at the same time as the previous one, and thus they form a double grave. Its northern wall is missing. The 'head cushion' is on the western side. Finds: bottom of a glass vessel, fragment of an iron knife. The double grave is plastered and painted. The painting can be best observed on the northern wall of G/1: painted vertical lines divided the length of the grave into three panels. The pattern is the same in each panel. Two lines painted in an 'X'shape join the vertices of the approximately square parts. The intersection of the X is the centre of a circle. In the circle there is a drawn horizontal diameter drawn that passes through the centre marked by the X. The rest is filled with flowers.

Commentary:

The grave goods do not support the idea that the grave belonged to a Christian. The pattern of the painting does not contradict this idea, but it also allows for a 'pagan' owner. In my opinion, the main points to be addressed are: which of the graves are 'pagan', which are Jewish Christian, which preserve Jewish customs, and which belong to Christianised pagans. We have to deal with all four types of burials in Pécs and in this very area. However, we have to admit that it is not always possible to determine which grave is which. Therefore, I would rather classify this grave according to its finds and objects.

32.II.j. Gem, fig. 246.

From grave G/17, opened in Geisler Eta Street, F. Fülep published a gem. From the illustration he discerned 'The Good Shepherd', and drew and published it accordingly in his book.[816] In reality, the gem depicts Artemis shooting an arrow, with a dog at her feet.

There are also further tombs, but they are so deep within the Roman cemetery that their early Christian use may be questioned. The detailed description of the cemetery can be found in Ferenc Fülep's work.[817]

32.I.h. Finger ring, fig. 247.

Provenance: Zsuzsa Katona Győr excavated a late Roman cemetery at 3, Székesfehérvár Street in 1982. **Location:** JPM - **Literature:** SCHEIBER 1983: 45-46; Here. Scheiber mentions KATONA GYŐR's article, which has not yet appeared.

Blue glass ring, with the illustration of a menorah on its head. Diam: 2.2 cm, diam. of head: 1.1 cm. I have to remark that it is possible that the ring proves the presence of Jewish people, or Jewish Christians, as there is other evidence confirming this; ☞ tomb G/4. The mere illustration of a menorah is not proof of Jewish presence as Christians also used this symbol (as a reference to light).

32.II.e. So-called tomb VII, fig. 248.

Literature: FÜLEP 1962a: 23; FÜLEP 1977b: 15-16; FÜLEP 1984: 53-55.

It was found in 1958 by F. Fülep, west of the cella trichora. In his opinion, walls 1, 2, 3 and 5 are Roman. The length of wall 1 is 287 cm; wall 2 is 445 cm. Its inner size is 380x180 cm, according to Fülep. Opening the area surrounded by walls 1, 2 and 5 (fig. 249), the following objects were found. I quote: '... we found many pieces of Roman bricks; white fragments of mortar with smooth outer surfaces; then, 2 m below, many animal bones (horse, cattle, sheep), ash and pieces of charcoal. 280 cm below the surface, there was a white, thin layer of lime. At this level there was also a spread of ash containing many animal bones. A large quantity of pottery sherds was found in the tomb with as well as the animal bones. Immediately above the layer of lime there were many charcoal pieces, some of them quite large.'[818] When they had broken through the white layer and removed the soil, they came across many more animal bones and pottery sherds. At 348 cm below ground, sections of a cast-in-situ floor were discovered. The base of the wall was found at 358 cm below ground level, after the floor had been pierced. Wall 3a is 275 cm long. Its base was found at 135 cm. 'We found a Roman brick grave under the lowest row of stones of wall 3a.'[819] Figs 250a, b. This is R32, the grave of a child; it provided no grave goods. It is oriented east-west. The grave was completely under the wall.[820] 'Its side was built of Roman bricks with and without rims; its base was also tiled with Roman bricks.'[821] The size of the bricks is 48 x 38 x 5 cm. The earliest pieces among the pottery sherds found in the so-called tomb - according to the above-quoted layers - are from the 9th century. One bronze mount is Roman.[822] Fülep mentions fragments of Roman

[816] FÜLEP 1969c: 16; FÜLEP 1984: 85, fig. 29.

[817] FÜLEP 1984: 63-76.

[818] FÜLEP 1962a: 23.

[819] FÜLEP 1962a: 27. (The given depth figures are sometimes uncertain or contradictory.)

[820] FÜLEP 1984: 55.

[821] FÜLEP 1962a: 34.

[822] FÜLEP 1962a after fig. 18.

glass. He notes that the amount of Roman bricks found was so small that they could not have represented the vault of the tomb.

Commentary: The grave is supposedly not Roman, and it was not destroyed when the wall was built, but it was built over. There cannot be a relationship between the grave and the building. Thus the grave, Roman or not, does not provide information to help date the building or indicate its function. The building is later than the grave.

The building: It is noticeable that the fragments of bricks appear only in the top layer, and they are few in number. Fülep discovered this too. The pile of animal bones and the ash simply show that the animals were cooked or roasted. At the bottom level, the bigger pieces of charcoal may lead us to suspect that the building was not vaulted, but that it was covered by beams. The deepening of the wall in several places, reconstructed as the spandrels of the vault by Fülep, may refer to this. The large amounts of pottery (9th, 10th and 12th centuries) found, and also the fact that mainly medieval phenomena were registered in the neighbourhood of this building, warn us that we should be careful when making conclusions. Even under the trichora, a very small section of Roman wall could be observed. Taking all this into account, I do not regard it to be a tomb; there are not even finds or phenomena referring to that.

32.II.k. Double grave under the northwest tower of the Cathedral

Literature: FÜLEP 1977b: 15; FÜLEP 1984:51, R/23-24.

It was found 1.5 m deep under the tower in 1922. It is oriented east-west. It was built of stones, and plastered inside. There were disturbed graves. Four cannon balls were also found, one in a grave, three around it. The balls are from the 17th century. The different descriptions were interpreted by Fülep as follows: '... this might have been a double tomb with painted walls, which was subsequently rebuilt. Originally, however, both were Roman tombs.'[823]

32.II.l. *Cella trichora* and its surroundings, cf. fig. 172a

Literature: SZŐNYI O 1926; NAGY L 1931b: 22; NAGY L 1938: 117; GOSZTONYI 1939: 97; GOSZTONYI 1941: 43-44; NAGY T 1941: 147 fig. 7; GOSZTONYI 1943: 14-16; GRABAR 1946: 106; DEICHMANN 1954 No 27; FÜLEP 1959b; FÜLEP 1969a: 152; FÜLEP 1977b: 15; FÜLEP 1984: 51-53; FÜLEP/DUMA 1972a; FÜLEP/DUMA 1972b; MÓCSY 1974: 333-335 fig. 53; PARRAGI 1976: 178; Severin, Katalog 8.5; VAES 1986 Abb. 80/8a; TÓTH E 1988c: 33; TÓTH E 1994b Abb. 8; Itinerarium 96-103.

The trichora, fig. 251.

This feature was found and opened by István Möller and Ottó Szőnyi during work in the Cathedral in 1922. In 1955, Ferenc Fülep and Aladár Radnóti excavated in the same area. The orientation of the trichora is approximately north-south. It was defined as a central building with three apses. The central hall is 5.2 x 4.6 m. The radii of the apses are not of the same length: the northern one is 2.3 m and the two on the sides are 1.95 m. The size of the porch is 4.3 x 1.5 m according to Szőnyi and Möller. The walls are 130 cm thick on average. There used to be four buttresses (80 x 80 cm) on the outer side of the opened northern apse. The eastern and the western apses have deteriorated. The remaining walls are approximately 130 cm high. The walls were made of stones and bricks. The level of the floor is 26 cm higher in the eastern apse than elsewhere. Stairs connected the different levels. On this higher floor, there was a sandstone slab (129 x 60 x 15 cm), which was defined as an altar base. During the 1922 excavations, as the floor was found already disturbed it was decided to break through it and deepen to the present level. The original pavement in the trichora was a 15 cm mixture of brick, tufa and limestone. The wall was originally painted in two layers. The lower layer was a fresco of white and black stripes, some of which has remained in the northern apse.[824] The fresco of the lower layer may be as old as the building itself. The upper layer has remained in patches, but we can see more of it. It depicts a curtain drapery. In the trichora, 'among the secondary debris, a piece of carved door-post and fragments of frescos from the second painting, including the illustration of a hand, were found.'[825] The floor of the narthex was simple ground, whitewashed in a thin layer. Its entrance was blocked up to the height of 45 cm.[826] The door leading to the porch is 1 m wide.[827] In the porch a grave was opened in 1922, and a similar grave was found next to it in 1955 (CT/1 CT/2).[828] These graves were made of stone and brick. In the second grave, a 15th-century ring was found, fig. 252.

The 1922 grave is regarded by Gosztonyi as from the 'Frankish Age'.[829] It was 44 cm under floor level. Inside the trichora, Szőnyi also found walls that he considered to be from the Middle Ages on the basis of their Gothic carving. The wall running from the west to the east at the northern apse is still observable at a point where the footings deepen to their present level. According to Fülep, this wall extends under the Roman floor.[830] (I have to note that Fülep also asserts this about a medieval wall extending under the Roman footing.) One of the outer buttresses of the eastern apse was recorded in 1955. A further observation was that a medieval wall was joined to the eastern apse, and that this wall was deeper than the foundations of the buttresses.[831] Radnóti states that there were two building periods. The first one is contemporary with the cemetery, which was traced back to the last decades of the 4th century by the excavators. The cella trichora was obviously considered an important cemetery building as it was surrounded by graves. The second period

823 FÜLEP 1984: 51.

824 It is very doubtful whether the part under the curtain-motif could belong to an earlier period. Its dark stripe follows the edge of the curtain so noticeably that it can hardly be accidental; this would still allow a multi-layer painting: only the main reason is missing. Furthermore, Ottó Szõnyi, in his report of the excavation, did not write that the fresco was Roman, but 'Roman-like'. - SZÕNYI O 1926: 179.

825 GOSZTONYI 1939: 16.

826 Ferenc Fülep has a different interpretation of this statement; he says that only 45 cm has remained from the block. - FÜLEP 1959a: 414; 1959b: 85.

827 GOSZTONYI says it is cca. 1 m, Fülep says it is 145 cm.

828 FÜLEP 1984: 52. Fülep marked the graves according to where they were found, CT stands for *cella trichora*.

829 GOSZTONYI 1939: 16.

830 FÜLEP 1959a: 415; 1959b: 85.

831 FÜLEP 1959a: 402.

postdates the Roman era; this view is supported by the paintings (fig. 253, 254) and the 20 cm layer of broken fragments on the threshold. Ferenc Fülep puts the second painting phase for the trichora in line with the reconstruction after the fire in 1064, and dates its final destruction to the Turkish occupation.[832]

On the ground plan based on Ottó Szőnyi's excavation (fig. 251), the destroyed parts of the walls are well noticeable. On the ground plan drawn at the excavation in 1955, the cut in the floor within the trichora, and the measurement of the later deepening, can be seen. The buttress of the eastern wall may be observed as well as the walls whose definition is impossible on the strength of this excavation carried out in a small area of 6.5 x 4.5 m. The complicated description of the excavations also shows that the correlation between certain parts of unearthed walls could also not be defined by the excavating scholars.

I attach Ferenc Fülep's sections; section *F-G* (fig. 255) shows us that the Roman wall was observable only on the lowest floor, and then only at a very insignificant height and length. Section *H-I* (fig. 256) showed an even smaller Roman wall on the lowest floor. Ferenc Fülep did publish some small finds, but they do not belong to the Roman period. Fülep dated the pottery fragments (fig. 257) to the 9th century. The Dutch tile and the tessera (fig. 258) can be dated to the 15th century. Section *A-B* has a north-south orientation, indicating the wall that was found under the footings by Szőnyi and his colleagues and which was defined as medieval, fig. 259. In section *B-C* there is a Roman floor at a depth of 284 cm. Section *D-E* shows the depth of the new cut, the footing of the western apse of the trichora, and the eastern apse's footings, which were partially concrete and partially a 'Roman' floor. The slab defined as the base of the altar can also be seen. These sections reveal that the Roman wall, or footings, is rather small and short and at a very low level. Only in one place, in the eastern apse, are the 'Roman' footings at their original height. But here the problem is that the medieval wall is extended just under its level. It is also important to point out that, according to Fülep, the length measurement of the porch directed from the north towards the south is 2.83 cm instead of the former 1.5 m. Thus the measurements of the porch are 4.3 x 2.83 cm, i.e. a little more than 12 square metres. He explained this by saying that the part defined as an outer wall by Szőnyi is a medieval wall. Nevertheless, he did find the outer wall, which he defined top a certain extent. What is remarkable is that, on the eastern wall of the part defined by Fülep as a narthex, there were also buttresses similar to the apses of the trichora. Fülep assumes that when the medieval walls were built the Roman walls were demolished. He has also found an underground room that he called a cellar; its dimensions are 230 x 280 cm. A buttress 125 cm wide divides the 230 cm north-south wall; its present height is 94 cm (see fig. 255, sections *F-G*). Everything is medieval. He states that the foundations of these buttresses are deeper than the Roman wall.

[832] FÜLEP/BACHMAN/PINTÉR 1988: 34-35.

Conclusions from the excavation reports. The most remarkable fact is that even from the 'secondary debris' no fragments of Roman objects have been unearthed. Moreover, it is also remarkable that the fallen fragments of fresco (plaster) remain from the second painting, although the plaster of the first painting has not deteriorated; its remnants can still be observed in the northern apse. We have no Roman trichora finds from either the 1922 or 1955 excavations, at least as far as we can know. It is notable that the earliest find can be dated to the 9th century. It is also worth mentioning that a layer, albeit a small one, from the Roman building was found and that it was the bottom level. The one walking level that has been defined as Roman is in the eastern apse: the only problem is that there is a medieval wall under it. Can we really regard the brick, tufa and limestone pavement as Roman under such circumstances?

Paying attention to these facts, we must ask ourselves is the building really from the 4th century, is it merely a central trichora, or is the trichora only a part of a naved church?

The excavation data at our disposal does not prove that there was a trichora here in the Roman Age, in the 4th century. However, we also have to state that all research so far has been conducted with a 'certain awareness' of the fact that a 4th-century trichora and its surroundings were being excavated. This 'certain awareness' has misled the study and the process of classification. As a result, the unearthed wall that seemed to be joined to the trichora was separated from it by Radnóti. The excavation of the porch was not continued southwards, which is unfortunate, as now we would know whether it was merely a porch or an aisle joined to the trichora. We can surely point out that there used to be some Roman structure under the trichora based on the unearthed wall-fragments; but it is not at all certain that this Roman antecedent was also a trichora. Rumour says that Saint Peter's Cathedral, standing next to it, also had a Roman antecedent. A Roman antecedent is not necessarily a trichora or a church. We do not know what type of building might have been here in the Roman period.

According to the descriptions, there are many graves around the trichora but this is not the case. In the area spreading south and east of the Cathedral, there are graves, but they have no relationship with the trichora. There are five graves in total around the trichora (R25-29). Their descriptions are as follows:

R25: 'A walled grave was found in 1954 ... 200 cm below the ground level of the Cathedral. ... Dombay J. collected the bones. There were no grave-goods in the grave.'[833]

R26: 45 cm below the previous one. 'The mutilated skeleton was also oriented W-E...'[834] This is connected to the trichora. In the reports the finds are not mentioned.

R27-29: 'Orientation W-E, heads to the W. The walls were made of stone and brick and most of the bricks are of medieval character, although there are also some Roman

[833] FÜLEP 1984: 53.

[834] FÜLEP 1984: 53.

ones. There were no grave-goods.'[835]

Of the above five graves, three (27, 28, 29) are surely medieval, as the ancients could not have used medieval brick, but vice versa is possible. Therefore, the Roman bricks in the graves do not convey any information concerning the age of the grave. The other two graves are also probably medieval. At the present state of research, it is not possible to confirm whether there is any connection with the trichora. It is impossible to form an opinion that concurs with the facts if we ignore the Middle Ages and deal only with the Romans; the relics of the two eras are so close that their assessment can only be carried out simultaneously by means of a thorough excavation.

The surroundings of the trichora

R31 was unearthed south of the trichora and its measurements could only be reconstructed. South of it, R30 was unearthed near three column bases, square/rectangular in cross-section. The two graves are oriented west east. Their finds are not mentioned.[836] The column bases can be seen in figure 172a, joined to grave 30. Their measurements are: 150 x 150, 160 x 135 and 150 x 150 cm; these column bases are not explained by Ferenc Fülep. In 1872, these columns were considered as the ruins of an early Christian basilica, but it turned out that they were not, although their ages could not be determined.[837] Szőnyi writes the following in relation to the columns: '... also surrounding the column shank was a floor of rectangular tiles...' Later it came to light that this 'floor' belonged to the paving of the square, after the building of the bishop's palace.[838] It makes the dating of the columns difficult; they may be modern, too. Szőnyi tends to regard them as pedestals, and he does not think they belong to the building. This opinion is not without basis, as there is no trace of any further building in the area. There are no other graves near the trichora.

Mention must made of the conduit running on the northern side of the Cathedral, fig. 260. Ferenc Fülep also conducted the newer excavation.[839] Szőnyi at the northwestern tower of the Cathedral carried out an excavation in 1922. He found the conduit at this time. According to his survey data, its height is 100 and width 60 cm. At this time, there was water in the conduit. Fülep opened the basin of the conduit, in which, below 165 cm, he discovered fragments of a Roman fresco; this layer reached down to 310 cm. The fragments do not have any figural ornamentation, only vegetal and geometrical. Colours like red, green, yellow, brown, blue, white and black were observable on the fragments. On one sole part of it, he could read a scratch, which he interpreted as 'ACTDΓ'.[840] I read it is 'ACTOR', fig. 261. Here a Valentinian I coin was also found. Fülep derives the fresco-fragments from a burial chamber. The rest of the finds are from the 16th and 17th centuries. Fülep thinks the conduit was built by the Romans to conduct water; it was found and used in the Middle Ages. There is a spring on the hillside northeast of the Cathedral that still operates and supplies Pécs with water. It is known as the Petrezselyem Spring. He assumes that the line of the conduit was the edge of the cemetery.

I have to add that research has accepted it as the remnant of a Roman *aquaeductus*. However, it is not clear where it conducted the water. Moreover, it has not been finally determined whether it really is an *aquaeductus* at all. It is true - be it an *aquaeductus* or not - that it marks the side of the cemetery, but does it mark the beginning of the settlement? Research has not localised the Roman city north of the *aquaeductus*, although the existence of the city itself is not without problems.[841] Are we really to seek Sopianae at Pécs?

Research of the Cathedral area would not be complete without a reference to Gyula Gosztonyi's theory. Its core is the assumption of an early Christian basilica under the Cathedral, whose ground plan would follow the one of the Cathedral. His argument is that the footings of the Cathedral crypt, the trichora's reconstructed foundations, and those of painted tomb I, are all at one level.[842] We should not dismiss his idea.

Summary of the trichora. At the present stage of research and publication, we may state that the trichora is not Roman: no Roman finds have been found in the trichora nor its neighbourhood. It might have been built at the end of the 7th century (at the earliest), and it existed until the late Middle Ages. All the alterations in the trichora were carried out in the Middle Ages. The neighbouring graves are also medieval, or at least may be suspected as so. The question is whether it is really a cella trichora. The 15th-century stove-tile is an extraordinary find. It raises the question whether the function of the building was the same over all the various periods. All this leads us to refute the reconstruction of the building.[843] The development of the area allowed for only a limited excavation, which was insufficient to provide a full interpretation.

---

835 FÜLEP 1984: 53.

836 FÜLEP 1984: 53.

837 SZŐNYI O 1926: 182.

838 SZŐNYI O 1926: 183.

839 FÜLEP 1961a.

840 FÜLEP 1961a: 63, fig. 7; FÜLEP 1962a: 45, fig. 21.

841 Cf. Endre Tóth's opinion in the historical evaluation of Pécs.

842 GOSZTONYI 1941: 43.

843 About the reconstruction works carried out at Pécs, we may generally say (as in the connection with the trichora) that they were led by the presumption that they were the ruins of early Christian churches. This issue automatically regards all the objects found as antique, and simplifies the early Christian building types, which, however, cannot be examined without using any geographical and chronological correlation. In the Hungarian scholarly works, we often observe the usage of the terms *basilica* and *martyrium* as synonyms, although they are not synonymous. The basilica is the scene of liturgy, while the martyriums are mostly central buildings, erected in cemeteries over the tombs of martyrs venerated by the local community. The building itself was regarded a symbol; the ordinary believer strove to be buried nearby (Kővágószőlős). Their beauty and form dazzled the contemporary observer. Based on the works of Paulinus of Nola, Ambrosius, and other antique authors, Mircea Eliade uses the phrase 'Städte außer Städte' to envisage such buildings (ELIADE 1983: 59-61). Finally, I have to note that none of the reconstruction of the Pécs relics is acceptable.

HISTORICAL EVALUATION PÉCS

Based on the above data at my disposal at present, the following historical summary can be prepared for the Roman Period, and Christian antiquity, of Pécs.

The area of the Roman city - if we can speak of a city at all – in relation to the size of the cemetery. The observable phenomenon of burials over previous graves brings our attention to the existence of different periods and a longer usage. The longer usage emphasises even more the small size of the city's territory, which is especially noticeable if we remind ourselves that it used to be Sopianae, the capital city of Valeria province. András Mócsy, who speaks moderately about Pécs's status as a city, and Endre Tóth, who does not consider Pécs to be a city in the early period, deserve more attention. The site-map prepared by Ferenc Fülep in 1984 shows that he concentrated on excavation of the cemetery first of all. The walls and wall fragments unearthed within the town are not enough to localise with safety the ancient city's public squares and main buildings. We may only infer the sites of the forum and the temple to the Capitoline trias based on the hypothesis that Pécs is identical with Sopianae. Houses have also been found, but their number is insignificant compared to that of the graves in the cemetery. Consequently, I have based my assessment on the cemetery.

The foundations unearthed in the cemetery should be considered as buildings belonging to the cemetery, as long as they are from the 4th century or from an earlier period. At Pécs, we have to exclude the possibility, observed in places like Aquincum or other cities in Pannonia, that with the shrinking of the city, the later burials are made within the former territory of the city.[844] The observable periods of the cemetery and the coins from the 1st century let us assume that there was an early settlement, although we have to admit that in this period it did not have the privileges of a city. The buildings of the cemetery show similarities not in their ground plans, but in the elaboration of their walls. What is even more extraordinary is the fact that there is almost no part of the 'city' where there are no graves.[845] The unusual archaeological phenomena and the geographical features (i.e. the cemetery lies on the hillside), require a unique interpretation of Pécs's objects. In other words, it differs from the common features of other Pannonian cities, and its interpretation must also be different. This uniqueness is borrowed from the cemetery's oriental style, which leads us to conclude that there were oriental immigrants, mainly Syrians, who may have been accompanied by Jewish, Hellenised or even Greek settlers.[846] The objects are so penetrated by oriental influence that we have to reckon with an early immigration and, to some extent, a settlement as well. The sole, certain parallel of tomb G/4 has been unearthed in the Holy Land, which leads us to the conclusion that the Pécs chamber is a creation of people of a Jewish origin. East of this chamber, in a nearby grave, the menorah ring was found. Although the menorah itself would not necessarily betray the presence of a Jewish population, since Christians used the symbol too, because of tomb G/4 we are forced to consider that the owner of the ring had a Jewish origin. The double graves, whose Christian nature is debatable, unlike their outstanding features compared to the other graves, are also in the vicinity of tomb G/4. The above facts make us conclude that in this area, namely the area east of Geisler Eta Street's bisecting line, together with the eastern block in Székesfehérvár Street, is located the graveyard of a population that does not, or cannot, use the rest of the cemetery. The Jewish, they may be Christian or not, belong to this category. Cf. fig. 172b.

We face different phenomena in the area stretching westward from Geisler Eta Street's bisecting line to the northern side of Szent István Square. We notice that the 4th-century pagan graves are never destroyed, but they are built upon. Sometimes we almost feel that the subsequent buildings include the previous graves. Such is the part of a building previously defined as tomb XIII, of which we can very probably suspect that it is a section of an *extra muros* basilica. This careful building upon the graves may perhaps refer to a certain reverence for the dead, though they were undoubtedly pagan. The pagan graves were oriented W-E; the divergence from this direction is so little that it is insignificant. The correlation of these buildings is not clear at the present stage of excavation, however, we can state that these parts of the buildings used to belong to one certain period, based on their building features and walling technique. The so-called mausoleum on the northern part of Szent István Square, which has been proved an *oratorium*, also belongs to this building period. It was erected on the Christian tomb below, but not necessarily with the purpose of raising a monument to the memory of the dead. Also in the chamber were pagan corpses; there were corpses at that time in the two sarcophagi that were older than the Christian frescos. We could observe alterations in this tomb that denoted the beginning of Christian use; these alterations did not mean destruction. Moreover, I would also raise the possibility that it remained the crypt of the same family who had buried their dead here, but adopted Christianity meanwhile. By means of painting Christian symbols and biblical scenes, they wanted to 'sanctify' their predecessors and create a resting place worthy of their faith. It as not they who raised the *oratorium*; it might have been built by their descendants. The walling technique of the superstructures was known in Byzantine architecture since the middle of the 5th century. Its main characteristic is that, after one or two courses of brick, there is a higher section of wall constructed of irregularly shaped stones and rubble embedded in mortar. I also consider it important that the size of the bricks is the same, 45 x 30 x 5 cm.

[844] Although Ferenc Fülep explains the presence of graves in all parts of the town by these historical events. ☞ Pécs, history of research.

[845] Vide FÜLEP 1984 Suppls. I-IV.

[846] Further finds will be needed to decide how to assess the other examples of the Philippi excavations. However, I would like to remind the reader of the fact already disclosed in the history of research, namely, that the Christians have Greek names, and it is also true of Sirmium's first bishop, Epaenetus. The first Christian settlers in the area may well have been Greeks. Their arrival at Pécs could have taken place quite early.

Although we could date the Pécs buildings to the middle of the 5th century theoretically, because this period was one of crisis in Pannonia, we think they were built at the very end of the 5th, or the beginning of the 6th, century. In this area, the Hellenised population had burials that included people from East Illyricum. We also have to allow for the existence of further burial chambers in Saint Stephen Square, south of the present chamber; their excavation would supply a more exact interpretation of this section of the cemetery.

The third, distinct part of the cemetery at Pécs is the one that includes the vicinity of the Cathedral, the area south and east of the Cathedral, and the western area next to it. This part of the cemetery is the most problematic, as it suffered most from subsequent reconstructions. It may also have been filled, partly artificially and partly in a natural way. Ottó Szőnyi writes: 'The mud and stones that are washed from Mecsek at times of showers, and that plough up the asphalt at any time, building real barricades, have filled it with a new layer. The medieval moat behind the Cathedral did not use to be there, so the water, with its stones, used to rumble down towards the flatlands above tomb I.'[847] This account explains the later infilling, which has buried the tombs (that were once above ground) below the surface. However, at the same time, the account warns us to look for those objects that would have protected the graves from this kind of destruction in the Roman period. It is inconceivable that the Romans exposed their own tombs to recurrent natural disasters. Such constructions would include the one defined as an *aquaeductus* at present. A short section of it has so far been excavated - the section joined directly to the Cathedral. Here the question is whether it really is an *aquaeductus*, or a conduit[848] flowing into a storage-lake. In this respect, we must be suspicious of the object designated tomb VII and that hides a walled shaft which can be followed almost up to the current surface. It was used as a refuse pit from the 9th century, which accounts for the presence of so many animal bones, pottery sherds and ash. It is probable that the hot ash was also tipped here after cooking, setting fire to any wooden contents and resulting in most of the charcoal fragments found amongst the ashes. In my opinion, the grave exactly under the wall is not Roman, but earlier.[849] The Romans built the wall onto it, walling the corpse with bricks. We can safely state that the ditch that caught and collected rainwater was ruined in the Middle Ages. It cannot be accidental either that Ferenc Fülep thought the edge of the cemetery was denoted by this *aquaeductus*. He was right, since the territory south of it was protected from natural damage. This object cannot be considered as an *aquaeductus*. All over the Empire these constructions used to extend high above ground, not under the surface. Naturally, I would not like to say that Pécs did not have an *aquaeductus*, but it was not this particular construction. In connection with the so-called tomb VII, and the so-called *aquaeductus*, we may interpret the remains of Roman walls found under the trichora. The excavations next to, and under, the trichora have unearthed many undefined wall sections, whose comprehensive excavation would shed light on the processes of water-collection and the building of cisterns. The trichora itself and its vicinity, together with their graves, are medieval.[850] I have dealt with the western part of the territory in question.

In the eastern part of this area is the *septichora*, the painted tombs I and II, as well as tombs III, IV, V, VI, VIII, IX and X, defined as unpainted tombs by the research. As for the tombs, the assumption was that all were early Christian, and there used to be a chapel above each of them. This axiom has been accepted in the Pannonian research, hindering the finds and the phenomena to convey information. The phenomena themselves are accidental, since there have not been any comprehensive excavations; the finds were uncovered during construction, canal digging, and other similar works. We have to be aware of the fact that there may be other tombs next to the known ones, or that they may be chronologically continuous. We may raise this question in terms of tombs II and VIII. These two tombs are older than the rest (tomb VIII may be dated to the 2nd century) and they are located along one linear line, which suggests that they may be parts of a tomb-street, and we do not know the other buildings of the lane because of the low level of excavation. In fact, the regularly arranged tomb-street is typical of the earlier period, while the later is characterised by tombs arranged in a network.[851]

The tombs – as regards their ground plans - belong to the same type, except for the painted tomb I. They are square, with an entrance on the southern side and a niche on the northern one. Superstructures used to belong to tombs II and VIII, but these were certainly not chapels; they were very common superstructures that occurred in many places of the Roman Empire denoting graves. Based on the data provided by the excavators, especially because of the data of its entrance, I classified the building over the painted tomb II as a tomb, although, as well as the measurements, all the phenomena suggested a *locus orationis*. In all probability, there was not a superstructure above the rest of the tombs. For the conditions of the

[847] SZŐNYI O 1907: 22.

[848] Ernő FOERK, when he was studying and reconstructing the *aquaeductus* of Aquincum from the viewpoint of an engineer, described those phenomena that did not belong to the *aquaeductus* but were parts of a conduit system. He found such a phenomenon at the fort in the area stretching towards the Danube, and wrote: '.. in all probability, there was only one ordinary sewer like those extending in Aquincum's streets 1 m deeper than the bottom of the edged brick drains, i.e. the gutters extended above them. (Az óbudai aquaeductus - The *aquaeductus* of Óbuda - BpR X 1923: 50). 'Ordinary' here refers to the simple gutters differentiated from the brick, vaulted, and high ones.

[849] Orientation E-W, so it differs from the orientation of the surrounding graves.

[850] Not being a medievalist, I would only note that the trichora may not be an independent building, but also the part of a church, as we may see in Salzburg - Wolfgang BRAUNFELS: Die Kunst im Heiligen Römischen Reich. Deutscher Nation. München Bd II (1980) 160-164. It is a well-known fact that the Episcopate of Salzburg envied Sirmium's greatness in the Middle Ages. It may serve with a correlation as for the building of the trichora.

[851] HESBERG 1987: 50-51.

ground features, the various means of approaching the chambers, and the probable existence of hypogeums, see the summary of ☞ painted tomb I (32.I.b)

We may regard the painted tombs I and II as early Christian. In the latter, the saint of the local community, Bartolomeus, may have been buried at the end of the 3rd century. The presence of ordinary graves among the chambers may be explained not in terms of an aspiration to rest near the saints, but simply because the site is a cemetery. These graves have no features that would refer to Christianity; no gravestones were used in this area. We know graves marked by gravestones from areas that, we think, used to belong to the territories of ancient cities or towns. This area, situated south of the Cathedral, was mainly used by Syrians and other oriental people. This can be shown by the Syrian styling traced on the fresco in tomb I. The grave marked with a chi-rho is also a unknown phenomenon in Pannonia; the chi-rho painted on a brick like this is unique. The skeleton's positioning, however, is not unique. Lajos Nagy discloses many examples of such burials in different Christian cemeteries. It was a practice to turn the dead to face each other.[852] ☞ This phenomenon is also observable in the cemetery of Szentendre (48.I.b). The vicinity of tomb X, as well as the mosaic, whose parallel was found in the mosaic of Djemila by Ferenc Fülep[853], can only be interpreted hypothetically. We have to raise the possibility that tomb X may denote a fourth part of Pécs's Roman cemetery, but only a new excavation could supply evidence of this.

We might also mention here the richness of some of the constructions and the inference of wealthy burials. Naturally, people were buried in burial chambers or in simple graves on an economics basis. The prosperity may be evidence of a Syrian population in the area. Epigraphic and villa researches have proved that in Pannonia the Syrians were amongst the wealthiest inhabitants.[854] We should also refer to the villa and the so-called mausoleum at Kővágószőlős. ☞ Kővágószőlős, for an interpretation of the frescos (26.II.a).

The peculiarity of the area is the *septichora*, which I have hypothetically classified as a *locus orationis*.

Due to the early medieval destruction, only disturbed burials were found in the chambers. We may point to the general destruction everywhere in Pécs, covering such a large area that it illustrates the merciless severity and brutality of a horde motivated by revenge. What survived the ravages of the horde suggests, or I dare say proves, the existence of a necropolis. The tombs, then following the destruction the buildings erected over the burial chambers, denote the necropolis, which is best characterised by Eliade's remark: 'Städte außerhalb der Städte'. In the necropolis the *extra muros* basilica has a peculiar significance, as it may refer to a stationary community that did not flee at the collapse of the Empire, thus representing a certain continuity of population. This building should not be interpreted in the context of a cemetery (as with the other superstructures as well), but as the church of a new settlement.

In the case of Pécs, we suspect an earlier settlement under the present inner town, beginning possibly at the turn of the 5th and 6th centuries, and later stretching over the previous cemetery too. The real question is whether the town of Sopianae can be identified with either one.[855]

The phenomena lead to a further hypothesis in connection with Pécs' medieval name, '*Quinque Ecclesiae*'. The five local communities being: (1) the Romanised native inhabitants and Italian settlers; (2) the Greeks or other Hellenised populations; (3) the Syrians; (4) the Jews; and (5) a community of the subsequent settlers who built the trichora. Together they would seem to form an *ecclesia* as communities (the original meaning of the term refers to the community), and they each possess an oratory according to their own rituals.[856] Thus '*Five*' does not relate to actual buildings, but rather to five living communities (i.e. *ecclesias*).

See references to Pécs at ☞ Pilismarót and ☞ Tokod. The features at these two locations have a link with Pécs, and it tends to suggest that Pécs was not the municipal centre of Valeria province.

33. PILISMARÓT-KISHEGY (CASTRA AD HERCULEM), fig. 262.

Literature: FINÁLY 1907; NAGY L 1938: 97; MRT 5 17/6 with further literature; MÓCSY 1974: 283 Fig. 47; SOPRONI 1974b; SOPRONI 1978: 46-48; SOPRONI 1985: 62-63; BARKÓCZI/SALAMON 1984: 152-155.

The site is presented here based on MRT 5 17/6. Its discovery as a Roman site is due to the work of Flóris Rómer. Róbert Fröhlich conducted the first survey in 1893; he prepared the ground plan based on certain features above the ground. Gábor Finály excavated in 1906. He opened a section of wall having seven semicircular turrets, as well as three buildings (*a, b,* and *c*) within the fort, fig. 263. The fort has an irregular ground plan (340 x 130 m). Building *a* is 33 x 27 m, it includes four rows of six pillars respectively; some of them in the northeast corner were ruined by subsequent constructions,[857] fig. 264. Building *b* is a rectangular one with an apse, it used to be joined to building *a* with three walls, fig. 264. Its size is 11 x 15.4 m, and the inner radius of the apse is 4.15 m. Building *c* also has an apse and it is completely independent from the other two buildings,[858] fig. 265. Its size is 9.45 x 14.7 m, and the inner radius of the apse is 2.9 m. Lajos Nagy considered the latter two to be early Christian churches. The MRT informs us of the available finds of the excavation, and it

[852] NAGY L 1938: 132.

[853] FÜLEP 1984: 61 No 137. There is also a parallel in Pannonia : WELLNER 1970: 364.

[854] ☞ Budapest III, Historical assessment, hypotheses on Aquincum's Christianity, M. Ant. Victorinus.

[855] No inscription has been found so far that would prove the name of the city here at Pécs, which suggests an expectant attitude. In fact, Romula's name was also presumed for a long time before an inscription was found.

[856] The communities may have had several buildings, since among the Syrians we know two different rituals.

[857] FINÁLY 1907 fig. 13.

[858] FINÁLY 1907 fig. 15.

emphasises the importance of the bricks with stamps. Stamps: *leg II ad., Quadriburgium, Terentius dux, Corta vicen.* The latter ones - in Sándor Soproni's opinion - are typical of the constructions under Valentinian between 370-372.[859] 'The rest of the finds are: green-glazed pottery and iron, bronze and bone objects typical of the 4th century. Among the numerous 4th-century coins, the ones of Constantine I, Constantinus II, Valentinian I and Valens, led Gábor Finály to the conclusion that the Marót fort was built under Diocletian, and it was reinforced with further turrets under Valentinian.'[860] The topography discloses the fact that during the site-survey Prehistoric and further 4th-century Roman finds were found. However, no evidence was found of the medieval tradition that reported by Simon Kézai. His remark is important as regards continuity in the Middle Ages.

Soproni identifies it with Castra ad Herculem. The name is present in the Itinerarium Antonini as a construction from Diocletian's period. In his opinion, its distance from Cirpi allows such identification. Thus, he also dates the building of the fort to Diocletian's reign, with alterations during Valentinian I. The identification of the name serves as a basis for all other conclusions. The name identification is due to the presumption that the object is a fort, so he writes as follows: 'Das Lager wurde in typisch spätrömischer Form unter Diokletian erbaut.'[861] Although he refers to the horseshoe shape of the turrets, I have to note that not all the turrets are horseshoe-shaped, but they have a semicircular shape.[862] András Mócsy dates the building of this type of turret to Constantine I, and, at the same time, he describes the type of fort associated with the Pilismarót edifice. These features include an irregular ground plan, enceinte walls following the geographical conditions, and its construction in such a way as to include hilltops. There are few gates - perhaps only one.[863]

Geographical conditions: the Malom brook flows at the foot of the hill, on its right-hand side is the top with the fort; west of the left bank is the Roman cemetery, cf. fig. 262.

The cemetery: The opened part is a section of an extensive cemetery.[864] Its dating is very ambiguous. The finds are evidently from the second half of the 4th century, but a more accurate dating is not possible. Even whether the cemetery was in use after 375 cannot be proved one way or another. László Barkóczi thinks that the cemetery was used under Valentinian I, and even after 375.[865] The beginning of its use could also not be firmly established. Some graves may refer to Christians. 'Two skeletons in grave No. 1 refer to Christianity, as well as at least two skeletons in grave No. 16. In connection with Pannonia, Lajos Nagy dealt with the cryptic usage of late Roman tombs assembled from stone slabs, which also points to an early Christian custom.'[866] He also gives a detailed analysis of the simultaneous presence of jugs and beakers in Pannonian cemeteries.[867] He mentions the frequency of a certain type of cup, of which four pieces were found, fig. 266.[868]

Commentary:

The absolute chronology of the building may be uncertain, but the two debated points of time still allow a certain interval, i.e. the reign of Diocletian and Constantine I. The descriptions take into account the different, and separable, periods of the walls, which means that we can establish a relative chronology. This is the only way one may clarify the relationship between the buildings, or their relationship with the wall, or its alteration. My own position is that the 'subsequent' construction referred to is later than building *a* (33.I.a), as it destroys the row of pillars of building *a*. However, this 'subsequent' construction does not destroy building *b* (33.I.b). On the other hand, the ground plan shows the section of wall belonging to the 'subsequent' construction above the apse of building *b*. This fact may make us think that building *b* is later than both building *a* and the 'subsequent' building, which were partly used, partly pulled down, when building *b* was erected. Thus, the presently observable and existing 'subsequent' walls do not reveal their original functions.[869] Reflecting the possibilities provided by the ground plan, the following order seems logical: building *a* is built, it is followed by the 'subsequent' building, which destroys the pillars in *a*. It is followed by the building of the fort-wall and the construction joined to it, its trace can be seen on Gábor Finály's map,[870] cf. fig. 263. Building *b* is erected after that, and at this time the earlier one is completely altered. Building *c* (33.I.c) may be considered as the latest one, built after *b*. We may suppose, perhaps, that the demolished building used to be a dwelling.

The reports do not mention whether there existed a Roman settlement on the mount, or in its vicinity, before the building of the fort. In connection with this, the uncovering of the cemetery, because of the low number of investigations, also provides no data. In spite of this, such a settlement is presumably similar to that of Keszthely-Fenékpuszta, where an early settling preceded the fortified settlement.

I also raise the issue that this object was actually not a fort but a settlement surrounded by a wall.[871] Even the

---

[859] SOPRONI 1985: 53-54.
[860] MRT 5 p. 285.
[861] SOPRONI 1985: 62.
[862] FINÁLY 1907 figs. 6-11.
[863] MÓCSY 1974: 282 fig. 47: further places that belong to this type: Budapest Esküter (this one is not on a hill), Hideglelőskereszt, Tokod, Visegrád. He also writes about the dating of the type of fort here.
[864] BARKÓCZI 1960: 127. He found 31 graves.
[865] BARKÓCZI 1960: 127.

[866] ibid. 127. This custom refers to the turning of the corpses to face each other. ☞ Pécs, summary, the third cemetery section.
[867] This theme is not rare in the Pannonian research, ☞ Pécs, painted tomb II. (32.I.c.).
[868] BARKÓCZI 1960 Pl. XXIV/4.
[869] We have to note that the ground plan is not based on the complete excavation. We cannot assess what proportion has been excavated.
[870] FINÁLY 1907 fig. 2.
[871] Đuro BASLER: Kasnoantičko doba (The Late Antique Age). In: Kulturna istorija Bosne i Hercegovina 1984: 309-369 I quote from his article (p.368): 'As for the fortresses, they assumed new forms in the 4th century. Besides a great number of refugiums erected by inhabitants of a certain area for defence in case of barbarian marauding raids; the central authorities built several fortresses which might be counted as *castrums*.' The towers of the refugiums listed by him that are surrounded by walls,

geographical conditions provide shelter here. If we consider the section surrounded by a wall as not only a military fort but also as a later settlement (like the ☞ Tokod fort), then we may regard the Öregek-dűlő Roman cemetery of provenance MRT 5 No. 17/31 as the cemetery for this settlement, whose dimensions - due to rather partial excavation and uncertain measuring - can only be estimated. Thus we may suppose: 'a cemetery which was approximately 130 m long, situated in a wide, 50-60 m strip of land with an East-West orientation'.[872] László Barkóczi lays stress on a certain type of cup and he writes in detail that the simultaneous use of the jug and the cup is characteristic of the cemetery.[873] 'At Pilismarót, glasses of this kind occur in the 60s and 70s of the 4th century.'[874] Eszter T. Szőnyi confirms the dating and offers further finds as examples when she puts the Győrszentiván cemetery between 360 and 380 on the basis of certain objects.[875]

The above-mentioned phenomenon, especially the simultaneous usage of the jug and cup, is not proof of Christianity, but it can be proof of the prevalence of earlier customs, possibly even of the Christians as well. '... Zugrunde liegt die auch in der christianisierten Bevölkerung festgehaltene Vorstellung vom leiblichen Fortleben des Bestatteten im Grab als seinem Hause, das mit täglichen Bedarfs - und Nahrungsmitteln ausgestattet wird, und das man Stücke des persönlichen Besitzes des Toten mitgibt.'[876] Jugs and cups actually belonged and belong to the objects used in everyday life. That is why their shape and elaborations have changed with times and fashions, and have consequently been of distinct value with regard to dating. The reference to the Eucharist cannot be accepted under any circumstances, but we may state that, as an item of everyday use (in a form matching the style of the age), it may have been used at agapes, in connection with the cult of the dead, to symbolise refrigerium, and, what is more, as a chalice too. The first chalices were also made of glass. Thus, in the first four centuries, a glass cup used at any time played acted as a chalice too, and therefore it could have been put into a grave as a symbol of having died with a purified soul, or alluding to the sacrificial meaning of the Eucharist. If the circumstances do not refer to the latter case, then we cannot prove that fact from an archaeological point of view. What is curious about the jug and cup illustration found in burial chamber II at ☞ Pécs,[877] is that the illustrated beaker has come to light also as a real archaeological find.[878] The jug must be looked for among the bronze dishes, since the colour discernible in the fresco brings to mind more the colour of bronze than that of a glass vessel. The shape of the handle also refers to the bronze dishes. This scene can be classified among such fresco types as the one that illustrates the feast in the Petrus and Marcellinus catacomb (with the inscription: *'Irene porge calda, Agape misce nobis'*), or the A3 cubiculum fresco in the Callixtus catacomb.[879]

SUMMARY

The definition *fort* is doubtful and 'walled settlement' seems more preferable. Due to the opened buildings and the nearby cemetery, a civil population may be imagined among whom there presumably were Christians as well. The idea about the presence of Christians in the cemetery is supported by the observations in connection with the position of the skeletons. Therefore, it may be supposed that there was an early Christian church (or house-church) among the buildings that was located within the walls. I would favour a house-church, but only at the very end of the 4th century. The possibility is supported by building *b*, mentioned so many times because of the reconstructions, in as much as the sections between *a* and *b* are considered to belong to it as well. At present, an assessment of *c* is impossible because of lack of correlation. Only excavation can confirm or disprove my hypothesis.

For other instances, see also ☞ Pécs and ☞ Tokod.

the thickness of their walls and the buildings inside the wall, are not analogues of that of Pannonia, but they are, in any case, analogues on an intellectual level.

[872] MRT 5 p. 294.

[873] BARKÓCZI 1960: 111-132. For more details on this topic see p. 127.

[874] BARKÓCZI 122-123. Here, he enumerates other provenances of this cup: Aquincum, Brigetio, Intercisa, Ulcisia Castra, Ságvár and Kisárpás. All of them can be dated to the second half of the 4th century.

[875] SZÕNYI E/TOMKA 1985: 109-110. Here the authors bring up analogues from Pilismarót, Pécs and Intercisa.

[876] DINKLER-VON SCHUBERT 1980: 157.

[877] Cf. GERKE 1952 we can read an argument about the early Christian references of these objects in this writing.

[878] PAŠKVALIN 1974: 125-126, Taf. VIII. 2.

[879] FÉVRIER 1978: 242, 250.

34. PORNÓAPÁTI (PORNÓ)

34.II.a. Tombstone

Location: It is lost. **Literature:** CIL III 4219; ILCV 352; NAGY L 1938: 96; NAGY T 1954: 512; THOMAS 1977b: 88. Steind 158; RIU 128; Onomasticon 1994: 340.

Its inscription, according to the Steindenkmäler (Steind) reads: *Aspaliae filiae caris/simae fideli ann(orum) XXV Gaudentius vet(eranus) ex p(rae) p(osito) si[l]/varum dom[i]nicarum / et Crescentia parentes / fideles vivi fec(erunt) et sibi.* The authors of the Steind date it to the 4th century, and, because of the names Gaudentius and Crescentia, they think it is Christian. The Onomasticon classified cognomen *Aspalia* as a Christian name.

35. PÖLÖSKE

35.II.a. Dagger-sheath, fig. 267.

Provenance: It was found in 1887 when the Szévíz (Seevíz) flowing into the Zala was controlled. There was also earthenware found together with it, Fig. 268. **Location:** LFM previous 53.311.1 and present Inv. No. 55.148.1. **Literature:** BELLA 1890: 362-364 fig. 85; BELLA 1891; NAGY L 1938: 140 fig. 85; KÁDÁR 1938-1939 fig. 20; KÁDÁR 193: 34; KÁDÁR 1942: 1-5; KÁDÁR 1969: 196; Severin Katalog 524.

Embossed metal plate; not gilded as described in the Severin catalogue or elsewhere. After restoration the brass evidently showed, but it has lost all its sheen since. Because of its shape, it is known as a dagger-sheath in the publications. L: 22.4 cm, W: 4.8 cm at the top, 2.2 cm at the bottom. On one side, the bronze plate is in one piece, on the upper part, the inscription VTERE / FELIC/ITER is in a beaded border, fig. 269. The rest is ornamented with a foliated scroll growing out of a cantharos with vine leaves, bunches of grapes and birds. The other side consists of two parts. On the upper part, there is an individual piece that seems to be a later application. On this, there is a male figure sitting on a throne, in front of him three standing male figures, all in a bead-bordered medallion, fig. 270. The latter figures are small compared to the sitting one and so the scholarly works interpreted the scene as 'Christ with the children', and considered it as the expression of triumphal art. This is why the sheath is regarded as an early Christian object. The sitting man is raising his right arm, his index finger and middle finger pointing out. His thumb folds over the little finger across his palm. This position of the hand was interpreted as blessing. This metal plate is thicker than the rest of the sheath. The piece under the substitution has three parts. In the middle, there are two parallel rows of pearls bordering a vine-tendril. This ornament is surrounded by unadorned stripes of plate, fig. 271. The sides consist of bronze plates decorated with rows of pearls, fig. 272.

Inside the bronze plates, remnants of wood were noticed, and even traces of iron were pointed out by the microscopic examination.[880]

The illustration on the substituted plate. The position of the sitting figure's hand and its gesture should be interpreted as a 'Redegestus'. A similar illustration is found on the casket-mount of ☞Császár, but there, above the scene, we see the chi-rho monogram and other biblical scenes clarify the early Christian references of the casket-mount. A similar illustration also exists on a glass medallion,[881] whose interpretation is debated. Cambi defined it as the scene of the 'Sermon on the Mount',[882] and declared it evidently Christian. These analogies suggest that such a metal plate was used for the reparation of the Pölöske sheath, and that its illustration may have obtained an early Christian interpretation in a Christian context, namely, the teaching or healing Jesus. However, the same scene on a dagger-sheath requires a different explanation, as its function does not account for an early Christian interpretation.[883] Based on the function, I would rather classify the scene as an *adlocutio*, or at least I would attribute the title of a *legatus*, an imperial representative to the sitting figure. It is true that the illustration of the *adlocutio* would require a standing *legatus*, but it is clear that the scene resembles it. This phenomenon supports the fact that the sheath needed repairing, and the craftsman thought this piece was the most suitable among the ones at his disposal. The gesture allows for this presumption. The elaboration and composition of the scene also remind us of the *tempora anni* illustrations of the pagan world.[884]

36. SÁGVÁR, figs. 273.

There are two well-separable sites, the fortification and the cemetery. Aladár Radnóti started the excavations and published his preliminary report in Archaeológiai Értesítő.[885] Alice Burger did the complete documentation of the opened graves in the cemetery.[886] The inner fortified settlement lies within the present village, the cemetery is farther out. Tömlöchegy, is east of the village; the Roman cemetery spreads on its south-east hillside. The excavations in the 1930s opened 259 graves, which were not located continuously one after the other; sometimes there was a distance of 300 m between the groups of graves.[887] The distance between the graves could not be explained by chronological differences as the excavator described the graves as contemporary. In Aladár Radnóti's opinion, the *cella memoriae* stood in the centre of the cemetery.[888] I would like to add that at best he could mean the centre of the known part of the cemetery, since the whole cemetery has not yet been opened. Further graves were excavated until 1942 and the total, number of the graves, including the more recent ones, is 342.[889] The verifying excavations

---

[880] I am grateful to Mrs. Nyári for this examination.

[881] CAMBI 1975: 54 fig. 4.

[882] I would rather think of the healing of the cripple, since above the twelve apostles there is a lying figure towards whom Jesus's hand is directed. This difference in the interpretation does not change the fact that the scene is really Christian. ☞Pécs, painted tomb I, frescos.

[883] In Roman art it is a common phenomenon that the same motif is used for the illustration of different persons and scenes. Later the illustration is explained in an inscription. In the present case the inscription is missing, so only the environment can serve as a clue.

[884] L. A. CASAL: Koiroi/Tempora anni. LIMC V. (1990) Kat.-No. 54.

[885] RADNÓTI 1939.

[886] BURGER 1966.

[887] RADNÓTI 1939: 151.

[888] RADNÓTI 1939: 152.

[889] BURGER 1966: 99.

dealt with the identification and publication of these graves. The correlation of the location of the cemetery, the settlement, and the today's village is taken from the documentation of the verifying excavations by Alice Burger,[890] cf. fig. 273.

Aladár Radnóti, overestimating the Christian finds, spoke of a cemetery rich in early Christian relics.[891] This idea was so widely adopted by scholars that later the whole cemetery was referred to as Christian. Alice Burger, the leader of the authenticating excavations, stated that there were graves in the cemetery without grave goods, and that they belonged to Christians and Jews. Ferenc Fülep agreed with this opinion to the extent that he spoke simply of a Christian cemetery: 'In the centre of the Early Christian necropolis ...'[892] The fact that Christians lived at Ságvár is indisputable; it is proved by the real Christian finds. But the presumption that the whole cemetery should be Christian is controversial. Even the buildings with apses cannot automatically be considered as Christian The apse, as an architectural form, was not a Christian invention, they simply became popular among these populations.[893] Alice Burger's observation that there were Jewish burials in the Ságvár cemetery is notable. In her opinion, mainly the west-east-oriented brick graves, unadorned sarcophagi, and graves without goods belong to the Jewish population.[894]

Burger defined the dating of the cemetery based on the coins found there: 'According to the coin finds, the early Christian-type cemetery of Ságvár was used between AD 340-375. Since the above discussed Jewish graves lack coin finds, they are certainly from an earlier time, but not sooner than the beginning of the 4th century.'[895]

The excavations of the fortification (fig. 274) were started by Endre Tóth in 1971, and were continued by him in the 1970s. He documented the working process regularly in the Archaeologische Forschungen column in Archaeológiai Értesítő,[896] as well as in the Régészeti Füzetek.[897] He also wrote a preliminary report,[898] a short summary of which is: The forts of this type make an inner chain of fortifications within the province - Ságvár, Kapospula-Alsóheténypuszta, Környe, Mórichida-Kisárpás, Fenékpuszta. The Ságvár fortification is located at the deepest point of the area, so it is only noticeable from the top of the nearest hills. Only one way leads to it, through the north-south valley.

Therefore, Endre Tóth thinks the area has the features of a refugium. Only the basements have remained from the inner buildings. The towers are round. However, there used to be an earlier period, too, whose traces may be observed at the northern gate, in the square towers; their dimensions being 6 x 6 m, their walls 150 cm thick. This period belongs to the tetrarchy. The inner buildings were erected in the second period. The fortification was built upon an early-imperial-period settlement. The settlement contemporary with the fortification lay north of it along the road leading out. In this territory he found grave-aediculas, pottery and coins datable to the 4th century.

As for function, Endre Tóth disclosed that Sándor Soproni's view of a military purpose should be reconciled with András Mócsy's point about the centre of a latifundium. The two theories are complementary. ☞ Keszthely-Fenékpuszta, history of research.

There are several points to be made concerning the excavations. Despite the fact that the present village was built on most of the fortifications, the line of its wall has been defined on the north, west and south. The system of side towers has also been marked. The extent of the fortification is 250 x 300 m, the wall is 200 - 220 cm thick. The side towers that joined the walls are 9 m in diameter inside. At the northeast corner, fragments of frescos, the remnants of floor and wall heating, as well as fragments of a mosaic floor have been unearthed. The mosaic was unicolour.[899] At the excavation of the southern gate, it turned out that a tower, which was square inside, round outside, used to protect the gate. The first period of the fortification was characterised by horseshoe-shaped towers, whereas the round towers were built in the second period. In one of the towers there was a refuse pit 90 cm in diameter. The pit contained a Christian casket mount with inscription, a 'white marble altar slab', a silver-plated capsule, a bronze lamp, an iron patera, and a bronze pan.[900] Endre Tóth considers these finds to belong to an early Christian basilica.[901] The excavations unearthed settlement material from the 1st and 2nd centuries.[902] In the second period of the fortification, a row of houses was joined to the wall of the fort, but it was demolished later. A dwelling-pit from the 5th century was also found.[903] In the second period of the wall, the first floor in that tower was supported by a wooden construction, which later burnt down. Consequently, there was a 50 cm thick levelling layer to be observed.[904] In the northeast section, two large buildings were found. One of them was 20 x 30 m, stone, decorated with a fresco, and furnished with a terrazzo floor. It could be heated throughout. Early imperial-period phenomena were also in evidence in this house. The other building is 30 x 60 m. Its walls are thick; the terrazzo is poor quality. It was heated.[905] In the same section, under the large farm-buildings, 'the painted adobe walls of an earlier Roman building'[906] were found in certain places. Six pillars of the horreum's western pillar row were

[890] BURGER 1966: 182. Fig. 2.
[891] RADNÓTI 1939: 156-157.
[892] FÜLEP 1980: 47.
[893] DEICHMANN 1954: 947-948.
[894] BURGER 1966: 162.
[895] BURGER 1966: 163.
[896] TÓTH E ArchÉrt 1972: 258-259; ArchÉrt 1973: 265; ArchÉrt 1974: 317.
[897] TÓTH E RégFüz Ser. I. No. 25 1971 77; No. 26 1972 79; No. 27 1973 70; No. 28 1974 78; No. 29 1975 70; No. 30 1976 57; No. 32 1978 72.
[898] TÓTH E 1975b.
[899] TÓTH E RégFüz Ser I. No. 25 1971: 77.
[900] TÓTH E 1975b 186.
[901] TÓTH E 1975b 187.
[902] TÓTH E RégFüz Ser I No. 26 1972 79.
[903] TÓTH E RégFüz No. 27 1973 70 p. 45.
[904] TÓTH E RégFüz No. 28 1974 78 p. 60.
[905] TÓTH E RégFüz No. 29 1975 70 p. 42.
[906] TÓTH E RégFüz No. 32 1978 80 p. 53

opened. The stone foundations of the pillars were made of red ashlars. A long corridor joined the eastern wall of the horreum. The horreum suffered from the Sarmatian invasion in 374-375. A two-room building was found under the horreum. Its walls are 120 cm thick in the basement, the building's size is 21 x 25-30 m. East of the horreum, next to the north-south highway, the excavators found 'a basilica divided into three naves by pillars'. Its inner width is 17.3 m, L: 45 m. Its walling is *opus incertum*. Pillars also divide the walls. 'On the western front of the building a grave was found containing a female skeleton. Bakers' ovens were built onto the north-western corner of the basilica.' When these were built, 'the wall by the corner of the basilica was partially destroyed.'[907]

As regards the dating of the construction, or alteration of such fortified settlements, Endre Tóth takes the following position: 'Der Umbau fast aller Militärlager am Limes in Pannonia in hufeisen- und fächerförmigen Türmen muss ... an die Zeit von Constantius II, möglicherweise an die Mitte des 4. Jh. datiert werden. Die drei Befestigungen in Valeria wurden gleichzeitig unter der Regierung von Constantius II. errichtet. Die inneren Festungen wurden nach 375. aller Wahrscheinlichkeit nach infolge der Verwüstungen des quadisch-sarmatischen Krieges und dessen Erfahrungen, umgebaut.'[908]

A rectangular mausoleum was also found at Ságvár, as Endre Tóth writes when classifying 'cult' buildings. Such buildings were also found at Pécs, Kapospula, Szentendre and Óbuda.[909]

Commentary: In connection with the fort, I am obliged to give my opinions in a restrained way; the fact that the excavator possesses a large amount of unpublished material, wholly unknown to me, causes an uneasy feeling. So, when I disclose my thoughts on the Ságvár and Alsóheténypuszta fortifications, I feel as if I were bound up in Plato's cave, expressing my opinions about shadows only.

We know only about the 4th-century sections of the Tömlöchegy cemetery, although the settlement had an early period, too. However, there is no report of the site of the early settlement's cemetery. The early settlement is hardly mentioned, and even then only to the extent that is needed for the description of the later periods. On the other hand, Radnóti really expected the further excavations to provide evidence of continuity.[910] He also writes that the coin finds begin with Constantine's dynasty and last until Gratianus. Further questions include:

- What was the extent of the Roman settlement, and what was its legal status?
- Why did a later settlement exist outside the walls in parallel with the fortification, while the fortification was built upon the previous settlement?
- Why were there three horrea next to each other, or, more exactly, were all three buildings horrea?
- Why did they build the bakers' ovens in the side of the basilica? How is it certain that the building called a basilica, really was a basilica?

My ultimate interpretations of Ságvár, formed through the shadows of my cave, are as follows: As Endre Tóth speaks about an early-imperial-period settlement, we have to suppose that those who migrated here in the early imperial period also settled down here, and lived together with the autochthonous population, who had been Romanised. We may search for Jewish people among these early settlers, who were pleased to find shelter in the valley for some reason. Besides these Jews, there were Christians as well; some of them had been converted from the pagans, some from amongst the Jews. We know nothing of the correlation between the early and late settlements - if there was a correlation at all. We cannot decide whether these Christians arrived as Christians, or whether they adopted their faith here. Only one thing is certain: the existence of Christians. Thus, we may try to find a house-church and, allowing for a Jewish population, we may also suspects a synagogue.

The early dwellers may have had burials along the road for a certain time, but we must seek a main cemetery on Tömlöchegy, where the cemetery must have a period. The late cemetery was used by the dwellers of the fortification regardless of their religious status, or any other factor. At most, the cemetery may be divided into sections based on religion. ☞Pécs, historical summary.

36.I.a. Casket-mount, fig. 275.

Provenance: It was found in the pit of the tower belonging to the southern wall of the fortified settlement; excavation 1973. E. Tóth. **Location:** MNM 73.1.195-196. **Literature:** THOMAS 1982b: 279, Abb. 23; GÁSPÁR 1986 Kat.-Nr. 924; THOMAS 1986a: 529-530; THOMAS 1988: 291 Abb. 11; TÓTH E 1989a: 218-220; TÓTH E 1995.

Two pieces of bronze sheeting with identical illustrations. The size of one is 23.8 x 9.4 cm. It depicts male figures, the two chief apostles and Timotheus, in a pearl-bordered metope. Inscription: PET/RVS TIMOT/HEVS PAVLVS. The Timotheus reference is the focus of the Pannonian research. In her comments on the question, Edit Thomas said that Timotheus was venerated in Ságvár's Christian basilica, or he was the patron saint of the basilica.[911] In one of his articles, Endre Tóth writes of the 'Timothy-feast' depicted on the casket-mount: 'Eine marginale Perikopenbemerkung (about the 'q' gospel of Munich) deutet auf das Fest des Timotheus und der sieben Jungfrau hin, das dem Martyrologium Hieronymianum zu folge am 15. Mai in Sirmium begangen wird.'[912] Endre Tóth's recent publication gives a detailed account of Timothy, who, he considers, was a martyr of Sirmium, and was venerated in the province.[913]

---

907 TÓTH E RégFüz No. 32 1979 80 p. 53

908 TÓTH E 1985: 130; TÓTH E 1988b: 32.

909 TÓTH E 1991a: 123.

910 RADNÓTI 1939.

911 THOMAS 1986a: 530.

912 TÓTH E 1989a: 218. Martirologium Hieronymianum writes the following on 15 May (p. 256): Timotheus Antiochenus est et virgines VII Sirmienses non sunt. Quam ob causam utrumque huc translatus sit ignoramus.

913 TÓTH E 1994-1995.

36.I.b. Casket-mount, figs. 276, 277.
Provenance: grave 54. **Location:** MNM 11.1938.1. **Literature:** KÁDÁR: Zur Frage der römerzeitlichen ägyptischen Elementen in der altchristlichen Ikonographie. Atti 1969: 578. Abb. 10; GÁSPÁR 1986 Kat.-Nr. 913.

Bronze. Fragmentary, 4.1 x 4.7 cm. The raising of Lazarus is depicted in a pearl-bordered metope. In the top right corner a chi-rho monogram, in the left one a star. Christ's figure is tall in comparison to Lazarus' grave. He holds a staff in his left hand, touching the grave with it. His bent right hand holds a pleat of his clothes, which may be identified with the pleat of the draped *pallium*. The depiction of this pleat is not very successful. Evidently, the craftsman, who embossed the sheet, misunderstood the magic staff. A similar composition may be seen on a fondo d'oro, see fig. 83.[914]

36.I.c. Fibula, fig. 278.
Provenance: It was found at Tömlöchegy from grave 42 during the excavations of 1937. I quote the description of the grave by Burger: '[Depth] 150 cm. Length: 171 cm. Female. Grave goods: 1. 'Silver ring' by right hand. Made of a flat band, which is wider at one part. In this section there are three bronze knobs welded to the ring. Diam: 1.9 cm, MNM 10.1937.5. Lost. It may be identified only on the basis of old photographs. 2. 'Silver ring'. On last vertebra. Half-round circle. The empty mounting is made of twisted silver wire with two buttons on both sides. Diam: 2.2 cm. MNM 10.1937.6. Lost. 3. 'Gilt fibula'.'[915]**Location:** MNM 10.1937.4 (At Burger under the same inventory number as the lost ring.). **Literature:** RADNÓTI 1939: 158-159 Abb. 128; KOVRIG 1937: 94 Taf. 33/1; BURGER 1966: 103 fig. 97, 42/3 pl. XCI, 2.

Bronze, crossbow fibula with three knobs. Pelta-shaped open-work on the pin holder. In the middle of the pin holder and on the spine of the stirrup there are niello leaves and a silver inlay. At the end of the pin holder there is a chi-rho monogram in silver inlay. Gilded. H: 7.6 cm, W: 5.5 cm.

36.II.a. So-called cemetery chapel with two apses, fig. 279.
Literature: RADNÓTI 1939: 151-153; BURGER 1966: 154 fig. 90; FÜLEP 1980: 47; TÓTH E 1990a fig. 8.

Dimensions: 'Its north-south oriented nave is 9.75 m long and 5.75 m wide. The entrance was probably in the south; the northern closure is linear. A semicircular apse is joined to both of its longer sides symmetrically (their radius is 3.4 m). Only the basement of the building has remained to us, and not completely. The thickness of the walls is 75 cm on average; their foundations are 95 cm. A part of the walls had been removed completely; they had to be reconstructed from their debris. The floor has remained in some areas only, it was roughly ground terrazzo with a thick layer of crushed brick below.'[916] Figs 280a, b.

Three graves were found in the building in their original place, but they were completely plundered. 'In the western apse, deep below the pavement, there was a brick-walled grave with a large sarcophagus slab.' Fig. 280c. In the eastern apse, however, 'the departed was laid to rest in a tomb composed of red sandstone and bricks.'[917] Fig. 280d.

In this two-apse cell nothing was unearthed that could have early Christian references; neither the carving of the sarcophagi, nor any other mark suggests such a relationship.

36.II.b-c. Cellae II and III, figs. 281-282.
Literature: RADNÓTI 1939: 155; BURGER 1966: 154-155 pls. LXXX.

'West of the burial chapel we unearthed the basement of two smaller cellae, we only managed to fix the last row of stones of the basement. They are simple buildings with a semicircular closing; one of them has a small hall. Despite the fact that we could hardly save anything from the walls, we found undisturbed graves in both of them.'[918] Alice Burger discloses the size of cella II: 5.5 x 6.75 m outside, the thickness of the walls is 50 cm. The entrance was in the south; its width is 1.4 m. Grave No. 192 is the undisturbed one mentioned by Radnóti.[919] Its size: 85 x 48 x 195 cm. Male. Goods: earthenware, an iron tool. The coins were lost.[920]

The dimensions of cella III have not remained, its apse is intact. In its middle, the grave was discernible, but its goods could not be defined.[921]

36.II.d. Buckle, figs. 283a, b
Provenance: Grave No. 127, male. **Location:** MNM 9.1939.5. **Literature:** BURGER 1966: 111, figs. 102/127, pl. XCI/4. THOMAS 1982b Abb. 24.

Bronze. Size: 6.4 x 5.7 cm. Its upper plate is ornamented. Above the waist, the figure of a man with outstretched arms, below the waist the shape of a fish. The incision moved a little, so its upper and lower part does not join perfectly. The buckle was classified as Christian because of the half-man half-fish body in the illustration, although we mostly come across such creatures in pagan tales. Its Christian nature is rather questionable.

36.II.e. Brick with incised pattern, figs. 284a, b
Provenance: Grave No. 203, male. **Location:** MNM 9.1939.29. **Literature:** BURGER 1966: 120, fig. 91.

There is an incision on one of the bricks in the grave. It was thought to be an illustration depicting the souls ascending to heaven. Brick size: 39 x 28.5 x 6.5. In the picture, there are two human figures, a man and a woman. The figures were shown with a wavy, undulating line. The female figure was drawn regularly, while the man was depicted by the reverse of the usual lines, as if expressing the negative sphere. The illustration demonstrates the polarity of male and female, and consequently has nothing to do with Christianity. An example of the same polarity in Pannonia, from the same cemetery, is the buckle in grave No. 177,[922] fig. 285. The plate of the buckle is square. The portraits of a man and a

[914] TESTINI 1966 Tav. II, Museo Sacro Vaticano. Inscription: ΣESVS CRISTVS.
[915] BURGER 1966: 103-104.
[916] RADNÓTI 1939: 152-153.
[917] RADNÓTI 1939: 155.
[918] RADNÓTI 1939: 155.
[919] BURGER 1966: 155.
[920] BURGER 1966: 119.
[921] BURGER 1966: 155.
[922] BURGER 1966: 117, fig. 107, pl. XCI/5.Location: MNM 9.1939.27.

woman in a bead border. Between the two portraits, there is a tree of life. Another parallel is a silver ring from grave No. 7 in the Keszthely-Fenékpuszta cemetery. On its quadrilateral head, there are two schematic human figures; the only difference between them is made by the marked sexual characteristics,[923] fig. 286. The classification of the brick as a Christian object is due to the misinterpretation of the scene. Because of such illustrations, Rudolf Noll's warning is very important, namely, that we cannot examine early Christianity without the religious mentality typical of the given area and period.[924] I have to emphasise that Tibor Nagy began his monograph on Pannonian Christianity by studying the antecedents.[925]

The significance of the fact that the illustrations highlight polarity can be a further aspect of our arguments. Although the statement that everything can only exist with its opposite is well known from Greek philosophy, still this polarity, which is observable in several graves from Pannonian cemeteries, so much emphasises thinking, that we must raise the hypothetical possibility that these were the graves of non-Christian Jews.[926]

36.II.f. Brick with incised pattern, fig. 287a, b

Provenance: Grave No. 203, male. The brick comes from the roof of the grave. **Location:** MNM 9.1939.28. **Literature:** BURGER 1966: 120 fig. 92 pl. LXXXV, 4; Severin Katalog 8.48.

The size of the brick is 48 x 33 x 10 cm. Alice Burger published it as early Christian because in the incision he discerned a chi-rho monogram. She interpreted the rest of the incision as the name of the dead person.

The previously discussed brick has already directed our attention to the fact that, in interpreting this cemetery, we have to reckon with a belief, or at least a 'philosophical' idea, which is not Christian, although in certain aspects it may very well not contradict the Christian sphere of thought. The brick's inscription might refer to the mysteries, a map allowing the dead to find the right way in the other world.

The brick is not edged, nor its surface smoothed, so the unevenness makes it difficult to interpret the pattern; there are cavities and lines, which were not made deliberately, just because of the roughness of the surface. The pattern was made with the finger. The order of the lines was well determinable; one could see which line crossed the other. If we hold the brick by its profiled thinner side, we see the following order of lines: 1. A line starts from bottom on the left, it proceeds upwards, and at the top, it returns. 2. At the bottom, from the midpoint, a line starts, continuing to the right. It breaks at the tierce point, and continues upwards in a curve. 3. A horizontal line in the middle, from left to right, crosses the first and second lines. 4. An approximately linear line that starts at the bottom, in the middle, and crosses the second line down. 5. A curve was drawn on this fourth line at the top, but not exactly at its end. This was interpreted as letter rho. 6. Above curve *5*, a parallel line was drawn in a section. 7. Between the two vertical lines, there is a shorter one, which bends to the left at the bottom; it crosses the third line. 8. It crosses line *7* at the top. 9. Small patterns, not touching any of the lines, for instance above line *3*. The next pattern is above lines *4* and *5*. These patterns are encircled by an oval line that does not close in the middle on the left. There is a line on the brick, which was caused by damages, and does not belong to the pattern. I do not find the interpretation of these lines as a chi-rho monogram a probable one, although the solution is still disputable at the moment. I would also like to note that in the grave the brick with the so-called chi-rho monogram was upside down. This fact also seems to reinforce the idea that in this case it is not a chi-rho monogram.

We know of a chi-rho illustration on a brick from another part of Pannonia, its find-spot is Daruvár,[927] fig. 288.

37. SÁRISÁP,[928] fig. 289.

37.II.a. So-called cemetery chapel

Two sites, Nos. 17 and 18, are significant in connection with the Roman research. At site 17, the traces of a Roman settlement were observed. Site 18 is on the southern side of the hill and here Albin Balogh opened an early Christian grave-chapel in 1926. The available finds were mostly pieces of frescos. Further finds: pottery sherd painted brownish red, a glass ring, two vessels and an iron knife. Albin Balogh briefly disclosed the excavation. The ground plan was also made by him,[929] fig. 290. 'There was a grave in the middle of the apse, barely forming a mound above the ground; it was made of tiles. It was painted inside, and vaulted at the top. The grave was empty, no bone was found in it. No Roman graves have been found around the chapel yet.'[930] The frescos show simple floral motifs, in other parts red, blue, yellow and green patterns; the inside of the grave is dark grey and black.... There are

---

[923] BARKÓCZI 1968: 279-280 pl. LXII/2.

[924] NOLL 1954: 4-16.

[925] NAGY T 1939: 5-30. Zoltán Kádár also wrote about the preliminaries - KÁDÁR 1972.

[926] On Jewish mentality and way of thinking, see BENYIK 1992: 9.

[927] A brick whose find-spot was identified based on the old Hungarian scholarly works. István Gyárfás wrote, referring to Ljubić that it could be seen in Zagreb. Literature: GYÁRFÁS 1889: 97-98; NAGY L 1938: 105 Abb. 67; MIGOTTI 1997 III.d.1. Katalog 1998: 7. When he defined the find-spot, Lajos Nagy wrote that it was Daruvár/Darova. This modern definition is incorrect, because Daruvár, whose modern name is Darova, is in the territory of Dacia; but since this piece could be seen in Zagreb in the 19th century, we had better presume that here Lajos Nagy was wrong when he defined the location and it is Daruvár between the Drave and Save Rivers. Albin Balogh knows about two bricks, which are preserved in the Zagreb museum. (1927: 198. These are: MIGOTTI 1997 III.c.1, III.d.1).

[928] The possibility of identifying it with Quadriburgium was also raised, but according to TIR L34 p. 94: 'Pann.Inf. Not. Dig occ. 33, 60. Unidentifierbarer Lagerplatz in der Provinz Valeria, ...' It is analysed not as a place, but as a type of building by MITROVA-DŽONOVA D: Das frühbyzantinisches castrum Stenos an der 'via a Singiduno per Hadrianopolim usque ad Constantinopolim' am Succi-Pass (Claustro: Stenos). Atti 1998 III 531-540. There is a reference to Pannonia on page 537 Note 5.

[929] MRT 5, 19/18, fig. 51.

[930] MRT 5, 19/18 p. 308.

subsequent incisions on many fresco-fragments. ... The pieces of very fine marble were found in the apse, which are ... the fragments of a *mensa*. The chapel used to have a terrazzo-floor, ... the thick layer of ash observed on the floor makes us conclude that the building was ruined in a fire.'[931] Albin Balogh disclosed its supposed early Christian references. He then defined the fragments of marble as the pieces of a *mensa*. The pattern of the frescos is possible in any Roman dwelling.[932]

Although it is known as an early-Christian cemetery chapel by the research, this fact is not verified by the finds. There was hardly enough space for a *mensa* next to the grave marked on the ground plan, and so I do not consider the marble fragments as pieces of a *mensa*. As the building was not erected in a cemetery, we have to suppose that it was a separate burial; its parallel is found at ☞ Kővágószőlős. Thus, the building belongs to the nearby villa. There is no evidence of Christian references. The ground plan itself is not an argument since the apse had also been used earlier. About the Diópuszta 'basilica' mentioned as an analogy, the verifying excavation proved that it was not a basilica, ☞Ugod-Diópuszta, 59.II.a. This fact reinforces the suspicion that the building at Sárisáp was not early Christian.

38. SÁRSZENTLŐRINC

38.I.a. Finger ring

Location: MNM 47.1906.55 = 91.10.16 purchase. **Literature:** unpublished.

Bronze ring with a round and flat head. There is a chi-rho monogram on the head, which is hardly discernibly encircled. It is not intact. Diam: 1.8 cm.

39. SOKLOS

39.I.a. Tombstone

In the CIL, its provenance is reported as follows: 'in oppido Soklos'. This place was identified with Siklós actually, according to its sounding, instead of certain data. If this identification was acceptable, the inscription would be included in RIU 4, but it is not there. **Location:** demolished. **Literature:** CIL III 3688; BT 1942: 316 (ALFÖLDI); RADAN 1973: 273; SCHEIBER 1983: 42-43.

I quote the inscription according to the text in the CIL: *d m /Septimae Mariae/ Iudeae quae vixit / annis XVII Actia / Sabinilla mater*. The dead woman is evidently Jewish.

40. SOMLÓVÁSÁRHELY

40.I.a. Fibula, figs. 291a, b, c.

Literature: TÓTH E 1994a figs. 15 and 17/A. It is not mentioned in the text, only its photograph and drawing are published.

Its type is B4a - this is the Sofia-group, which indicates the oriental occurrence of the objects. The characteristics of this type are its big size and trapezoidal cross-section.[933] In the drawing and photograph it seems as if there were depictions of the chi-rho monogram on the pin holder and stirrup.

41. SOMOGYSZIL

Pottery with incised pattern and smoothing

Location: RRM. **Literature:** BURGER 1979; Severin, Katalog 8.70-8.73.

Alice Burger excavated a late Roman cemetery on the site. On some of the pottery unearthed here there are engravings illustrating different patterns. Alice Burger, who was the excavating and publishing scholar, did not mention these engravings in her writing. Edit Thomas, however, spoke of these incisions, considering them as Christian motifs in some of her lectures. She thought the majority of these motifs were crosses. As a result of her lectures, the Christian reference of these objects became widely known. I regard the identification of these engravings with crosses or chi-rho monograms, as stated by Edit Thomas, rather far-fetched.

41.II.a. Grave 126, Inv. No. 68.49.2.[934] Fig. 292.

41.II.b. Grave 123, Inv. No. 68.46.2.[935] On this object, ligature of a rho and a cross is also conceivable, fig. 293.

41.II.c. Grave 86, Inv. No. 68.14.1.[936] Figs. 294a, b.

41.II.d. Grave 84, Inv. No. 68.13.17.[937] The pattery is partly smoothed, partly incised. Fig. 295.

41.II.e. Grave 80, Inv. No. 68.9.2.[938] This might be interpreted as a chi-rho monogram; the question is which meaning is acceptable. Fig. 296.

41.II.f. Grave 54, Inv. No. 66.100.4.[939] Fig. 297.

41.II.g. Grave 41, Inv. No. 66.91.1.[940] The drawing is smoothed. It is not the ligature of X and P, but only that of X and I, i.e. i-chi, which was not used only by the Christians. For example, this motif is the decoration on the altar of Esztergom; between the acroteria on the front capital. The altar was consecrated to *Silvanus Domesticus: [Silv]ano / Dom(estico) / Liciniu[s] / V S L M.*[941] As there are no further Christian objects in the cemetery, the interpretation of the illustration as Christian seems to be a premature conclusion, although the fact that the smoothing in was deliberate is evident. Fig. 298a, b.

41.II.h. Grave 41. Inv. No. 66.91.3.[942] Fig. 299.

41.II.i. Grave 40, Inv. No. 66.90.3.[943] There is an illustration on its side, which might be a chi-rho monogram. There is another illustration inside at the bottom. We should actually ask the question what these signs might mean. Besides the chi-rho monogram of grave 41, no other mark is so clear and evident that we might suspect a religious content behind it. They are rather stamped signatures. Fig. 300.

931 MRT 5, 19/18: 309.
932 BALOGH 1934: 50-51.
933 TÓTH E 1994a: 150.

934 BURGER 1979 Taf. 21.
935 BURGER 1979 Taf. 21.
936 BURGER 1979 Taf. 17.
937 BURGER 1979 Taf. 16.
938 BURGER 1979 Taf. 15.
939 BURGER 1979 Taf. 31/6; Severin, Katalog 8.73.
940 BURGER 1979 Taf. 9/1; Severin, Katalog 8.72.
941 MRT 5 8/4, 66 Table 2.
942 BURGER 1979 Taf. 9/3.
943 BURGER 1979 Taf. 9.

41.II.j. Grave 33, Inv. No. 66.84.5.[944] Fig. 301.
41.II.k. Grave 6, Inv. No. 66.62.4.[945] Fig. 302.
Besides the smoothed in depiction, it is difficult to suppose a symbolic interpretation of the incisions. If they are deliberate marks at all, they were perhaps put there when the vessel was made, and they served practical purposes. Furthermore, the marks do not allow us to decipher their message. The vessels betray a uniform making, which may mean that the function of the marks is also uniform. Let me cite the example of the incisions of the vessels from the sigillate deposit found at Tác. In the deposit there are early sigillates, the latest piece is from AD 250.[946] None considered that the objects or incisions might have been Christian, although the superficial observer, or hasty interpreter, might have found 'suspicious' patterns.[947] Fig. 303. Examining the Tokod sigillates and their incisions, Dénes Gabler came to the conclusion that the incisions on the vessels were made during the 1st century, or in the first decade of the 2nd century at the latest.[948] Among the incisions, we may discern X-es, crosses and even ligatures called *i-chi*.[949] Their interpretation is not easy, however. They represent such a specialized genre of epigraphy that is a rather neglected field, as Dénes Gabler complained when examining the Tokod sigillates. The signs might stand for numbers, and thus cubic measure, or they might give the name of the owner, etc.[950] With this knowledge, I dare say even more that the incisions on the Somogyszil potteries are not Christian.[951]

42. SOPRON (SCARBANTIA) ☞ Szombathely

Although archaeological research has continued for decades in the town, early Christian finds, or at least supposedly early Christian finds, have hardly ever been found.[952] The excavations have determined Sopron's city wall and forum; they have unearthed evidence of its religious life.[953] We know that there was a kiln near the city. A settlement of potters has also been reported.[954] Thanks to János Gömöri's efforts, we have quite a good knowledge of Sopron's early Roman topography and the condition of the city. Despite this, János Gömöri only mentions Christian finds later. The exact site may be seen on his excavation plan,[955] fig. 304: 1: the pavement of the forum. 2a: base of an equestrian statue. 3: conduit. 4: pillars, 150 cm wide. 5: footing level with brick fragments. 6: wall. 7: column base. 8: doorway with stairs. A: a room in the south, where the doorway marked *8* leads. B: room west of the yard. 9: wall, divides *B*. Heating-flues with a coin of Constans. C: room east of the yard, with a subsequently built hypocaustrum. 10: collecting-box in the hypocaustrum. 11: roof-tile to cover the stoke-hole. 12: the southern wall of the room. 13: earlier hypocaustum-remnants in *A*. 14: houses. 15: (no description). 16: well. 17-18: the southern and northern walls of the same building. It has three building periods. Its floor is one metre higher than that of the forum. Inside plaster painted red. 19: the western side of the forum, western conduit-section. 20: inn. 21: the western wall of a building on the eastern side of the forum; eastern conduit-section. Thickness of the wall 90 cm. 22: (no description). 23: Capitol. 24: road, 480 cm wide, decumanus. X: exedra. Y: exedra. P-Q axis: crosses the Capitol. 25: building in a ruined area, 3rd century, the find-spot of the Minerva-statue. 26: intersection of cardo-decumanus (today's Storno House). 27: eastern road crosses the cardo.

The size of the forum is 45 x 45 m. It was paved with stone slabs. A gutter runs around the forum for rainwater. The Capitoline shrine stands in the north; symmetrically with it there is a Hercules shrine. There is a colonnade to the south; its roof was tiled with tegulas and imbrexes. From here stairs lead to the forum. On the western side, there is a building erected on a podium; it was probably a basilica. The forum was approached from the south. At the end of the road there was a shrine to Silvanus. At the crossing of the roads, an altar was found built for Isis and Bubastis.

At *25*, not only a Minerva-statue was unearthed, but also a monumental white marble statue, perhaps from the Capitol. On the walls of room *C*, a newer house was built in the 5th century, which was ruined a century later. In the demolished layer of the building, a Lombard 'S-fibula' was found. The Lombards appeared in Pannonia in the autumn of 526, and they conquered any suitable settlements. The brick with the incised figure, to be described later, was found in this house.[956]

42.II.a. Fragment of a jug, fig. 305.

Provenance: 'Die früheren Heizpfeiler aus Krosbacher Steinblöcken hatten regelmäßigen, rechteckigen Querschnitt, und ihre Tiefe entsprach dem des Forumsniveaus. Die Pfeiler des späteren Heizsystems bestanden aus Lehm zusammengeklebten Ziegelbruchstücken unter denen wir Krugfragmente mit eingekratzter Inschrift TEO fanden.'[957] It was found in room *A* on the forum.[958] **Location:** LFM. **Literature:** GÖMÖRI 1985: 21; GÖMÖRI 1986 Abb. 15.

944 BURGER 1979 Taf. 6/5; Severin, Katalog 8.71.
945 BURGER 1979 Taf. 2/3; Severin, Katalog 8.70.
946 GABLER/KOCZTUR 1976.
947 GABLER/KOCZTUR 1976: 88. Taf. VII.
948 GABLER 1967: 61.
949 GABLER 1967 fig. 1/3, 5. fig. 2/13, 15.
950 The interpretation and solution of the inscriptions were studied by János HARMATTA. Pannoniai edény-feliratok. AntTan XIV 1967: 67-101; Inscriptions on Pottery from Pannonia. ActaArchHung XX 1968: 247-274 pls XLIX-L. In these articles J. Harmatta studied the inscriptions on the sigillatae published by Gabler (1967).
951 By this I do not mean to exclude the possibility of such a cross on the vessels that can have a Christian background. Cf. TÖRÖK L: Abdallah Nirqi 1964. Finds with Inscriptions. ActaArchHung XXVI 1974: 367-393. The inscriptions were put on these vessels before baking.
952 GÖMÖRI 1986 Abb. 15 and Abb. 21.
953 BELLA 1911; RADNÓTI 1953: 23-30; ALFÖLDI 1959; HOLL/NOVÁKI/PÓCZY 1962; NOVÁKI 1964: 240; PÓCZY 1967; PÓCZY 1971b; PÓCZY 1976a: 24; PÓCZY 1976b; PÓCZY 1977; PÓCZY 1980; GÖMÖRI *RégFüz* Ser. 1. No. 35. 1982 93/2; No. 36 1983 81/3; No. 41 1988 74/1.
954 GÖMÖRI 1992: 5.
955 GÖMÖRI 1986 Abb. 1. I have prepared my explanations to the map outlining Gömöri's description: GÖMÖRI 1985; GÖMÖRI 1986.
956 The whole passage is the outlining of GÖMÖRI 1985.
957 GÖMÖRI 1986: 350.
958 GÖMÖRI 1985: 11.

A fragment with part of an inscription: TEO. On the left edge of the vessel there is the beginning of a vertical hasta of a letter. The inscription should also be considered as a fragment. There are also fragments of letters in the second line, which should perhaps be read as XIT. It is hardly probable that we should regard the first letters of the first line as the Latin form of the Greek Θεοζ. Considering it as Christian is expressly disputable. No other patterns or illustrations have been found on the fragment.

42.II.b. Brick fragment, figs. 306a, b.
Provenance: the house built above building C, the 6th deterioration layer. **Location:** LFM. **Literature:** GÖMÖRI 1985: 19 Abb. 21; GÖMÖRI 1986: 353 Abb. 21.

The incised brick is 46 x 31 cm. János Gömöri speaks about a haloed orans figure, which is a Christian relic. The height of the figure is 19 cm. He considers it proof of the fact that the arriving Lombards dissolved the Arian Christianity they found here, which again strengthens Gömöri's belief that Episcopus Vigilius's seat was Sopron. He participated in the Council of Grado between 572-579, but 'in 568 he left for Italy with the Lombards.'[959] Gömöri treats the presence of the bishop as a fact, and thus considers the presence of Christians verified.[960]

Commentary:

- The description of the halo resulted from a misinterpretation of the hair.[961] We simply have an orans figure.
- As for its date, the fact that it was found in a building burnt down in the 6th century is not a determinative condition, especially not if the building had previous periods as well. The brick may not have been produced in the 6th century. There is an incision on the ☞ Kisdorog roof tile, but regarding the method of illustration there is a more similar example from Bátaszék.[962] We have no basis for defining the figure on the brick as early Christian. The 'orans' illustrations themselves are not proof of early Christianity; they are too ancient to be interpreted as exclusively Christian relics. Let me quote Hans Liniger: 'Adorant (Planet) zeigt steigende Sonne an.'[963] The 'eternal' feature of the orans motif in Pannonia and Hungary was dealt with by Győző János Szabó.[964]

42.II.c. Fragment of marble slab defined as an altar table, fig. 307.
Provenance: It was found in the southern part of the forum, in the southeast corner of the exedra, in a period earlier than the 6th century. **Location:** LFM. **Literature:** GÖMÖRI 1985: 21; GÖMÖRI 1986: 355. No illustration is published. Hence, my special thanks are due to Dr Gömöri and Dr Gabrielli for permission to reproduce the photograph.

I quote Gömöri's description: 'We found a fragment similar to the Fehéregyház marble altar table in the southern part of Scarbantia's forum, in the SE corner of the exedra, a few metres away from the house with the incised brick. The two fragments of the red marble *mensa* and the white marble pillar of the monopodium are one period earlier than the 6th-century building with the incised brick.' For the Donnerskirchen parallel ☞ Csopak.

The unsatisfactory description does not allow proper evaluation. The classification as altar table seems suspicious. Those slabs that really are altar tables - in Pannonian relations - were all made of white marble. What is more, a question has emerged whether this marble fragment is Roman at all. There is no decision at the moment.

According to the archaeological evidence and observation, we cannot point out early Christianity at Sopron.

Written sources. Quirinus writes of Scarbantia: *'Cumque deductus fuisset beatus Quirinus ad primam Pannoniam, et per singulas civitates vinctus catenis, ad Praesidis Amantii judicium traheretur, si quidem ad ripam Danubii ad singulas civitates (ducebatur), Amanto eodem die revertente de civitate Scarabetensi, offertur ei beatus Quirinus Episcopus, quem Praeses ad urbem Sabariensem ad audiendum censuit repedari. Tunc ingredientes ad sanctum Quirinum Episcopum Christianae mulieres, cibam, potumque obtulerunt ei. Quarum fidem intuens beatus Episcopus, dum ea quae offerunt benedicit, catenae quibus ligatae manus eius et pedes fuerant, ceciderunt. Accepta igitur esca, regressis mulieribus, hi qui eum custodiebant Sabariam duxerunt.'*[965]

Nagy Tibor mentions Scarbantia's Christians; he managed to prepare the following summary of Sopron from the evidence at his disposal: 'The populous city of Scarbantia, north of Savaria, would even have been suitable for the foundation of an episcopate. However, the Christianity of the city is not dealt with in the written sources except the Quirinus Passion. At best only a presbyter or a deacon might have directed the community.'[966] József Török last dealt with this passage from the passion in the following interpretation: 'The bishop was escorted in fetters from town to town, because Amantius had ordered that the interrogation should take place in the City of *Savaria*. Christian women took food and drink for the prisoner. Then the interrogation began.'[967] In an earlier translation, we read that he was taken from town to town 'along the banks of the Danube'.[968]

Until 1974, there had been no headway made in the matter of Sopron's antique episcopate, or any other Christian relations, when Endre Tóth published his

959 GÖMÖRI 1985: 20.
960 GÖMÖRI 1986: 355.
961 The depiction of the halo has well-defined iconographic conditions and methods, see Agneta AHLQVIST: Tradition och rörelse. Nimbusikonografin i den romerskantika och fornkreistna konsten. Helsinki 1990.
962 I am grateful to headmaster József Sümegi for showing me the brick incision kept at Bátaszék, which also illustrates an 'orans' figure. LAKATOS 1993.
963 LINIGER 1976: 25 Tabelle 2, 15.
964 SZABÓ J Gy: A keleti kereszténység egyik ismertetőjegye temetkezéseinkben. - A typical characteristic of Orthodox Christianity in our burials. JPMÉ XXVIII 1983: 83-97.
965 RUINART 1859 Passio Quirini IV, p. 523.
966 NAGY T 1939: 211.
967 TÖRÖK J 1991: 30.
968 RADÓ Polikárp: A vértanúság emlékei - The relics of martyrdom. Budapest 1944: 237, the analysis of the passion JARAK 1994: 26-27, 164-166.

article.[969] He wrote that among the signatories of the Council of Grado, in the second half of the 6th century, *Vigilius episcopus Scaravaciensis* was Scarbantia's bishop. Scholarly works already knew of his existence earlier, and he had been written about in many places interpreting and explaining the attribute *Scaravaciensis* in different ways.[970] Thanks to Endre Tóth's articles since then, research has respected 'Vigilius episcopus Scaravaciensis' as the 6th-century bishop of Sopron - even if with a slight degree scepticism.

Commentary: With reference to Sopron's Christian antiquity, not only the archaeological sources are scant, but written sources as well. As a matter of fact, there is only one, the cited Quirinus acta. The here-mentioned *Scarabetensis* can really be identified with Sopron. Previous readers misunderstood the references of the *acta* and went on to misinterpret it. Let us examine the passage.

• The civitas Scarabetensis features in the clause relating to Amantius' journey. The translation is: '*Amantius, having returned from the City of Scarabetensis on the same day* ...'. Like this, the passage only tells us that Amantius has been to the City of Scarabetensis. However, this tells us nothing about Christians, whether there were any or not. Nevertheless, we get to know that Amantius obviously met the bishop in Savaria, who therefore - according to the above - was not in Scarbantia.

• Consequently, the bishop was not taken to Sopron and thus the interpretation that the Christian women who lived there came to meet him is wrong. But from which town did these Christian women come to see him? The order was given by Maximus in Siscia. Then comes the road journey along the Danube to Pannonia Prima. Next, the focus is on Savaria, where Quirinus arrives on the same day Amantius returns. This is followed by the appearance of the women. Whatever the antecedents of this part of the story were, this passage is a unit whose time and place is given by *Tunc*, a word that refers to the time when Maximus gives the order. The word *Tunc* is just preceded by Maximus' order, which thus draws the attention back to Siscia as the location. Therefore, the women - we may say his followers – have come to see him in Siscia. This section ends: AFTER HE HAD EATEN, AND THE WOMEN HAD ALREADY GONE AWAY, THOSE WHO WERE KEEPING HIM TOOK HIM TO SAVARIA. Consequently, it was the Christian women of Siscia who came to see the bishop.

The translation: '*When Venerable Quirinus was taken to Pannonia Prima, they dragged him in fetters from town to town to the interrogation of Praeses Amantius; he was just led through the towns along the bank of the Danube; Venerable Bishop Quirinus was taken to Amantius, having returned from the City of Scarabetensis on the same day. Praeses (Maximus)*[971]*ordered that he should be taken to Savaria to be interrogated. This time Christian women came to meet Saint Bishop Quirinus, and they brought him some food and drink. The saintly bishop, seeing the women's faith, while he blesses what they are giving him, the chains with which his arms and legs were tied, fell away. After he had eaten, and the women had already gone away, those who were keeping him took him to Savaria.*'

• The interpretation of the text is that Quirinus was taken along the Danube and that he did stop at Sopron. He had already met Amantius in Savaria. The women, regarded as Sopron's Christians, are, in fact, the inhabitants of Siscia who make their own farewell to their bishop. It is therefore certain that in the *acta* they are not women of Sopron. As a consequence, the only written source that might have been seen as proof of Sopron's Christianity, was only deemed so as a result of previous misunderstanding and misinterpretation. It is not such a proof.[972]

• The figure of Praeses Maximus. *Acta II*[973] comments on the activities of Maximus: *'Inter multos autem qui in Christi exercitu triumphabant, beatus Quirinus Episcopus Siscianus, a Maximo Praeside jussus est comprehendi, quem cum studiose quaererent, et beatus id sensisset Episcopus, egressus est Praeside interrogaretur, quo fugeret? Quirinus Episcopus respondit: Non fugiebam, sed jussum Domini mei faciebam, Scriptum nobis est: Si vos persequentur in una civitate, fugite in aliam. Maximus Praeses dixit: Quis hoc praecepit? Quirinus Episcopus respondit: Christus, qui verus est Deus. ... Tunc Maximus praeses jussit eum fustibus caedi cui et dixit: Respice et agnosce potentes esse deos, quibus Romanorum servit imperium. Propter quod consentiens, eris sacerdos magno deo Jovi, alloquin ad Amantii Praesidis primae Pannoniae iudicium dirigeris. ... III. Post triduum autem Maximus Quirinum Episcopum jussit ad Amantium Praesidem, ad primam Pannoniam deduci, ut pro contumacia, quam in leges Imperatorum exhibuerat, ultimam sententiam sustineret.'*

There is an allusion to this order in caput IV later, which was misunderstood. On the official rank of Maximus, Jenő Fitz has written a summary.[974] He studies the administrative problem of why it was not Maximus himself who sentenced Quirinus. The solution is really an administrative issue, which is associated with the division of Pannonia.[975]

• The interpretation of *Episcopus Scaravaciensis* as the bishop of Sopron on the basis of the archaeological and written data is as follows: We may state that there is

[969] TÓTH E 1974b; TÓTH E 1977c. In one of his earlier papaers, he had already written about Vigilius - TÓTH E 1971-1972: 235.

[970] This earlier literature (e.g. POLASCHEK 1937: 1027) is to be found in the articles TÓTH E 1974b, 1977c together with the quotation from the debated concilliar documents. On the assessment of the variations in connection with the attribute (*Scaravansiensis, Scarabanciensis, Caravaciensis)* and its sources BERG 1985: 85. Here, Heinrich Berg reflects on András Mócsy's sceptical opinion (MÓCSY 1977: 398-399) too. Berg's opinion is that Vigilius was the Lombards' bishop.

[971] Here the text of the *acta* was misunderstood. The two praeses were taken for one person. '... quem (i.e. Quirinus) Praeses (id.e. Maximus praeses) ad urbem Sabariensem ad audiendum censuit repedari.'

[972] Cf. MIGOTTI 1997: 6.

[973] RUINART 1859 p. 522. III. p. 523.

[974] FITZ 1994 No. 903.

[975] FITZ 1994 pp. 1181-1183. Cf. ibid. No. 894 - Amantius, No. 896 - Valerius Catullinus, No. 905 with Aurelius Iustinianus' cursus and task.

neither archaeological, nor written evidence of Sopron's Christianity. This, however, does not mean that there could not have been Christians in the town, but we cannot prove such a fact on the strength of the data at our disposal at present. If we compare Sopron's data to the data of those sites (towns) where archaeological finds or written information prove the presence of Christians, then the difference is remarkable.[976] In places where there were Christians, we always find some of their relics; and where there was even an episcopate, the available relics are much more numerous, as in Sirmium. Regarding Sopron, there is a great silence in this respect, which raises the suspicion that we can hardly reckon with a considerable Christian community. Therefore, the episcopate of Vigilius Scaravaciensis in Sopron becomes rather vague.

Let us suppose that the first Christians appear in the territory only at a very late time, in the course of the 5th and the 6th centuries, and that is why the episcopate exists only in this late period; we should even then have found their relics either within the city or in its environs. Such late data are provided at Keszthely-Fenékpuszta. In Sopron, there are no such relics or traces. Consequently, we have to face the much-asked question: which town was Episcopus Scaravaciensis bishop of?

• The identification of the city: The opinions so far - besides Sopron - have identified the City of Scaravaciensis with places outside Pannonia. It is obvious that the names Scarbantia and Scaravaciensis sound similar enough for them to have been combined into one, perhaps due to a mishearing or a superficial identification.

We must not forget that we are already in the last quarter of the 6th century, since the Council of Grado, among whose signatories we see Vigilius, was held between 572-577. In this period, the phrase *scara* was already used with an individual meaning as the anterior constituent of compounds: *acies, cuneus, copiae militares.* Hincmarus in Epist. ad Dioecesis Remensis Episcopos cap. 3: *Bellatorum acies, quas vulgari sermone scaras vocamur.*[977] If we divide the word into two, i.e. *scara* and *vaciensis*, the analysis will be easier. The TIR[978] discloses even allonyms under Bassiana I, among these are the two parts of Cod. Just. Tibor Nagy writes about this Codex-passage, and he states that there is no mentioning of the episcopate.[979] In this place, Iustinianus lists the provinces belonging to the new patriarchate. His decree was written in 535. I quote: *'Multis et variis modis nostram patriam augere cupientes, in qua primo deus praestitit nobis ad hunc mundum, quem ipse condidit, venire, et circa sacerdotalem censuram eam volumus maximis incrementis ampliare, ut primae Iustiniane patriae nostrae pro tempore sacrosanctus antistes non solum metropolitanus, sed etiam archiepiscopus fiat, et ceterae provinciae sub eius sint auctoritate, id est tam ipsa mediterranea Dacia, quam Dacia Ripensis, nec non Mysia Secunda, et Dardania, et Praevalitana provincia, et Secunda Macedonia,* ***et pars secundae etiam Pannoniae, quae in Bacensi est civitate.***[980] We can read the interpretation of this passage at Konrád Szántó as follows: 'Bassiana (Petrovci), or 'Civitas Bacensis' as it is called elsewhere, which lies near, south-east of Sirmium also played an important part in Pannonia Secunda's Christian life. In the reign of Iustinianus I (527-565), the metropolitan of the province was no longer Sirmium's, but Bassiana's bishop.'[981]András Mócsy writes about the identification of Baciensis civitas and Municipium Bassiana.[982] However, if Bassiana was called Baciensis in the 6th century, then it seems logical to treat this city as the seat of Vigilius.

Some indirect data may support Konrád Szántó's opinion that Sirmium was joined to Thessaloníki together with the prefecture at the beginning of the 5th century. This was useful for Thessaloníki from the aspect of prestige, but it was not useful for Sirmium. And, indeed, in Sirmium there was a certain decline, as a result of which another city could take over her role - in certain respects - by the beginning of the 6th century.

Reverting to Quirinus' story,[983] I would like to add that Quirinus was Siscia's bishop who was captured while he was leaving his seat. Afterwards, he was taken through the towns along the Danube. This report presents Bassiana as the more obvious one, considering the identification of the city under discussion instead of Scarbantia. It is much closer to Sirmium in fact, it is easier to reach the bank of the Danube from there, thus the mentioning of the towns along the Danube also becomes more obvious by the help of the identification Bassiana = Baciensis. Donji Petrovci, ☞ Szombathely.

• The question of spelling: the orthographic differences between *Vaciensis* and *Baciensis* in the middle of the words are the letter-groups 'vaci' and 'bat'. The exchange 'b' - 'v' was quite common, it also occurred in inscriptions, for instance BICTORINVS was the name of a fossor.[984] Graffiti from Trier hill: ΒΟΗΘΕ ΒΙΚΤΩ[ΡΙ.[985]

As for the meaning of *scara*, the explanation might be given by the Sirmium inscription (Sirmium) where the *cuneus* is already included according to its Christian meaning. But the late language usage of the

[976] E.g. Dunaújváros, Pécs, Sremska Mitrovica.

[977] DU CANGE vol. VI. 1954 p. 336 - *scara*.

[978] L 34 Aquincum, Sarmizegetusa, Sirmium. Budapest 1968. Some remarks on Bassiana: András Alföldi writes about Bassiana's (Donji Petrovci) role played in Pannonia's Christian life. Its other name is Bacensis civitas, as referred to by Justinian's decree subordinating it to the archiepiscopacy founded in his native town. As for its episcopacy I quote András Alföldi: 'We may find Bassiana's name even in Hierocles' Synecdemus; it suggests a bishop's residence as our author compiled his town-lists with the help of lists of bishoprics. It is curious that the Byzantine court still kept the claim of the archiepiscopacy of Iustiniana prima to the churches in South Pannonia when the territory had already been conquered by the Gepids.' (ALFÖLDI 1938: 156.) Other summaries ZEILLER 1932; GRBIĆ 1936; GRBIĆ 1937; NAGY T 1939: 22, 209, 210; MÓCSY 1968; TIR 34 Bassiana; MÓCSY 1973: 375.

[979] NAGY T 1939: 209 No. 73.

[980] Cod. Just. Nov. XI. CIC P. 3. p. 84. Ed. E. OSENBRÜGGEN Lipsiae [15]1875. Cf. Note 1461.

[981] SZÁNTÓ [2]1987: 242.

[982] Baciensis civitas. RE Suppl. XI. 346.

[983] Though Episcopus Vigilius has nothing to do with it.

[984] GUYON 1974 No. 2.

[985] REUSCH W, Frühchristliche Zeugnisse im Einzugsgebiet von Rhein und Mosel. Trier 1965: 258, No. 55 III 20.

Sirmium inscriptions and their style referring to the East generally explain the occurrence of *scara* with such a meaning. The Christian usage of *cuneus* is also shown by an inscription from Sirmium.[986]

• The problem of the Scarabetensis gate is still to be discussed. For details, see *Szombathely*.

Summary based on the written data. In the Quirinus passion, the Christians having been considered as women from Scarabantia so far, were not the inhabitants of Scarbantia, but they were from Siscia in fact. Amantius and Quirinus met in Savaria where Pannonia Prima's praeses and the bishop of Siscia arrived on the same day.

Episcopus Vigilius was not Sopron's but Bassiana's bishop;[987] the town was already called Baciensis in the 6th century.

Hypotheses based on the knowledge of a part of the unpublished archaeological material.[988] Despite the counter-arguments that have been listed so far, I still suppose there were Christians in Sopron, but not in a vast majority. I base this supposition on the material of unpublished cemeteries. We can see in other examples elsewhere that the Christian relics are spotted more easily in the cemetery than in the settlement, ☞ Aquincum, canabae. Cemeteries have been found in Deák Square, at the Szent Miály Kapu (Saint Michael Gate) or in Csengery Street.[989] Further sites are the old Lutheran cemetery, the Catholic cemetery, the Hofäcker, Bécsi domb (Vienna Hill), Rákóczi Street and Mátyás Király utca (King Matthias Street). Other sites in the centre (besides the forum) are Templom Street, Új Street and Orsolya Square.

In Deák Square a lamp with a suspension, a gilded bronze fibula and a silver spoon were found. A silver spoon was also found in Rákóczi Street.

Among the lamps published by Dóra Iványi - also with the help of similar examples -, we can find ones that we may perhaps consider Christian.[990]

We must not neglect the finds from sites around Sopron since there is no guarantee that the Christians would necessarily live in the centre of the town. Moreover, we should also not neglect the Germanic cemeteries, although they already date from the 6th century on the one hand;[991] on the other hand, these people did not belong to the population of Scarbantia. We have to note that Klára Póczy depicts the 5th century at Sopron as a restless one, due to the earthquake in 456 and harassment by the Germans (barbarians).

The complete study of these cemeteries (focusing on the 4th century) will also provide more reliable information about the Christians. Klára Póczy states that, similar to other German settlers' dwellings, the dwelling place of Marcomanni Queen Fritigil was also in this territory. [992] The queen was under the influence of Ambrosius, bishop of Mediolanum, and so we cannot be indifferent to her dwelling place from the aspect of Christianity in the region. At the very end of the 4th century, Fritigil settled down in Pannonia Prima,[993] but we are not sure whether it was in Sopron or not. Tibor Nagy suggests that the seat of the Christian Queen and her Marcomanni was at Mórichida-Kisárpás.[994] The latter hypothesis - given the present excavation results at our disposal - is more realistic.

HISTORICAL SUMMARY - SOPRON

In spite of the fact that - according to our present knowledge - we do not have Christian relics from Sopron, as the women coming to meet Quirinus were not the inhabitants of Scarbantia, we still have to assume that the unpublished material and the territories that have not been excavated so far will bring in some evidence in connection with this question. This assumption may be realistic on the basis of the similarity that is shown between the material of finds published on a larger scale from other towns, and that

[986] CIL III 14340^{3}; BRUNŠMID 1908-1909: 193 No. 403. A white marble fragment, 9.3 x 15 x 2.9 cm. According to Hytrek, it was found next to Synerotas' tomb. I quote its inscription from Brunšmid: ...] r? cuneus t[... / ...] ne Sirmi[... / ...]us o[... . Brunšmid gives Domaszewski as a reference, who proposes *ecclesiae* as an addition. Brunšmid considers a further addition possible in the first line: letters 'e' and 'f, r'. What is under discussion is the military interpretation of *cuneus*, which might be possible if *cuneus* occurred together with *equitum*. As a 'Truppenform', it is first included under Constantine the Great, (HOFFMANN D 1969: 273. In Pannonia - SOPRONI 1974B). The expression in the inscription is difficult to classify among the regular military forms, since these forms and names are defined, (see details in HOFFMANN D 1969: 247-279). Domaszewski must not have proposed a non-military interpretation of the inscription by chance; he involved the expression *ecclesiae Sirmiensis* with his proposal. It is reasonable to use *cuneus* as a metaphor, for example as the *cuneus veritatis* of the Sirmium Church, (BLAISE 1975 headword *cuneus* Tertull, Marc I, 21). The expression *ecclesiae Sirmiensis* explicitly refers to the owner who holds possession of something. We can reckon with such expressions coming from other Pannonian territories as well, because it belongs to the same legal category as the *domus ecclesiae*. The *ecclesia*, like the *synagoge/a* may refer to a certain community, e.g. *ecclesia ex circumcisione* and *ecclesia ex gentibus*, (WILPERT 1916: 495 Taf. 73, 74/2), by which the Christian and the Jewish communities were to be separated. The iconographical symbols of these are Jerusalem and Bethlehem, as Wilpert expounds in his quoted work. Bethlehem symbolises the Church of the peoples, while Jerusalem stands for the Christian Church of the Jewish people. We can reckon with both here in Pannonia. Sirmium and its neighbourhood, the area between the Drave and the Save, belonged to the peoples' Church; that is why they participated in the heretical movements. On the other hand, in the northern part of the province, due to the Christianising of the Jews who had already come here earlier, the *ecclesiae ex circumcisione* were established. ☞ Tokod, ☞ Pilismarót, ☞ Pécs. The contemporaries knew which town belonged to a certain Church, therefore they omitted explanatory comment and wrote down only the names of the town, e.g. *ecclesiae Sirmiensis*. Thus if the missionaries took away objects owned by these churches, then wherever they went the ownership 'sheet' on the objects also disclosed what kind and exactly which Church they had belonged to.

[987] Cf. JARAK 1994 - Bassiana's bishop is unknown although we know that he existed.

[988] T.-SZŐNYI Eszter: A preliminary report on the excavation of the cemetery in 1 May Square. SSz XXXI 1977: 340-345. 1 May Square = Deák Square.

[989] BELLA 1895: 396.

[990] IVÁNYI 1935 No. 4353 LFM old Inv. No. 53.355.1 new Inv. No. 55.125.9.

[991] Hegykő: BÓNA 1960; 1961; 1963; BARKÓCZY/LENGYEL 1971: 191-199; ☞ Keszthely-Fenékpuszta, Fertőszentmiklós; TOMKA 1980: 5-30.

[992] PÓCZY 1977: 38-40.

[993] MÓCSY 1974: 345.

[994] NAGY T 1983.

is suggested by the unpublished material of finds from this town. On the basis of the data given at present, Vigilius was not Sopron's bishop. Furthermore, nothing confirms the assumption that there was an episcopate in Sopron at all. Those finds that have been published as early Christian ones by the research so far, could not survive the 'crucible'. A future task is to publish the complete find material and to decide whether the objects referring to Christianity belong either to the surviving Romanised population or to the new settlers. We cannot suppose a total destruction or a total migration either. We could rather accurate and define the shift in the size of the city or the change in the life of the city. There was certainly a kiln near the city, which reminds us of the shrinking of the ☞ Tokod settlement and its nearby kiln. The research has not as yet focused on such changes to the settlements in the 4th century and on the workshops appearing near them that indicate occupational changes. We cannot expect an alteration in the architectural structure as shown in Emona's example. Following the fire in 238, the restoration followed the pattern of the old structure and was maintained even in the later periods as well.[995] The comparison of Emona and Scarbantia is justifiable because both were established in the early period, although it is true that Emona gained the rank of a city under the Iulius-Claudius dynasty, when it belonged to Pannonia, whereas Sopron became a city in the Flavian period. An inscription found in Savaria is proof of this.[996] The inscription was first seen and described by István Schoenvisner: *municipivm / Flavivum / Avg / Scarbant.*[997] It reads as Municipium Flavium August(ale) Scarbant(ia). The historical assessment of the attribute *Augustale* would also clarify Amantius's journey to Scarbantia.

43. SÜMEG fig. 308.

Two buildings near to each other have produced a very wide range of opinions as regards their date and function. For the sake of simplicity, let me call them I and II. The first point to be raised is that it is No. II that is thought to be an early Christian church.

The history of research is extremely short, as the last excavation took place at the end of the 19th century. According to Bálint Kuzsinszky, 'the area whose whitish surface indicates all kinds of Roman walls underground is about two acres.'[998]

43.I.a. Building I, fig. 309.

Literature: ÁDÁM 1882; THOMAS 1964: 111-114 Abb. 51; MRT 3. 54/23 fig. 68; SARNOWSKI 1978: 65 figs. 86-87; D GÁSPÁR: Römische Villa bei Sümeg. Balácai Közlemények III 1994: 343.

The Topography writes the following about Sümeg as a site: 'The pieces of tegulas, imbrexes and earthenware found between Sümeg and Sümegcsehi on the hillside edging the Sarvaly brook, near the ruined Berki Mill, in the plough-land, indicate a considerable Roman settlement. According to the earthenware, the settlement existed all through the Roman Age. The Director of Secondary School, Iván Ádám, excavated here, and as a result, the picture of a 39.4 x 43 x 39.43 m Roman villa emerged, with osculatory square corner turrets whose sides were 5 m. Ádám thought that the walls and heating-flues found in front of and under the building used to belong to an earlier Roman building. In his opinion, the villa with corner turrets was inhabited even in the Migration Period, which is indicated by the bricks whose shape and quality differ from the usual Roman material.'[999]

Edit Thomas writes that it was built from limestone and Roman bricks; the thickness of the turret-walls is 73 cm. The archaeologist does not mention bricks with stamps although they are various and in good condition. There are some, 'deren Breite fast gleich der Stärke ist.'[1000] These prove a subsequent alteration and renovation. However, hypocaustum bricks were also found, which indicated heating. We have no data about heated rooms. Finds from the building include Roman storage vessels, silver coins with illustrations of clover and lily, and coins belonging to the Archiepiscopacy of Salzburg (in Edit Thomas's opinion). The different building materials and walling techniques betray several building periods. A conduit system was discovered south of the building. The villa was constructed in the 3rd century.

Tadeusz Sarnowski dealt with parallels of the building. Rectangular buildings with corner turrets are common in several provinces: Thrace, Dalmatia, Pannonia, Germany, and Gallia; in North Africa, too, round turrets were found. The buildings can be as long as 60 m, and as wide as 30/40 m. He dates the Sümeg and Pogánytelek buildings to the end of the 3rd and the beginning of the 4th centuries, and thinks that they match the type of the Orlandovci, Mogorjelo villas, or Diocletian's palace, at Split. This similarity in the type provides a basis for dating.[1001]

Among its various bricks, a type also appears in the building that is thought to be an early Christian basilica.

43.II.a. Building II (So-called early Christian basilica), fig. 310.

The Topography writes the following: 'The same bricks appeared in the early Christian basilica excavated by Iván Ádám, 150 steps from the above-mentioned building with four corner turrets.' The church facing exactly east is 27.8 m long and 21.6 m wide. On the eastern side, there is a 9.6 m-long rectangular projection of wall. On the northern and southern parts of each longer wall, the basements of two buttresses have remained. Between the two northern buttresses are the traces of a 1.4 m-wide

[995] PLESNIČAR-GEC 1976: 29.

[996] According to Steindenkmäler. 61. Municipium / Flavium / Aug(ustum) Scarbant(ia) - the reference to Schoenvisner is not punctual. RIU 134.

[997] SCHOENVISNER 1791 p. 12, 51, Table 9.

[998] KUZSINSZKY 1920: 115.

[999] MRT 3. 54/23.

[1000] THOMAS 1964: 112.

[1001] SARNOWSKI 1978: 65-66.

doorway. The main entrance used to be on the western side. The western side was joined to porches divided into several parts; Ádám could not explore this area satisfactorily because by then the basement walls had been mainly excavated. There were few bricks in the walls. Limestone from Sümeg was mostly used together with a little basalt. In the basilica, the inner buttresses had also been excavated previously.[1002] Tibor Nagy thinks the church is similar to the one at Kisdióspuszta (Ugod-Dióspuszta) whose subsellium was made of wood.[1003] E. Thomas supposes a three-aisled basilica.[1004]

No one has conducted an excavation in the area since then, so the basis for different views and opinions is provided by the above-listed data. The different ideas mainly concern the dating and the building of the church. Its dating is ambiguous; Ádám puts it to the 8th or 9th century,[1005] T. Nagy to the period after the collapse of Roman rule,[1006] L. Nagy also to after Roman times,[1007] E. Thomas considers it as Roman,[1008] while K. Sági and Miklós F. Füzes say it is from the 8th or 9th century.[1009] The Topography takes a standpoint as follows: 'Based on traces on the ground surface ... the place of Ádám's excavations could not be determined as the settlement had a large territory.' A great number of Roman tegula and imbrex fragments could also be collected northwards along the Sarvaly brook. It is clear that it was also a part of a loosely built up large Roman settlement.[1010]

Sarnowski[1011] also supplied further examples of the so-called early Christian basilica, fig. 311. These ground plans do not seem to be Sümeg's parallels at first sight, but in some details we may recognise its similarity to the Sümeg 'basilica' - about which we must add that its accurate and complete ground plan is unknown. The orientation of the Holbury villa (e) is different from ours, but its ground plan is very similar. There are four columns in the rectangular room. Furthermore, there is a projecting part with a linear closing, though it cannot be seen because of the built-up corners. In the case of Clanville (g), there is also a side entrance on the eastern side. The projecting part with the linear closing is also unobservable from the outside. The inner hall is divided by several columns, but we have the further rooms opposite the 'apse', whose location is very uncertain in the Sümeg 'basilica'.

On this type of building, Tadeusz Sarnowski writes as follows: 'On peut supposer qu'elles abritaient des locaux d'habitation, alors que le rez-de-chaussée était occupé par une écurie ou une grange. Nous sommes enclins à croire qu'il s'agit d'un genre de bâtiment que nous ont fait connaître les provinces européennes de l'Empire, auquel A. Greiner donne le nom de "type basilical", tandis que les chercheurs britanniques l'appellent "aisled house". En Afrique, des constructions proches de ce type par leurs plans sont signalées dans la région de Sétif.'[1012] According to Sarnowski, such buildings are depicted in Tabarca's mosaics; he dated these mosaics to the beginning of the 5th century.[1013]

We have to note that site No. 54/24-25 lies very close to this area, which supports the opinion of the Topography that the Roman settlement had a very large territory.[1014] Cf. fig. 308.

Since the publication of the Topography (1970) no new information has been disclosed about the area, thus we can only form opinions based on other examples. Moreover, these opinions have to be verified by concrete data in case of new excavations.

Commentary: Kuzsinszky's experience and the opinion of the Topography agree that we have to suppose a larger Roman settlement, and the excavated building or buildings form a part of this settlement. No one questions the fact that building I was erected in the Roman Age, but in the case of building II the archaeologist himself suggests a later point of time. The peculiar bricks are indeed medieval but it only betrays a medieval usage. As for building II, we can suppose Roman basement walls, but we have to consider its medieval dating very seriously.

The analogies: Building I. Sarnowski mentions the constructions of Mogorjelo, Split and Orlandovci.[1015] In Mogorjelo's case, it is clear that the surroundings give the impression of a building with corner turrets. The villa itself is not a parallel. The case is similar to that of the Split palace. Orlandovci is more similar to the Sümeg villa,[1016] fig. 312. Two of its corner turrets are rectangular and two are round. In its centre, however, there is a part of a building, which is identical with the one joined to the NW turret of the Sümeg villa. The identical parts are thus the rectangular room with a closing (with apse), joined by a two-part rectangular room. In the Sümeg villa this 'partition' may have been created by joining the other side and the room with the apse to the already existing corner turret. Considering the sizes of these three rooms compared to each other, we may observe the same solution in the Pforzheim villa,[1017] fig. 313. In the case of Pforzheim, there seems to be a building complex where a villa comparable to the one at Sümeg is situated in the centre.[1018] These comparisons allow the conclusion that in dating the first phase of the building with corner turrets to Diocletian's reign, we may suppose that only the four square turrets were erected then together with the inner buildings. The rooms in question were joined to the NW turret in the second phase; their date of building, based on other villas

[1002] MRT 3. 54/53, p. 221, fig. 69.
[1003] NAGY T 1941: 146.
[1004] THOMAS 1980: 196-197.
[1005] ÁDÁM 1882.
[1006] NAGY T 1939: 212, Note 99.
[1007] NAGY L 1938: 82-83.
[1008] THOMAS 1964: 114-115.
[1009] SÁGI/FÜZES 1967: 82.
[1010] MRT 3. 54/23, p. 221-222.
[1011] SARNOWSKI 1978 fig. 97; RIVET 1969 fig 2,6 d: Mannsfield Woodhouse, e: Holbury and g: Clanville
[1012] SARNOWSKI 1978: 67.
[1013] SARNOWSKI 1978: 11.
[1014] General plan MRT 3. 54.
[1015] SARNOWSKI 1978 figs. 90-91: Mogorjelo, fig. 92; Split, fig. 96: Orlandovci
[1016] SARNOWSKI 1978 fig. 96.
[1017] SARNOWSKI 1978 fig. 98-99.
[1018] SARNOWSKI 1978: 66-67.

used as parallel examples, can be put to the beginning of the 5th century.

Assessing the two-part room created by the alteration and the neighbouring part with apse, it is feasible that it is the equivalent of a part of building that may have been used as a house-church in Pannonia. Therefore, although it has not been studied in connection with early Christianity, I classify this building among the 'suspicious' examples; its case will have to be decided by further excavations. I have to note that one of the buildings belonging to the villa at Szentkirályszabadja is often mentioned as a similar example, but that building has not been excavated yet either, ☞ Szentkirályszabadja.

Building II. The parallels listed by Sarnowski are dwelling houses. Rivet assesses this type of building as follows: 'As to the form of the house, there is no doubt that the nave was roofed and was not an open courtyard, for no arrangement for draining off rain-water is ever found to exist. The entrance is sometimes, though not always, in the middle of one side. In the more developed form of the building, internal partitions are added, mostly in the aisles, but not infrequently in the nave also, cutting off the space into individual rooms.'[1019] His further analysis reveals that in such buildings the owner of the villa lived together with the workers and that therefore there was a main building.[1020] In our case, there is also a main building with traces of subsequent alterations (building I). If we regard both buildings as dwelling houses, we cannot separate the function and building time of II from that of I. The only question is whether it is possible to connect buildings I and II, thus, classifying them as a winged-house.

The ground plan of building II: The western entrance is generally mentioned, whereas the northern entrance is rarely discussed. The porch 'could not be clearly identified by Ádám as the basement walls had almost been quarried. ... Inside the basilica the buttresses had also been quarried.'[1021] The interpretation of these inside pillars supposes that the building was either once a three-aisled basilica or, under other circumstances, a dwelling house. What seems reasonable is the supposition of four pillars, which acted both as a space divider and supported the roof. However, the currently suggested ground plan is unusual in antiquity and therefore a medieval dating for the building should be considered.[1022] The coin of the Archiepiscopacy of Salzburg that was found nearby, despite the fact that it was located in the other building, does suggest a church - moreover, a medieval church.

Both ideas are acceptable - they are mere hypotheses. The unexcavated building II does not provide any clear answers, and therefore I must add it to the list of uncertainties.

SUMMARY – SÜMEG. Because of the limited excavations, only hypotheses are possible. It seems probable that the second phase of building I may be put to the beginning of the 5th century, when, as a part of the building, a house-church was created. As for building II, although it would be a fascinating possibility to interpret it as an antique construction, the more probable dating is medieval.

44. SZABADSZÁLLÁS .fig. 314.

44.II.a. Glass beaker

Location: KJM 65.21.3. **Literature:** H.-TÓTH 1969; H.-TÓTH 1971: 115-138; GÁSPÁR D: Ein sogenannter christlicher Glasbecher aus dem Barbaricum. Atti 1998 III 309-316. Let me attach a corrigendum to Atti Abb. 1; it is correctly: Ein weinroter kegelförmiger Glasbecher.

In 1966, Elvira H.-Tóth opened a single grave at Szabadszállás. According to the publication,[1023] research of the grave's environment did not provide finds other than the beaker and the final description was that of a single grave with a male skeleton. (The man's head was cut off before burial and the body wrapped in a winding-sheet; the head was also buried.[1024]) Purplish-claret glass beaker H: 16.5 cm, Diam: 10.5 cm, fig. 315. Elvira H.-Tóth's described of the glass beaker as follows: 'Der Glasbecher hat eine einigermaßen unregelmäßige Form, das Abschneiden und Schleifen des Becherrandes gelang ein wenig schief. Die Wanddicke ist unterschiedlich, sie beträgt am Rand etwa 3 mm, verringert sich 4-5 cm unterhalb des Randes an der Grenzlinie zwischen Inschrift und Bilderfeld bzw. an dessen unterem Saum und verdünnt sich an letztgenannter Stelle ganz besonders stark.Hier erreicht die Wandstärke kaum einen halben Millimeter, um sich dem in Tropfenform endenden Urteil zu von neuem bis zu 4 mm zu verdicken. Die Farabe schwantk der Wanddicke entsprechen zwishcenb ins Violette spielendem dunkelweinrot bis zurm durchsichtigen lila-rosarot. Unter dem Rand erblickt man aus der Bearbeitung stammende, spiralförmig aufsteigende längliche, im mittleren Teil runde Luftbläschen, die in den Glasstoff eingeschlossen sind. ... Als wichtiger Umstand sei noch erwähnt, daß auf der Außenwand im Mitteilteil des Becheres eine 7-8 cm lange, nahezu vertical verlaufende Naht zu sehen ist, die abe nicht von einer zweiteiligen Form stammen kann, ...Unter dem Becherrand verlaufen parallel zu einander zwei tief eingeschliffene Linien, an die sich nach unten zu einem ca. 2,5-3 cm breites glattes, unverziertes Band

[1019] RIVET 1969: 65.

[1020] RIVET 1969: 67.

[1021] MRT 3, 54/23, p. 221.

[1022] Walter Muir WHITEHILL: Spanish Romanesque Architecture of the Eleventh Century. Oxford 1941. 220-221: Deals with the dating of the Church of Santo Tomé de Zamora; its ground plan is fig. 104.

[1023] H.-TÓTH 1969: 6. The excavation of the environment of the grave brought to light only insignificant Bronze Age remnants. The Grave: 150x285 cm, oriented W-E, its bottom is 65 cm wide, it contracted to the depth of 170-175 cm. Male skeleton, stretched, lying on its back; fig. 1. Goods: iron hoe, L: 17.5 cm, fig. 2. a. Shovel-shoe 14.5 x 13.5 cm. Fig. 2 b. Glass beaker. H.-TÓTH 1971: 115.

[1024] H.-TÓTH 1969: 24-25: Burial. It took place in the Barbaricum, although there was nothing in the grave that would support a burial in the Barbaricum. The discolouration around the skeleton allows the possibility that the corps was buried in a winding-sheet. The skull was cut off the body, according to Note 56: 'The grave was undisturbed, the displacement of the skull together with the vertebras is impossible; because of the intactness of the vertebrae we have to suppose a careful severing.' The presence of iron objects is explained by the gravediggers' superstitious fierce of the dead. This rite is unique both in the Barbaricum and in the province. H.-TÓTH 1971: 117-120.

anschließt, unterhalb dessen zwischen zwei Linien eine aus grichischen Majuskeln bestehende Inschrift in griechischer Sprache angeberacht ist.'

Under the inscription, there was an approximately 6.3 - 7 cm band with the following depictions:

- In an emblem framed by a double concentric, slightly pressed circle there is a middle-aged man riding a caparisoned horse without stirrups; the horse is gracefully prancing. The figure is turning slightly left, raising his right hand in a salute. His garment is a *chyton* gathered up with a fold. Over his shoulder there is a flying *clamys* fixed with a crossbow-fibula. A ribbon, as a wreath, holds his short curly hair. In front of the horse's legs, there is a well-discernible illustration, which is, however, difficult to interpret. In Elvira H.-Tóth's opinion, it is the heads of the defeated. The background is decorated with an oscillated pattern. Under and above the emblem there is a fertility symbol, fig. 316.
- In the area between the emblems there is a standing female figure. She is wearing a long, richly pleated *peplos* pinned up on her left shoulder and gathered in a fold. She holds something high in the air. In her left hand, close to her body, she holds a twisted '*tyrsos'* decorated with ribbons that extend behind her, fig. 317.
- 'Rider in an emblem.' The gilding is worn and the glass damaged, therefore the illustration is hardly discernible. A riding figure similar to the first one. He is a young man with long hair and riding a prancing horse. He has a *clamys* slung over his raised right arm. The tail of the harnessed horse seems to be tied up in a knot, fig. 318.
- A figure, without clothes, standing on tiptoe, raising its right arm high, turning its head left; its curly hair is held down with a ribbon. There is a *pedum* in its left hand and a loose veil (*nebrys*) on its left shoulder. It is stepping forward lightly with its left foot, there is a bunch of palm-leaves under its feet. *Satyr*, fig. 319.
- Young rider; the position, the clothing and the harness of the horse and its rider respectively are completely the same as in the previous representations. A part of the shield worn on his left arm can be seen behind the horse's head. The background is filled with an oscillated pattern from the bottom to the top, as well as from the centre outwards in front of the horse. There is a fertility symbol above the emblem, under which we may suppose the figure of a snake in an oval field, fig. 320.
- 'A clothed female figure turning left in a position, slightly bending ahead; her garment is a long, richly pleated *peplos*, which is tied up on her waist, and pinned with a disc fibula. Her hair is held down by a ribbon. In her right hand she holds the '*tyrsos'* decorated with ribbons (as seen on the second figure), while she stretches her left hand behind the first emblem, quasi holding it. There is a bound up palm or acanthus-leaves.'[1025] Fig. 321. The Greek inscription is as follows: ΠΙΕ ΖΗΣ ΕΝ ΟΝΟΜΑΤΙ ΘΕΟΥ ΜΕΤΑ ΤΩΝ ΣΩΝ ΠΑΝΤΩΝ ΙΕΧΝΙΤΟΥ (sic),[1026] fig. 322. In connection with the inscription, Elvira H.-Tóth asked for the assistance of István Hahn, András Mócsy and György János Szilágyi whose opinions are: 'Die Inschrift enthält eine häuffigverwendete spätantike-frühchristliche Formel, die man auf Trinkgefäßen sowohl in lateinischer wie auch in griechischer Sprache antrifft: ΠΙΕ ΖΗΣ ΕΝ ΟΝΟΜΑΤΙ ΘΕΟΥ ΜΕΤΑ ΤΩΝ ΣΩΝ ΠΑΝΤΩΝ ΙΕΧΝΙΤΟΥ. Die Auslesung wird lediglich dadurch erschwert, daß die Worte ΜΕΤΑ ΤΩΝ ΣΩΝ hinter ΖΗΣ und vor ΕΝ stehen sollten. Nach Vornahme dieser Korrektion lautet der entsprechende lateinische Text folgendermaßen: *bibe, vives cum tuis in nomine Dei omnium creatoris*. Die meiste Wahrscheinlichkeit hat die Annahme für sich, daß der Text von einem Glasbecher abgeschrieben wurde, auf dem die Inschrift nicht kontinuierlich ringsum verlief, sondern auf den leer gebliebenen Flächen einer bildlichen Darstellung angebracht war, vgl. beispielweise ActaAntHung. 16 (1968) I, 1: in zwei Reihen und dreimal unterbrochen: *viv(as) Innoc cum tuis / enti in Deo*. Auf der hypothetischen Vorlage dürfte die Inschrift mithin folgendermaßen angeordnet gewesen sein: ...

Eine andere, weniger wahrscheinliche Möglichkeit besteht darin, daß das durch raumfüllende Ornamente von den vorangehenden Wörtern getrennte ΙΕΧΝΙΤΟΥ im Originaltext fehlte und der Inschrift erst nachträglich zugefügt wurde, als man feststellte, daß hinter dem Wort ΠΑΝΤΩΝ noch genügend Platz verblieben war. Ursprünglich gehörte ΠΑΝΤΩΝ zu ΤΩΝ ΣΩΝ im Sinne von *cum tuis omnibus* und gewann im neuen Zusammenhang erst nach Hinzufügung des Schlußwortes ΙΕΧΝΙΤΟΥ seine gegenwärtige Bedeutung.'[1027] In the publications there are many analogies with regard to the figure, the illustration and the inscription of the beaker.[1028] The author also mentions a Pannonian illustration where we see a beaker shaped like this.[1029] After the very thorough and accurate examination of the find - mainly based on the inscription - the author classifies this glass find among the

[1025] H.-TÓTH 1969: 6; H.-TÓTH 1971: 120-123.

[1026] I am grateful to Dr Elvira H.-Tóth for her assistance in my work in many ways, and for placing her drawing at my disposal.

[1027] H.-TÓTH 1971: 134.

[1028] H.-TÓTH 1969: 14-18. Cone-shaped cups, analogues: 1. Dunapentele grave No. XXVII, 6.1926.88 lost, H: 19.3 cm, diam: 12.5 cm. According to István Paulovics: 'Reverse cone-shaped, funnel-like, its base is rounded, there is a little deepened line running around the rim, smooth. Its colour is light green, translucent, made of 3 mm glass. Around the mouth, then downwards, there is a 3.3, then 4 cm lower around its waist, an approximately 5.5 mm wide ribbon running around. The ribbons are made of 10-12 finely incised close lines which are today hardly discernible.' BARKÓCZI 1988 No. 159.
2. Mosdós, KV Inv. No. 6805. H: 16 cm, diam: 11.4 cm. Crystalline, translucent. 'Under the mouth, there is a surfacing running around in the upper and middle thirds; in the upper part between the surfacing there is a surfaced pattern of hexagons.
3. Ságvár, grave 262. Green. H: 18.8 cm, diam: 12.6 cm.
4. Ságvár, grave 32. H: 13 cm, diam: 7.8 cm. According to BARKÓCZI 1988, the following cone-shaped beakers are there with inscriptions: 105 Ságvár Grab 270, 108 Ságvár Grab 26, 109 Ságvár 32, 110 Ságvár Grab 262, 158 Intercisa Grab 95, 159 Ságvár Grab 212, ΠΕΙΕ ΖΗCΗC, 167 Mosdós H.-TÓTH 1971: 124.

[1029] H.-TÓTH 1969: 18. József Hampel 1903 No. 10 originally published it: A 7.3 x 7.9 cm bronze plate. The riding youth with a Phrygian cap is holding a beaker shaped like this in his right hand, which he is handing over to the woman standing in front of him.

early Christian objects, and about its illustration - according to its comparison with the Cologne blue bowl - he says the following: '… haben ihre Gegestück in der frühchristlichen ägyptischen Kunst, vor allem auf Geweben, bzw. unter den Kaiserbildnissen und dionysischen Figuren und Szenen, die als antike Elemente, ihrem ursprünglichen Sinn entfremdet in der koptischen Kunst auftauchen.'[1030]

This extraordinary find has not been re-considered since then, although everybody accepts that it offers further secrets and that there are more questions to be asked. These may be summarized as follows:

The inscription is a common, 'good-wishes' formula, which very often appears in Latin or Greek, not only on Christian objects but also on so-called *Spruchbechers* and other drinking vessels. The inscriptions, however, differ in their verbalisations; they have many variants (as we ourselves use different words to express our best wishes according to the situation). Our task is then to choose the occasion that best corresponds both with the illustration of the object and with its inscription.

What conclusions do the scenes on the beaker provide?

To what time can we date the find?

The dating raises the question whether the burial is connected with the beaker or not. Does the fact that this grave was found between the Danube and the Tisza Rivers have a special importance?

The problem with the inscription. Based on analogous examples, it could be pointed out that the inscription was a common good-wish formula, and that the beaker could not be classified as an early Christian object simply because of the inscription.[1031]

I would not alter the word order of the inscription on the beaker. In the present word order, the IEXNITOY is clearly separated from the rest of the inscription, so I would treat it individually. In this form, the translation is: Drink, you will live in the name of God together with your ones - I (am) the master's. Thus, the term τεχνιτου is interpreted as the expression of the genitive referring to the object. This solution is also a common verbalisation of the owner of an object.

The illustrations. They can be divided into two groups, the riders in the medallions and the figures among the medallions. The figures among the medallions belong to a Bacchanalian procession, two maenads and a *satyr*. The order of the illustration starts according to the inscription, under the word '*pie*'. This part of the beaker is in the worst condition, it is rather weathered, so it is difficult to interpret the illustrations. The standing clothed female figure, cf. 317. The riders symbolise the soul itself, man at different ages. The first rider after the genius is the young, energetic soul. He is also followed by a standing figure, who - in my opinion – is not unclothed, but wearing close fitting, very tight leather trousers: they seem to end above the ankle. Its profile shows a very characteristic face, which suggests that the figure is wearing a mask. The second rider is a strong man in his prime. The bacchante with the *tyrsos* may already refer to happiness in the other world. Finally, the third rider, the oldest of all, on the verge of old age, symbolises the fourth phase of man's life. If he is a triumphant rider, his reward will be happiness in the next world. The victory may be symbolised with a military image: the defeat of enemies. Thus the beaker, with its symbolism, belongs both to life and death. In life, it keeps the memory of one's genius alive; while in death, it refers to happiness in the other world. The latter could be taken for granted if the person were either initiated, or if he gave himself deliberately for sacrifice. Perhaps the term *master* may refer to the first possibility.[1032]

The question of dating. In Elvira H.-Tóth's article, the most elaborated question is the one of technical parallels and dating. She enumerates all those glass beakers that have the inscription ΠΙΕ ΖΗΣΕΣ with an incredible thoroughness. These beakers - according to their inscription - really show similarities with the one found at Szabadszállás, but are not suitable for more accurate dating. In other aspects they are not examples of the Szabadszállás beaker. Moreover, at our present state of knowledge, and according to the archaeological finds at our disposal, the Szabadszállás beaker could be said to have parallels in Pannonia only with regard to its shape. One was found in grave 143 in the Bátaszék-Kövesdpuszta cemetery. It is a brick grave with a child's skeleton,[1033] fig. 323a. The cemetery was used from the first half of the 4th century to the end of the 4th century.[1034] The other example comes from Ságvár,[1035] fig. 323b. The cemetery can be put to between 340-375.[1036] László Barkóczi dates the Ságvár glass beaker to the second half of the 4th century.[1037] He found several variations of the type, but these variations, based on their parallels, must have originated in the Mediterranean and their production was also in that

[1030] H.-TÓTH 1971: 137. H.-TÓTH 1969: 25: 'Besides the few grave-goods and the W-E orientation typical of Christian burials, the Greek inscription of the beaker also supports the Christian feature of the burial to a large extent, …' The solution of the inscription is the work of András Mócsy, István Hahn and György János Szilágyi. Their opinion: The inscription is a common late antique - early Christian formula. The word order is not right - ΜΕΤΑ ΤΩΝ ΣΩΝ - this is a mistake due to oversight. Considering this, their translation is clear: 'bibe vivis cum tuis in nomine Dei omnium creatoris, i.e. drink, and you live together with your ones in the name of God, the creator of everything.' The IEXNITOY was written in the ansata because the place of the inscription had been measured improperly. 'The Christian feature of the Szabadszállás beaker's inscription is out of question.'

[1031] I dealt with the problems of the inscription in my lecture at the early Christian congress, therefore I wish to outline my lecture here briefly.

[1032] The Bacchic motifs suppose Dionysos' presence, which, on the beaker, is only possible through the riders. In this case, Dionysos has to be understood as the god of vegetation and revival, vide AUGÉ Ch / LINANT DE BELLEFONDS P: Dionysos (in periphera orientali), commentaire LIMC III/1 529. We can see his illustration with the seasons LIMC III/2 117, 118, 119 Dionysos/Bacchus.

[1033] PÉTERFI 1993 Taf. 36/11.

[1034] PÉTERFI 1993: 96.

[1035] BURGER 1966: from grave 32 a glass beaker, fig. 96, 32/1 is its drawing, the description of grave 32 is on p. 103.

[1036] BURGER 1966: 163.

[1037] BARKÓCZI 1988: 84, No. 109.

territory.[1038] Considering its technique of production, such a glass product was rare even in Imperial relations. The gold foil is burnt into the formed, and still hot glass. This enables the pattern to be seen even if the gilt is lost.

Elvira H.-Tóth also cited the sole parallel that could serve as a starting point for the making of the beaker. This analogue is the Cologne beaker.[1039] The beaker has a different shape and illustration, but the scales used for filling the space between the illustrations is the same. Moreover, its production method is also the same as regards the annealing of the gold foil on the Szabadszállás beaker. The Cologne beaker illustrates the sons of Constantine in small medallions (there are biblical scenes in the larger medallions). The portraits of all the four sons of Constantine I leave no doubts as to the dating of the Cologne beaker, and, in addition, solves the dating of the Szabadszállás beaker as well. We may put the appearance and first phase of its implementation to the reign of Constantine I (324-337), and the Szabadszállás beaker was also produced in this period.[1040]

The relationship between the burial and the beaker. The above-discussed arguments have already elucidated the fact that Pannonia is unconnected with the beaker, and, as a conclusion, there could be even less connection with the Barbaricum. Thus, only one solution offers itself to interpret the burial and the grave-goods between the Danube and the Tisza Rivers, and this is the geographical location; it has always been a determining factor throughout our history. The Carpathian basin is in the middle between east and west. Those who migrated in either direction had to cross our territory. The person buried here might also have been a traveller who died suddenly. He was buried here together with his personal effects. The glass beaker had also been 'acquired' by him, by grave robbing perhaps or other means. It did not belong to him originally. Therefore, there is no relationship between the funeral rite and the beaker.

SUMMARY

The beaker cannot be classified as a Christian object based on its illustration or inscription. It fits well with the religious mentality and burial customs of the 'pagan' world. The inscription is a good-wish and was commonly known in the 'pagan' world. All the formal parallels of the beaker were found in pagan environments. Its sole technical analogue depicts the members of the Constantine dynasty together with biblical scenes, where this matching of scenes can be interpreted as an announcement of an administrative measure for the benefit of the Christians. Thus, had they wanted to express their Christianity they could also have illustrated biblical scenes instead of, or among, Bacchic ones.

The beaker is datable to the time of Constantine I, and there is no connection between the presence of the beaker and the history of the province. Its appearance in the Barbaricum is accidental.

[1038] BARKÓCZI 1988: 80-85.
[1039] FREMERSDORF 1958: 53 Taf. 119.
[1040] I am grateful to Frau Dr Gudrun Gomolka and Dr László Török for the advice, literature and consultations concerning the Szabadszállás beaker.

45. SZÁZHALOMBATTA-DUNAFÜRED (MATRICA) fig. 324.

In the 1940s, some ruins of a building were found in a canal project. K. Sági and A. Radnóti visited the site and documented the structures found.[1041] In 1943, Ádám Dormuth had an excavation in the southern part of the fort.[1042] Éva Bónis and András Mócsy conducted an excavation in 1953. They opened three dwelling-pits and some stone buildings. They dated the dwelling-pits, based on the material fills, to the reign of Marcus Aurelius.[1043] 'Building I was a larger, villa-style house among the stone buildings. During the rescue excavation, wc could only open its southern part. We were searching for its further extension with the help of trial trenches, but we could not gain a comprehensive picture.'[1044] The buildings are marked I-VI; Károly Sági opened two rooms of VI in 1942. András Mócsy determined six periods, observing the traces of repeated alterations. At building III, a narrow wall was found made of a course of dry stonewalling. The thickness of the wall is 25-40 cm. Among the stones he found a small bronze coin from the Valentinian period and a bone tessera. He dates the building to the second half of the 4th century based on the coin.[1045] Edit Thomas conducted a rescue excavation in the 1960s. In the 1970s, Sarolta Tettamanti and Judit Topál worked in the area and excavated the cemetery. No finds from here refer to early Christianity,[1046] but the bone *tessera* seems to have Christian features. Naturally, we cannot draw major conclusions from this sole object and, perhaps, further excavations will supply more Christian finds.

45.I.a. Bone tessera, fig. 325.

Provenance: In building III. a narrow wall was found made of a course of dry stonewalling. Among the stones a small bronze coin from the Valentinian period and a bone tessera were found. **Location:** MNM 53.44.6. **Literature:** MÓCSY 1955: 64-65, fig. 5; MÓCSY 1964: 265 Abb. 141.

On the bone tessera there is a chi-rho monogram. Diam: 2.5 cm.

46. SZÉKESFEHÉRVÁR, ☞ Tác

46.I.a. Pillar or mullion, fig. 326.

Provenance: Arnold Marosi thinks it was found in a demolished house at Székesfehérvár. The majority of the Roman stones was taken to the town as building material. As to the forms of this type of material transport, see for more detail ☞ Tác, as the newer research - not mentioning Székesfehérvár - gives Tác-Gorsium as the site of the screen. **Location:** Tác, Szabadtéri Múzeum, Inv. No. 9904. **Literature:** MAROSI 1936 107; NAGY L 1938: 61 fig. 26; ALFÖLDI 1938: 166; FITZ 1967-1968b; Taf. 6; BOJÁR 1967-1968; TÓTH E 1974a: 172; FITZ 1976: 388 Note 16; Severin 8.29; THOMAS 1982b: 290-291; KISS Á 1987 No. XXI.8; TÓTH E 1990a: 26-27 Abb. 14.

[1041] THOMAS 1964: 261.
[1042] MÓCSY 1955: 59.
[1043] MÓCSY 1955: 62.
[1044] MÓCSY 1955: 62.
[1045] MÓCSY 1955: 64.
[1046] TOPÁL 1981.

Size: 98 x 40 x 24 cm. The stone was damaged and split. The edge of its front panel is broken off and the pattern is missing from the back panel. Front panel: The middle of the stone is filled with an encircled chi-rho monogram. There is a vine plant below, and above there are two crossing dolphins, between them a peacock feather. In the left corner, there is a fish. Under the dolphins there are a crayfish, a snail and a cuttlefish. Bojár, objecting to the idea of the tree-of-life motif, as others defined it, defined the pattern under the chi-rho monogram as a vine plant. Back: intertwined rosettes fill up the entire surface. Bojár's iconographical interpretation is the following: the fish is Jesus Christ as the monogram and the motif of the vine plant below it show. The rest are sea creatures. Even the small fish thrown back by the fishermen (who catch souls) are considered unclean. Others reach Christ through justification. The back panel, judging by the back panel, may be interpreted as a Sol-Lux symbol.

Iván Bojár dated the stone to the 8th century and provoked a debate. The first reaction came from Jenő Fitz, who put the stone to the end of the 4th century or rather to the 5th century. Endre Tóth joined the debate dating it to the 5th century. The other disputed question was the find-spot of the stone. At the time it was found, Arnold Marosi immediately thought it had been taken to Fehérvár from the 'early Christian basilica' of Tác. The problem is that at that time, the building was considered an early Christian basilica that later proved to be a bath belonging to a villa. Therefore the solution that then seemed to be evident has vanished. However, the theory saying that the stone could only have been transported to Fehérvár from Tác still prevails.

Commentary:

Provenance. The secondary provenance is obviously Székesfehérvár. A question might be raised about the original place from where it had been brought. Two possible versions are Tác and Aquincum. We certainly know that stones were brought to Székesfehérvár from Aquincum.[1047] It has not been proved that stones were brought to Székesfehérvár from Tác as well.[1048] Therefore, it may be assumed that the object considered as a screen had been brought to Székesfehérvár from Aquincum.

Style analysis and function. What struck Iván Bojár was that the technique of carving was actually equivalent to the chip-carving technique. He consequently dated the stone to later than the Roman Age. It is true that the carving of the stone denotes a later period, but not as late as Bojár claimed. The more accurate definition of this later period is helped by a further example. A basilica was excavated by N. Drandakis at Kyparissos (its antique name was perhaps Καινήπολις). The researcher dated this basilica to the 5th century. Woodcarvings, a part of a rood screen, and pillars that used to hold the arches of windows were found.[1049] One of the pillars is a close parallel. On the upper part there is a chi-rho monogram surrounded by alpha and omega. The part below is filled with a vine-tendril growing out of a pot. The representation of both the bunches and the leaves is similar to that on the Székesfehérvár stone. It verifies Bojár's opinion that there is a vine-plant and not a tree of life on the stone. The analogue is informative as to the function as well. The Székesfehérvár find has so far been defined as a screen, although it is higher than a usual screen. The analogue was not used as a screen but as the mullion of an arched window. A screen was also found in the same basilica. It is rectangular and in the middle of the profiled field there is a cross that has an ornament in a vertical stripe on one of its shorter sides. The rectangular screen was set on its longer side. This example shows that we have to define the Székesfehérvár stone as a pillar, and not as a screen. It may surely be dated to the 5th century.[1050]

Thus there are finds in the territory of the province that have significance beyond antique Pannonia, such as the building with apses in Aquincum, 4.I.g. It also supports the fact that the pillar was taken to Székesfehérvár from Aquincum.

The Saint Stephen (the first King of Hungary) sarcophagus[1051] should also be mentioned here. It was originally a Roman sarcophagus and its subsequent carving and medieval usage account for the fact that it has exclusively been dealt with by medievalists, although its motif could be a link between antique Pannonia and evolving Hungary.[1052] Furthermore, we should note those relics from the 6th and 7th centuries published by Branka Migotti,[1053] and those stone fragments from Siscia and the Muč area that are described by Hampel.[1054]

47. SZEKSZÁRD

47.II.a. Sarcophagus fig. 328

Location: MNM 23.1849.1. **Literature:** KUBINYI 1857a Tables I-II; KUBINYI 1857b; HENSZLMANN 1869 I. 311; CZOBOR 1879 170; GYÁRFÁS 1889 93; PULSZKY 1897 I. 253 Table XCIX; WOSINSZKY 1896 744-746; NAGY L 1938 48-51; NAGY T 1958 423; CERMANOVIĆ 1965; KÁDÁR 1969 188; NAGY T 1971 130; BURGER 1973 42-49; MÓCSY 1974 334; ERDÉLYI G 1974 68-70; FÜLEP/TÖRÖK 1977 312 Abb. 392; BARKÓCZI 1982 41; KOCH/SICHTERMANN 1982 327; THOMAS 1982b 274-275, GÁSPÁR 1986 § 89; BURGER 1991 Kat.-Nr. 97.

Marble sarcophagus whose cover and box are not contemporary. The research considered it early Christian because of the illustrations of the basket with bread and fish in the acroterions of the cover. On the box there are pictures of Eros and Psyche, as well as the competition between Marsyas and Apollo. One of the shorter sides is ornamented with a vine-tendril growing out of a *kantharos*.

1047 ALFÖLDI 1939; ALFÖLDI 1940; ALFÖLDY 1977: 229-230 and Notes 50, 53.

1048 They were brought to other places, but not to Székesfehérvár, ☞ Tác

1049 Chroniques des fouilles en Greece en 1958, Kyparissos BCH 88 1959 : 631-632. Figs. 8, 9, 10.

1050 A capital executed in a similar style was found in Aquincum and in Ulcisia Castra's cemetery.

1051 BOGYAY Tamás, Stephanus Rex. Wien, München 1976, figs. 54, 55, 56.

1052 The survival of pagan motifs is spotted by Alán KRALOVÁNSZKY - Contribution à la question du sarcophage de Székesfehérvár dit de Saint Étienne. Aregia VIII-IX 1967-1968: 85-91, pls 23-24.

1053 MIGOTTI 1997 II.2, II.3.

1054 HAMPEL 1894.

Koch and Sichtermann wrote: 'Der Kasten aus Alisca könnte noch im 2. Jh. oder um 200 n. Chr. entstanden sein.' This dating is significant, not because it could preclude the Christianity of the box, but because the Pannonian research dated it to the 4th century, only to declare it Christian based on the above-mentioned illustrations. The illustrations do not have a Christian theme. In my opinion, even the articles in the box, including the *vas diatretum,*[1055] preclude the possibility of a Christian association.

47.II.b. *Vas diatretum* fig. 327.

Provenance: In the above-mentioned sarcophagus. **Location:** MNM 23.1894.2. **Literature:** KUBINYI 1857a Table III; KUBINYI 1857b; HENSZLMANN 1869 I. 311; CZOBOR 1879: 170; GYÁRFÁS 1889: 93; KUZSINSZKY 1895 L; WOSINSZKY 1896: 744-746; PULSZKY 1897 I. 260; NAGY L 1930a: 116; NAGY L 1938: 49-51 fig. 17; KÁDÁR 1969: 199; MÓCSY 1974 pl. 35a; FÜLEP/TÖRÖK 1977: 312 Abb 388a; THOMAS 1982b: 274 Abb. 18; BARKÓCZI 1988: 556.

Size: diam: 16.5 cm; L: 10.5 cm. It was considered Christian because of its inscription: LEIB[E] T[ω] ΠΟΙΜΕΝΙ ΠΙΕ ΖΗ[CA]IC. Lajos Nagy first noted that the Christianity of the inscription is uncertain.[1056] ☞ Szabadszállás, 44.II.a.

48. SZENTENDRE (ULCISA CASTRA / CASTRA CONSTANTIA) fig. 329.

Roman sites are the fort,[1057] the cemetery,[1058] the building opened on Hunka Hill[1059] and the recently excavated villa[1060] in the territory of the village museum, ☞ Aquincum, the *cella trichora*, 4.II.a.

Flóris Rómer started the excavations at the fort. He was followed by Lajos Nagy, and then in 1939, 1940 and 1942 Tibor Nagy,[1061] who clarified the ground plan of the fort.[1062] In 1959, Sándor Soproni conducted an excavation, crossed the southern wall of the fort and opened the wall-section parallel with the wall of the fort. Next to the section he found fragments of frescos. Another excavation was directed by Márta Kelemen H. in 1968, she crossed the western wall.

The fort was built under Trajan and altered in Hadrian's reign. It was re-built after the Marcomann wars. It was in ruins in 270, putting an end to the first period of the fort. Although we may suspect a rebuilding by Diocletian, the horseshoe-shaped towers denote the second period. Under Valentinian, an elevation of the floor was executed as well as restoration. Even smoothed pottery was found in the *castellum*, leading S. Soproni to suppose the usage of the *castellum* in the 5th century as well.[1063] L. Nagy describes traces in the fort that refer to the fact that 'the continuity of the fort's settlement can be demonstrated until the Magyar arrival.'[1064]

The size of the civil settlement next to the fort is 500 x 400 m.[1065] The settlement was first excavated by Tibor Nagy. In the 1970s, Sarolta Tettamanti excavated the settlement and the cemetery. A summary of the civil settlement was given by the Topography. The houses are quadrilateral. The walls are 50 cm thick with frescos imitating marble-inlay. The material comes from the 2nd and 3rd centuries. Tibor Nagy also supposes the usage of the building in the 4th century, although 'in the remaining material there is no 4th-century piece.'[1066] A large (10.5 x 16 m) *villa rustica* style of building, with a 60-cm-thick *opus incertum* walling, terrazzo floor, a *hypocaustum*, frescoed walls and a westward closing with an apse, raised the excavator's suspicion that he had found a 4th-century Christian basilica.[1067] 'But we can only repeat that late material was not found in the *vicus*, which was, in our opinion, gradually abandoned at the end of the 3rd and at the beginning of the 4th century, getting closer to the fort, and in its territory they opened the late cemetery',[1068] fig. 330.

The late cemetery - based on the Topography's information - belonged to the fort.[1069] Flóris Rómer again started the excavation here. Lajos Nagy conducted a major excavation in 1928. In 1923 and 1924, several late Roman graves were found at 1-3 Római temető utca (Roman Cemetery Street). 'In the same area, a brick grave was found whose goods - casket-mounts and three bronze bracelets - were saved by Head Physician, Henrik Péchy.'[1070] The casket mount is decorated with biblical scenes and leaves no doubt as to the Christianity of the object.

48.I.a. Casket mount with biblical scenes, fig. 331.

Location: MNM 68.8.1. **Literature:** BUSCHHAUSEN 1971 A 62; GÁSPÁR 1986 Kat-Nr. 947.

The illustrations are composed in medallions on bronze plates: the miracle of the loaves and fishes, Daniel among the lions, the raising of the dead, Jupiter enthroned, the passage of ships, a portrait Constantine, Hercules with a shield and butt. This scene's legend is INVICTO CONSTANTINO FELIC ... ATI. The diameters of the medallions are 1.5, 3.2 and 7.2 cm. The smallest medallion portraying Constantine is next to the large medallion of Jupiter. This may even make us think that the successor of the attribute *Iovius* is the one appointed by Jupiter the Great (God). He, the INVICTVS, is offered the military service of the possessors of the attribute *Herculius*, with shields in their hands. A frieze with running animals in rectangular fields also belongs to the mount. The edge of

[1055] GÁSPÁR 1986 Kat-Nr. 946 and § 89.
[1056] NAGY L 1938: 50.
[1057] MRT 7. 28/1. For the disclosure of the history of research belonging to the certain areas the relevant passages of MRT were used.
[1058] MRT 7. 28/2.
[1059] MRT 7. 28/11.
[1060] MRT 7. 28/12.
[1061] NAGY T 1943: 392.
[1062] NAGY T 1943: 392-399; here he refers to his earlier publications that study the results of excavations at Szentendre.
[1063] SOPRONI 1978: 68-71; even BARKÓCZI/SALAMON 1984: 158-161.
[1064] NAGY L 1939: 147.
[1065] MRT 7. 28/1 p. 252.
[1066] MRT 7. 28/1 p. 253.
[1067] TÓTH E 1991a: 123: in the summary concerning antique Christianity the building is mentioned as 'a rectangular building with an apse.'
[1068] MRT 7. 28/1 p. 253.
[1069] MRT 7. 28/2.
[1070] MRT 7. 28/2 p. 255.

the bronze plates is bordered by mouldings; the zone is ornamented with a waving line and leaves. We rarely hear of the symbolic interpretation of this wavy line decoration, although it is to be found not only on casket mounts, but also on gravestones and other finds. We usually define it as an undulating line (albeit a spiral), and we should interpret its existence according to its iconographic message. Assessing different scholars' opinions to an ancient meaning, the spiral is the symbol of 'life-giving'.[1071] Opinions differ as to the origin of the form; it may derive 'from the waves of the sea (Okeanos), from the *Octopus* or *Nautilus*, ... or *tendrils* (scroll) of germinating plants, ... from the holy palm-tree (*phoenix dactylifera*), or - leaving earthly considerations entirely - from the 'spiral twist' of the planetary orbits, already noted by Plato.'[1072] The illustration of the spiral on this mount harmonises with the passage of the ships;[1073] we are drawn to the waves of the sea. And in the scene, the name of the *Nautilus* aligns perfectly with the *navis*. Together with the Christian scenes, they make an interpretable program of pictures we may describe as follows: Constantine is also represented as Iovius, INVICTUS. To him, the invincible, the men known as *Herculius quasi* offer themselves in service. This Constantine is the one who brings life by declaring Christianity a legal religion. This attribute of the one who brings life is supported by the biblical scenes: the raising of the dead, the survival of Daniel, and the multiplication of the loaves and fishes.

48.I.b. Cemetery[1074] fig. 332.

'At the end of 1928 and the beginning of 1929, twenty graves and a quadrilateral grave-chapel were opened by Lajos Nagy on the plot at 16 Római sánc út (Roman Fort Street - the so-called Reichert-plot, No. 1448). ... A small (7 x 5 m) double-room tomb. The walls below are probably the remains of the earlier *vicus militaris*. The larger room is 5.35 x 5.2 m. and a terrazzo type plaster covered its smaller inner sides. Its lower part consisted of a grave that was found open. There were no graves in the bigger room, but its elaboration may have been rather decorative, suggested by a capital with leaves and a fragment of a cornice ornamented with *kyma* and acanthus-leaves. The building has become well known in the scholarly literature as an early Christian *cella memorise*, although we have no primary evidence proving this fact. The orientation of the surrounding graves was W-E and, according to the marking of the earth graves, they were lying with their heads pointing westwards.'[1075] The height of the capital is 32.5 cm. The size of the cornice-fragment is 94 x 114 cm.

Lajos Nagy thinks the capital[1076] is late Roman according to its style; he draws a parallel between this and the two capital fragments from Aquincum,[1077] (4.I.h). He also adds that the bigger room is the same age as the graves, and is in close connection with the cemetery. The forms of graves are various: sarcophagus, sprinkling with lime, and simple burial in the ground or brick graves. Sometimes the tombstones were re-used. One bears the name of *Castra Constantia*.[1078] Lajos Nagy also mentions that two burials took place in one sarcophagus. The dead were placed either next to, or facing, each other.[1079] He dates the usage of the cemetery to between 369 and 375.

Commentary: The tomb is similar to tomb IX at Pécs as far as its ground plan and size are concerned. In the chamber at Pécs, too, burials only occurred in the 'small' part. I do not regard this building to be anything other than a tomb, as is the one at Pécs. Its larger part is the *memoria* belonging to the grave and the chamber: nothing more and nothing else. I do not think it has a relationship with the cemetery generally or with the rest of the burials. There is no find or phenomenon referring to early Christianity. The capital found next to it might have theoretically belonged to the tomb, although the usage of pillars is not supported by the size of the building. Its surroundings do not help date the capital, only its style and the stylistic marks of the above-mentioned two capitals at Aquincum may serve as a clue. Two capitals - one from Aquincum and one from Szentendre - display the chip-carving technique, a method observable on other capitals dated to the 7th century.[1080] It is evident that our dating cannot differ from it considerably in Pannonia, but we may say, at the present stages of excavation and publication, that the find does not offer more than a general hypothesis. I present my hypothesis in the form of a question, as the period is beyond the limit of this study: What correlation can we recognise between the strap-end, long or short buckles, or mounts, and these capitals?

The evaluation of the data from Szentendre. We may state that one of the excavated graves bears Christian characteristics, made evident by the illustrations of its casket mount. Considering the program of the pictures on the mount, we will see that its Christian reference is even stronger, but, on the other hand, it connects the change of the name from *Ulcisia Castra* to *Castra Constantia* to the person of Constantine I. Thus, in the territory we have to reckon with Christianity (in various forms) even though the data at our disposal makes this difficult. Although the grave and the *memoria* are close to each other physically, I would not say there is a relationship between the two. Accepting

---

[1071] BARB 1953: 207-208.

[1072] BARB 1953: 208.

[1073] This scene is not unique in Pannonia. The same motif appears on another casket mount in the same form. GÁSPÁR D, Kästchenbeschläge und Kästchenbestandteile aus dem Komitat Fejér. Katalog II. Aregia XXVI 1997 Kat.-Nr. 2192. The research tried to identify this ship scene with a concrete historical event. I would rather attribute a symbolic meaning to it.

[1074] NAGY L 1938 fig. 82. The general drawings of Christian cemeteries - including this cemetery part at Szentendre - see MÓCSY 1974 fig. 52. TÓTH E 1990: 21, fig. 4.

[1075] MRT 7. 28/2 p. 256.

[1076] NAGY L 1938: 130 fig. 83; BT (NAGY L) 1942: 777; KISS Á 1987 No. XIV.2 - the definition of its find-spot is not correct; it was from the fort not the cemetery. As to the cornice fragment KISS Á 1987 No. XIV.6.

[1077] KISS Á 1987 No. XV.73-74.

[1078] NAGY L 1938: 131; CIL III 15172

[1079] Pécs, grave 13, 32.I.a.

[1080] H. SCHLUNK: Iberische Halbinsel. In: PKG Suppl.I. 1977 Nos. 331, 332, 333, pp. 277-278. The mentioned phenomena is said by Schlunk o be due to an oriental influence.

the opinion disclosed in the Topography, I dare say that this *memoria* is not Christian. Or, to be more exact, it is not Christian if it is the same age as the graves. The probable dating of the capital found next to it reaches beyond the time limits of the present essay, but it can be informative as to the meeting point of antiquity and the Early Middle Ages.

As in Aquincum, we may also note here the strange fact that the burials - few as they are - supplied Christian finds, whereas the settlement did not. This means that the Christians were represented by a smaller population, perhaps also belonging to the *nobiles*. At present, however, it is not possible to say where they lived.

49. SZENTKIRÁLYSZABADJA-ROMKÚT, figs. 333a, b.

49.II.a. 'Hall church'

There was an excavation in the territory, but it was not published.[1081] Edit Thomas, attaching a general drawing of the buildings, disclosed a brief summary of it.[1082] This was the occasion Edit Thomas described building II as a hall-church.[1083] The Topography offers a re-evaluation of this excavation and publication.[1084] The result of this re-evaluation is the general drawing of the Roman villa based on the excavation.[1085] I and II published by Edit Thomas are not present in the drawing of the Topography as 'the buildings were not opened, their places were marked next to the other buildings according to traces on the surface, and so we cannot draw conclusions as to their functions merely from their ground plans.'[1086] That is why it is regrettable that the reconstruction has become well known in the scholarly works that also serve as a basis for serious conclusions. ☞ Sümeg.

50. SZILÁGYPUSZTA-MALOMALJA

50.II.a. Bronze sheet, fig. 334.

Location: JPM. **Literature:** DOMBAY 1957 Abb. 62; THOMAS 1964: 299 Abb. 156.

The size of the bronze sheet is 5.3 x 1.7 cm., with a chased pattern of dots on one side. Edit Thomas interpreted it as a chi-rho monogram with A-Ω. If we wish to see A-Ω in the dots, the chi-rho monogram is upside down. The interpretation is certainly problematic.

51. SZOMBATHELY (SAVARIA) ☞ Sopron, Donji Petrovci

History of Research

Contents: 1. Quirinius's basilica, 2. Saint Martin's native house and birthplace, 3. Savaria as an episcopal seat, 4. Savaria's topography.

The research of the town goes back to the 18th century. István Schoenvisner wrote the history of the town and a monograph of the archaeological finds.[1087] The special importance of his work lies in the fact that he enumerates, describes and illustrates finds that have been destroyed since then, which means that his documentary of the history of the town is of considerable value.

1. The Quirinus basilica. Gyula Géfin[1088] and István Paulovics Járdányi[1089] contributed a greatly to the archaeological excavations and research of the town. They directed the excavation of the Romkert (garden with ancient ruins) and found the building that was documented as Quirinius's basilica by István Paulovics.[1090] This theory persisted for a long while,[1091] until 1973, when Endre Tóth proved that the representative building thought to have been Quirinius's basilica had actually been an imperial palace.[1092] From this time on, however, the question has been: Where then is the basilica mentioned in the *passio*? Today's research suggests that the basilica is located at the gate to Sopron (presuming that the term *porta Scarabetensis* means the gate to Sopron, as suggested by the earlier research. ☞ Sopron.)[1093] Quirinus - at the beginning of the *translatio* - was later taken by the fugitives to Rome, and was buried in the catacomb on Via Appia next to Saint Sebastian, where a church was later erected.[1094] The earthquake in Savaria on 7 September, 455, destroyed the Szombathely basilica.

2. The birthplace of Saint Martin of Tours, or the dispute concerning the place, as both Szombathely and Győrszentmárton rivalled for the birthplace; the dispute had been going on for at least 150 years by 1944.[1095] The problem seemed to have been settled when István Paulovics disclosed the results of his previous excavation. This excavation had been conducted in the village of Szent-Márton (Saint Martin), approximately 1.5 km east of Szombathely, fig. 335. In fact a tradition existed, and still exists, at Szombathely, whereby the medieval Saint Martin's church was supposedly erected over the saint's native house. It was thus important to determine whether there were any Roman antecedents of the church. István Paulovics thought he had found this Roman antecedent under the church,[1096] and the excavation seemed to justify the tradition.

In Tibor Nagy's opinion, Saint Martin's birthplace is evidently Savaria. The saint's biography mentions the

1081 MRT 2, 44/5.

1082 THOMAS 1964: 118-122 Abb. 56.

1083 THOMAS 1964: 118-119.

1084 MRT 2, 44/5. The concordance of the origins of the buildings: T I = Th IV, T II = Th V, T III = Th III. T=Topography, Th= Thomas.

1085 MRT 2, p. 188, fig. 38.

1086 MRT 2, p. 189.

1087 SCHOENVISNER 1791.

1088 GÉFIN 1992.

1089 PAULOVICS 1943a; 1943b; 1944.

1090 PAULOVICS 1938a.

1091 BALLA/KÁDÁR 1958.

1092 TÓTH E 1973: 117-137; TÓTH E 1975: 121-139; TÓTH E 1978-1979.

1093 The so-called Sopron basilica, i.e. the basilica next to the gate to Sopron. GYÁRFÁS 1889: 63.

1094 GYÁRFÁS 1889: 64; NAGY T 1946: 244-257; TÓTH E/ZÁGORHIDI CZIGÁNY 1994: 24-24 Appendices a, b. Cf. ŠAŠEL 1974. The mausoleum of St Quirinus – the Platonia, see A. FERRUA: The Basilica and the Catacomb of Saint Sebastian. Rome 1978: 38.

1095 PAULOVICS 1944: 3; MESZLÉNYI 1977; With the list of more recent literature and the disclosure of the dispute TÓTH E /ZÁGORHIDI CZIGÁNY 1994: 27-28, No. 16.

1096 PAULOVICS 1944: 12-16.

presence of *sacerdotes* in Savaria.[1097] In the same question, Endre Tóth brought up new arguments in favour of Szombathely.[1098] It still seemed necessary to conduct a verifying excavation within Saint Martin's parish church. In the first phase, the archaeologists stated that there were parts in the crypt that were earlier than the Baroque church.[1099] The new excavations, however, proved that Savaria's early Christian church and Saint Martin's native house should not be sought there. The church, as it were, had no Roman antecedents. 'Before the medieval constructions, all through the Roman period, there was a cemetery in the territory, proved by numerous stray finds as well as the, unfortunately, robbed 4th-century sarcophagus.[1100] The excavation also led to the fact that Saint Martin's native house and the church erected over it should not be looked for under Saint Martin's parish church, but under another medieval church, the Franciscan one. The excavation, therefore, could not settle the problem.[1101]

3. Episcopate and bishops. According to Salagius,[1102] we do not know exactly the location of the seats of Bishops Migdonius and Megasius, who both took part in the Arian struggles, but presumably one of them was the bishop of Savaria. This idea is based on Sulpicius Severus's information in Saint Martin's biography, saying that Saint Martin was exiled from the country of the Arian bishops.[1103] In Endre Tóth's opinion, Savaria's bishop is the Arian Gaius.[1104] Tibor Nagy thinks that the archaeological finds do not allow us to localise the Episcopal seat here, even if we reckon with the existence of a baptistery, as Canon 38 of the Council of Elvira (c. 300) expresses that, because of a shortage of priests, baptism could be administered by anyone.[1105] He does not even consider the (already mentioned) term *sacerdotes* in the biography as evidence, as he thinks the plural refers rather to priests.

Gaius's seat is thought by Salagius to be in the City of Cibalae, whereas Tibor Nagy suspects it was in Iovia. He argues that 'Gaius used to possess a respected position in the Pannonian Arian clergy, as shown in the 5th chapter of Athanasius's *Historia contra Arianos*. Accordingly, to the age, his seat must also have been significant. We know that Iovia was such a place between the Drave and the Save Rivers. The importance of the community, from the viewpoint of orthodoxy, may be deduced from Amantius's mission, or also the fact may be concluded that the dominant denomination was Arianism for a while. Our impression is that Gaius's Episcopal seat was most probably here.'[1106]

Actually, the Episcopal seat was also considered very seriously because everybody was led by the presumption that there used to be a huge basilica whereas, as it later transpired, there was just an imperial palace.

4. Savaria's topography, fig. 336. Savaria was the capital of Pannonia Superior/Prima, it was reported on by Ammianus Marcellinus. He mentioned the earthquake in the city as well as the fact that there used to be a *lavacrum regium* here, where the emperor, during his visit, had a bath.[1107]

Recent excavations have rendered it possible to define the extension of the city as the opened cemeteries around the area that used to belong to the settlement. Terézia Buocz prepared its earlier topography.[1108] Besides the dwelling houses, road-sections and conduit parts, one of the greatest excavation works was the opening of the *Iseum*.[1109] Near the Iseum, a *horreum* was also found, about which we know nothing but a single remark.[1110] The further excavations have brought other religious relics to light as well; 'a large building complex partitioned with inner pillars' was opened at 3 Alkotmány Street. Here, several altars were found. Among these was an altar to Mercurius and Jupiter, as well as votive statuettes to Mercurius, a Mercurius-shrine and a statuette to Minerva.[1111] The excavation on the schoolyard at 8 Hollán Street began in 1983. The 5 x 5 m. southern gatehouse belonging to Savaria's western gate was found here. From this place, the city wall extended in a southeast direction.[1112] Outside the city wall, a workshop for bronze casting was also found.[1113] The research has defined two periods of the city; the city was extended in the second. Until the most recent excavations, however, the direction of the city wall and the situation of the gates had been debated. A brief summary of the history of the city can be read in the first volume of the RIU, from which I quote the following passage: 'Die Gründung durch Claudius ist ausdrücklich bei Plin. n.h. III 147 erwähnt; der Beiname *Claudia* wird auf Inschriften regelmäßig ausgeschrieben. ... Collegien: *fabrum et centonariorum und genii p(rovinciae) P(annoniae) S(uperioris)*. Letzteres wird wohl mit der Ara Augusti von Oberpannonien zusammenhängen, die in Savaria stand. ... Savaria war eine station des *publicum portorium Illyrici* und hatte auch eine Benefiziarstation. ... Die Blütezeit der Stadt war die Zeit von den Flaviern bis zu den Severern. Die überwiegende Mehrzahl der Inschriften stammt aus dieser Zeit. Im 4. Jh. war Savaria Sitz des *praeses Pannoniae Primae* und erlebte eine nochmalige Blüte, davon zeigen die in Pannonien sonst ziemlich seltenen

[1097] NAGY T 1939: 210-211. - Sulp. Vita s. Martini c. 2. CSEL 1 ed. HALM.
[1098] TÓTH E 1972b: 244-252.
[1099] MEDGYES/TÓTH E RégFüz Ser. 1. No. 38 1984 83/3.
[1100] KISS G/TÓTH E RégFüz Ser. 1. No. 44 1990 167; KISS G/TÓTH E 1993: 185.
[1101] The debaters were not influenced by the fact that the biography clearly states that the birthplace is Sabaria, see later.
[1102] SZALÁGYI 1784: 223.
[1103] Vita s. Martini c. 6. CSEL 1 ed. HALM.
[1104] TÓTH E 1991: 121.
[1105] 'lavacrum suum integrum habet' NAGY T 1939: 211.
[1106] NAGY T 1939: 210.
[1107] Amm. Marc. 30. 5. 16 ed. SEYFARTH 1971.
[1108] BUOCZ 1962; BUOCZ 1967; A Romkert úthálózatáról - On the road system of Romkert BUOCZ 1971-1972: 201-215.
[1109] SZENTLÉLEKY 1960; SZENTLÉLEKY 1965.
[1110] BARKÓCZI/SALAMON 1984: 172; TÓTH E 1972b: 243: This area was built up later than the *colonia*; the road system was also different.
[1111] BUOCZ/SZENTLÉLEKY RégFüz Ser. 1. No. 34 1980 78/1; No. 37 1983 92/1.
[1112] MEDGYES RégFüz Ser. No. 38 1984 83/2.
[1113] SELLYE 1990: 51-52.

altchristlichen Grabsteine. ... Das Territorum der Colonia war wahrscheinlich ziemlich groß, wenn auch seine Grenzen nur im Norden, gegen das Municipium Flavium Scarbantia hin ... Größere Siedlungen sind im Territorium bisher nicht bekannt, die Itinerarien zählen als Station der Bernsteinstraße *Arabo* (Körmend?) und die Station der Straße Savaria-Arrabona *Bassiana* (um Sárvár?) auf.'[1114]

As to the last sentence, it should be noted that the research knows of two sites named Bassianas in Pannonia. One of them is called *Baciensis civitas* and is near Sirmium,[1115] and other is between Savaria and Arrabona,[1116] fig. 336. The research of this last road section has now taken place,[1117] fig. 338. The results may be outlined as follows: Leaving the City of Savaria through the eastern gate, one takes the road that proceeds straight to Bassiana. The road is well observable in the area. The question of the road is linked with the identification of the ancient Bassiana with a modern settlement. In the beginning Sárvár and Körmend seemed probable, but the road leads to a Roman settlement that was 26 km from Savaria (as written in It. Ant. in connection with Bassiana. XVII, XVIII MP - 26.64, 25.126, 28.082 km). This Roman settlement lies along the Sárvár-Celldömölk railway-line, north of the Zalabér junction today.[1118]

The results of the most recent research (regarding the city walls and city gates) have been collected by Magdolna Medgyes.[1119] As a supplement to her summary, she discloses the defined city walls and gates, fig. 339. To see the changes caused by the new results see fig. 337.

Commentary, disclosed in points I and II.

I. CHRISTIANS IN SAVARIA AND QUIRINUS

'Siscia was the most significant, and as we presently know, only episcopate in Savia. Its bishop represented this province at the Councils of Serdica and Aquileia (?).[1120] András Mócsy shares this opinion: 'The only martyr in the province of Savia was the famous Quirinus, bishop of the provincial capital at Siscia. Quirinus was executed in Savaria, ...'[1121] 'The story of his sufferings is told by *Passio Quirini,* whose author is unknown, as follows: On the announcement of the edict ordaining the persecution of the clergy, Quirinus flees Siscia. However, the soldiers sent out to find him arrest him and present him to Praeses Maximus. Despite all the persuasions and tortures of the procurator, the aged Bishop of Siscia persists in his faith insistently. Maximus imprisons him, then in three days he sends him to Amantius, Pannonia Prima's proconsul in Savaria. However, Amantius is also unable to break the resolve of Quirinus and he condemns him to death. They bind a millstone to the bishop's neck and drown him in the Sibaris (today Perint) brook.'[1122] I quote: '*Tunc ... jussit sancto dei Sacerdoti vel famulo molam ad collum ligari, et in fluvii Sibaris undas demergi. Cumquae de ponte praecipiatus fuisset in fluvium, et diutissime supernataret, et cum spectantibus locutus est, ne suo terrerentur exemplo, vix orans ut mergeretur obtinuit. Cujus corpus non longe ab eodem loco, ubi demersum fuerat, inventum est, ubi etiam locus orationis habetur. Sed ipsum sanctum corpus in basilica ad Scarabetensem portam depositum, ubi major est pro meritis eius frequentia procedendi.*'[1123]

Translation: Then ... he ordered that a millstone be bound to the neck of God's holy priest, like that of a slave, and to sink him in the water of the Sibaris brook. As he was pushed into the brook, he floated on the surface for a long time, and spoke to the spectators not to recoil at his example. Hardly had he prayed for sinking, when his wish was granted. His corpse was found not far from the place where he had sunk, where their oratory stands today. Yet, the sacred corpse was buried in the basilica at the Scarabetensis gate, where, because of its merits, the possibility of visiting it is better.

The questions concerning the place cited are as follows: 1. Which town should be identified with *civitas Scarabetensis*? 2. Where is the *porta Scarabetensis*? 3. How can we reconstruct the site described in the *Passio*? 4. What is *locus orationis*? 5. How should we mean by *basilica*? 6. Why was Quirinus taken to Savaria?

1. Civitas Scarabetensis

I must refer back to Sopron where the question has already been partially examined and where *civitas Scarabetensis* is identified with Bassiana, more exactly with the Bassiana next to Sirmium. Furthermore, we should remind ourselves of the fact that Quirinus was transported from town to town along the Danube. It seems probable, therefore, that he was taken to Arrabona by boat, and then by road to the other Bassiana II. The author of the passion did not differentiate between the two cities, regarding their name or any other respect, as those who knew this did not confuse the circumstances. Thus, in this case we are talking

---

1114 RIU 1 p. 13.

1115 GRAF 1936: 52, 57, 58, 111. ☞ Sopron.

1116 GRAF 1936: 52, 57, 89.

1117 TÓTH E 1977b. We have to note that from this road another one branches off, along which, next to Katafa, a sentry post was excavated. The sentry post was used in the 2nd and 3rd centuries, and a treasure of 18 Marcus Aurelius coins was found in it. No mark of burning or any human destruction was observed. It was identified with the sentry post of 'Arrabone'. CSERMÉNYI V/TÓTH Endre: Eine römische Strassenstation und die Strassenstrecke zwischen Salla und Arrabona. Savaria 13-14 1979-1980 171-203. PALÁGYI S: Adatok a Savariából Arrabonába vezetõ római út kemenesszentpéteri szakaszának kutatásához. In: A Dunántúl település története, IX. Város - mezõváros - városiasodás. - Details as to the research of the Kemenesszentpéter section of the road from Savaria to Arrabona. In: The History of Settlements in Transdanubia, IX. Town - market-town - urbanisation. Ed. SOLYMOSI L / SOMFAI B. Veszprém 1992: 27-34.

1118 TÓTH E 1977b: 73.

1119 MEDGYES 1977; MEDGYES 1998. Both works contain detailed literature of the recent research. I have to note that I had known about the new results before the publication of the articles as they had been disclosed in a lecture at an international colloquium, and Magdolna Medgyes had also informed me about her own results personally, and had put the documentation at my disposal, for which I am grateful.

1120 NAGY T 1939: 226.

1121 MÓCSY 1974: 326. ŠAŠEL 1974: 738 writes about the uncertainty around the date of his death, as well as his surrender to Pannonia Prima's preases and finally the *translatio*.

1122 NAGY T 1939: 65-66. In the same place, Tibor Nagy studies in detail the literature of Quirinus and the sources containing data as for the martyr. There are three such sources: '1. Hieronymus's *Chronicon*, 2. Prudentius, 3. the Martyrologium Hieronymianum.' He also deals with their chronological order and authenticity in detail, 1939: 66-73.

1123 Passio Quirini V, RUINART 1859 p. 524.

about the Bassiana (II) between Arrabona and Savaria, and not about Sopron. It may explain why nothing mentioned in the passion can be found around the actual Sopron (northern) gate. In the present case, enough support is given by the *argumentum ex silentio* to prove that the *civitas Scarabetensis* is not Sopron. Because of its similar name with the South Pannonian Bassiana I, the Bassiana II near Savaria becomes more easily identifiable. Another argument for this city may be the fact that the author of the passion did not use the name *Scarbantia* alone, but the name *civitas Scarbetensis*, as he was obviously influenced by the other Bassiana.[1124]

2. Porta Scarabetensis

The previous determinations conclude in the result that in the story of Quirinus' sufferings, not the gate to Sopron is mentioned, but the one to Arrabona. The survey of the road between Savaria and Bassiana II makes this possibility even more probable. This was the shortest way from the Danube, only 26 km. This road, however, does not lead to Savaria's northern, but to its eastern gate. Consequently, the *porta Scarabetensis* is supposedly identical with the town's eastern gate.

3-4. Site and the *locus orationis*

Endre Tóth has already attempted to reconstruct the site, fig. 340, where the marks are the following: 1. the imperial palace, 2. the Sopron gate, 3. Saint Martin's hypothetical native house and the hypothetical Christian cemetery, 4. the place of Quirinus' martyrdom, 5. the western gate, 6. the *amphitheatrum*, 7. the southern gate.[1125]

Let us have a look at the allusions to the site in the text: The interrogation took place in the amphitheatre.[1126] The order was in all probability also given here which is reinforced by the reference to other sufferings as well *(inter ceteras quas pertulit passiones)*. According to the command he was to be thrown into the Sibaris brook. The Sibaris flows west of the city. Quirinus was pushed into the water from a bridge. Its site is well determinable, as pillars of a bridge have been found, marking the place of execution. Further information provided in the text is that his body was found not far from where he had sunk. The whole event must have taken place near the bridge as he clearly spoke to the spectators. Following this there is a remark that may be more significant to archaeological research; it relates to where he was found (*locus oratoris habetur*). It is clear that the Christian community gathered for prayer here, outside the city walls, in a place known as *locus oratoris*. It must have been at least part of a private house, or a whole house, similar to the medieval *oratoria* (a private house of prayer).[1127]

In the general plan (fig. 341), I have indicated where the corpse was found. Opposite, on the left bank of the brook, is the *locus orationis* that may have been erected. There has not been an excavation here but the site survey and other observations have spotted traces of buildings. The verification, or refutation, of my theory needs an excavation. The place is marked (2), and it is certain that in the 3rd century Christians were already living in the area. The place on the right bank marked (3) is where I suppose the dwelling places of the Christians were in the 4th century; the community born as a result of the events.

5. The basilica

The word *basilica* has several meanings. It is usually a building but its form may vary.[1128] It is necessary, therefore, to determine its form in this case.[1129]

The events discussed took place during Diocletian's reign. The Christian bishop is executed because he is not willing to offer a sacrifice with incense. He was buried after his body had surfaced (if we select this meaning of *depositus*). The definition of the *locus orationis*, mentioned by the Passion, as being an existing building was certainly more suitable in the era of persecutions.[1130] However, Quirinius is not buried near the *locus orationis* and the fact that he was buried in the basilica next to the Scarabetensis gate is emphasised as a contradiction by the author of the Passion. He also adds that this was 'so that he (the bishop) could be visited more easily.' The stress here is not on the word *basilica*, but on the clause '*so that he could be visited more easily*'. That is, they specifically chose such a burial place for him, not a simple earth grave but a marked tomb, a *martyrium*.[1131] Such buildings in the Middle Ages could have been the *tituli minores* churches, where priests, not bishops, undertook their duties, subordinated, however, to their bishops.[1132] From the text it seems as if the execution, finding of the corpse, and his burial all occurred on the same day. We have to suppose that the basilica already existed and that, therefore, it could not have been a showy, representative building. This basilica, nevertheless, had been erected in the cemetery, denoting which part of the cemetery belonged to the Christians. It is certain that the Roman cemetery was also in this area and that the gravestones were found here.

There is one more event in the Quirinus story, his

[1124] This possibility raises the question of at what time Bassiana was called *civitas Scaravaciensis*. In Iustinianus' decree, the name seems to be well accepted.

[1125] TÓTH E 1973 fig. 2.

[1126] Passio Quirini IV, RUINART 1895 p. 523.

[1127] Aclim. Ep. 7. (6th century) In: BLAISE 1954: *basilica*. Cf. ADAM 1985: 304.

[1128] ADAM 1985: 302. Cf. note 1440.

[1129] In Tibor Nagy's opinion, the Passion of Quirinus was written in its present form between 380 and 395, based on previous sources: NAGY T 1939: 66. This idea resulted in the conclusion that the existence of the basilica should not be put to the time of Diocletian, but to the period of the compiler, when even a representative basilica may have been erected: TÓTH E 1975a: 421-422. This assumption has no archaeological support at the moment. In the same place, Endre Tóth clearly discloses the fact that 'around the Scarbantia gate there is no Christian find'.

[1130] Endre Tóth supposes that the basilica was erected at the end of the 4th century, so we can reckon with a representative building.

[1131] BLAISE 1954 basilica: ce mot désigne aussi des sanctuaires moins importants; BLAISSE 1975 basilica: tombeau royal, chapelle funérarie, mémoire de martyr. For us it is important that the basilica can be a house of prayer, a private building. It also requires consideration in connection with the basilica in the text of the passion.

[1132] SZEREDY 1883: 573 with reference to C. Ticin. c. 13, a. 850: 'Propter assiduam erga populum Dei curam singularem plebibus archipresbyteros praeesse volumus; qui non solum imperiti vulgi sollicitudinem gerant, verum etiam eorum prebyterorum, qui per minores titulos habibant, vitam jugi circumspectione custodiant.'.

translatio.[1133] The opinion that Scarbantia's Christians, as they flee, take Quirinus' body to Savaria still prevails in the Pannonian research even though Tibor Nagy has demonstrated that this could not have been so.[1134] He puts the date of the *translatio* to between 405 and 420. The year 405 is provided by Prudentius's stay in Rome; in 404, he is there and writes a hymn to Quirinus.[1135] At this time the body could not have been in Rome. However, the *Calendarium*, written in Rome around 420, already mentions Quirinus and so the remains of the Saint must have been there by that date. It is certain that the building in which his body had rested before being transported to Rome was also respected later, and this is the point where Saint Martin's story has a link. István Paulovics was right when he considered it important that 'Saint Martin's Church, the parish church at Szombathely, was not in the territory of the town but one kilometre east, where Savaria's late Roman Christian cemetery lay. ... The continuation of medieval Sabaria's E-W main street leads across the Gyöngyös brooks, about 210 m away to the east. The church ... was erected approximately 630 m east of the medieval town gate. Its distance from the castle is exactly one and a half kilometres. ... In the described topographical situation, the most important peculiarity of Saint Martin's church is that in 1237 the church of Virgin Mary was also standing south of the castle. ... Two factors enhance the uniqueness of the City of Sabaria-Szombathely. One is that Saint Martin's church was erected on Sabaria's late Roman Christian cemetery. ... In this case ... there is a cultic continuity, insinuated by the surviving name of Sabaria and the continuity of the cemetery. The inhabitants of Szombathely have used the area around Saint Martin's church as a cemetery (over the old Roman cemetery) since the Middle Ages.'[1136] The archaeologists found the 9th-century church and the graves immediately around the church; the continuity of the graves is supposed until the 11th century. This was the time the church was extended.[1137]

My hypothesis regarding the story of Bishop Quirinus is that he was buried in a building called *basilica*, which was located in a cemetery but served as a place of prayer. After his *translatio* the basilica continued his memory and veneration.[1138] In the Middle Ages it was known that the place had been the resting place of a saint and bishop. Thus, in the 9th century, the church was erected in a special place and the burials took place in its vicinity. They kept the memory of a saint and bishop but they did not remember his name. Because before the 11th century towns did not have parish churches, and because this place met the legal and formal requirements necessary to have a *presbyter de plebe* (parish priest),[1139] Savaria's first parish church was built here. Owing to the saint, it continued to preserve this significance. By the 11th century, however, the tradition of Saint Martin had become more dominant, and he was known as both a bishop and a saint. For the Middle Ages, it was not the name that was important but the notion of sainthood, and especially that of a saint/bishop. In this respect Quirinus and Saint Martin were the same and from this time on Martin was venerated. Because he was not a martyr, only his native house could explain the exactly defined place of the building, and this is how Martin's native house was built on the site of the tomb of Quirinus.

6. Why was Quirinus taken to Savaria?

The literature so far has thought, and supposedly with good reason, that Quirinus was taken to Savaria because the community wanted to set a precedent for all Christians who refused to offer a sacrifice with incense. This argument implicitly involves Tibor Nagy's careful opinion that in Savaria, contrary to all appearances, there was no bishop (or at least not at this time). Amantius was Savaria's procurator (*praeses*), which was also a reason for the sentence to take place there. However, Quirinus was the bishop of Siscia which raises the question why Savaria's *praeses* had to pass a judgement over him. The problem can be solved by explaining the contemporary administrative system, as Jenő Fitz does.[1140] The relationship between Siscia and Savaria already existed in early Roman times. Savaria was Pannonia Superior's, later Pannonia Prima's capital, as well as the location of the provincial diets in the territory. In my opinion, Savaria was of a higher rank than Siscia, and consequently their *praeses* were subordinate to one another. I quote an inscription that was found next to the amphitheatre in Savaria. It is a marble table devoted to Nemesis: Nemesi / Aug(ustae) sac(rum) / L(ucius) Val(erius) Valerian(us) / dec(urio) col(oniae) Fl(aviae) Sisc(iae) /duumvir i(ure) d(icundo) flamen /divi Cl(audii) duumvir q(uin)q(uennalis) / sac(erdos) p(rovinciae) P(annoniae) S(uperioris) ex vot(o).[1141]

Lucius Valerius Valerianus, the character in the inscription, is a member of the *ordo* in Siscia, although he is still *duumvir quinquennalis* in Savaria, in the centre of Pannonia Superior's religious life. The *ara*, at which the *concilium provinciae* gathered annually, was here. The *sacerdos* was charged with its leadership. In the present case, it is Lucius Valerius Valerianus, whose name appears on one more marble table in Savaria. According to the data, Lucius Valerius Valerianus was a citizen of Siscia but his office as *sacerdos* was in Savaria.

II. CHRISTIANS OF SAVARIA AND SAINT MARTIN

The above hypothesis does not wish to infer that

[1133] BULIĆ F: St. Quirinus, Bishop of and Martyr of the City of Siscia (Sisak). rep. In: Catalogue 1994: 13-15, 151-154; JARAK 1994 : 168.
[1134] NAGY T 1946: 249-250.
[1135] Aurelii Prudentii hymnus de martyrio S. Quirini episcopi. Disclosed by RUINART 1859 p. 524.
[1136] KISS G/TÓTH E 1993: 178-179.
[1137] KISS G/TÓTH E 1993: 185.
[1138] This tomb may have been destroyed, that is why the excavation could not supply Roman antecedents.

[1139] SZEREDY 1883: 573.
[1140] ☞ Sopron, Quirinus' route. FITZ 1994 pp. 1181-1183, Nos. 894, 896, 903.
[1141] RIU 20 with further literature, the solution of the abbreviations were taken from here. The inscription was solved and completed with a commentary by FETTICH 1920-1922; Pannonia 1990: 58-59 (FITZ).

Saint Martin was not born in Savaria. It is clearly stated by Sulpicius Severus that he was from Sabaria. He was a child of pagan parents, converting his own mother when he returned to her. This was the occasion when Sulpicius Severus writes that Arian *sacerdotes* exiled him from the town (*de civitate*). To substantiate this further we must look into the main events of the life of Saint Martin.

There are many disputes over his date of birth and other events in his life. The main details come from Justinus Uttenweiler.[1142] He was born in Savaria in 316 or 317,[1143] and he died 8 November, 397 in Gallia. He was the son of a tribune from Pavia who wanted the fifteen-year-old boy to join the Roman army and consequently Martin served in Gallia. It was at this time that the much quoted scene of him sharing his cloak took place. He was a catechumen from the age of ten and baptised at eighteen. Two years later he left the army and went to Hilarius, who ordained him exorcist. He then returned home, converted his mother, and was excommunicated. These events in Martin's life, at the ages of ten, fifteen, and eighteen, are provided by Sulpicius Severus. The differences in opinion are caused by the difficulty in reconciling this information with the historical facts. If we accept that he was born in 316, then we have to assume that he became a catechumen in 326, joined the army in 331, was baptised in 334, and left the army in 336. This would mean that all the events took place in the reign of Constantine I. However, Sulpicius Severus writes that he left the army during the Gallian campaign of Julian, and so could not have been born in 316 or 317, but in 335, as claimed by other scholars.[1144] If we accept the latter year, then he was a catechumen in 345, became a soldier in 350, was baptised in 353, and left the army and found Hilarius in 355. Hilarius was then the Bishop of Poitiers, exiled by the Arians in 356. Sulpicius Severus also writes of this event. Therefore, we may define the absolute chronology of Saint Martin's life in a way that reconciles them with the well-known facts provided by the biographies of Hilarius and Julian.

1. Hilarius: If Martin had been born in 316 or 317, he would almost have been the same age as Hilarius.[1145] Nevertheless, according to the biography, they were not the same age. Hilarius was not ordained bishop under the reign of Constantine I but Constantius II,[1146] thus Martin could not have left the army in 336. Nevertheless, if he was born in 355, then all the events in Martin's life could have occurred at the ages given by Sulpicius Severus, and he could have visited Hilarius when he was a bishop.

2. Julian: On 6 November, 355, Constantius II created him Caesar. He travelled to Gallia in December the same year. Constantius also personally escorted him to Ticinum. He was in Vienna in the second half of December. Constantius ordered Ursicinus and Marcellus to be with him. The latter was a *magister equitum et peditum* at the time. Julian was appointed consul for the first time on 1 January, 356, while Constantius II became one for the eighth time. This was the period when Julian started to prepare for his Gallian campaign that was to last from 356 to the spring of 359. He was in Autun on 24 June, 356 and he went from there to Auxerre,[1147] and thence to Reims by way of Troyes. Here he united his army with that of Marcellus. This united army marched to Metz, and onwards to the Lower Rhineland through Dieuze and Zabern. He was in Sens, in 356, during the Alemann raid, which lasted for 30 days.[1148] Reconciling these events with those in Martin's life, it may well have been that he left the army in the December of 355, assuming that Julian was already present on that occasion a s well. Julian's presence seems to be confirmed by the fact that Martin asked for his exemption, referring to his Christian faith; it is a well-known fact about Julian that, until he became *augustus*, he always inclined to pretend to favour the Christians as he considered this to be the way to guard his physical safety.[1149] However, the dialogue of Martin and Julian cannot be regarded as a real historical source in the form it has remained;[1150] it rather calls attention to the power and example of the saint. Nevertheless, it has to be accepted that Martin did ask for exemption. The reformation of the army in the late period of the Roman Empire could have made this possible.

Although the duration of the then compulsory military service is uncertain, we know that it lasted for at least 5, and at most 15, years.[1151] After five years of service certain members of the family could gain exemption from taxation.[1152] It should not appear unusual that Martin joined the army in his nonage, as the system for compulsory military service introduced the habit of children being taught to carry arms by their fathers.[1153] The tribune mentioned in the biography was a *contubernium familiare*

[1142] UTTENWEILER 1934. These data are in accordance with Antal Meszlényi's data. The research of Saint Martin's life - due to the difficulties in understanding the sources - is an individual theme and an individual volume. That is why I do not wish to study it more than necessary. See also FONTAINE.

[1143] The date of his death is certain. His date of birth was calculated from this; they subtracted 80 from 397, because he died at the age of 80 according to the sources.

[1144] We also have to mention the other opinion of the debaters, namely, that the enumerated events of Martin's life did not take place at the ages defined by Sulpicius Severus.

[1145] F. WERNER: Hilarius von Poitiers. In: Lexikon der christlichen Ikonographe. Begründet von E. KIRSCHBAUM. Rom, Freiburg, Basel, Wien Bd 6 1974: 533: he puts the date of his birth to around 310.

[1146] J. G. CAZENOVE: Hilarius Pictaviensis. In: A Dictionary Biography, Literature, Sects and Doctrines. Ed. W. SMITH / H. AWCE. London, Vol. III 1882: 55.

[1147] The ancient name of Auxerre remained in several written versions, see GRAESSE J. G. Th. / BENEDICT F.: Orbis latinus oder Verzeichnis der wichstigsten lateinischen Orts- und Ländernamen. Berlin 1909: 20.

[1148] E. V. BORRIES: Flavius Claudius Iulianus. RE X.1. 30-35, (26-91). We can read about the preparation for the campaign in Ammianus Marcellinus: XVI. 2.

[1149] He was baptised in Macellum; he wanted to be a priest. It was a deceit, but this way he had hopes for avoiding Constantius' suspicion and thus the execution.

[1150] J. WORDSWORT thinks it is only a legend: Iulianus (103) In: A Dictionary Biography, Literature, Sects and Doctrines. Ed. W. SMITH / H. WACE. London, Vol. III 1882 485.

[1151] VÁRADY 1961: 27.

[1152] VÁRADY 1961: 28.

[1153] Cf. 15.II.a. the interpretation of the *biarchus*.

for Martin,[1154] who can be interpreted concretely, i.e. he was also the relative of his tent-mate. This fact explains the benevolence of the tribune who warned Martin that he had to stay in the army for at least five years because he could not quit after three years. However, after he had completed his time, Martin could leave as a *honesta missione*, therefore Julian did not give him exemption but, at the same time, he did not violate the law either. Thus, it was absolutely normal that Martin should ask for exemption. His request was legal. Martin may have been a soldier in Marcellus's army, together with the above-mentioned tribune.

Summarising Martin's life story, I accept the year 335 as his date of birth, as well as all the information supplied by Sulpicius Severus and referring to his age. He was a catechumen at 10, a soldier at 15, baptised at 18, and asked for exemption successfully at 20. This series can be well reconciled with the historical facts. Nevertheless, he was not 80 when he died. It is fairly probable that the span of 80 years given in the sources cannot be treated as a historical fact, but merely a symbolic number[1155] expressing that Martin had already left this Earth and was between Heaven and the world, ready for death, and therefore the time was right for the saint to enter Heaven.[1156] Sulpicius Severus mentions several historical events that are not ordered chronologically, although there is no reason to expect them to be so. He does not write history but a biography: the biography of a saint, an example to follow. The latter is more essential.

Martin did not stay long with Hilarius; therefore we may conclude that he visited his mother at the age of 21 or 23 (i.e. the period between 355 and 357).[1157] By this time, Constantius II was monarch,[1158] and his reign proved the most difficult period for Arianism in the history of Illyricum.[1159] In all probability, Sulpicius Severus alludes to this when he characterises the heat of the Arian struggles (in Illyricum) by his use of the adverb *maxime*.

He writes: '*Igitur Martinus Sabaria Pannoniarum oppido oriundus fuit, sed intra Italiam Ticini altus est, parentibus secundum saeculi dignitatem non infimis gentilibus tamen.*'[1160] (Thus Martin arrives from the Pannonian City of Sabaria, but his parents, themselves not of humble origin and valuing dignity, had him raised in Ticinum, Italy.) '*Nec multo post admonitus per soporem ut patriam parentesque, quos adhuc gentilitas detinebat, religiosa sollicitudine visitaret, ex voluntate sancti Hilari profectus est, multis ab eo obstrictus precibus et lacrimis ut rediret. ... Itaque, ut animo ac mente conceperat, matrem gentilitatis absoluit errore, patre in malis perseverante, plures tamen suo salvavit exemplo. Dehinc cum haeresis Arriana per totum orbem et maxime intra Illyricum pullulasset cum adversus perfidiam sacerdotum solus paene acerrime repugnaret multisque suppliciis esset adfectus - nam et publice virgis caesus est et ad extremum de civitate exire conpulsus -, Italiam repetens, cum intra Gallias quoque discessu sancti Hilari quem ad exilium haereticorum vis coegerat, turbatam ecclesiam comperisset, Mediolani sibi monasterium statuit.*'[1161] (Not long after, he received a command in a dream to bring religious zeal to his homeland and parents because they kept to their pagan religion. He set off with the consent of Saint Hilarius, but the latter, by means of pleading and tears, made him promise to return. ...Being so committed, he prised his mother from her paganism but his father still persisted in his evil ways. However, his good example led many to salvation. Because Arian heresy was raging all over the world, and especially in Illyricum, he struggled almost alone, and very fiercely, against the perfidious priests. He was maltreated many times,[1162] flogged, and finally exiled from the city. On his return to Italy, he learned that the Church had also been disrupted in Gallia soon after the departure of Saint Hilarius, who had been forced into exile. He chose *anchoretism* in Mediolanum.)[1163]

What we learn from the text.

His parents are not of humble origin, but where they come from remains vague. The reference to Ticinum suggests that they must have had Italian relations there and that they looked after Martin. It should also be mentioned here that there is a further opinion suggesting that Martin's parents were Celts.[1164]

1154 Sulp. Sev. Mart. (FONTAINE 1967) III, 5.

1155 We must not forget that, in antiquity, numbers had not only quantities, but qualities as well. As for quality, the 8 (80) expresses that he had far from Earth (as number 4), but he had not yet reached Heaven.

1156 The previous philological analyses have not resolved the contradiction of the data at our disposal. I have chosen the least contradiction, which was indifferent to Pannonia's history, but at the same time, according to my supposed hypothesis, the contradiction itself can be resolved.

1157 MESZLÉNYI 1977: 103: Meszlényi puts the event to 355 while Endre Tóth dates it back to 354 - TÓTH E/ZÁGORHIDI CZIGÁNY 1994: 27.

1158 In 337, when the Empire was divided between Constantine I's sons, Pannonia was given to Constans, who spent time in Illyricum. We know that he also stayed in Sirmium. He did not support Arianism.

1159 See chapter 'A History of Events'.

1160 Sulp. Sev. Mart. (FONTAINE 1967) II, 1.

1161 Sulp. Sev. Mart. (FONTAINE 1967) V, 3, VI, 3-4.

1162 Cf. Suet. in Galb. 12: supplicio capitis afficere.

1163 The phrase *sibi monasterium statuit* was translated as *anchoretism* because the word *sibi* clearly expresses the fact that it was him, on his own. We do not have to dwell on the significance of the text as to the history of eremitism, or coinobite lifestyle. The translation of SZUNYOGH X. F.: He established eremitism for himslef. - Az elsõ remeték és szerzetesek világából - The world of the first hermits and monks. Chapter II. Az elsõ szerzetesek nyugaton - The first monks in the west. [Keresztény remekírók 7] Budapest 1944 p. 20. Cf. FONTAINE 1967 p. 267. The following translation was based on Fontaine: SULPICIUS SEVERUS The Life of Saint Martin. Pannonhalma ²1998. Translated by: BORIAN E /BŐRCSÖK T /KOVÁCS B /REICHARDT A /SZITA B: The volume contains the letter to Eusebius, Deacon Aurelius and Bassula. The biographical interpretation in the volume is Fontaine's interpretation, to which I wish to remark in connection with only one aspect. It is the Roman military affairs, whose interpretation is ambiguous because he did not consider the military reorganization conducted in the late imperial period. Thus the conclusions based on it are also uncertain. For the sake of completeness, I have to mention a new translation of Saint Martin's life, with which I cannot deal here as it was published following the completion of the present work: VANYÓ L: A III-IV. század szentjei - The saints of the 3rd and 4th centuries. Budapest 1999: 151-197.

1164 MESZLÉNYI 1977: 96. Among the debaters of Saint Martin's life, there is an issue saying that his parents have perhaps a Slav origin. It is

After his dream, he sets off with the consent of Hilarius. In this detail of the *vita*, amongst others, the difference can be observed between the modern, scientific way of thinking and the ancient way of thinking: a thinking not detached from God, and a thinking according to 'ontological anatomy'[1165]. This nuance, today, warns that exaggerated criticism can deprive the *vita* of its value as a source.

It is undisputed that he was born in Savaria, brought up in Italy, and was ten years old when he became a catechumen at his own request. At the age of 15 he joined the army, at the behest of his father; at the age of 18 he was baptised and he was 20 when he successfully asked for exemption. He went on to become the disciple of Hilarius and departed from Gallia after returning to his parents. These are historical facts.

He had an eventful journey. In his account, Sulpicius Severus follows the tradition of legends. He was captured by highwaymen, one of whom he converted. In Mediolanum, Satan tempted him in human guise, but Martin resisted and Satan left him. Martin continued his journey and at this point one might expect to read of his arrival, but it is missing; after his struggle with Satan we see saintliness in his person and the final destination of his journey is left unspecified. The word Savaria does not even appear in the text, as we might expect. The research assumes that he converted his mother in Savaria.[1166] He arrived in his homeland (*patria*), but it could just as well be Pannonia as Italy. It is ambiguous whether the town mentioned (*civitas*) was Savaria.[1167] In Pannonia, Arianism was not a matter of faith, but rather a political loyalty to the ruler.[1168] This theological controversy took place mainly between the Drave and the Save Rivers. This geographical place may be discovered in the expression '*intra Illyricum*' in the *vita*. It is true, however, that we cannot exclude the fact that the heresy was known throughout the province.[1169] The centre of the controversy, however, was in the south. The phrase '*de civitate exire*' does not provide more concrete conclusions. We may think of his exile from Savaria, but Pannonia's southern cities are feasible too. Moreover, we cannot exclude the Italian city, Ticinum, either.

The city itself - the destination of his journey - is almost immaterial; the emphasis is on the saint's behaviour, fighting strongly against heresy (*acerrime*), and that explains why he was exiled from several towns by the followers of Arianism. Although the phrase '*repetens Italiam*' seems to support Pannonia, he left for Pannonia from Gallia and wanted to return to his master as he had promised. Meanwhile, however, he learnt that the Arians had also exiled his master, Hilarius, and so he returned to Italy instead. This latter assumption supports Ticinum, as it was where he had converted his mother. The possibility or impossibility of identifying the actual city has, however, nothing to do with the interpretation of the word *sacerdotes*. As regards the saint's life, the actual locality is also insignificant, as is the fact that priests or bishops beat him. What is essential is that he struggled against those who were in *perfidia*, regardless of where they lived and who the *secundum saeculi dignitatem* actually were. Our pragmatic modern research would be off-target if we took for granted (based on this passage of the *vita*) that in those times an Arian bishop led the Savarian *ecclesia*. This passage does not even tell whether there was a bishop in Savaria at this time or not. I refer again to Tibor Nagy's above opinion that the word *sacerdotes* does not refer to *bishops*, but it is only a reference to the clergy. I would endorse this idea by adding that the *sacerdotes* were not even necessarily priests of the actual civitas (perhaps Savaria) but, in general, all Arian priests present in the province.

We have to remind ourselves of Abbot Leonianus, who went to Gallia with Attila's Huns. He was active in Augustodunum and Vienna.[1170]

the more rational side that has held to the issue of Celtic parents, though it does not mean that the Celtic origin is commonly accepted by the researchers.

[1165] The expression was borrowed from Valery LEPAHIN (Az óorosz kultúra ikonarcúsága - The Icon-like Features of Old Russian Culture. Szeged 1994). On page 115, he explains what we should understand by the term. Let me summarise his explanation: 'the fight against passion, temptation or the evil living in ourselves and the world has to be started in the church, in the religious service, where we can gain strength for this struggle.' It is promoted by the sermons. He also demonstrates the fact by the sermons of Saint John of Damascus (ca 650-750), where 'the dogmatic, catechetic, iconographic and didactic features are together.' It is also a characteristic of the icon, the devotional picture or the legend. He studies the icon-like features of the latter on pp. 117-127, whose main idea is that we have to accept the strange and seemingly impossible statements as true, because if we do not the picture will be deformed.

[1166] József TÖRÖK is of another opinion. 1991: 32-37.

[1167] TÓTH E/ZÁGORHIDI CZIGÁNY 1994 No. 16, p. 27.

[1168] HEUSSI 1958 p. 101 § 24 q.

[1169] In a very detailed way see NAGY T 1939: 127-156.

[1170] ALFÖLDI 1938: 163; STADLER / HEIM III 1869: 774 Leonianus.

III. FINDS

51.I.a. Tombstone / Loculus cover, fig. 342.

Location: SM 67.10.11 **Literature:** CIL III 4218; ILCV 554; RIU 77; THOMAS 1975-1976 Abb. 16; THOMAS 1977b: 56-57.

István Paulovics writes that this was found in the cemetery in Saint Martin suburb.[1171] Size: 69.5 x 84x 11 cm. Chi-rho monogram / Bon(a)e memoriae. / in Deo vivatis / Aurr Fla(v)ini et Leonis. / conlactanei fideles et / innocentes. Aur Flavinia/nus et Nemesius par(iter) memo(riam) / fec(erunt).

The interpretation of the inscription: The names of the foster brothers are *Aurelius Flavinius* and *Aurelius Leonis.* The double 'r' indicates that the name of both brothers is *Aurelius*, cf. 51.II.c. Their name is in the vocative in the inscription, in accordance with the call *in Deo vivatis.* The inscription later tells us that the two foster brothers are believers/faithful and innocent. The abbreviation *par* is interpreted as *pariter*, in contrast to the interpretation in RIU as *parentes*, because the two masculine names before it exclude the possibility that it relates to parents. I would, rather, think of comrades who had had the memorial erected in the same proportion.

The translation, therefore, reads as follows: To good memory. Live in God, Aurelius Flavinius and Aurelius Leonis! Foster brothers, believers/faithful (this meaning cannot be made certain here) and innocent. Aurelius Flavinianus and Nemesius had the memorial erected in the same proportion.

51.I.b. Grave inscription, fig. 343

Location: MNM R-D 207. **Literature:** CIL III. 4220; ILCV 2201; Steindenkmäler 160; RIU 78; THOMAS 1975-1976 Abb. 18; THOMAS 1977b: 58-59; Severin, Katalog 8.42.

A marble slab; size: 61 x 82 x 7.5 cm. Its inscription, according to the authors of the Steindenkmäler: Bon(a)e memori(a)e in Deo / vivas Aur(elius) Iodorus civ(es) / Graec(us) ex reg(ione) La(o)dic(ena) q(ui) / vix(it) an(nos) L et Aur(elio) [F]ron/toni an(norum) II et Aur(eliae) Celsi/n(a)e m(ensium) VIIII Aur(elia) Domnica u/xor con(iugi) car(issimo) et fil(iis) dul(cissimis) / mem(oriam) viva fecit.

Translation: To Good Memory. Live in God, Aurelius Iodorus, a Greek citizen from the Laodicea area, who has lived 50 years. And to the two-year-old Aurelius Fronto and the eight-month-old Aurelia Celsina. Wife Aurelia Domnica has made the memorial for her beloved husband and beloved children in her life. Interpretation: The phrase *vivas* - as it is singular, instead of plural - emphasises Iodorus's situation. I interpret it that, among the persons mentioned, he is the only Christian. It is confirmed by the fact that Iodorus was from the Laodicea area. Probably, *viva* refers also to the Christian status in connection with the wife. Otherwise the the word has no meaning.

51.I.c. Tombstone

Provenance: It was walled into Saint Martin's church. Location: lost. **Literature:** CIL III 4186; ILCV 474; Steindenkmäler 162; RIU 81; THOMAS 1977b: 60-61.

Inscription: *Bene memorandae co/niugi dulcissimae / que vix ann XX M V Fl Pomenti/us Aur Iustinae maritus /⁵ et vector nam et protecto/r contra votum memo/riam posuit.*

Line No. 5: Despite the previous interpretations, I regard the ligature as a VE-connection.[1172] We read of *vectura* offices and *vectures* as officials, both in a symbolic meaning and in the meaning of a prescribed range of duties[1173] (shipping in this case).[1174] Translation: To the wife of good memory, the beloved, who lived for 20 years and 5 months, from Flavius Pomentius, husband of Aurelia Iustina, her director and guardian, erected a memoria instead of a votum.

The contrast between *votum* and *memoria* raises the suspicion that the husband was a Christian.

We have to note that inscription 51.II.d is also on an Aurelia Iustina tombstone.

51.I.d. Tombstone

Location: lost. **Literature:** RIU 82; THOMAS 1977b: 61.

Its inscription (according to the CIL): *Hic positus est Fl/orentinus infans qui / vixit annos septem / et requiem adcepit in Deo / patre nostro Hristo eius* (sic). The phrase *Hristo Christus* has an interpretation, but as the stone itself has been lost, the correctness of the interpretation cannot be checked. We may, however, question whom the word *eius* refers to. It either refers to the dead person, or God the Father. I prefer the latter solution, i.e. *Hristo (filio) eius.*[1175] Translation: The Florentinus child was buried here, who lived seven years, and gained rest in God, our Father, and in His Son, Christ.

51.I.e. Grave inscription / Loculus cover, fig. 344.

Location: SM 67.10.13. **Literature:** RIU 83; THOMAS 1975-1976 Abb. 17; THOMAS 1977b: 57-58.

Fragmentary; size: 53x75x7 cm. Chi-rho monogram / *Memoriam pictoribus / duobus pelegrinis (sic) Launioni[s] / vixit annis L et dies XI / et Secundinus vixit anni[s] XXV et dies VII / ec erunt collegas e[...*

Reading: 7: according to the RIU *[f]ecerunt collegas e[ius].* I interpret the whole line differently. The initial F is not on the stone; what appears to be F is actually the result of damage. The line begins with E. It would also be curious if they had left a space within a word, i.e. they wrote the word as F EC ERVNT. Other parts of the inscription do not repeat this occurrence. Moreover, the *collegas* cannot be the subject related to *fecerunt* as it is in the accusative. The *eius* would have enough space, but as

[1171] PAULOVICS 1943b: 46. I would like to note that this is the cemetery already mentioned above in the first point.

[1172] It is mostly read as *lector*.

[1173] Hieronymus Commentarii in Ezechielem CETEDOC Cl 0587, lib 8, cap (s.s.) 27, linea 1398: quando tyrus ciciderit, gubernatores illius turbabuntur descendent que de classibus, et remiges vectores que ac proretae - sive nautae et universi gubernatores maris -, iactati fluctibus, stabili aliquando consistent gradu.

[1174] Pl. Petrus Chysologus: Collectio sermonum. CETEDOC Cl 02279 M, sermo 32, linea 88.; Hieronymus Commentarii in Ezechielem CETEDOC Cl 0587 lib. 8. cap. (s.s.) 27, linea 1386.

[1175] VICTORINUS Commentarii in Apocalypsin XI, 1 (CSEL 49 Ed. HAUSLEITER J) p. 96, 13. ☞ Historical evaluation in The History of Events.

two names are mentioned, a singular pronoun cannot be used to replace their names. The last visible letter might be *e*, *f*, or a *p*, but it cannot be firmly identified. My hypothesis is: *e(orum) c(ausa) erunt collegas. P(osuerunt).* They became colleagues thanks to their common case, perhaps being collegium members, or there may be an even simpler explanation. Lactantius writes '*maximum pietatis officium peregrinorum et pauperum sepultura*', and the practical application of this sentence is analysed by Pasquale Testini in his description of the duties of *fossors*.[1176] In the course of his analysis, a simpler explanation arose, namely, that the Christian community regarded the two wandering painters as fellow-believers who were obviously staying temporarily in Savaria on account of their work. Here they died, explaining why it is not essential to know who actually had the *memoria* erected.

Translation: A memoria was erected for the two wandering painters. Launionis lived for 50 years and 11 days, Secundinus lived for 25 years and 7 days. They become colleagues because of their case. A Đakovo (Deakovár) inscription may be cited by way of comparison.[1177]

51.I.f. Tombstone, fig. 345.

Provenance: it was found in the modern cemetery next to the Dominicans. **Location:** MNM R-D 135. **Literature:** CIL III 4190; ILCV 401; Steindkmäler 165; RIU 84; THOMAS 1975-1976 Abb 19; THOMAS 1977b: 65.

Size: 47 x 69 x 5 cm. Inscription: *L. Maximine vivas / in Deo Aurelia Hernilla coniux / fecit Maximimo (sic) / centurioni caro suo/ benemerito viro p/<n>io suo mem(ori)a(m) posu/it a[nn(os) vi]xit XLV.* Translation: L. Maximinus, live in God. It was made for Maximinus by his wife, Aurelia Hernilla, for her dear centurio, for the pious man of good memory, for hers, she erected a memoria. He lived for 45 years.

Commentary: 2: Maximino, 5: Paulovics offers the solution *virginio*.[1178] The word is used for a man in an unconsummated marriage.[1179] It was classified as a Christian relic because of the reading *virginio*. On the strength of this, it could be both Christian and non-Christian.

We may find a similar composition and errors in an inscription from Sirmium: in Brunšmid's transcription: chi-rho monogram / *Aurelia Urbic(a) / Fla(vio) Martiniano, / cumpare (sic) suo <Inacae> / carissimo et (Inacae) dulcissi/m(a)e filiae mater piissima / dol<l>e(n)s (f)ratri (sic) et filiae / m<a>emoriam posuit.*[1180]

Another inscription from Sirmium gives an example for the usage of *viro*: [A] chi-rho monogram Ω/ *[ego Aur]elia Aminia po/[sui] titulum viro meo /[F]l(avio) sancto ex n(umero) Iov(ianorum) pr(o) tec(tori)/ bene merentius qui vixit* ⁵/ *ann pl(us) m(inus) L qui est defunct/us civit(ate) Aquileia. Titulum posuit ad beatu(m) Synerot(am) ma/rture(m) et infante(m) filiam* ¹⁰/ *suam nomine Ursicina, / qui vixit annis n(umero) III.*[1181]

51.I.g. Lamp, figs 346a, b.

Location: SM 61.13.43. **Literature:** THOMAS 1975-1976 Abb. 7-12; THOMAS 1977b: 53-55.

Bronze. The so-called 'Agnus Dei' lamp. L: 11.7 cm, diam: 5.8 cm, H: 4.8 cm without the lower part. The lower part and the cross are subsequent complements. Edit Thomas dates it to the 5th century, ☞ Zalaegerszeg, 60.I.a. Two pieces are known from South Pannonia.[1182]

Clay lamp. It was thought to have been found at Szombathely; in fact its provenance is unknown. It has a menorah illustration. ☞ Finds of unknown location, 62.I.i.

51.I.h. Menas ampulla, fig. 347.

Provenance: It was taken to the museum as a local find.[1183] **Location:** SM 54.775.365. **Literature:** NAGY L 1938: 96. fig. 60. KÁDÁR 1969: 199; THOMAS 1975-1976 Abb. 5-6; THOMAS 1977b: 47 fig. 49; THOMAS 1980: 200 CXLVIII/1; THOMAS 1988: 293 Abb. 16.

We have no data as to its exact site and surroundings. H: 10 cm, diam: 7 cm. Saint Menas is standing with his arms outstretched. His robes flow around his elbows. There is a camel on each side of him, and crosses (with arms of the same length) next to his head.[1184] Zsolt Kiss classified pieces of this type as belonging to the second group, datable to between 560 and 610.[1185] A very close parallel is mentioned in the catalogue of an exhibition.[1186]

We have to accept Zsolt Kiss's dating as regards the Szombathely piece. This find has so far seemed unlikely in Pannonia because research has shown no continuity.

51.II.a. Marble screen, fig. 348.

Location: SM 67.10.1. **Literature:** PAULOVICS 1943b: 43; PAULOVICS 1944: 21, FIG 8; Stendenkäler 235; THOMAS 1977b: 78; KISS Á 1987 No. I.37.

It was made of white marble. Size: 100 x 80 x 15 cm. Paulovics writes that 'we do not know its exact provenance.'[1187] He connects it with the act of confession. It has been regarded as Christian ever since, even though commentaries to the finds are mere hypotheses. It was the essence of confession that the martyr's relic could be seen from the outside through the pierced wall.[1188] However this

[1176] TESTINI 1966: 224.
[1177] CIL III 4002; BRUNŠMID 1908-1909: 166 No. 361; NAGY L 1938: 110, fig. 74. Marble; fragmentary; 50.5 x 39 x 4.5 cm. Above the inscription there is an XP, which has remained fragmentarily. Inscription: Fl(avio) Mauro f[ilio ...]/nis, bene[merenti]/ fedeliq[ue iuveni]/ an(norum) (viginti duorum), c.../ m[emoriae causa]/ coll[egae ...] /f[ecerunt ...
[1178] PAULOVICS 1944: 36.
[1179] BLAISE 1954: 850: uirginius.
[1180] CIL III 10237; ILCV 3611; BRUNŠMID 1908-1909: 199 No. 376; NAGY L 1938: 101 fig. 63.
[1181] CIL III 10232; BRUNŠMID 1908-1909: 187 No. 393; ILCV 2182; NAGY L 1938: 101.
[1182] MIGOTTI 1997 VI.b.1-2.
[1183] NAGY L 1938: 96.
[1184] THOMAS 1977b: 47 defined it as a rosette.
[1185] KISS Zs 1973: 146.
[1186] L'art copte. Petit Palais Paris, 17 juin - 15 septembre 1964 Paris. No. 137. Provenance: Abu Mina, location: Städtische Galerie Liebighaus, Frankfurt am Main.
[1187] PAULOVICS 1944: 21.
[1188] An example is Saint Felice's tomb in the Basilica nuova di Paolino. - TESTINI 1985: 340 figs. 8-9. Another piece ornamented with a chi-rho

section is unpierced and it is unlikely that it would have been used for this purpose. (But it has also been described as a screen elsewhere and it may well not have been pierced.) The authors of the Steindenkmäler do not mention these opinions and do not touch upon the Christian references of the carving. Edit Thomas classifies it as a Christian object, together with the fragment built into a modern wall at ☞ Egyházashollós (13.II.a). The Szombathely or Egyházashollós fragments may also be screens, but this cannot be proved. A similar section was found during excavations of an early Christian church,[1189] fig. 349. Despite the analogy, I do not regard the Pannonian pieces as Christian because the same (and similar) carvings were used generally as balusters, even for imperial orders, e.g. Constantine's triumphal arch.[1190] The same motif (with piercing) may be observed in Pompeji, in a non-Christian environment.[1191] At Szombathely there is an imperial palace that was formerly considered an early Christian church,[1192] and we should allow for the possibility that these pieces come from there. ☞ Egyházashollós, Smerska Mitrovica.[1193] The same baluster found at Sirmium seems to support the idea that this object was, in fact, a component of an imperial building.

51.II.b. Loculus cover? fig. 350.

Provenance: Bitnitz says its provenance is Szombathely.[1194] **Location:** MNM R-D 235, Petrédy collection. **Literature:** RIU 75; TÓTH E 1972c; THOMAS 1975-1976 Abb. 13-14; THOMAS 1977b: 48-51.

Marble, the size of the fragment is 18 x 19 cm. Thickness: 35-37 mm. The remaining part of its inscription reads NVS VIP, but NVS VLP is also possible. Endre Tóth saw the illustration as a representation of the Good Shepherd. However, the authors of the RIU do not consider it as early Christian: 'Fragment einer Marmortafel mit vertiefter figürlicher Darstellung: rechter Fuß und Mantel eines sitzenden(?) Mannes in langer Hose. Typische spätantike Arbeit.'

The stone was classified among early Christian objects largely because of the above association with the Good Shepherd. It may also be interpreted in other ways, and is, perhaps, best described as debateable. In any event, because of its uncertain provenance it is of no real value in terms of a connection with Pannonia's Christianity.

51.II.c. Grave inscription of Aelia Kalendina, fig. 351.

Provenance: István Paulovics says it was found in the cemetery next to the Dominicans.[1195] Its exact find-spot is unknown. **Location:** MNM R-D 208. **Literature:** RIU 76; THOMAS 1975-1976 Abb. 20. THOMAS 1977b: 64.

Size: 51 x 84 x 4.5 cm. Inscription: *In D M Bon(a)e Memori(a)e Aeliae Ka/lendin(a)e fideli matri dulcissim/ae qui vixit annos n(umero) LX Aelii M/ucianus, Bassus, Paulinian/us et Paulus fili(i) pientissimi/ memoriam posuerunt.*

István Paulovics adds *D(o)m(ino)* to the letters D M in the first line, which is accepted by the authors of the RIU. It is classified as early Christian partly because of this completion, partly because of the tradition of Saint Martin connected to its find-spot. *D M* is a common abbreviation on tombstones. The *IN D(is) M(anibus)* is an ablative. There is a linguistic simplification in the inscription. Third line: The plural form of the name *Aelii* indicates that all of them are Aelii, cf. 51.I.a. Translation: Among the Spirits of the Departed, to the faithful mother, Aelia Kalendina of good memory, the much beloved, who lived exactly 60 years, Aelius Mucianus, Aelius Bassus, AeliusPaulinianus and Aelius Paulus, her sons, the most pious erected the memoria.

The meaning of *fidelis* cannot be *believer* here in any way. The phrase *bonae memoriae* does not occur in Christian use exclusively; therefore, we do not have to consider it as Christian.

51.II.d. Tombstone of Aurelia Iustina, fig. 352.

Provenance: Szombathely, 15 Fő tér (Köztársaság tér) - Main Square. **Location:** SM 67.10.12. **Literature:** CIL III 4197; Steindenkmäler 138; RIU 79; THOMAS 1975-1976 Abb. 24; THOMAS 1977b: 62-64.

Partially because of its find-spot, partly due to the expression *bonae memoriae* on it, ithe stone has been thought to be early Christian in a secondary usage. The stone was originally produced in the 2nd century, and as evidence of its pagan inscription, the existence of the *Dis Manibus* was mentioned. The authors of the RIU stated that the habit of the present inscription and the style of the stone were not of the same period. According to the authors of the Steindenkmäler, the inscription is: *[Bona]e memori/[ae] Aureliae Ius/tinae an(norum) XL / coniugi caris/simae Aurelius / Gaianus et Caius / et Iustinus / fili ma/tri pientissime v(ivae) f(ecerunt) / et filia Surica / D(is) M(anibus)*. Size of the stone: 160 x 75 x 16-20 cm. The tombstone may be divided into three parts. The inscription panel is in the middle, bordered by columns on the two sides. The columns bear a frieze, on which there used to be a tympanum. The tympanum and the frieze are now fragmentary. There is an animal in the preserved corner of the tympanum, as well as a tree with grazing ungulates on both sides of the tree in the frieze. Under the inscription is a relief sculpture, where two geniuses hold a garland. An eagle is sitting on the garland. The letters *D* and *M* of the *Dis Manibus* can be seen on the frame of this relief sculpture. The ninth line of the inscription was also incised in the upper part of the frame: here one may read *pientissime* instead of *pientissimae*, and the Steindenkmäler solved only one of the two F-s of the inscription - f(ecerunt).

No mark of any subsequent carving is noticeable, although it is true that the habit of the inscription does not match the style of the stone. I think the tombstone was not used when it was carved, only later, and so the inscription must also have been made later. The letters DM do not show an earlier habit of writing either. We have to suspect

comes from another excavation - Marcel LE GLAY: A la recherche d'Icosium Antiquités Africaines 2 1968 No. 19, pp. 50-51 fig. 37.

[1189] NIKOLAJEVIĆ 1978: 684-685 fig. 6.

[1190] R. BRILLIANT, Scenic representations. Age of Spirituality No. 58.

[1191] ESCHBACH 1978: 239.

[1192] TÓTH E 1973: 117-137.

[1193] BRUNŠMID 1910-1911: 98 No. 668; MÓCSY 1974 pl. 41a; KISS Á 1987 116/3.

[1194] TÓTH E 1972c.

[1195] PAULOVICS 1944: 36 fig. 17.

that the carving of letters DM took place when the whole inscription was carved.

Its addressee is Aurelia Iustina. This name, in an unchanged from, also occurs on tombstone 51.I.c, however no relationship between the two stones may be assumed.

51.II.e. Tombstone with *carmina*, fig. 353.

Location: MNM R-D 126. **Literature:** PAUR 1859 226; CIL III 4185; ILCV 296; Carmina Epigraphica Latina 578; PAULOVICS 1944 35-36; RADNÓTI 1954 493; Steindenkmäler 161; RIU 80; THOMAS 1975-1976 Abb. 21; THOMAS 1977b 61-62; ADAMIK 1983; AnnÉp 1984 722; EDH001731.

Limestone, size: 59.5 x 100.5 x 5 cm. Marks of red paint are observable. The inscription is in an un-profiled border. According to the authors of the Steindenkmäler, it reads: *Fl(avius) Dalmatius v(ir) p(erfectissimus) ex pro[tectore] / qui vixit an(nos) XL et Aur(elia) Iulia conp(ar) dulc(issima) / qui vixit an(nos) XXXV ideoque hunc titulum / scripserunt quisquis he(ris) post me d(o)m(inus) Laris huius / et (h)orti vicinas mihi carpe rosas mihi lilia pone / [can]deda(s) q(uae) viridis dabit (h)ortulus. I<s>ta beatum! / .... Volussius et Sacatia lib(erti) posuerunt.* According to the RIU, the 1. *ex pro[tectore Aug(usti)].* In Adamik's opinion, the 6. *can]deda, q(ue) viridis dabit ortulus ista beatum...*

The inscription has two parts, one in prose and one in verse. The gravestone has been classified as early Christian because such a context was felt in connection with the verbalisation of the inscription. Diehl also shared this opinion. T. Adamik, in his article, refutes this false notion with the following arguments: 1. the word *hortulus* does not mean a cemetery, but a private graveyard. 2. The indication of a person's social status is a pagan, and not a Christian habit. 3. In the epitaph we read about strewing flowers onto the grave, which is never mentioned in early Christian death poems as 'pagans lumped the circus and the brothel with the funeral.' Consequently, wreaths of flowers were considered inappropriate by Christians.

Tamás Adamik translated the poem into Hungarian, however, the English translation is in prosaic form: Whoever you are, owner of the house and garden after me, pluck a rose, strew it on me with bright lilies that our small garden grows. flowers belong to happy people.

Translation of the prosaic parts: Flavius Dalmatius, the very perfect man from the guards, who lived 40 years, and Aurelia Iulia, the dearest, who lived 35 years. Therefore this poem was written: .... (the memoria) was erected by Volussius and Sacatia, freed slaves.

It is not Christian.

51.II.f. Tombstone of Namius Quintus, fig. 354.

Provenance: Saint Martin's suburb, in 1904. **Location:** SM 67.10.69. **Literature:** Steindenkmäler 166; RIU 85; THOMAS 1975-1976 Abb. 23; THOMAS 1977b: 59-60; THOMAS 1980: 200 pl CXLVIII/2; Severin, Katalog 8.41.

Size: 29.5 x 60 x 4.5 cm. Inscription: *Namius Quintus / custor cymeteri / an(norum) LXXXVI et Aurelia / Irene coniux an(norum) LXIXI (sic).* It was classified as a Christian tombstone due to the expression *custor cymeteri*, but as this also occurs on pagan stones, there is no certainty as to its true classification.

51.II.g. Fragment of tombstone, fig. 355.

Provenance: In the suburb of Saint Martin, in the present cemetery there. **Location:** SM Inv. No: ? **Literature:** CIL III 10935; Steindenkmäler 167; RIU 86; THOMAS 1975-1976 Abb. 22; THOMAS 1977b: 65-67.

Only a very small fragment remains (26 x 31.5 x 2.7 cm), marble. Inscription: *Bonae m[emoriae].* It is impossible to tell where it originated.

51.II.h. Roof tile, fig. 356.

Location: SM 54.435.18. **Literature:** THOMAS 1975-1976 Abb. 15; THOMAS 1977b: 52.

Edit Thomas connected the lines on the tile that are finger-drawn to the written letters. Thus, the letter T became a cross, and the O, next to it, became the sun (Christ's symbol). It was consequently classified as a Christian object. However, there are only two letters – 'T' and 'O' on the tile. The two letters can hardly be separated in the photo, but it is easy to do on the tile itself. I do not consider it Christian.

SUMMARY- SZOMBATHELY

The latest topographical researches have clarified the town wall of Savaria, as well as the situation of its gates. This work has made it possible to re-evaluate the scenes and places mentioned in the Quirinus passion. As a result, it seems that the *porta Scarabetensis* is identical with the city's eastern gate and therefore the *basilica* of the Passion should be sought there. The examination of the meanings of *basilica* and *locus orationis* have led to the conclusion that they had the function of private houses of prayer. Furthermore, the *martyrium* gained the function of a church during the Early Middle Ages, as shown in the address of *tituli minores*.[1196] This basilica does not mean a representative building with several naves. Therefore, I think the marble 'screen' belongs to the imperial constructions.

The place of the *locus orationis* was definable based on the text of the Passion; it lay outside, west of the walls, on the bank of the Sibaris (Perint). As to its dating, we may state that it was already built during the course of the 3rd century. This fact suggests a Christian community (probably communities) in Savaria from at least the 3rd century onwards. The existence of the *basilica* refers to several communities, or the larger number of Christians. Quirinus' Passion relates the migration of a part of the population as a real event, but the presence of the *menas ampulla* betrays the continuity of Christianity. The Christian tombstones belong to Romanised Christians, who were partly converted locals, partly previous settlers. The inscriptions reveal foreign newcomers: the itinerant painters (51.I.e) and the Greek Iodorus (51.I.b). Compared to the presumed large Christian community, there are few small finds. (The very large quantity of small finds brought to light by the latest excavations have not yet been published.)

There is a need to mention the lamp with the

[1196] SZEREDY 1883: 573. *Tituli minores* stood below the Ecclesia baptismalis. The hierarchy followers: Basilicae, ecclesiae baptimales, tituli minores.

menorah illustration. It is not valuable from the viewpoint of Pannonian history; its find-spot is unknown and it is not certainly a Pannonian piece. Here in Savaria - and in this respect Savaria is unlike other areas in the province - we cannot assume Syrian Christians. It was North Italy[1197] that had an impact on Savaria's Christianity, and therefore this is an opportunity to evaluate Aquileia's role - as is reflected in the history of the research.[1198] I think we may also put the Quirinus *translatio* down to the influence of Aquileia. The origin of the family of Saint Martin also strengthens North Italy's role. As the family did not belong to the class of *humiliores*, according to Roman law, it is hardly imaginable that they did not settle in the city.[1199] Thus, we cannot expect Saint Martin's family home to be found outside the walls of the Roman cemetery. The problems concerning the medieval church, the title, and the existence of Szombathely's first parish church, are intended to be solved by the suggestion that veneration of the two saint-bishops were merged in the Middle Ages. Saint Martin's church was erected on the site of Quirinus' tomb. The term *sacerdotes* is interpreted to denote not a bishop, but priests, who belonged not only to Savaria but also to the whole of Pannonia.

52. SZŐNY / KOMÁROM (BRIGETIO)

Brigetio is a town lying along the *limes*, whose military importance is studied primarily. Families from North Italy, such as the Caesernii,[1200] settled down here quite early. It became a part of Pannonia Superior, when Pannonia was divided at the beginning of the 2nd century; in 214, it was connected to Pannonia Inferior. It has an auxiliary fort and also a legionary. No. 1 is the place of the I Adiutrix station. Shrines to Dolichenus and Mithras were found in the canabae that had developed around the fortress. Its civil settlement lies west of the canabae. It achieved its rank as a *municipium* from Caracalla, which is referred to by its name: *Municipium Brigetio Antoninianum*. Later, during the 3rd century, it gained the rank of a *colonia*. It is significant that the second largest Syrian colony was here. They had an isolated life just like in Intercisa. 'This isolation is all the more striking since the Syrians in Aquincum, and especially in Brigetio, had worked their way up to membership of the *ordo decuriotum*.'[1201] Their migration did not stop; even in the 4th century Syrians kept arriving in Brigetio in considerable groups.

The RIU 2 has lately disclosed a summary of the identification of Szőny with ancient Brigetio (as well as other sites in the area with ancient names).[1202] Although some of the discoveries made by local excavations were lost in the war, the objects at our disposal make it evident that Brigetio was much more important than can be deduced from present publications.[1203] The summary of the RIU 2 also shows where the research history may be investigated,[1204] and a general map of the site is included as an Appendix,[1205] fig. 357. Because of its Christian reference, the historical event should not be forgotten (involving the Brigetio region), when Fritigil's Christian Marcomanni settled down in Pannonia around 396-397.[1206]

A fish carved of bone,[1207] mentioned as a Brigetio find by Béla Czobor, is also referred to later by József Hampel.[1208] Lajos Nagy did not regard the dolphin-shaped amulet as Christian.[1209] It is not reasonable to classify it so, as the fish was also a common symbol among the followers of other religions in antiquity.[1210]

52.I.a. Clay lamp, fig. 358.

Location: Komarno, Podunajské Múzeum.[1211] **Literature:** IVÁNYI 1935: 14-15, 106 No. 990, pl. XL/9; NAGY L 1938 fig. 51.

Clay lamp. Size: L: 10.9 cm, w: 3.2 cm, H: 4.5 cm. Chi-rho monogram on the discus. Lajos Nagy thinks that there is a schematic picture on the discus of the twelve apostles. It is a mere hypothesis. It could also be said that they are the symbols of the twelve zodiac signs. The illustration is not very discernible. Such decorations are much better characterised by J. Garbsch's term 'Sigillatadekor', which refers to recurring geometric figures produced with a stamp.[1212] Those lamps (to which group this piece also belongs) display various patterns; only their shape is similar. Next to the filling holes there are a tied iota and a chi. Type: Abbiani 1-a.

52.I.b. Clay lamp, fig. 359.

Location: MNM 59.1883.1 presented by János Marossy. **Literature:** CZOBOR 1883: 245-246 fig. 31; IVÁNYI 1935: 107 No. 994, pl. XLVI/13 and LXXVIII/1; NAGY L 1938: 86 fig. 52.

Fragment of a red clay lamp (which has been restored). There is a chi-rho monogram on its discus, but the *P* faces to left. Its holder is solid and smooth. Diam: 7.7 cm, H: 4.5 cm. On the bottom, there is an encircled four-leaf rosette. Type: Abbiani 1-a.

52.I.c. Bronze suspension for a lamp, figs. 360a, b.

Location: MNM 4.1933.27 Milch collection, purchase. **Literature:** PAULOVICS 1938a: 60-61 fig. 104. NAGY L 1938: 143-144 fig. 88; TÓTH E 1977a: 143-148, 150, 152 Abb. 1-2: scale drawing of all details; TÓTH E 1980b Abb. 3.

A cross-shaped suspension unit for a lamp. Lajos

[1197] MÓCSY 1974: 120: Aquileian families settle down in Savaria.

[1198] BRATOŽ 1994. It is a part of the town's history that the migration from Aquileia begins early, e.g. the Canii family. MÓCSY 1974: 120.

[1199] Although we might reasonably ask who would live in the cemetery.

[1200] MÓCSY 1974: 120-122.

[1201] MÓCSY 1974: 228. ☞ Aquincum.

[1202] RIU 2 p. 89.

[1203] The latest excavation are convincing: BORHY L/ SZÁMADÓ E: A Szõny-Vásártéri ásatás leleteibõl - From the finds of the Szõny-Vásártér excavation. Budapest, Komárom 1996; BORHY L, Glasgemme mit antoninischem Frauenporträt aus Brigetio (Komarom/Szõny, Ungarn). Archäologischer Anzeiger 1997: 97-100 with further literature.

[1204] RIU 2 p. 91-92.

[1205] RIU 2 p. 90.

[1206] NAGY T 1983: 113-121. ☞ Sopron.

[1207] CZOBOR 1883: 245 fig. 31.

[1208] HAMPEL 1880: 173.

[1209] 1938: 86.

[1210] WESSETZKY 1974.

[1211] I am grateful to the archaeologists and assistants of the Podunajsko Múzeum for helping my work.

[1212] GARBSCH 1982: 103-104.

Nagy writes that on the right and left sides the letters A and Ω used to hang; the lamp was hung in the middle. L: 22.6 cm, w: 17.3 cm. Its site has never been questioned, although it is not typical of the Pannonian material, and is different. Endre Tóth, who dates the piece to the period between the second half of the 5th century and the middle of the 6th, writes the following about its historical evaluation: '1. Es würde sich nicht lohnen den Gegenstand wegen seines geringen Rohmaterialgehaltes zu rauben. Als awarisches Beutegut kann an den Gegenstand höchstens nur als Rohmaterial auffassen. 2. Was zumindest so wichtig ist: die Datierung steht im transdanubischen christliches Denkmalmaterial dieser Epoche nicht allein, obwohl es zur Publikation oder Umwertung dieser Gegenstände erst in den letzteren Jahren gekommen ist.'[1213]

Commentary: It is a generally accepted view that the lamp was suspended from the bottom of the cross, whereas the letters A and Ω were suspended from the two side arms. This arrangement of the letters is, indeed, a usual custom with crosses. A Latin cross is meant here and not a cross-shaped object. We have to make a distinction between the two, because the suspension for a lamp is a cross-shaped object, and not a cross. It can have a symbolic value, but it is never identical with the cross itself as an object. Consequently, it is not at all evident that there used to be an alpha and an omega on the horizontal bars. It is also conceivable that there were lamps suspended from them as well. It is, in fact, quite likely, as this would achieve a greater amount of light. Therefore, in contrast with the general view, I support the idea that on this suspension there were never A and Ω letters, but three lamps. Furthermore, it is only a cross-shaped object, and not a cross.

Endre Tóth considers this piece invaluable for the Christianity of the province because of the neglect of Pannonia's 5th-century history experienced so far. It is true that we have generally thought that the Romanised population as having fled, and so the 5th century can only supply the heritage of peoples in the Migration Period. That is why we have not reckoned with the continuity of the Romanised. Provided we suppose that this find occurred here accidentally, due to the migration, then the object does not convey any information as to the Christianity of the province. However, assuming a consciousness in carrying the object here, we have to deal with the continuity of the Christians and the Romanised in the Migration Period also. In any event, we may certainly say that the new finds show a tendency that suggests the survival of the Roman population. Furthermore, it is also indisputable that we have finds referring to the Early Middle Ages (☞ Aquincum, ☞ Keszthely-Fenékpuszta, ☞ Pécs, ☞ Székesfehérvár, ☞ Szentendre), which explain the presence of the suspension for a lamp. These finds prove that Christianity did not cease to exist on the transfer of Roman Pannonia. This Christianity, even if the descendants preserve it, conveys a strong oriental (Byzantine) influence.

[1213] TÓTH E 1977: 150.

52.I.d. Silver finger ring, fig. 361.

Location: Kunsthistorisches Museum, Wien. **Literature:** BOROS 1899: 149-150; NAGY L 1938: 85 fig. 50.

A cornelian stone was set in the ring, decorated with a chi-rho monogram.

52.I.e. Bar Kochba coin.

Location: MNM 4.1940.17. **Literature:** R.-ALFÖLDI 1956; BARKÓCZI 1957-1958: 19; SCHEIBER 1960: 57-61; THOMAS 1964: 267; THOMAS 1977a: 22; SCHEIBER 1983: 62-63; SOLIN 1983: 762.

Only one piece was found, in 1940, from a cremation grave under the Roman villa on the eastern side of the fortress. Edit Thomas provides the ground plan and the excavation plan of this villa.[1214] A squad of the *legio I adiutrix* stationed in Brigetio took part in the Jewish war; one *vexillatio* belonging to the *II adiutrix* too.[1215] The opinion that the coin was brought here by soldiers is certain. Heike Solin thinks the soldiers brought it as an amulet. The coin, therefore, lacks significance as regards Pannonia's Christianity, or Jewish population.

52.II.a. So-called augur's staff / so-called crosier, fig. 362.

Provenance: the first grave of a six-grave cemetery. Male grave. Barkóczi puts the cemetery to between 270-320. He thinks that it was used by those oriental people who had participated in Aurelian's eastern campaign and who had come to Pannonia. Here they preserved 'their native burial customs' (1965: 257). **Location:** MNM 58.12.33. **Literature:** BARKÓCZI 1965: 215-257; THOMAS 1974: 144; THOMAS 1982a; THOMAS 1982b: 265 Abb. 62, b, c; GÁSPÁR 1989: 103-111.

This peculiar object was found in a cemetery in the area between the Brigetio fortress and the canabae. The cemetery consisted of only six graves. Edit Thomas wrote that the cemetery and its grave goods were early Christian. The silver staff with gold inlay belonged to the first grave. On one end it ends in a spiral; here the cross-section is rectangular, 1.5 x 0.3 cm. The other end, virtually the shaft section, ends in a quadrilateral pyramid, 0.6 x 0.6 cm. L: 33.1 cm. It has a niello ornament with incision and gold inlay. The stylistic analysis of the ornament and its dating was disclosed by László Barkóczi,[1216] who defined the object as an augur's staff. Edit Thomas defined it as a crosier and this explains why it is now placed with Christian finds. I have conducted a full examination and functional analysis of it and it proves not to be Christian, although it certainly had some religious background. It seems to have been used at 'pagan' rituals, especially by oriental people. Its name is possibly *virga ritualis*.

52.II.b. Inscribed roof tile, fig. 363a, b.

Location: MNM 62.208.1. **Literature:** CIL III p. 962; RÓMER 1886: 154; R-D 429; THOMAS 1974: 139-146; THOMAS 1977b: 39; MÓCSY 1977: 396 Note 448; THOMAS 1987c; THOMAS 1980: 194-195; THOMAS 1982b: 261-262; JARAK 1996: 267.

The tile is broken; its present size is 25.5 x 35.5 x 8 cm. It does not have an edge. Flóris Rómer, who documented the tile, defined the stamp as the one

[1214] THOMAS 1964 Abb. 142.

[1215] KUZSINSZKY 1934 No. 285: *Aelius Silvanus ... domo Syria Palestina;* BARKÓCZI 1957-1958.

[1216] BARKÓCZI 1965: 250.

belonging to the *legio I Adiutrix*, and described the inscription as follows: hoc die felice / eunt per secul/ utes quorum / nones ..... ic / longius i. Edit Thomas supervised this reading of the inscription, altered it, and completed the missing parts: hoc die felices fratres / sunt persecuti morie/ntes quorum anima / non est victa in deo / longius iubilabit. She displayed the tile as a Pannonian early Christian relic in her exhibitions and lectures. In her article, E. Thomas gives a more detailed summary of early Christianity in Pannonia based partially on this tile and other objects (crosier, glass beaker, jug). We find the interpretation of these objects as early Christian rather premature. Concerns and objections were expressed in connection with the interpretation of the inscription by András Mócsy, who noted that the word *persecuti* could not be interpreted as passive. Furthermore the completion of *victa* was impossible as the initial *f* was certain. He also raised objections to the Christian interpretation of the other objects. It is a real problem that the rest of the objects - the augur's staff / crosier, the glass beaker (grave 2), the jug and the beaker (grave 3) - do not refer to Christianity individually.[1217] I would complete the first three lines of the inscription as follows: *hoc die felice militave/runt persecuti fugie/ntes quorum.../ non es...fic....../longivs ...* Interpreting the word *felix,* according to the *utere feliciter* inscriptions, I do not think it refers to heavenly happiness, but rather to this world. Consequently, I would translate the first three lines as *on this lucky day, the pursuers besieged the pursued, whose...* From this point the reading is very uncertain, but we could perhaps finish it: *whose lines or troops never arrayed again.* The tile is fragmentary. The reading of *est* must be wrong as there are four spaces between *non* and *fic*, where the fourth one is covered by the stamp. I agree with András Mócsy that it is not a *v* but an *f*, as is made evident by the word *felice.* I do not think that the inscription is Christian. I raise the hypothesis that Pannonian tiles had a function similar to that of those Egyptian papyri on which the *pridianums* were written. Thus, I assume that these inscribed tiles are Pannonian *pridianums*. We know several inscribed tiles from Siscia,[1218] whose texts, or the scheme of the texts, are similar to the scheme of military reports on papyri.[1219] Therefore, there is a strong suspicion that the tile is similar.

52.II.c. Votive plaque, figs. 364a, b.

Location: lost. **Literature:** CIL III 10998; BARKÓCZI 1951: 250; RIU 440; TÓTH I 1995: 181-184.

The inscription is rather fragmentary. The RIU transcription: Deo M[agno] / Aeterno /gam pr [...] / Claudia p[...] / que pecun[ia sua] / fecit ex v[oto]. Later István Tóth studied the inscription and completed it as follows: Deo M[agno] / aeter[no synago]/gam pr[oseucham] / a [solo] / Claudia p[ia filio]/que pecun[ia sua] / fecit ex v[oto]. He also attached a historical summary to the completion whose essence is that there was a Jewish community and a synagogue in Brigetio. He thought the inscription from Intercisa (☞ votive plate, 12.II.c) also supported this opinion. Furthermore, he interpreted the usage of *aeterno* - based on his references - as an attribute of Yahweh. Supporting the Jewish interpretation of the inscription, he mentions the Bar Kochba coin (52.I.e) found at Szőny.

Commentary:

• *The completion and interpretation of the inscription.* Analogues of the first line (partly the second line): an inscription is established to *Deo magno aeterno* by Flavius Donatus and his wife, Tertulla. This inscription is the establishment of a *votum.*[1220] An inscription is erected to *Deo aeterno et Iunoni et angelis* by Marcus Procilius Aphrodisius with his daughters.[1221] An inscription to *Iovi aeterno Dolicheno* is established by C. Iulius Apollinaris,[1222] and to *deo magno* by Marcella.[1223] Quintianus decurio alae erects an inscription to *Silva/no deo / aeterno.*[1224] The listed examples show that the usage of the epithet *aeterno* is not exclusively Christian or Jewish. It could be the attribute of the Sun, the Moon, the signs of the zodiac or the Phoenix as an Alexandrian influence.[1225] According to Saine's study, the attribute *aeternus* is due to Syrian gods, where Silvanus is an exception.[1226] The term itself also derived from Syria. It is adopted in the figure of Aion in the Mithras cult. Overall, the *aeternitas* expresses a conception that cannot be limited to Yahweh, although it is a usual attribute of his.[1227] Second line: So that we could interpret the epithets of the addressed god together, I connected the *aeterno* from the second line to the first one. The dedicated object follows it. In István Tóth's completion it is the *synagogam.* As the completion is based on the presumption that we are dealing with the Jewish/Christian God, the *-gam* ending suggests this kind of completion. Besides the fact that there is no criterion as to dividing the Christian and non-Christian, Jewish and not Jewish inscription,[1228] the *-gam* ending allows other possibilities as well, for instance *aedem longam, domum longam, bigam* or *quadrigam.* In the present case, I propose the completion as *[dom(um) lon]gam.* Third line: I think the completion of the letters *PR* is proper as *pr[o se et suis]*, since it is a common - although not compulsory - custom on votive inscriptions that the purpose of erection is stated. Fourth line: The *A* could also be an initial letter, as the fraction does not proceed in the way we can see it in István Tóth's drawing. Therefore, I think this letter belongs to the name *Claudia*

[1217] For a more detailed reflection see ☞ Pécs and ☞ Pilismarót.
[1218] LJUBIĆ 1876: 16-21, 23.
[1219] FINK 1971 morning report of the Coh. XX Palmyrenorum, No. 47. p. 185, or No. 49. p. 191, from the middle of the 3rd century. Here we can find names and numbers listed according to a completely similar scheme as on the tiles in Siscia.

[1220] AnnÉp 1907 3.
[1221] AnnÉp 1914 106. Cf. SAINE 1968: 1102.
[1222] AnnÉp 1911 215.
[1223] AnnÉp 1956 225; 1961 348: Peremustae *Deo / magno Araca / Marcella pro/ salute sua et suorum / v s l m.*
[1224] see 12.II.c.
[1225] SASSE H: Aion RAC 1 1950: 199; dedications to *Deo Magno*, see ALFÖLDY 1977.
[1226] SANIE 1968: 1093.
[1227] SANIE 1968.
[1228] SOLIN 1983: 657.

based on the female names at Szőny, where these names do not consist of only a cognomen: Aurelia Aeliana,[1229] Aurelia Vera,[1230] Aurelia Verina,[1231] T. Claudia Aemerita,[1232] Petronia Severina,[1233] Pompeiae Secundae,[1234] Aelia Valeriana,[1235] Ulpia Martina,[1236] Valeria Valentina Ova[1237] or Aurelia Felica.[1238] As a consequence of the listed names, the completion as *A[urelia]* matches the type of the female names at Szőny. I accept the completion of the text from the fifth line. My version is as follows: *Deo M[agno] / Aeter[no dom(um) lon]/gam pr[o se et suis] / A[urelia] / Claudia P[ia filio]/que pecun[ia sua] / fecit ex v[oto].*

• *The historical evaluation of the inscription.* The appearance of the epithet *Deo Aeterno* on ☞ the votive inscription from Intercisa (12.II.c) can really be mentioned as an parallel. However, it cannot serve as evidence of the Jewish interpretation of the Szőny inscription, since, in the evaluation and interpretation of the Intercisa plate, H. Solin has already disclosed that the person who erected the Intercisa plate was not Jewish.[1239] Thus, the usage of the attribute *aeterno*, or the inscription itself, does not support István Tóth's completion as *synagogam* or his interpretation of the inscription. The combined usage of the epithet *magno aeterno* does not allow us to interpret the *aeterno* independently from the other one. The presence of the two attributes together is more typical of the mysteries than of Jewish or Christian inscriptions. The Bar Kochba coin, which was found in the excavation of a villa, is an individual piece, and is invaluable as regards the question of whether Jewish people lived in Brigetio. We know that in Hadrian's reign, several Pannonian *vexilatios* were stationed in Palestine as mentioned in a Samarian inscription.[1240] Therefore, the inscription does not prove the fact that there were Jewish people in Brigetio, or that there was a synagogue there.

At the present stage of research, we can state that the presence of a synagogue in Brigetio is not confirmed.

53. TÁC-FÖVENYPUSZTA (GORSIUM - HERCULIA) fig. 365.

The site of the excavation is approximately 2 km from the Szabadbattyán road.

The definition of the function of the buildings has been intensively challenged by recent research. We not only have to examine the question of the 'Christian basilica', but also the question of the building complexes.[1241]

The site - in contrast to others - has an advantage in that it had been a pasture before excavation. No damage was done to the below-surface Roman and medieval relics by ploughing or other activities. The excavation began in 1934, when a building with an apse was found.[1242] It was defined as an early Christian basilica because of the apse. Arnold Marosi disclosed its finds; these were *vas diatretum* fragments, pottery sherd, a spindle whorl, an iron object, and the lead plaque of the Thracian Rider-god. The presence of the apse was decisive in the definition, although Marosi himself admitted that there were ambiguous phenomena, and he disclosed the definition of the basilica very cautiously.[1243] Lajos Nagy wrote about the usage of 'a large Roman *villa rustica* for cult purposes,' noting that the excavation of the basilica had not yet been completed.[1244] Edit Thomas wrote of a three-aisled basilica with apse, and of the fact that the excavations had unearthed an altar slab.[1245] The later excavations completely opened the buildings in question and the constructions gained new identifications.[1246] The excavations proved the fact that the buildings that had been thought to be early Christian basilicas were not actually so. However the continued excavations after 1954 unearthed buildings that are today classified in the literature as early Christian churches. Disputes arose not only over the buildings classified as early Christian, but over nearly all of the other constructions, whose classification, function, and even chronology, had been defined based on the presumption that the *ara Augusti provinciae Pannoniae Inferioris* used to be at Tác. The buildings had been given names like *palatium,* provincial assembly hall, the Emperor's main temple of worship, Capitoline temple, etc.[1247] The names of buildings also changed as their functions were redefined.[1248]

The name of the settlement. The antique name of the settlement used to be identified with the *Gorsium* that appeared in the *Itinerarium Antonini*[1249] (as opposed to previous opinion that searched for Gorsium in Székesfehérvár). The research concluded by saying that Gorsium was not Székesfehérvár, but it was not determined to be Tác. The dating of the ancient name, as well as its change, caused a new debate. One issue was that Gorsium's name was changed to *Herculia* in the late period, so Tác's Roman name was first Gorsium, and then Herculia.[1250] The other issue refutes this alteration of the name[1251] at the same time. It also dismisses the identification of Tác with

[1229] RIU 571.
[1230] RIU 535.
[1231] RIU 536.
[1232] RIU 544.
[1233] RIU 573.
[1234] RIU 574.
[1235] RIU 575.
[1236] RIU 588.
[1237] RIU 598.
[1238] RIU 627.
[1239] Vide the elaboration of the whole problem at 12.I.c.
[1240] BARKÓCZI 1957-1958 p. 19. No. 22.
[1241] I wish to thank Dr Jenő Fitz, leader of the Tác excavations, for the permission to examine and use the small finds from the buildings. I also admit that their examination has not yet been finished, so I know only a part of the material.
[1242] MAROSI 1934: 60-71. - I note that the opening and research of Tác was so regularly and systematically documented in publications that only a selection of the most significant ones will be outlined here for the reader to find further literature and references.
[1243] MAROSI 1934: 69.
[1244] NAGY L 1938: 125.
[1245] THOMAS 1980: 197. Mihály Nagy has proved these altar slabs to be monopodiums, ☞ Aquincum 4.II.h.
[1246] FITZ 1961: 147-156.
[1247] TÓTH E 1989c; TÓTH E 1991. Reply: FITZ 1993; FITZ 1997.
[1248] TÓTH E 1989c: 49-50; TÓTH E 1991: 104-105.
[1249] FITZ 1962a: 155-158.
[1250] FITZ 1976; FITZ 1976a; FITZ 1976b: 14, 100.
[1251] TÓTH E 1982: 55-72. The reply to this: FITZ 1984: 279-281.

Gorsium on the grounds that the unearthed objects do not match those associated with the idea of a town.[1252] Tác was identified with Gorsium, a *municipium* etc., by the 'euphemistic and exaggerated choice of words'.[1253] The alteration of the name also brought about a dispute over the roads, as the protagonists concerned themselves with this subject as well. Outside the debate, Gyula Fülöp expressed his opinion concerning the question of the roads.[1254]

Subsequent transport of Roman stones for building purposes. Several inscribed Roman stones were found at the excavation of the medieval basilica in Székesfehérvár. These inscriptions repeatedly referred to Aquincum, so it was taken for granted that the stones had been taken from Aquincum for the building of the basilica. The site is described by Árpád Dormuth data,[1255] ☞ Székesfehérvár. At the beginning of the Tác excavations, a Roman settlement appeared that could have served as a closer source of stones and this led to a further presumption that stones came from a site at Tác.[1256] The arguments were that it was nearer than Aquincum, and that there was no Roman site in the Fejér region that could have provided the material. However, apart from rescue excavations, there had been few planned excavations in Fejér County. Arnold Marosi had started a systematic and continuous excavation in Intercisa.[1257] Only in recent years have new and systematic excavations been started by Gabriella Nádorfi, which have produced an unexpectedly large amount of material.[1258] Thus, the argument that no nearer source of stones existed apart from Tác has been refuted. According to Jenő Fitz, 'the abandoned stone material of Aquincum and other forts along the Danube was mainly used for building local churches, monasteries, castles, and palaces in the Middle Ages. ... We can confirm only one transportation of stones from Aquincum to Székesfehérvár, which were required for Saint Stephen's Basilica.'[1259] ☞ Székesfehérvár.

The exact route of the transportation of the inscribed stone material (and therefore its site of origin) was proved by Géza Alföldy. A part of it must have come from Aquincum as the inscriptions themselves refer to Aquincum.[1260] On the other hand, it can also be confirmed that the stones were taken from Tác to r. Sárpentele.[1261] The question of this transportation of stones is essential as it forms part of the Roman interpretation of Tác.

The study of excavations. The systematic excavations were documented in annual publications, disclosing the latest results, hypotheses, and the more consolidated opinions.[1262] The historical conclusions deduced also appeared very systematically.[1263]

1992 was the 33rd year of the Tác excavations. By this time many objects had been brought to light, and scholars had tried to define their functions with the help of ground plans, inscriptions and small finds. The newer results sometimes led to changes in the concepts, which were revealed in the *Forschungen* by the archaeologists. Their work could be divided into three phases: 1. The excavation, 2. The definition of the function, 3. The compilation of Tác's history. Two of these three phases produced constant debate. Undertaking the disclosure of the issues related to Tác, I will outline the archaeologists' opinions, limiting the debate to the problems regarding Christianity, and then I will disclose my own opinion, again focusing on Christianity.

Tác is situated at a Roman crossroads, one road connecting Aquincum and Sopianae, the other Brigetio and Sopianae. The earliest finds from the territory are from the 1950s: two bricks with a stamp belonging to the *ala I Scubulorum*. Another early find was a Vespasian denarius. The building once thought to be a basilica is actually the wing of a villa's bath, under which an early period horreum was found. Near the villa, a smithy was opened. East of the smithy, the researchers came across a 1st-century shrine (25.3 x 11 m). It was covered with terrazzo. It had burned down, but some of its wooden door remained. In the 2nd century, there was an Eraviscus settlement here that was ruined during the Marcomann wars. We cannot follow the Eraviscus culture from the end of the 2nd century. Thracian settlers came, witnessed by the Thracian names in the inscriptions or the lead plaques of the Thracian Rider-god. The restored shrine to Jupiter Dolichenus was consecrated during Septimus Severus' Pannonian stay. This was a period (until 260) when oriental people came to the territory (the time of the Roxolan invasion). Gorsium was rebuilt in the 4th century, and its name was changed to Herculia. There were extensive building operations in the reign of Valentinian I, denoting a prosperous era. The graves found in the buildings already mark the decline to follow; the settlement was demolished, but life continued in

[1252] TÓTH E 1991: 100. Cf. his opinion of Pécs, see Pécs, history of research.

[1253] TÓTH E 1989c: 45; TÓTH E 1991: 101.

[1254] FÜLÖP 1978: 281-285.

[1255] András ALFÖLDI (1939: 101) writes as follows: It was found from the basement of 'the building from Saint Stephen's time'. 103: In the reign of King Matthias, some building material from the demolished Óbuda church was transported to Fehérvár.'

[1256] FITZ 1958 6.

[1257] The possibility that the stones had been transported from here was not even raised. In addition, I have to note that the recent research has raised the opinion that, in the Roman period, such a stone carver's workshop used to exist in Gorsium and that Intercisa's inhabitants might have been customers. This means that the Roman stone relics found in Intercisa had been carved in Gorsium. BARKÓCZI 1984. From Trajan's time, oriental craftsmen worked in Gorsium: BARKÓCZI 1982. Cf. Note 1289.

[1258] NÁDORFI 1992 No. 83-117.

[1259] FITZ 1958.

[1260] ALFÖLDY 1977: 231.

[1261] ALFÖLDY 1977: 229-230, and Notes 50 and 53.

[1262] The Alba Regia volumes, *Forschungen* entry, from 1964 to 1990, by Zsuzsa Bánki, Jenõ Fitz and Vera Lányi. The excavations, buildings and epigraphic finds were disclosed by Jenõ Fitz, the coins by Vera Lányi, the rest of the finds by Zsuzsa Bánki. As I have to refer to the excavations more often, this part of the *Forschungen* is referred to by Jenõ Fitz's name in the notes of abbreviations.

[1263] FITZ 1961; FITZ 1962; FITZ 1962a; FITZ 1967-1968a; FITZ 1970; FITZ 1972; FITZ 1972b; FITZ 1972c; FITZ 1975; FITZ 1976; FITZ 1976a; FITZ 1976b; FITZ 1978; FITZ 1979; FITZ 1980b; FITZ 1981; FITZ 1983; FITZ 1985a; FITZ 1985b; FITZ 1986; FITZ 1987; FITZ 1990; KOCZTUR 1972.

the area as Germanic tribes arrived.[1264]

From a military aspect, and in terms of initial expansion, Tác's historical significance is analysed by András Mócsy, who writes that internal strife made it necessary to establish an inner military fort north of the Drave River.[1265]

This fort served as a basis for a later settlement where we now find the buildings discussed here. As already mentioned, the names of these buildings is debateable but the terminology will still be used for the sake of easier identification.

The general plan of the excavation is shown in fig. 366. The overall map of the excavations, indicating conditions in the 2nd, 3rd, and 4th centuries[1266], is presented as fig. 367. The ground plans of the buildings from the 2nd and 3rd centuries[1267] are shown in fig. 368. The ground plans in the 4th century[1268] are shown in fig. 369.

PALATIUM or building I.

Size: 65 x 49 m. It is a building with a *peristylium*, its main entrance lying on the southern side. The rooms had underfloor heating supplied from the outside - the *praefurnium* was also found. The most attractive section is the hall with an apse on the western front. The southwest part constitutes an independent group of rooms. The halls with apses on the northern side indicate the bath. In the eastern part of the palace there are marks of a previous construction, which was defined as a horreum (25.3 x 11 m). The archaeologists compared its walls to those of the palace, and stated that the walls belonging to the eastern parts were thicker, of better quality, and had deeper footings. The floor was covered with a good-quality terrazzo under a thick layer of soot, burnt fragments of beams and remnants of frescos. Because of its good quality and frescos, it might have had a cult function.[1269]

The sigillatae found between 1958 and 1961 were studied by Dénes Gabler. The sigillatae with relief can be dated to the period between the third quarter of the 1st century to the first quarter of the 3rd century.[1270]

FINDS

53.II.a. Basilica maior, or building III,[1271] fig. 370

Size: 12.5 x 20 m, the radius of the apse is 4.2 m. Its walling is *opus incertum* and in the walls there were 2nd-century gravestones built in. Its floor was very bad quality. The fresco fragments found on the floor were very poor. The coin finds are from the Constantine and Valentinian dynasties, consequently its building is dated to the middle of the 4th century.[1272] Based on its ground plan, and especially because of its apse, the archaeologists defined it as a Christian basilica, recognising a font in the quadrilateral construction next to the basilica. The latter construction is actually a small, brick annex between the large common well (C) and the '*basilica maior*'.

In the studied area, the E-W street, the palatium, the *tabernae*, the *basilica maior*, the *basilica minor*, and the *area sacra* constitute a unit.[1273] On the southern side of the E-W street is the *peristyle* of the *tabernae*, on the northern side is the *basilica maior*. The building with an apse belongs to the 4th-century layer; below it, the remnants of earlier buildings were found, dating to the Severus era. The building was destroyed by fire, perhaps connected with the Roxolan invasion that devastated Gorsium in 260.[1274] This early building was decorated with frescos.

53.II.b. Basilica minor, or building V.[1275] fig. 371.

Size: 18 x 13 m. The diameter of the apse is 6.3 m. It has only one room. The brick floor is poor quality. Coins of Trajan, Diocletian, and Valens were found here, but as a result of the obvious disturbance they were useless for dating. The basilica-style building may be dated to the turn of the 3rd and 4th centuries, thus it is earlier than buildings I and III.[1276] It was probably 'Herculia's first Christian church'[1277] It was fenced from the west and south.

Epigraphic debate. Although the further buildings are not Christian, I have to refer to them as they have religious references and have recently been included in the debate.

In 1970, Jenő Fitz published a brief paper[1278] in which he arrived at the conclusion that Pannonia Inferior's *ara Augusti* was in Gorsium, i.e. the annual provincial councils would have taken place here. He referred to two previous opinions. András Mócsy thought that the *ara* was either in Aquincum or in Gorsium, while Árpád Dobó did not consider Gorsium to be a likely candidate. Jenő Fitz took issue with Dobó and asserted his own opinion that the provincial *ara* was in Gorsium. In his argument, he cited building VII as the hall for the council, while building IX was the temple of Emperor worship. The basis for these definitions was - in his opinion - a brick stamp, on which the ligatures TE and PR might be read: in his interpretation: te(mplum) pr(ovinciae). From this time on, all finds were interpreted in connection with the function of the *templum provinciae*, and - as years passed - the *area sacra* developed.

Endre Tóth revised the above argument, and the Tác findings, publishing a polemical article[1279] in which he precisely criticised the method of defining the excavated objects, as well as the *a priori* presumptions and the way

1264 Fitz 1962a. The same historical outline was published in Hungarian and English, with different illustrations. In Hungarian: Fejér Megyei Szemle 4 1967: 38-47; in English: Acta Archaeologica Carpathica 10 1968: 287-300.

1265 MÓCSY 1971.

1266 BARKÓCZI/SALAMON 1984: 172 Abb. 26: these drawings were first published by Jenõ Fitz.

1267 TÓTH E 1991: 98 Abb. 1.

1268 TÓTH E 1991: 99 Abb. 2.

1269 FITZ 1976b: 26.

1270 GABLER 1972.

1271 FITZ 1976b Abb. 6; HAJNÓCZI 1987 Abb 10.

1272 BÁNKI 1965: 213.

1273 FITZ 1972; FITZ 1972c: 4-24, 52; FITZ 1976: 386; FITZ 1986: 365-367.

1274 FITZ 1976b: 32.

1275 FITZ 1967-1968a: 219 Abb. 2. - FITZ 1972c: 14-16.

1276 FITZ 1967-1968: 218-219.

1277 FITZ 1976b: 38.

1278 FITZ 1970.

1279 TÓTH E 1991: 97-112.

Jenő Fitz had taken them as facts. At the same time he provided counter arguments for each statement. He proposed different interpretations for the TE.PR brick stamp, i.e. te(gularia) pr(…), te(gula) pr(…). He connected all his interpretations to the fact that the inscription was on a brick, so the solution had to be a derivation of the Latin word *tegula*. He further remarked that such brick stamps had been found not only at Tác, but also at Tés (Veszprém County).[1280] Géza Alföldy also joined the debate supporting the argument 'wonach die Ziegelstempel mit der Marke TE.PR für die These von Fitz ebenfalls[1281] keinen gesicherten Beweis liefern.'[1282]

Another issue of the epigraphic debate is the inscription of an altar, which is read by Jenő Fitz as follows: [---] / pro salute /templenisum / L(ucius) Virius L(ucii) fil(ius) Mer/cator sacerd(os) / v(otum) s(olvit) l(ibens) m(erito).[1283] The first line of this inscription, based on the remaining letter fragments, was completed by Géza Alföldy as follows: Dis Magn[is].[1284] Starting from the interpretation of *Dis Magnis*, he proved that the *area sacra* really did deserve this name: it was not the place of the *concilium provinciae Pannoniae Inferioris*, but the sacred district of gods belonging to unidentified oriental mysteries. The *templum*[1285] was not the temple of the *Divi Augusti*, as it had been interpreted by Fitz, but the temple of the *Deorum Magnorum*. Tác was indeed a religious centre, but not the place of the provincial council, Emperor worship, or the *templum provinciae Pannoniae Inferioris*.[1286]

As we see, the epigraphic debate not only reinterprets the inscriptions in question, but also the role of Tác in the Roman period.

Fitz's response to the debate.[1287]

The dedication *Dei magni* is unacceptable as it does not occur in Pannonia, and especially not in the period to which Jenő Fitz dated the building connected to the inscription by Alföldy. Furthermore, the restoration of the *templum*, according to other examples, is always an imperial construction, or at least a construction on the emperor's command, therefore the completion of *divi Augusti* is better. Gorsium remains at the centre of Emperor worship.

The new names in Pannonia during the tetrarchy, such as the province of *Valeria*, support the idea that Maximianus Herculius might have given the restored Gorsium a new name - Herculia. Endre Tóth's objections to the dating are weak; he considered only a few selected buildings, not the whole building complex. In addition, lacking a parallel example, the presumption that 'militärischen Zeremonien' took place in the territory is groundless.[1288]

In connection with the transportation of the stones, Jenő Fitz touches upon Barkóczi's opinion,[1289] to which he adds that 'womit die Forschung bisher nicht viel anfangen konnte: drei Steinfragmente aus Gorsium ind Intercisa gehörten zusammen.'[1290]

Debate on the so-called early Christian buildings III and V. Endre Tóth criticised the functions of almost all the buildings, including the ones belonging to the so-called early Christian basilica.[1291] 'What consideration can make us think that the shed joined to the northeast corner of building III is a font? Why are buildings III and V early Christian basilicas (moreover, possibly an Episcopal residence)? Why do we call the two buildings with approximately the same inner territory *basilica maior* and *basilica minor*? These are technical terms, and do not refer to 'Tác-type' buildings in this period. It is inexplicable why the latest publication marks pillars in the two 'basilicas' while neither the excavation report, nor the earlier publications (and ground-plans) did so.'[1292]

Jenő Fitz did not mention these objections in his responses.

Commentary:

• We cannot avoid adding our remarks to Tác as a whole, although it does not belong closely to the present subject. The debate has clearly showed that there are problems with the interpretation of Tác in the Pannonian research. The historical image that has previously been revealed, based on the excavation reports and the writings by Jenő Fitz, is the sequence of statements that has been questioned in nearly every regard. Once a suspicion has emerged in connection with the statements, it cannot be dispelled in the tense atmosphere of written argument and counter argument. The fact must be faced that only a comprehensive reinterpretation can help, which would require the publication of the whole material of small finds. If by any chance the suspicion as to the identification of the name proved true, it would not at all reduce the significance of Tác.

Starting with László Barkóczi's statement on the stone-carver's workshop, we have to accept that oriental stone-cavers were already working at Tác in Trajan's reign.[1293] Under such circumstances, however, we cannot wonder when we find traces that suggest worship of oriental gods (and shrines suggesting oriental rites) in Trajan's time. Although the Pannonian research always highlights the relationship between the oriental and the

---

1280 TÓTH E 1989c: 53; TÓTH E 1991: 106.

1281 He refers to the altar inscription, which he reinterpreted in his article, and interpreted Tác's role in Pannonia accordingly.

1282 ALFÖLDY 1997: 232.

1283 FITZ 1972c: 41; 1976b: 74-75, No. 19; Pannonia 59; For further literature see ALFÖLDY 1997: 225 Note 3. Reply: FITZ 1997 especially 68.

1284 1997: 226.

1285 CIL III 3342; ALFÖLDY 1997: 235, further literature Note 50, where Alföldy mentions his own previous interpretation (templum Dianae) as well.

1286 ALFÖLDY 1997.

1287 FITZ 1993; FITZ 1997.

1288 FITZ 1997: 68.

1289 Vide Note 1257.

1290 FITZ 1997: 69.

1291 Without any details: TÓTH E 1989c: 45.

1292 TÓTH E 1991: 100.

1293 BARKÓCZI 1982; FITZ 1997: 69.

Septimius-dynasty, it does not preclude the possibility that oriental people appeared in the province in other periods and under different emperors. When András Mócsy studies the Flavian-Trajanian period, he writes that although the 'Italians, Gauls, Spaniards, and Noricans' served best, 'the composition of the auxiliary units was even more varied ... Syrians, Thracians, Dalmatians, Illyrians, natives of the Alps, Britons, Spaniards, and Germans.'[1294] The *ala I Ityraeorum* was composed of Syrians. The *legio II Adiutrix* was joined by Macedonians and Syrians when this legion was sent to oppress the Jewish rebellion under Hadrian.[1295] As for Tác, the *ala I Ityraeorum* is essential as this unit was stationed in Intercisa, a few kilometres from Tác. Craftsmen, therefore, could have settled at Tác while their customers lived in Intercisa. It might also have been that the auxillary units were placed farther from the limes, in a more protected area. Naturally, the incoming craftsmen and their family were allowed to keep their culture and religion.

- The dedication *Deo Magno* is known from Brigetio, ☞ Szőny, 52.II.c. Although the inscription is fragmentary, the *M* is still discernible. Thus the dedication is certain. We also know that there used to be a Syrian settlement in Brigetio.

- We can almost certainly say that the two buildings are not basilicas. The finds disclosed so far do not support their Christian interpretation. There is a well next to building III, but it is so close that a miniature room could hardly have been joined to the eastern side of the apse. This was interpreted as a font but it cannot be one. The location of baptisteries was defined and they were mostly octagonal.[1296] There are, however, examples of reusing earlier Roman buildings and creating octagonal fonts in them. Charles Bonnet describes such a building.[1297] 'Le baptistère présente un plan rectangularie tripartite, la pièce principale étant séparée des annexes latérales par de grands arcs. On accédait da le nef par trois portes, alors qu'un petit escalier permettait de descendre dans l'annexe nord, depuis la galerie supérieure des cryptoportiques. La cuve primitiv a été modifiée, elle était circularie ou polygonale (figg. 11, 12 et 13). Le deuxiéme état, cruciforme, est très rare puisque seuls les exemples de Dion en Grèce et Barcelone en Espagne nous sont parvenus. Une nouvelle transformation du bassin lui donne une forme octogonale plus habituelle.'[1298] From this we learn that in rare cases the font may have different ground plans, but there then have to be other circumstances to authenticate the object. Such circumstances are missing from the Tác building, which neither meets the iconographic requirements of a font, nor provides any information for such use.

We also have to note that we have found no baptisteries in Pannonia. Emona is the only place where we know one;[1299] at that time Emona was not part of Pannonia. The baptistery can be substituted by a simple *lavacrum*, which was possible to build in Pannonia. In Tác's case, however, even this cannot be proved.

The main characteristic of building V is the shape of its apse, which forms almost a full circle. There is also an example of such an arrangement where the continuous alteration of one single building is observable through different periods, from the 4th century to the 11th century, within a building complex that is clearly an independent unit.[1300] As a comment, I quote Charles Bonnet again: 'There is a great deal of archaeological material that establishes the date of these first Christian buildings as being in the middle of the 4th century. This religious complex was no doubt erected between 350 and 370. However this may be, it is certain that a Christian community existed in Geneva well before this time.'[1301] The difference between Geneva and Tác, notwithstanding the similarities of ground plan, is the number of finds referring to Christianity, or at least to the building of a baptistery. There are many in Geneva but not one in Tác. Therefore, we cannot prove that Christians used the Tác 'basilicas'.

- With reference to the terms *basilica maior* and *minor*, the archaeologists only wished to express that there were two buildings, and No. III was found first. It was, however, an unfortunate choice of names as these terms refer to hierarchical churches, and their priests, in ecclesiastical law.[1302] ☞ Szombathely.

- There is, however, one illustration that might be regarded as Christian. The illustration is a common one, and, no doubt, more usual among Christians. However, in view of the circumstances, I do not consider it to be Christian.

53.II.c. Roof tile

Location: IKM 73.19.1. **Literature:** FITZ 1975: 322; Severin, Katalog 8.49.

'Bruchstück eines Leistenziegels (tegula) mit Ritzzeichnung und Resten eines Graffito. Die Zeichnung stellt einen stilisierten, langgestreckten Fisch neben einen viergeteilten Brot dar. Fisch und Brot sind durch verschiedene Kerben stark gegliedert.'

SUMMARY

Based on the material presently at my disposal, I do not think we can prove the Christian usage of the buildings that have been thought to be Christian. However, we may consider several shrines and workshops in the area. Tác's shrines and consequently Tác's role in Pannonia's religious life should not be underestimated. The presence of the shrines does not confirm the religious thinking of the

[1294] MÓCSY 1974: 154.
[1295] MÓCSY 1974: 155.
[1296] LASSUS 1968: 88.
[1297] BONNET/PERINETTI 108 fig. 11: Cathédrale Notre-Dame. La cuve du baptistère. Fig. 13: Pans des états successifs de la cuve. BONNET 1986: 485 figs. 6-7.
[1298] BONNET/PERINETTI 99-100.
[1299] PLESNIČAR-GEC 1983 Taf. 57.
[1300] BONNET 1985 fig. 4; BONNET 1986a figs. 1-5; BONNET 1987: 332-333 -We can see Geneva's map, and within the city, the situation of the Episcopal buildings, in the 'corner' of the city, near the walls. BONNET 1988; BONNET 1991: 222-223.
[1301] BONNET 1986a: 23.
[1302] SZERÉDY 1883: 571-575; ADAM 1985: 302.

locals, but rather that of the settlers and the religiousness of the *nobiliores*. We may only suspect a relationship between Tác's inhabitants and Christianity rather than prove it. My assumption is that this territory may have been a *refugium* for a long time, where the *pagani*, in critical periods, took shelter, keeping their ancient religion. That is why we cannot suppose that the oriental population, having adopted Christianity, erected their church in this area. I would even put forward the view that Pécs was the 'reservoir' of the Christian oriental; whereas Tác was that of the pagans, or in the Christian era, the stronghold of emperor worship. We may determine the length of paganism after study of the finds. I presume that in this area no Christian churches were erected, but that pagan shrines were transformed into churches.[1303]

54. TÁP

54.II.a. Fragment of architrave, fig. 372.
Literature: TÓTH E 1980b: 99 Abb. 5.

The stone has been lost and its dimensions are unknown. Endre Tóth estimates its height to be around 50-70 cm. It was classified as a Christian object due to its similarity to the carvings of the Theoderich memorial in Ravenna; and Endre Tóth dated it to the 5th or 6th century accordingly. A similar fragment of architrave was found at Au am Leithaberg.[1304] Despite this parallel find, and because of the uncertain nature of the object, I regard the item as controversial.

55. TATA

55.I.a. Golden finger ring, fig. 373.
Location: MNM 103.1889.1 purchase, Jakab Krausz. **Literature:** NAGY L 1930b: 243-246 fig. 153; NAGY L 1938: 88; TÓTH E 1979 Kat.-Nr 14.

Multiangular, of small size. Its inscription runs around it. Between the letters there is a cross with arms of the equal length. Diam: 2.1 cm, H: 3 mm, weight: 3.4 gr. The inscription was read and interpreted by Lajos Nagy: 'chi-rho monogram PAXPOP/TANTIDORINVI/DENTI (sic) = pax portanti dor(mitio) invidenti - Be peace in Christ for the owner of the ring, eternal rest for the malicious.'[1305]

[1303] BANDMANN G.: Mittelalterliche Architektur als Bedeutungsträger. Berlin 1951: 144-159, writes about sacred districts and that pagan shrines are transformed into Christian churches. There is a reference to Pope Gregory I's letter to Mellitus, where the subject is that one of the *daimons'* temple passes into the Church of the God. The archaeological observations belonging to this written tradition is disclosed by Branka MIGOTTI 1992b.

[1304] SCHOBER 1914; NAGY L 1938: 126. A cemetery with Christian features, where a building was found with an architrave and two capitals inside. The architrave fragment seems to be analogous to the one at Pola, which helped Egger define the function of the building. Schober quotes Egger's opinion as follows: 'Nach Egger wurden solche Säulen als Tisch- oder Ziborienstützen in den Presbyterien der frühchristlichen Kirchen verwendet.' (SCHOBER 1914: 256 fragment of the architrave: fig. 211; the two capitals: figs. 212-213.) The building was defined as a Christian basilica based on the finds; and based on its analogy to the Pola dome, it was put to the 5th century.

[1305] NAGY L 1930b: 243, 245-256.

56. TIHANY-SAJKÓD

56.I.a. Fibula , figs 374a, b, c.
Provenance: Graves were found at 11 Mályva Street. The owner sold this object to the MNM in 1966. **Location:** MNM Inv. No. 66.14.3. **Literature:** TÓTH E 1994a figs. 2-3.

Crossbow fibula. L: 7.5 cm. Gilded bronze. A chi-rho monogram decorates at the pin holder. It can be dated to the second half of the 4th century.

57. TOKOD

Among Tokod's sites, the Erzsébet-shaft and Várberek are dealt with here. The general map of the sites, as well as their correlation, was published in MRT Vol. 5[1306] fig. 375. The excavations and data concerning the area are summarized and based partly on Topográfia and partly on the Tokod monograph.

The site has been known since the 19th century. The new excavations were initiated by mine-works in the area. A. Mócsay led an excavation in the 1960s to try and resolve problems that arose at previous excavations.

As a result of rescue excavations and a site survey, the settlement of the original Celtic inhabitants, the early and late Roman settlements, the cemetery and the 'fort', have been brought to light. The Roman settlement was established in the 1st century, and was damaged during the Marcomann wars. After the wars, the building activity was reduced to the reconstruction of already existing buildings. There are few facts concerning the life of the settlement in the 3rd century. It survived, however, and in the 4th century, there were industrial activities in the area of the former settlement. In connection with these activities, most is known about the pottery.[1307] The settlement continued into the 5th century; according to Mócsy's assumptions, the 'fort' might still have existed in the 6th century as well.

The fort and its name

Among the remarkable finds are the red marble fragment,[1308] and the especially great number of early sigillatae from North Italy[1309] and from the Po valley.[1310] The column whose fragment has been found belonged to an early building.[1311]

At Tokod, the Roman settlement, found in separate but correlating sites, was already started in the 1st century, and, what is more, can be regarded as a significant one. Moreover, since there are available finds up to the 5th century, we have to postulate its continuous existence. The destruction caused by the Marcomann wars, as in most places of the province, did not mean a complete desolation or devastation.

The fort lies south of the settlement on the hillside, fig. 376. The corners of the approximately rectangular fort are directed into the four cardinal points. The thickness of

[1306] MRT 5, Site No. 22; p. 325.

[1307] BÓNIS 1991: 87-150.

[1308] Tokod 1981: 13-22 (KELEMEN)

[1309] GABLER 1964: 98.

[1310] GABLER 1964a: 15.

[1311] MRT 5 22/16, p. 338, pl. LXVIII,3.

the walls is 160 cm; the thickness of the tower-wall is 110 cm. Its size is 122 x 140 x 115 x 149 m.[1312] 'The fort was surrounded by a 6-metre ditch at a distance of 21 m from the walls. The horseshoe-shaped corner-towers are 7 and 10 m in diameter; the linear sides of the semicircular side turrets of the north-east and south-west walls are 8 m; the linear side of the only large semicircular turret of the south-east wall is 13 m. The quadrilateral gatehouses are 7 x 6.5 m. On each longer side, at different distances from the turrets, there is a 125 cm doorway with a carved stone slab covering on the two sides, and with a door-stone with the engraved hole for the axle of a single-door. The doorway on the north-east side was closed by dry-wall_construction of stone.'[1313] In the southeast turret a terrazzo floor was found, and a pillar of 110 x 110 cm in the middle. The renewal of the footings was recorded in two of the turrets. Earlier, 2nd- and 3rd-century gravestones were also used for building the 'fort'. No buildings were found in the middle of the fort, only alongside and joined to the walls. The only exception is the *horreum*, which has four independent walls. The *horreum* is also an exception because of its deep foundations; the other buildings had no foundations and consequently were not so durable. The only place where reconstruction was recorded was at the entrance of the *horreum*. The walls of the buildings are 60-65 cm thick.

Between the Roman fort and the already excavated late Roman cemetery, people found graves when ploughing. Besides the damaged graves, the archaeologists opened a disturbed brick grave, which can be dated to the age of Valentinian I, according to the marks LUPICINI TRB and OLIMPV.[1314]

András Mócsy, who directed the excavation, defined the object as a 'fort'. The historical assessments have been using this term since then, and their key question is the name of the fort, or its absence from the Notitia Dignitatum register. To prove these two facts, A. Mócsy presented his arguments and evidence.[1315] Concerning the previous period, however, he stated that 'daß Tokod im 3. oder frühen 4. Jh. kein Lager hatte.'[1316] That is why Tokod may have been left out from the list. Alternatively, if we accept S. Soproni's identification of Gardellaca-Cardabiaca and Tokod, then Tokod is present in the Notitia Dignitatum list as a garrison.[1317] Soproni's main argument defending the identification of names is the name itself. Because Gardellaca-Cardabiaca appears in several sources, it must be a fort, and must be located near Tokod. Since no other fort has been found in any excavations or perambulations, we must see Tokod behind the Latin name in the sources.[1318] S. Soproni, too, has to admit that certain circumstances make the object a unique fort: 'Mit der Identifizierung entfällt jene seltsame Annahme, daß das Lager von Tokod, obwohl für das Militär gebaut, unvollendet geblieben und nur als Zivilsiedlung oder als Nachschubstation benützt worden sei, ohne militärische Besatzung. Dies widerspricht gänzlich der römischen Praxis.' To prove the fact that it is still a fort, he has the following argument: 'Es widerspricht aber nicht dem militrischen Character, daß das Lager 3 km von der Donau entfernt liegt, wie das Lager Pilismarót, das bereits in seinem Namen - Castra ad Herculem - einem klaren Hinweis auf die militärische Nutzung zeigt. Seine Lage entspricht auch sonst der von Tokod: Beide liegen hinter der Donau und sperren ein Tal.'[1319] He then considers all the arguments that have been brought up in the heat of the debate to prove that the object at Tokod was not a fort; but he always has a response to them. His last argument is as follows: 'Ob man die Anlage als Lager oder befestigte Siedlung bezeichnet, ist im 4. und 5. Jh. unerheblich. Wenn Gefahr drohte, suchte auch die Zivilbevölkerung innerhalb der Mauern Zuflucht.'[1320] The fact that the question of Tokod, from the historical point of view, had never been satisfactorily settled, despite the many publications, was brought up by László Barkóczi and Ágnes Salamon, who examined the whole object in a larger context comparing it to similar ones.[1321]

Dating, and Anthony of Lérins

The authors of the Tokod volume unanimously agree on the fact that the fort was built during the reign of Valentinian I. The cemetery was used in the 5th century as well. What is more, B. Lőrincz refers to the 5th century very firmly in connection with the brickstamps that have been found in the cemetery: 'Sicher ist aber, daß die Belegung des Gräberfeldes viele Jahre nach dem Tod von Valentinian I. begann und die Ziegel daher erst erheblich später in die Gräber gelangten.'[1322] The fragments of early Christian vessels are dated to the second third of the 5th century by V. Lányi; consequently, she thinks that there might have been Roman life in the settlement at this time in its last period.[1323] A. Mócsy shares this opinion: 'Das Alpha-Omega-Keramik datiert den spätesten Teil der grauen Keramik in die zweite Hälfte des 5. Jh. Da barbarisches Material in der Festung nicht gefunden wurde, wird man die völlige Entvölkerung der Festung spätestens in den Anfang des 6. Jh. datieren können.'[1324]

In connection with the analysis of the cemetery (interpreting anthropological results), A. Mócsy mentions that adults migrated away, and refers to the biography of Anthony as follows: 'Schriftquelle, die von Romanen in Valeria in der zweiten Hälfte des 5. Jh. berichtet, ist die Antoniusvita des Ennodius, die von der Abwanderung des in Valeria geborenen Hl. Antonius von Lirina (Lérins) und implizite auch von der Abwanderung des Lauriacenser Bischofs Constantius berichtet. Beide haben ihre Heimat in

[1312] MRT 5 pp. 341-343.
[1313] MRT 5 pp. 341-343.
[1314] KELEMEN RégFüz Ser. 1. No. 35 1981: 101.
[1315] Tokod 1981: 42-45 (MÓCSY).
[1316] Tokod 1981: 43 (MÓCSY).
[1317] SOPRONI 1978: 170-171; SOPRONI 1985: 21.
[1318] SOPRONI 1985: 24.
[1319] SOPRONI 1985: 24.
[1320] SOPRONI 1985: 26.
[1321] BARKÓCZI/SALAMON 1984: 161-167.
[1322] Tokod 1981: 125. (LŐRINCZ).
[1323] Tokod 1981: 85. (LÁNYI).
[1324] Tokod 1981: 45. (MÓCSY).

der zweiten Hälfte des 5. Jh. verlassen.'[1325] He also mentions another remark by Ennodius saying that Pannonia was destroyed in the second half of the 5th century.[1326]

Anthropological analysis

The anthropological interpretations are remarkable. The parallels to the Tokod population are shown in the 13th chart (Tokod, 1981). According to this chart, the late Roman parallels are brought from Baranya County and the city of Pécs.[1327] The number of the population was reconstructed by Kinga Éry. She based her results on the number of buried people and the finds, and she showed them in the 5th chart. She calculated that the cemetery was used for 30, 60, 90 and 120 years, which means 1, 2, 3 and 4 generations. Based on the reconstructed data, she stated that the settlement was of medium size.[1328] The analysis showed that the majority of the buried people belonged to a race in which both the Mediterranean and the northern element can be traced. 'Soviel kann aber festgestellt werden, daß der Anteil der Nordoiden den veschiedenen mediterranen Elementen untergeordnet ist.'[1329]

Commentary:

- Was it a fort? The strained nature of Soproni's arguments in interpreting the object as a fort clearly proves the fact that it was not one. The fort theory has been forced by the strongly militaristic tendency of the Pannonia research. If we accept the non-fort point of view, then the phenomena, which are very real, become interpretable. The other question is whether the object at Tokod can be considered an inner fortified settlement similar to Fenékpuszta and other places. A civic nature is probable; a walled establishment, where temples are built with the character of an acropolis.

The object at Tokod is not a fort or an inner fortified settlement. If we compare it with inner fortified settlements, it turns out that it shows no similarity to them, so it cannot be considered one of them, at least not one like Fenékpuszta, Környe, Ságvár, Kisárpás, or Alsóheténypuszta.[1330] Thus, we have to find a solution from the other walled establishments. It cannot be a villa estate because the finds do not reinforce this idea. The possibility of a castle must be excluded at once, since it is refuted by the ground plans themselves, not to mention the lack of foundations, or the shallow foundations that are typical of the whole object. Hypothetically, I would propose the idea that we are dealing with an 'acropolis', which was, or may have been, inhabited to a certain extent and which was erected exactly there because of the natural features. In this case, naturally, the identification of place-names (Tokod = Gardellaca-Cardabiaca) becomes uncertain, but many difficulties mentioned above caused by the fort-theory can be ruled out.

The available finds from the settlement, and the grave goods from the cemetery, are naturally equivalent. In contrast, the amount of finds from the 'fort' is notably poorer than that from the settlement, and it becomes even more highlighted if we think of the 'industrial' establishments. It should seem evident that the goods produced near the settlement or in the canabae end up with the troops, as can be shown in the examples of other forts in Pannonia. Consequently, the stock of finds from the 'fort' should be as rich as the one from the settlement. The few pruning knives, hoes, hand-bells, or cattle-bells do not allow us to think of any agricultural activity. The sole dagger and the few arrowheads do not necessarily mean the presence of troops. It is true that they are weapons, but they could have been used for hunting as well as for military purposes. Let us not forget the hunting scenes on the marble slabs at Csopak.[1331] The notable lack of finds, the difficulty proving that it was a fort and that there were military forces here, make it, at last, clear that the object at Várberek was not a fort. In other provinces there are examples of walled villa estates,[1332] palaces, acropolises, and other larger buildings. Thus an object surrounded by walls is not necessarily a fort. (Walled towns do not have to be discussed here as their territory is incomparably bigger than those of the above-mentioned building complexes.)

- Dating, and Saint Anthony of Lérins. The information provided by Ennodius may be significant concerning the prevalence of Christianity in Pannonia, but is not at all relevant for the staying or migrating of the Christians themselves. It is known of Anthony that he was born in Valeria Civitas, and that he was nine years old when his nobleman father died.[1333] Then, for the sake of his education, the child was taken to Saint Severin.[1334] After Saint Severin's death, Anthony was taken to his uncle Constantius, Bishop of Lauriacum,[1335] who him 'in aller Wissenschaft und Tugend unterrichtete.'[1336] Ennodius expressly says that Constantius was his uncle, and after his

[1325] Tokod 1981: 45. (MÓCSY).

[1326] Cf. PEKÁRY 1955: 19-20. Note 2 on page 19. Analysing the sources concerning destruction and exodus, Tamás Pekáry states that 'we can make allowance for a certain tendentious exaggeration.'

[1327] Tokod 1981: 226, 238. (ÉRY).

[1328] Tokod 1981: 224. (ÉRY).

[1329] Tokod 1981: 224. (ÉRY).

[1330] There are more details in MÓCSY 1974: 304-305.

[1331] See at Csopak, 7.II.a.

[1332] The list of things that could, or had to belong to a villa estate, was legally determined; see BUCK 1983. Walled villas have been found at Nadar - SARNOWSKI 1978 fig. 73. -, Mogorjelo - ibid. fig. 90. -, or Pforzheim - ibid. fig. 98. An example of walled palaces is Diocletian's palace at Split - ibid. fig. 92. We could also illustrate the matter with those mosaics, bone-carvings, metal or glass vessels, etc., which depict walled building complexes, and there are hardly any forts among them. We may find pictures of walled forts exactly in the Notitia Dignitatum.

[1333] ENNODIUS, vita B. Antoni CSEL 6. Ed. HARTEL 1892. p. 384, 23-26.: qui servum suum Antonium tanta virtutum dote sublimen circa Danubii fluminis ripas in civitate Valeria Secundino patre lucis huius ianuam iussit intrare. On the life of Anthony and his bishopry in Lauriacum vide BERG 1985: 66-68.

[1334] ENNODIUS, vita B. Antoni CSEL 6. Ed. HARTEL 1892. p. 385.

[1335] Although Constantius is brother of Antonius's father, Secundinus, the assumption is that nothing proves the fact that Constantius, too, was born in Pannonia. - BERG 1985: 67. Moreover, there are opinions saying that Anthony himself was born in Noricum, too. - BERG 1985: 67, n. 58. The Hungarian research, because of Valeria province, locates Valeria Civitas in Valeria province, and suspects it is Tokod.

[1336] STADLER/HEIM 1858 I. S. Antonius, 28 December pp. 260-261.

death Anthony settles down in Weltlin, with Father Marius, who was his tutor from this time on, and with whom 'machte er schnelle Fortschritte auf der Tugendbahn.' From here, he went into mountains of the Alps and lived as a hermit until he was finally found. Crowds of people thronged to his cave and took him to the convent of Lérins, 'wo die Mönsche an ihm ein vollendetes Muster der Vollkommenheit fanden.' Two years later he died. Guttler dates his death to 525, Baronius and Ruderus to 488.[1337]

From the biography, it emerges that it was not an exodus. At the same time it is worth noting that here in Valeria in the 5th century there were people of noble origin, who happened to be Christians. It is also notable that the Christianity of Noricum and Pannonia kept in touch, at least through family relationships.[1338]

- Tokod's Early Christian history. Based on the above data, observations and assessments, we are able to draw the early Christian history of Tokod in some respects. There are chapters, however, which may only written of hypothetically.

We may state with some certainty that the settlement existed continuously from the 1st century up to at least the 5th century, despite war and barbarian raids. In fact, it may be said that it reached its prime in the later centuries, since the turning of the settlement into an 'industrial centre' can be dated to the 4th century.[1339] This change must have brought prosperity to the settlement, which, thus, may have become more attractive. At this point, the anthropological analysis supplies important information that indicates a parallel to the population of Baranya and Pécs. We must not forget, however, that this parallel equates Baranya's and Pécs's 4th-century population to 5th-century Tokod. From this we may draw the conclusion that the settlers from Baranya and Pécs arrived in the second half of the 4th century, attracted by Tokod's 'industrial' possibilities. The immigrants were most probably Christians, and, anthropologically, they might have had a Mediterranean character.[1340] This late immigration explains the fact that the first early Christian finds at Tokod have remained from the 5th century. These finds include pottery with illustrations that also appeared on earlier Christian objects, but the type of pottery sherds can really be only dated back to the 5th century. As for classifying the pottery, I agree with András Mócsy, who recognised simple pottery objects, and not with those who classified these fragments as terra sigillata chiara.

The industrial establishments raise another question, namely, the fact that the inhabitants, and thus the settlement, needed a wider territory. If the old dwelling houses were turned into workshops that can mean two things: either a loss of population, or new territories connected to the settlement. A loss of population would support A. Mócsy's theory of transmigration, but it leaves unexplained the growth of the settlement defined as canabae by the authors of MRT 5.[1341] Canabae lay to the east and south of the earlier settlement, where broken pieces of Roman earthenware were collected during the perambulations.

and to the immigration to Tokod at the beginning of the 4th century. Another common feature is that the Christian, or Christian-like, objects only appear in Pécs and Pilismarót from the second part of the 4th century onwards (especially in the period of Valentinian I), while they appear in Tokod in the 5th century. In any event, the documents at hand today suggest the trend that the research in these three places **Links between Pécs, Pilismarót and Tokod.** Because of the great similarity to the building complex at Pilismarót, we have to assume that the function of these buildings was also the same. The existence of pieces of earthenware related to Christianity, albeit few in number, arouses the suspicion that Christians lived at Tokod and that they needed a church. On the basis of the Mediterranean character detected by the anthropological analysis, as well as its reference to the relations at Pécs, we may suppose some relationship between the inhabitants of Tokod, the Hellenised population of Pécs, and the Hellenised Christians of Pécs. The Hellenised influence shows not only in the fact that the settlement and the cemetery were separated (as was compulsory in antiquity), but also in the existence of an acropolis, which is also separated from the settlement and is on higher ground. We must not disregard the fact that the Roman sites at Pécs, Pilismarót and Tokod are on hills, or near hills. At Tokod, we have all three features: settlement, cemetery, and acropolis. The walled object at Pilismarót, by reason of its similarity to Tokod, should rather be considered a type of inhabited acropolis. This means that the site of the settlement is unknown, even if some superficial finds suggest traces of settlements.[1342] At Pilismarót, we have a part of the cemetery and the acropolis. At Pécs, we know the cemetery, we suspect the settlement, and we have never searched for the acropolis. The main aim of the research at Pécs (as with Sopianae) has always been to open the Roman city first: to locate the centre, forum, city walls, etc). However, as I have mentioned, the Roman town will not emerge despite regular excavations. On the basis of the examples of Tokod and Pilismarót, the real question in connection with the problems at Pécs is perhaps whether

---

1337 STADLER/HEIM 1858 I. S. Antonius, 28 December pp. 260-261.

1338 As for Christianity in West Pannonia, let us just allude to some finds from Carnuntum. Fragments of vessels, Museum Carnuntinum, Inv. 3393, GABLER 1988: 25-26, 38 Abb. 7/3, 8: 'Wandfragm. einer Schüssel der Form Chenet 320. Rollstempel, Schrägstichmuster, Andreaskreuz mit P (Christogram) und complizierte Bogenmotive. Mustergruppe 7 und 8 nach Hübner.' The first quarter of the 5th century. Lampe mit Christogram, Museum Carnuntinum, Inv. No. 6581. Severin, Katalog 8.58: 'Aus ziegelrotem Ton mit dem Christogram Chi-Rho im Diskus, flankiert von zwei stilisierten Palmzweigen.' Silberlöffel mit Christogram, Museum Carnuntinum, Inv. No. 11.671, Severin, Katalog 8.55: 'Silberlöffel; am Schaufelansatz Namenskartusche MARIA, in der Schaufel Christogram; Serienguß. Das Christogram eingestanzt.'

1339 MRT 5 p. 339.: In this part of the settlement, industrial establishments came into being in the 4th century. The houses, which were only partly used for dwelling, gave home to the glass-making workshop; the bronze-smith settled down here; the kilns for baking pottery were produced here; and iron making took place here.

1340 Cf. the summary of Pécs.

1341 MRT 5 p. 339.

1342 See the general map of site 17 in MRT 5.

Pécs is really the Sopianae, the centre of the public administration in Valeria, or is it, rather, a Syrian and Jewish civitas the life of which was disturbed by the immigration of the Carps who settled down around Pécs in the 4th century. And, because every settlement needs some space, it may have caused some movement within the population. This might have been the cause of the northward shift and led to the building operations during Diocletian's reign, should not continue independently of each other.

The definition of the buildings at Tokod

The rectangular building divided into three by a row of columns has so far been defined as a horreum, fig. 377. According to the excavation records, this was the only building at Tokod on which reconstruction could be noticed. It should be added that here, as in Pannonia, because the research has mainly dealt with military history and the limes, neglecting almost all other features of life, the buildings divided by rows of columns have automatically been defined as horrea,[1343] as food had to be stored for the troops. Such ground plans are not exclusively characteristic of horrea[1344] cf. fig. 132. At the IX Congressus Internationalis Archaeologiae Christianae, several lectures dealt with Christian and Jewish cult buildings, their development and ground plans. N. Duval mostly examined the origins of basilicas as well as the influencing factors with regard to ground plans. He asks, 'La ressemblance est même parfois telle qu'on a pu hésiter dans l'identification de certains monuments: synagogues, églises de judéochrétiens?'[1345] B. Brenk compares plans of synagogues to those of profane buildings. He is not afraid to compare ground plans of pagan temples with those of synagogues.[1346]

These lectures convey the suggestion that here, under the mask of a 'horreum', we are dealing with a Christian church, which became one following a reconstruction. Alternatively, it is the case referred to by N. Duval's question. Thus, the functional possibilities of the Tokod 'horreum' are synagogue or church.

For the sake of completeness, it may be mentioned here that long ago the research raised the issue that the horreum, if the need arose, might be transformed into a Christian church.[1347] However, quite apart from anything else, the ruins do not always give the necessary information. 'Souvent, il est très difficile de qualifier les ruines qui peuvent être aussi bien les vestiges d'établissements militaire (burgi, castella), que de châteaux-magasins (horrea), d'auberges d'étape (mansiones) ou de villas.'[1348]

The buildings at Pilismarót give exact parallels to the octagonal buildings joined to the turrets of the walls. Considering the building complex, we may hypothetically raise the issue that the so-called 'horreum' was, in fact, a synagogue, and the two buildings with apses next to it were Christian churches. At Pilismarót, we have no further traces of buildings. At Tokod, the traces of buildings should also be considered as the ruins of cult buildings.

See more references at ☞ Pécs, ☞ Pilismarót, ☞Keszthely, and ☞ Ságvár.

FINDS

57.I.a. Pottery sherd, fig. 378.

Provenance: It was found near the kiln. **Location:** BBM 70.650.466. **Literature:** Tokod 1981: 80. Abb. 22/1, 25/1. (LÁNYI); Severin, Katalog 8.69b Taf. 64; TÓTH E 1994-1995: 135.

The inside is decorated. Base intact, wall broken. Motif on base: Greek cross, on each horizontal arm there is a palm-leaf, under these the letters A and Ω in reverse order.[1349] The motifs on the walls show the triangular base, which was the origin of the arms of the cross, and which can be found at the arms of the cross on the bottom of the vessel. There used to be four crosses on the wall of the vessel, divided from each other by a palm-leaf respectively (there were altogether four palm-leaves). The bottom is 6.4 cm in diameter. A parallel to the technique of the image has been found in a Christian settlement in Aigina.[1350] Its type is Hayes Form 77, 5th century.[1351]

57.I.b. Pottery sherd, fig. 379.

Provenance: It was found in the settlement. **Location:** BBM 74.52.1. **Literature:** Tokod 1981 80. Abb. 22/2, 25/2. (LÁNYI); Severin, Katalog 8.69a Taf. 64; TÓTH E 1994-1995: 135.

Also decorated inside. It has lost a part of its base. The Latin cross used to be encircled. Both the circle and the cross are ornamented. Staurogram with letters A and Ω in the usual order. The lower arm of the cross is 3.4 cm, which shows that vessels *a* and *b* were approximately of the same size and shape; obviously with the same function. A close parallel to the technique of the staurogram is to be found on one of the vessels from Kôm el-Dikka.[1352]

57.I.c. Pottery sherd, fig. 380.

Provenance: It was found near the kiln. **Location:** BBM 75.650.471. **Literature:** Tokod 1981: 80. Abb. 23/2. (LÁNYI); Severin, Katalog 8.69c.

A shoulder-fragment. Painted with a fish. The maximum width is 5.4 cm.

[1343] If a building is a horreum, we may well expect a considerable amount of seeds to be found in the excavations. Our excavation records, however, make no mention of seeds.

[1344] LECLERQ 1907: 341. Figs. 1-5: ground plans of synagogues.

[1345] DUVAL 1978: 519. Cf. SAUSER 1966: 463-511; DEICHMANN 1983: 68-88.

[1346] BRENK 1978.

[1347] 'Die alte '*horrea publica*', die in der Regel ein kleines heidnische Sacellum vor dem Genius loci hatten, wurden jetzt zu '*horrea ecclesiae*' mit einem christlichen Oratorium.' VAES 1986: 322.

[1348] SARNOWSKI 1978: 26.

[1349] In M. GUARDUCCI's opinion, the reversed image has a message to the religious man: the progress from the end, which is death, to the origin, which is the origin of eternal life. 1987: 1765.

[1350] FELTEN F, Die christliche Siedlung. In: Alt-ägina. I,2. Mainz/Rhein 1975 (55-81), 71. Taf. 23, No. 113.

[1351] HAYES 1972: 125.

[1352] RODZIEWITZ 1976: 34-35. forme B 1-3. pl. V, 16.

57.I.d. Pottery sherd, fig. 381.
Provenance: It was found near the pottery workshop. **Location:** BBM 70.650.468. **Literature:** Tokod 1981: 80. Abb. 22/3, 25/4. (LÁNYI); Severin, Katalog 8.69d Taf. 64.

The remaining fragment of ornamentation shows parts of two birds on two sides of a palm-leaf. The whole vessel was most probably decorated with plants, flowers, and birds. The maximum width of the sherd is 8 cm.

57.I.e. Casket-mount, fig. 382.
Location: private collection. **Literature:** GÁSPÁR 1986 Kat.- Nr. 1324.

Bronze. The size of the fragment is 10.8 x 7.5 cm. The scenes are in a metope framed by a pearl border. In one of the metopes there is Daniel among the lions, in the other is the scene of Moses striking the rock.

Summary of the above finds. The fragment in Kôm el-Dikka may be grouped with the pottery sherds as a general parallel, that might, perhaps, allow for the conclusion that the idea, or the sample, originated from a vessel from here. The B form is present in Kôm el-Dikka from the middle of the 3rd century up to the 7th century, continuously. In Alexandria, the most long-lasting forms are B 2c and B 3. But in North Africa this is a common form.[1353] For us it means that the sample form may have derived from Alexandria or another North African city.

Although we can see no early Christian symbols on them, the lamps found either in the settlement or near the kiln should be mentioned here. V. Lányi lays emphasis several times on the fact that no lamp was found in the fort (nor fragments).[1354]

We do not regard all these lamps as early Christian, but attention should be drawn to one piece: Tokod 1981 Abb. 16/1 (LÁNYI)[1355] (Fig. 383.) Lányi writes that 'Eine auf hohem walzenförmigem Fuß stehende Lampe.'[1356]

58.TÓTVÁZSONY

58.I.a. Bronze tablet
Location: ? **Literature:** Katancsich 1826 I. 439 No. 497; CIL III. 208*; NAGY L 1938: 84.

Inscription: A - chi-rho monogram - Ω / *Claudia H/elpis cum fi/lia et filio / in pace Avi/a f*(ecit). In the CIL volume, Mommsen comments that it is not genuine, because Christian inscriptions are very rare in this region. Lajos Nagy has already refuted this argument.[1357]

59. UGOD-DIÓSPUSZTA

59.II.a. "Church", fig. 384.
Literature: NAGY L 1931a:29; BALOGH 1934: 51; NAGY T 1941: 145; MÓCSY 1962: 756; KÁDÁR 1969: 183; TÓTH E 1991a: 123; MRT 4 75/5.

A similar building was found at ☞ Sárisáp 37.II.a, and T. Nagy mentions them as analogues. Nagy thinks the inner parts were made of wood which have deteriorated, explaining why nothing now remains of them. In contradiction, Endre Tóth does not think Christians used the building since no furniture items have been found.

Sylvia Palágyi writes: 'the excavation in 1968 only partly confirmed the ground plan made with the help of Rómer's data. The opened building had a porch and several premises; its longitudinal axis is directed west east. Its size is 22 x 13 m. Its western side was joined by a 5.6 m apse.'[1358] 'Its poor set of finds does not reinforce the function of a cemetery chapel. Moreover, the dross, the glass paste, and the domestic pottery suggest an economic usage for the building.'[1359]

60. ZALAEGERSZEG (Kaszaháza)

60.I.a. Bronze lamp, figs. 385a, b.
Provenance: 'In 1927 at Kaszaháza, which is a small village, people found a broken bronze lamp while ploughing.'[1360] **Location:** GM 53.33.1. **Literature:** ALFÖLDI 1931-1932: 5-6. Abb. 2; NAGY L 1938: 83 Abb. 48; THOMAS 1975-1976 Abb. 37-40; TÓTH E 1977a: 148; THOMAS 1977b: 81-82 figures on pp. 80 and 82; TÓTH E 1980b: 98 Abb. 4; THOMAS 1980: 198.

Thomas suggests that it symbolises the Lamb of God, and first proposed this meaning. Endre Tóth dates the object to the 5th century. Gramatopol, who supplies a very close parallel from Dacia, dates the type to between the 4th and the 6th centuries.[1361] Length: 12.1 cm, its discus is 5.6 cm in diameter; its height at the discus is 3.1 cm, at the chi-rho monogram, it is 9.4 cm. It was found broken; the stem and chi-rho monogram broken off. It should be noted, however, that, in the photo taken when it was found, the lamp does not have its base. It was either reconstructed later, or, as with the base of the Szombathely piece, was later re-made. ☞ Szombathely 51.I.g. ☞ Nagyberki-Szalacska 31.I.a., analogue from South Pannonia.

61.RUSOVCE/OROSZVÁR[1362]

61.II.a. Roof tile, figs. 386a, b.
Location: HM 58.287.1. **Literature:** BELLA 1914; SÁGI 1968b: 400 Abb. 5; PUSZTAI 1973: 26 fig. 14; THOMAS 1980: 200 Fig. 4; KOLNÍK 1984: 22; THOMAS 1986a: 518.

It was found in a late Roman cemetery. Thomas thinks it is Christ carrying the cross; K. Sági suggests it is Christ's caricature, the reaction of the pagans. Size: 30 x 46 cm. The figure must have a symbolic meaning. It is the shape of a man, but the mouth is not human. Its protruding tongue resembles the tongue of a snake. Its right hand is totally missing, and the left hand is not human either. We make no decision on the object interpreted as a cross. Jesus carrying the cross has its iconography, as seen on Passion sarcophagi.[1363] They used to depict the cross, but it was the *crux invictus*. The carriage of the cross appears on the

[1353] RODZIEWICZ 1976: 33.
[1354] Tokod 1981: 79. Abb. 16. (LÁNYI).
[1355] BBM 72.37.380.
[1356] Tokod 1981: 79. (LÁNYI).
[1357] NAGY L 1938: 84.
[1358] MRT 4, 75/5 p. 245.
[1359] MRT 4, 75/5 p. 245
[1360] NAGY L 1938: 83.
[1361] GRAMATOPOL 1982: 264 Taf. XXVIII/21
[1362] Although the site used to be in the territory of the Barbaricum, the object is kept in Hungary. The object is often mentioned and quoted, and it must be dealt with here, too.
[1363] GERKE 1939: 6-22.

Lateran 171 sarcophagus, but there it is being carried by Simon of Cyrene and not by Christ. Antiquity was unaware of the image of a cross being carried (and so often depicted in the Middle Ages), consequently, the carving on the tile cannot be Christ.

The picture is probably a caricature, and perhaps makes fun of a certain part of everyday life, or a certain character. Thus, we should, rather, find this person.

The drawing shows only the characteristic features; that is why the right arm is completely missing. This warns us that there might be other inaccuracies in the picture, so we are not to look for 'artistic' accuracy in terms of depicting the real world, but rather for an abstracted 'essence' in the caricature-style image.

If we compare the drawing with the representations of gravediggers, we notice some similarities, figs. 342-343.[1364] If a gravedigger puts on his protecting garment (covering his head) he might resemble the strange face of this caricature. In any event, the implement carried on his shoulder (sketched roughly) might give the impression of a cross. The fossor is an 'exclusivmente cristiano',[1365] but only its term. There were gravediggers in the pagan world as well, only the term used for them differed: *artifex*.[1366] Their spades were developed by practical needs; the shape did not depend upon whether a Christian or non-Christian used them. The figure in the drawing might well be a Christian, but it cannot be proven. The object itself should not be considered a Christian relic; only the 'cross' suggests it is one. The item thought to be a cross is surely not a cross.

62. FINDS OF UNKNOWN LOCATION

The finds classified here may presumably have been found in Pannonia,[1367] but we do not know their origin exactly. There are, however, Christian finds in the Déri Museum, or the Museum of Fine Arts, surely originating from outside Pannonia. These finds have been omitted from the catalogue.

62.I.a. Suspension of a lamp, fig. 387.

Location: MNM 1.1874.635;purchased, the Ráth collection. **Literature:** CZOBOR 1879: 180 fig. 5; CZOBOR 1883:242-243 fig. 28; NÉMETHY 1886: 243; GYÁRFÁS 1889: 100 fig. 12; NAGY L 1938: 143 fig. 86; TÓTH E 1977a: 146 Abb. 5.

Bronze, length: 14.5 cm. Shape: cross with rho. Its loop of chain that may have connected it to the fixing part has the form of figure eight. There used to be an A and an Ω hung on the arms of the cross; the marks of their fixing chain have remained. The stem of the letter P is shaped in a way that the lamp can be hung from it.

62.I.b. Suspension of a lamp, fig. 388.

Location: MNM 1.874.363; purchased, the Ráth collection. **Literature:** CZOBOR 1879: 179 fig. 4; CZOBOR 1883: 242-243 fig. 27; NÉMETHY 1886: 242; GYÁRFÁS 1889: 100 fig. 11; NAGY L 1938 text on page 142, the picture without a number on page 148; TÓTH E 1977a: 148 Abb. 4; THOMAS 1982b: 271, Abb. 12.

A bronze disk with a chi-rho monogram open work. Diam.: 11 cm, its height with the lower and the upper ring is 13.6 cm.

62.I.c. Suspension of a lamp, fig. 389.

Location: MNM 1.1903, purchased. **Literature:** NAGY L 1931b: 39 fig. 25; NAGY L 1934: 303, fig. 13; NAGY L 1938: 47, 142, fig. 12; PAULOVICS 1938b: 131 Abb. 95; ALFÖLDI 1942 XL. Table 3; TÓTH E 1977a: 146. Abb. 3; THOMAS 1977b: 42; THOMAS 1980: 198 pl. CXXXVIII; THOMAS 1982b: 271, Abb. 13.

Regarding the accuracy of the find-spot, here is József Hampel's description: 'Its site was impossible to localise, only the trustworthy merchant, from whom we got it, said that he had found it in a scrap-collector's store-room at Bonyhád.' Consequently, we had better take it as of unknown location.[1368]

Lamp suspension part. Cast bronze. Diam: 23.3 cm, Th: 8 mm. The chi-rho monogram is composed in openwork technique in the round part, fig. 390: a reconstruction. The pigeon-shape lamp in the reconstruction is not a Pannonian piece, it was just placed there to demonstrate the structure. A very similar piece was found in a village called Biertan (Berethalom) whose tabula ansata is also at our disposal. The inscription on the latter is: 'Ego Zeno/vius vot/um posui,'[1369] fig. 391. This bronze object is only one part of the 'chandelier', which held the lamps. The lamps were put on these dividing parts.[1370] We know similar suspensions of lamps from between the Drave and Save Rivers.[1371]

62.I.d. Clay lamp

Provenance: MNM 88.1925.3, gift, E. Spigl. **Literature:** not published.

H: 10.4 cm. On its discus there is a Latin cross, the longer arm of the cross is flanked by the two filling holes. There is a mark on the base that cannot be distinguished because of an irremovable stain. Abbiani type 2.

62.I.e. Clay lamp

Provenance: MNM 53.50.185. **Literature:** not published.

H: 10.5 cm. The representation of the Good Shepherd on its discus. Its handle is perforated. It has one filling hole. This type of lamp is published by Dóra Iványi at No. 902.[1372]

1364 TESTINI 1966 fig. 101: a graffitto of a fossor

1365 TESTINI 1966: 221.

1366 Thesaurus vol. II, fasc. 3. p. 697/47: artifex.

1367 Any information suggesting a site in another province is given in the description. These finds are not present in the assessment since they are by any means uncertain.

1368 HAMPEL 1903b: 434.

1369 NAGY L 1938: 142; ALFÖLDI 1942 pl XL,2; Römer 1969 Kat. H52, Taf 13; Erdély 1988: 128.

1370 On the suspensions of lamps found in various places in the empire, see DE ROSSI 1871 tav. V-VI.

1371 Site: Rogoznica near Poetovio. Location: Kunsthistorisches Museum, Wien. Inv. No. VI 727-728. Literature: CIL III 4098; DE ROSSI 1871 tav. V-VI; KUZSINSZKY 1895 CCXLVI, CCXLVIII; DOBSTAL 1914: 188-189; NAGY L 1938: 110 figs. 71-72; P. KOROŠEC 1980; KNIFIC/SAGADIN 1991 Nos. 1-2. Two pieces. Diameter 6.1 cm and 8.2 cm. Bronze. The chi-rho monogram is encircled, its inscription is: VOTUM PVSINO POSVIT and INTIMVS MAXIMILIANV(s fra) TRES CRISPINO POSVERVNT. The completions were taken from ROSSI.

1372 IVÁNYI 1935.

62.I.f. Clay lamp

Provenance: MNM 53.50.189. **Literature:** not published.

On its discus a squared chi-rho monogram. Nozzle has burnt black. H: 12.2 cm. Abbiani type 2 var. i.

62.II.a. Finger ring. Roman? Fig. 392.

Location: MNM 10.1951.225. From an unknown site. **Literature:** NAGY L 1938: 112, fig. 76, the second in line 1 (right above).

Golden, Diam: 2.2 cm, weight: 3.64 g. A very small Greek cross carved in the head.

62.I.g. Clay lamp, fig. 393.

Location: MNM 10.1951.226, the Ráth collection. **Literature:** CZOBOR 1879: 177 figs. 2-3; CZOBOR 1883: 244 fig. 29; NÉMETHY 1886: 242; R-D 539; PULSZKY 1897 I. 259 fig. 111; NAGY L 1938: 91. Note 1; THOMAS 1980: 198 PL. CXXXVII; THOMAS 1982b: 271, 273 Abb. 14 - here the information of the site is not clear.

Chi-rho monogram on the discus. Rim: series of alternate circles and triangles. Chi-rho monogram flanked by two filling holes. Its handle is marginally higher than the object itself. Not perforated. As for its provenance, L. Nagy writes that it may probably have come from Rome, and was acquired by the MNM from the Ráth collection; Edit Thomas describes it as 'Mit Christogram verzierte Tonlampen (Abb. 14) sind uns aus Aquincum, Brigetio und Savaria bekannt.'[1373] Abbiani type 1 var. a.

62.I.h. Finger ring, fig. 394.

Location: MNM 10.1951.236. **Literature:** NAGY L 1938: 112, fig. 76, the first one in line 2 (left below).

Bronze, the ring is narrow, the head is round and decorated with a chi-rho monogram. There are dots between the body-lines of the letters. Diam: 2.7 cm.

62.II.b. Gem. Roman? fig. 395.

Location: unknown. **Provenance:** unknown. Inaccurately, the find is always recorded with location and provenance Szombathely; but it is false. Its location is said to be Szombathely, and though the inventory number indicated in the literature exists, it refers to another object. There is no correction in the inventory. The inv. number published in Scheiber 1983 is also false: 5476485. **Literature:** SCHEIBER 1960: 55-57 with an outline of the previous literature; THOMAS 1975-1976: 105-160 Abb. 3; THOMAS 1977b: 47; THOMAS 1977a: 24; SCHEIBER 1983: 50-51.

The gem is made of blue stone, with the picture of a menorah. Size: 1.5 x 1.2 cm. Its shape is unusual, an ellipsis cut in half. Such pieces did not appear in the Roman age. If it had been broken, the fracture could be seen. In Scheiber's interpretation the figures at the ends of the arms are flames. This is a very logical assumption, but we have to add that the flames have the shape of birds. The birds are clearly distinguishable. I do not think the object comes from the Roman period.

62.I.i. Clay lamp, fig. 396.

Provenance: unknown. Dóra Iványi, who first classified the lamps, indicated an unknown provenance. Because of the mentioned Szombathely location, a Szombathely provenance is ascribed. **Location:** SM 26. **Literature:** Nagy L 1934: 306, fig. 14; IVÁNYI 1935 No. 986; NAGY L 1938: 96, fig. 61; SCHEIBER 1960:53-55, No. 9a with further literature; THOMAS 1975-1976 Abb. 4; THOMAS 1977b: 47, fig. 48; THOMAS 1977a: 24-25; SCHEIBER 1983: 48-49, 51; SOLIN 1983: 761.

There is a menorah on its discus, in Sándor Schreiber's opinion, surrounded by palm-branches (luláb) on both sides. L: 11 cm, Diam: 8 cm, h: 3.8 cm. Abbiani type 1 var. a.

62.I.j. Emperor's bust, fig. 397.

Provenance: unknown. **Location:** MNM 10.1951.23 from Ferenc Kiss's collection. **Literature:** THOMAS 1983: 305-314 Abb. 19; THOMAS 1986a: 509-511 Abb. 12-13; THOMAS 1988: 292 Abb. 13.

Iustinianus's portrait. H: 11.3 cm, w: 6.9 cm. Its further description in Edit Thomas's words: 'Die ovale Scheibe der Fibel trägt ein eingetieftes Kreuz. In seiner Rechten, die von einem dicht gerafften Hemd bedeckt ist hält der Kaiser den Kreuzglobus (Sphaira), das durch ein (Kreuz) viergespaltete Reichssymbol.'[1374]

Although Christianity in the age of Iustinianus has to be interpreted in a very different way from that of the first four centuries, I suppose it is improbable that the mark on the fibula is significant regarding the Christianity of the province.

[1373] THOMAS 1982b 271.

[1374] THOMAS 1986a 510.

VI. HISTORICAL EVALUATION

1. CHRISTIAN COMMUNITIES IN PANNONIA

The first question to be asked is the very familiar one: When did Christians live in Pannonia? The easiest answer, as I have stated in the research history, is the 3rd century. This answer may have seemed correct, because the early Christian life of Sirmium[1375] (cf. Zeiller's scholarly activities[1376]) was doubtful, we may say refuted.[1377] Criticism of the chapters on Pannonia was written by András Alföldi[1378] immediately after Zeiller's book appeared. His main issue was that Zeiller could not read the Hungarian literature on the subject, and did not check the meticulous data published in different sources in different languages; in fact he ignored them. He also referred to Egger, who expressed very similar concerns in connection with Austria. The idea, therefore, that the early Christian life of Sirmium begins in the late Roman period is due to Zeiller's research.

I cannot agree with Zeiller, and do not propose to re-examine his statement that all the accounts of the episcopate in the 50s AD are without exception later interpolations, falsifications or legends.[1379]

The importance of Sirmium began even before Christ, and it lasted continuously until the invasion of the Avar tribes.[1380] Ireneus, who is said to have been its first bishop, died a martyr.[1381] Among Pannonian towns, Sirmium contributed most to the Arian dispute,[1382] 'since the most different ecclesiastical policies and theological views met and were confronted here.'[1383]

According to Vladislav Popović, regular excavations began only in 1957;[1384] the previous ones had been only occasional. These previous excavations were reported by Petar Milošević, who thinks Hytrek's excavation in 1882/83 was the earliest.[1385] Let me refer to Smiljka Kašić's dissertation, in which the author exhausts and assesses all the achievements achieved by excavations and all the publications that appeared before 1975, including the early Christian sites and finds and bishoprics of Pannonia Secunda and Savia.[1386]

During the regular excavations, they found remnants of the southern and north-western walls of the town located on the left bank of the River Save, as well as parts of the town centre. The centre is presumed to be the old forum. Next to it was the public bath, the remains of the then *basilica* (we may perhaps suspect that the *curia* had been struck by lightning). The horreum was the best preserved construction.[1387] The quality of the remains found at the southern wall suggests an imperial residence, which is reinforced by the remnants of a hippodrome found in part N/10-11.[1388] Christianity in the city is described by Popović: 'The probable formation of the Christian districts near the walls, as a result of the cadastral situation, including the Cathedral with annexes, has not yet received archaeological confirmation.'[1389] Christian graves were found in all the cemeteries surrounding the city walls, but in the city itself only few finds have appeared so far that refer to Christianity. As the written sources describe the city as a significant one regarding Christianity, the only explanation may be that, because of modern development, the land has been built upon. Even regular excavations have been so far insufficient in gaining an archaeologically precise and satisfactory picture of the settlement. Maja Parović-Pešikan writes about a chi-rho monogram found while opening a villa.[1390]

The evaluation of the city, from the archaeological point of view, and especially from the viewpoint of Pannonia's history, has been summed up by András Mócsy. In his summary he writes: 'The imperial city of Sirmium was given most of its monumental buildings under the Tetrarchy. This conclusion is based not only on the disturbed conditions preceding the accession of Diocletian, which would not allow of large-scale building, but above all on the choice of Sirmium as a permanent imperial residence by Galerius. Licinius ordered baths to be built there, for which the column-capitals were fetched from a quarry on the Dalmatian island of Brač (CIL III. 10107 = ILS 3458). Whether these thermae Liciniane can be identified with the truly huge remains of baths uncovered by the new excavations is as yet undecided. These baths at Sirmium were constructed in building-titles, as were the other large buildings recently discovered at Sremska Mitrovica; among them were a large horreum, part of a palace-like building, a hippodrome probably connected with the Imperial Palace and, in the south-west corner of the excavated buildings, there are dwelling-houses, among them one of a distinctly upperclass character. Buildings constructed of tiles are very rare in Pannonia, and this is all

[1375] An accepted concomitant of modern research is that when making a catalogue, the present borders are kept on view instead of the ancient ones. Thus, the present work deals with sites in the territory of Hungary. However, we must not ignore those sites that are today located in foreign countries, but used to belong to the old Pannonia province. Among these one of the most important is Sirmium (Sremska Mitrovica) the capital city of Pannonia.

[1376] ZEILLER 1918.

[1377] A summary of this in NAGY T 1939: 31.

[1378] ALFÖLDI 1920-1922: 128-129.

[1379] Sirmiun is a very significant town regarding not only its importance in public administration, but also in Christianity. In all probability, its Christianisation began in the middle of the 1st century, which is supported by arguments discussed elsewhere (GÁSPÁR 1989-1990: 99-105; GÁSPÁR 1993b: 5-22), as well as by the logical conclusion that such a cultured community, so firm in its Christian belief and one that played such a notable part in the life of not only the province, but also Christianity from the 4th century on, could not have come out of nowhere in the end of the 3rd century. This significance must have had its antecedents. Its importance is undoubted from the 4th century, which is proved by the five councils held here between 347 and 359.

[1380] MIRKOVIĆ 1971.

[1381] FERJANČIĆ 1969: 56.

[1382] NAGY T 1939: 127-157, 160-169.

[1383] NAGY T 1939: 169.

[1384] POPOVIĆ1971: 120.

[1385] MILOŠEVIĆ1971.

[1386] KAŠIĆ 1975. For me, the fact that I could see and read it was a great help. My special thanks are due to the author, Smiljka Kašić.

[1387] POPOVIĆ 1971: 119-133 together with Appendix I of maps; the mentioned part is indicated by numbers 29 and 30: Loc. 29-30.

[1388] POPOVIĆ 1971: 126, 130.

[1389] POPOVIĆ 1971:129.

[1390] PAROVIĆ-PEŠIKAN 1973: 1-44.

the more striking since in many areas no building-stone is available. There are large districts where stones for building had to be fetched from quarries 50 km away from the building-site. Presumably the reason why, exceptionally, tiles were used in Sirmium was that the very precise technical execution of the public buildings was only possible in tile. Of great significance are the Christian cemeteries of Sirmium. These were excavated in the nineteenth century, unfortunately with little care (but perhaps with too much imagination). They have yielded a large number of grave-inscriptions and an extraordinary variety of subterranean buildings... Here it need only be mentioned that similar cemeteries are equally characteristic of the other administrative centres.'[1391]

In the Saint Synerotas cemetery a reliquary has been found.[1392] Bronze, its length is 10.2 cm, its diameter at the bottom is 28 mm, and at the top, it is 25 mm. Its two equal parts are fitted together with golden and silver thread. It contains remnants of human bones. It could be carried around the neck. It was made in the 4th century.

In 1976, The Archaeological Institute of Belgrade, The École Française in Rome and the Museum in Sremska Mitrovica undertook an excavation in the north-east cemetery. They found Saint Ireneus' basilica, where Macedonius had erected a *memoria*. The inscription found here confirms this.[1393] The historical value of the inscription is shown by the usage of the particular grammatical form, and the undoubtedly oriental influence present. The latter is important because it coincides with those observations and findings, which were listed by Gamber.[1394] As a final result of his argument and identification, he pointed out that the origin of several Latin liturgical texts, bearing an oriental influence, are Sirmium, or, at least, from its environs. Their author, therefore, must also have come from Sirmium, or its vicinity.

There is much more information about Sirmium's early Christian life than we would first think. The data partly comes from written sources.[1395] It is also shown in Klaus Gamber's article,[1396] where he analyses some written sources to demonstrate the oriental impact that can be very easily traced down in Sirmium. In his summary he writes: 'In Sirmium selbst, das schon relativ früh christianisiert worden war, scheint dagegen, wie wir sahen, von Anfang an der griechisch-syrische Einfluß vorherrschend gewesen zu sein.'[1397] Referring back to the statement in the research history, according to which the Christians in Pannonia have mostly Greek names. This statement may be augmented by stating that the Christians in the province were mainly immigrants. In the western part of the province, first in Savaria, the Christians were these Romanised inhabitants, whose birth-place was not Pannonia. They are immigrants from Italy or descendants of such immigrants, or people from the east that arrived here via Aquileia. Besides Savia and Pannonia Secunda, inscriptions were made in Savaria only, which indicates the Romanisation of the inhabitants. The inscriptions show that they are not local people: Iodorus' tombstone (51.I.b), for instance, or the *pictores pelegrini* tombstone (51.I.e). Saint Martin's parents also have an Italian connection. The presence of these immigrants might have resulted in the baptism of the local inhabitants, but we cannot speak about a deliberate mission, or at least we are not in the possession of such information. The existence of a Christian community has been proved, and its influence probably increased following Bishop Quirinus' execution here.

The community shows an Aquileian influence, which is highlighted by Quirinius' *translatio* at the beginning of the 5th century. Although there are written sources proving the presence of the community, no church has as yet been opened by the excavations. The usage of the *locus orationis* built in the 3rd century can also be regarded continuous after the reign of Constantine. We have no data about the bishop of the Savaria community. As a Christian community cannot exist without a bishop, we have to suppose the existence of itinerant bishops. This issue is supported by the fact that we know names of bishops, but we cannot localise their residences, obviously because they did not even have such seats. This observation mainly applies to the bishops of Valeria and Pannonia Prima; we know the residing places of several bishops in South-Pannonia, i. e. Savia and Pannonia Secunda. The possibility should not be ignored that, in the early period, more bishops may have come to Pannonia

[1391] MÓCSY 1974: 213-313.

[1392] GUYON 1975: 419-424 fig. 2.

[1393] In basilica domini n/ostri Erenei, ac mem/oriam posuit Maced/onius unacum m/atronam suam a/ammete evenati/ a leaf, chi-rho monogram and a palm-branch. (MILOŠEVIĆ/PRICA 1979: 41-43). The inscription is not quoted in the transcription of the authors because I interpret the reading differently. *Unacum* is the preposition, which appears in the late texts. *Cum* (and so *unacum*), although mostly takes the *ablative*, does not always go with it. The Church Fathers already used the accusative with it. (BLAISE 1954 *cum* entry 1. Cf. CIL VIII 21638: una cum filios; H. PETERSMANN: Zu christlichen Inschriften für das in Afrika gesprochene Latein. Lecture 22 April, 1994, Universitas Nominata de Eötvös Lóránd, Facultas Filosophica. They use the accusative instead of different cases, which is a sign of grammatical simplification. The word *Ammete* can only be a Greek place-name, which would end in *eta*, thus the iota subscript of the word transcribed in Latin cannot be seen at the word in the ablative because of the preposition 'a'. The word *evenati* comes from *evenio* (to come out of or from somewhere), and is the plural masculine nominative of its participium perfectum passivum. It is made to agree with the boy's masculinity, and the usage of the plural is justifiable. Consequently, the translation is the next: 'In our lord, Ireneus's basilica, Macedonius, with his widowed mother, who both came from Ammete, also erected a memoria.' Cf. No. 233.

[1394] GAMBER 1982b.

[1395] The other part is made up of finds and objects coming from archaeological excavations, such as Saint Demeter's memoria. Saint Demetrius' basilica was found in the centre of Mitrovica in 1978. It is in an eastern-western direction. Its length is 20.5 m, its width is 12.5 m. It can be dated to the first half of the 5th century (MILOŠEVIĆ/PRICA 1979: 45-48). There was also an excavation on the right bank of the River Save, in Macvanska Mitrovica, from where the report speaks about four Christian churches. These churches are surrounded by cemeteries (MILOŠEVIĆ/PRICA 1979: 45-48; on the phrases of the church BOŠKOVIĆ et al. 1974: 631-632).

[1396] GAMBER 1982b, the analysed sources are the next: *Fragmenta Ariana, Codex - Liber sancti evangelii, Istructio ad competentes* by Nicetas.

[1397] GAMBER 1982b: 84.

than are known of. They may have come with the Christians and wandered in the province (or wandered further or back). The case of Saint Anthony of Lérins shows that there really were Christian communities, but their principals are unknown. Further Christian communities were detected at Kékkút (21.I.a-c) and Sümeg (43.I.a.). These two sites differ from the others in that here a house-church was found, whereas in the graves, or in the inscriptions, there are no traces of Christianity. In the other places the situation is quite the reverse; there the majority of the Christian finds are grave goods, and we have no (6.I.a; 8.I.a-b; 17.I.a-b; 29.I.a; 48.I.a-b; 52.I.a-e), or very few, finds related to Christianity (4.I.a-g; 12.I.a-k) from the settlements. The inner fortified settlements, such as Kapospula-Alsóheténypuszta, Keszthely-Fenékpuszta or Ságvár, deserve an individual observation and evaluation. The case of Pécs and Pilismarót is totally individual.[1398]

While in western parts the Christians are Italian and Greek immigrants, in the eastern districts, even if they are few, Syrians appear early in the province, namely in Aquincum and Pécs. This explains why in these places the presence of Christians is assumed from the 1st century, but certainly from the 2nd century onwards. Their numbers did surely not exceed the populations of Hausgemeinde or Mahlgemeinde. In Dunaújváros (12.I.c), the presence of Syrian Christians is also possible from the 2nd century. In Pannonia, the Syrians constitute such a decisive and determinant part of Christianity that we may talk about a special Christian community. Let me raise the issue here that there is probably a connection between the believers of Mithras, the Syrian communities and the Christians.[1399] As in Dura-Europos the mithraeum, the synagogue and the Christian house-church are present at the same time. In Pannonia, therefore, the three are also present,[1400] but in a certain territorial distribution, so that in Aquincum there are all three present, in Intercisa the Syrians and the mithraeum, in Pécs the Syrians and communities of Jewish origin (the latter two with a considerably Hellenised community). It may also be supposed that, in Pécs, they came from a Greek territory. This territorial distribution is only relevant to the later Valeria. Naturally, we include Brigetio here, whose attachment from Pannonia Superior to *Inferior* under Trajan might have been caused by reasons other than military interests. From this viewpoint we may regard the territory detached from the former Pannonia *Inferior* to the later Valeria as a unit, which raises the suspicion that in Diocletian's reign the method of the division, besides geographical reasons, was due to ethnic and religious differences; call them differences in loyalty.

Those territories, where the eastern population with its mysteries and disposition to magic could give the semblance of loyalty, were integrated. The Jewish also lived in this territory. Their religion had never been *illicita* and they had always found a solution to avoid fatal conflict. This fact explains why there were no victims of persecutions of Christians in this region, and also how the predecessors of Anthony of Lérins, and similar Christians, could survive critical periods in the valleys of Tokod and Pilismarót. The latter phenomenon (that Pannonia keeps its early Christian principle, i.e. that the Church of God is a living community rather than a man-made institution, or buildings) may also be accounted for by starting out from this situation. Pannonia keeps this principle longer than in other places. This is explains why no church (in its present sense) is found in the former Valeria from the period before the early Middle Ages. However, I attribute the 5th-century finds (and to date considered by research as stolen) to the permanent, albeit small-scale, eastern connections; these include the pillar at Székesfehérvár (46.I.a), the suspension for a lamp at Szőny (52.I.c), the capital found in the cemetery at Szentendre (48.I.b), and the capital found in the excavation of the Aquincum synagogue (4.I.h). These finds presuppose a religious continuity. The Romanised Italian population does not adhere to Christianity because they decide to flee.[1401] The descendants of the oriental population and the Christianised local people represent the group of those who belong to antiquity, moreover, the Christian antiquity of the Migration Period (or perhaps from the time of the Magyar conquest). It is necessary to revise the cella trichora at Pécs (32.II.1), so as to demonstrate and assess precisely these relationships (cf. József Deér.[1402]).

The situation of Pannonia Prima is slightly different since a part of its territory used to belong to Noricum, whose role in the history of early Christianity is much more modest than Pannonia's.[1403] Its belonging to Noricum, as well as the Amber Road leading through it, considerably enhanced its relation to the western half of the empire and to Italy, together with the level of Romanisation.[1404] Obviously, this is why Savaria, the capital of Pannonia Superior (Pannonia Prima), proved to be an acceptable location for Italian traders and other Italian families to settle down. In such a place the stranger, even if he is an itinerant, missionary bishop, is not extraordinary.[1405] If the Christians kept a low profile in this territory they could lead their life undisturbed, since the officials, too, chose to neglect the edicts silently. The sufferings of Quirinus, Bishop of Siscia, serve as evidence of this. It proves the existence of the Savaria community on the one hand; on the other hand, the locals survive cleverly by means of setting an example. The significance of Scarbantia is less than Savaria's, and, because of its closeness, it could also be regarded as a resting place for the emperor and magistrates;

[1398] The individual characteristics are dealt with in the topographical part.

[1399] Cf. 12.I.c; the historical analysis of Aquincum; and the historical analysis of Pécs.

[1400] As it was referred to in the introduction.

[1401] This flight is continuous; it begins (378) soon after the *receptio* of the Germans (376), and finishes approximately in the third decade of the 5th century.

[1402] József DEÉR: Die heilige Krone Ungarns. Wien 1966.

[1403] NAGY T 1966.

[1404] For instance, the achievements of stone cutting: a hexagon in Savaria, or the statues of Jupiter, Juno and Minerva in Savaria and Scarbantia. The luxury living mirrored in the rich set of finds. The imperial residence in Savaria, etc.

[1405] In Church history, we know conditions where the only possible solution is the employment of itinerant, missionary bishops, cf. HAUCK 1954 II 337-339.

for instance, Praeses Amantius might arrive even from this direction. The authorities were in the capital, not nearby Scarbantia. The Christians could live here peacefully if they did not attract attention. This explains my earlier statement that the publishing of the small finds could prove the pesence of Christianity in Scarbantia.

The peculiar set of finds of Keszthely-Fenékpuszta (and the inscription that includes a name) makes it evident that some of the Romanised inhabitants had been spared until that time.[1406]

In the territories of Savia and Pannonia Secunda we may point to Christianity on a larger scale; here there is evidence of a deliberate and 'official' mission absent in the previously studied. The names of the Christians and their bishops are Greek. They are the ones who arrived and stayed here. The bishops may localised and are not itinerant missionaries. The local conditions of the two territories make them more similar to Illyricum Inferius, linking them to the group Bratož calls an 'illyrisch-alpine' phenomenon; its characteristic is that: '... ungewöhnlich oder sogar Ausnahmenfälle sind, ...'[1407] Such a characteristic cannot be found in Pannonia Prima or Valeria, with the exception of Pécs. In other words, the two parts possess different basic characters in terms of their culture and mentality. A political analogy best elucidates the difference between the two: Pannonia's southern part was conquered by the Romans the rest was simply annexed. In Sirmium, we find the term *ecclesia Sirmiensis* in an inscription, which can refer to the Sirmium community or to Sirmium's *ecclesia ex gentibus* community.[1408] This meaning becomes intelligible if we take Pannonia's Jewish and Jewish-Christian communities into account as well. The latter comprise the community called *ecclesia ex circumcisionibus*. In Pannonia, Sirmium represents the link to the Age of the Apostles.[1409]

Moving on to the finds, the chi-rho monograms, or crosses on fibulas, are not of a significant value from the viewpoint of Pannonia's Christianity. They are official symbols and actually refer to the era when Christianity was a legal religion (*religio licita*), and therefore this fact was also expressed in the public administration. It is rather a 'sign of the times', not the Christian feeling of the population. Cf. finds 2.I.a; 19.I.a; 36.I.c; 40.I.a; 45.I.a; 56.I.a; 62.I.j.

There are sites where only a few objects have been found. Reasons for this might be that the place has not as yet been searched thoroughly, or that we cannot hope to find any more Christian finds from a place where these are accidental, e.g. the Bar Kochba coin (52.I.e.) in Brigetio. These finds are invaluable in dealing with Christianity. Cf. 10.I.a; 27.I.a; 31.I.a; 38.I.a; 55.I.a; 58.I.a; 60.I.a.

The finds and objects that are at the moment uncertain may either be classified otherwise, or their present classification may be reinforced. These are: 4.II.a; 4.II.c; 26.II.a; 32.II.h; 37.II.a.; the excavation in site 28; and the reassessment of site 52. The objects from site 42 are still to be studied

No one has yet been able to evince the presumed existence of Christian life in Sopron; Vigilius Scaravaciensis was not Sopron's bishop. Despite the negative results, and because of the large quantity of unpublished material, further research may bring its reward.

Finally, we also have to discuss the possibility of liberated slaves. To be more exact, the possibility that slaves were liberated because their owners were Christian. This means taking into account a social grouping in the province, who could, and also wanted, to be actively involved in freeing slaves.

This question has been dealt with by András Mócsy.[1410] Examining the epigraphic material, he came to the conclusion that the number of slave-owners who liberated slaves is quite high. He located on the map the places of those inscriptions that are documents of such liberation, together with those that refer to freed slaves. His map demonstrates the fact that, in Pannonia, we have plenty of these documents in Pannonia, but his research does not include which of these liberations were the result

---

[1406] From the cemetery next to the horreum, a ring with the inscription BONOSA; Onomasticon 1994; cf. notes 110, 544.

[1407] BRATOŽ 1994 41.

[1408] Even if there is no direct link between ☞ the Bishopric of Sirmium to be studied later, and the expression found in the inscription in Sirmium, we can presume that there is some relationship between the beginning and the conclusion. In connection with this, it is necessary to quote Saint Paul, who wrote about his own task, Gal. 2, 7-8.: 'Sed contra cum vidissent quod creditum est mihi Evangelium praeputii, sicut et Petro circumcisionis: (qui enim operatus est Petro in Apostolatum circumcisionis, operatus est et mihi inter Gentes) ...' Cf. BRATOŽ R: Die römische Synode 680 und die Frage der Kirchenorganisation 'in gentibus' im 7. Jahrhundert. Atti 1998 II 587-602.

[1409] The evangelisation necessarily begins with the apostles since there is no other source. It also follows as a matter of course that the assistance is provided by the disciples and the seventy-two appointed by Jesus. It is unusual, but a fact, that in this period our main source is the Bible (Acts). I accept the early evangelisation conducted by Andronikos and Epaenetos authentic, who were not bishops in the area in the modern sense, only the principals or priests of the community, born as a fruit of their work. We can only estimate the number of the people belonging to this community, but it makes clear that it was a smaller one. The two arrivals shortly after the other suggests that one person was not able to cope with the task, on the other hand 'die ältesten Quellen reden immer nur in der Mehrzahl von den Bischöfen eines Ortes. (KARPP 1954: 400). In the opinions so far the question has been in what kind of hierarchic order the two bishops might have been, arriving so close in time. The answer has mostly been that only one of them was the bishop of Sirmium, the other was the bishop of the whole province. As long as my opinion is concerned, I do not think Andronikos and Epaenetos should be ranked in this way. I think it is much more likely that one of them stayed in Sirmium to direct the community, while the other one went on evangelising. The Greek names that have remained in inscriptions speak about their efforts. The research has considered these inscriptions as the documents of how Christianity spread in the country. It is an undoubted fact that those who are surely Christian have without exception Greek names, and they speak Greek better; as for example Victorinus, Poetovio's bishop, who could speak Greek better than Latin, as Hieronymus witnessed. (For more details see NAGY T 1939: 35-52.) In possession of these facts I feel the need to re-interpret the previous theory that emphasised Aquileia's role in the evolution of Christianity in Pannonia. Aquileia's role is evident in Savaria, and will presumably be shown in Brigetio, too (in the course of further studies), where some of those coming from North Italy move from Savaria. Aquileia's role is important, but it is limited in territory and time. Its general role is that the missionaries, travellers and traders arrived here via Aquileia. This means a route, but not a starting-point.

[1410] MÓCSY 1956: 221-250.

of economic or religious factors. At the same time, however, religious reasons do not necessarily mean Christianity. We should not forget M. Ant. Victorinus and the freedmen making the inscriptions in the mithraeums.[1411] So, at this stage of research, we may only state that the liberation of slaves for religious reasons is certified, but we cannot decide whether it was caused by Christian belief, or faith in Mithras.

The inner fortified settlements may be regarded as the typical phenomena of the transition from antiquity to the Middle Ages. Despite all the difficulties, the fort at Keszthely-Fenékpuszta is the most appreciable. The basilica opened here (22.I.a) cannot be dated earlier than the 7th century (in the plan last registered). Its earlier periods belong to different periods in time, although their exact dating is uncertain. The function of the near horreum (22.I.b) is dubious, as analyzed by László Barkóczi, because there is a cemetery next to it, which follows the building. Otherwise, the accounts do not mention findings of seeds, or such remnants of food that would certify its function as a horreum. This observation proves that it was not a horreum and makes it necessary to supervise, and if necessary, reinterpret those buildings already considered as horrea.

The end of antiquity, and the beginning of the Migration Period, is dated to the end of the 5th century in Pannonia. This does not mean the disappearance of Christianity, only the meeting of two Christianised cultures. Antiquity is represented by the Menas ampulla in Szombathely (51.I.h) in the 7th century, or the cruciform bronze suspension for a lamp (52.I.c) in the 6th; whereas the Christianity of the Migration Period is represented by the finds of Závod, Deszk, etc.[1412]

The saints of the later period are Leonianus ☞ Szombathely; Martinus Bracarensis, who became the bishop of Braga; Bishop Ioannes, who fled to Italy from the Avars; Paulus, who struggled with the sect of the Bonosiani; Abbess Ioanna, who left for Salona from the Avars.[1413] The special choice of words in the symbolum, drawn up in Pannonia, gives a clue to Martinus Bracarensis's activities in Hispania, to the point that he had connections with the Mozarabic liturgy.[1414]

The German influence deserves special attention. Their presence and influence have been assessed focusing on the fact that they converted Pannonia to Arian faith. It is clear that Pannonia's role in Arianism is great. Arius was exiled here, thus the heretic faith must have inevitably spread in the province; then Bishops Ursacius and Valens were staunch adherents of Constantius II, so they would never shrink from disseminating Arian dogmas by force. But in the period when these events took place (351-361), the Germans were out of the picture. They became conspicuous in the empire from 376 onwards, and in Pannonia a little later. Thus, attributing Pannonia's Arianism to the Germans is a simplification of the matter. The problem is complex; its first component is the Arian heredity completely independent from the Germans, the second the adoption of Christianity by the Germans, and the third their Arianism. As a fourth component, we have to mention the rather unfriendly terms between the Germans and the people of Pannonia.

The Germanic tribes adopted Christianity at different times and became Christian to different extents. Becoming a Christian marks a big change: 'Aber bei wie vielen Germanen hat ein solcher Bruch wirklich stattgefunden?'[1415] The Germanic religious mentality was analysed by Walter Baetke. Its essence is that they did not differentiate religions and so built Christianity into their religious ideas. In Baetke's words they 'Germanised Christianity'. The group of Visigoths, who adopted Christianity when paganism was the legal religion, may be described as Christian. Some of them, in the reigns of Traianus Decius and Diocletian, died as martyrs. Wulfila and his Visigoths, the 'sogenannten Kleingoten,' are not in the same category.[1416] Their Arianism is quite different, as described in their easy-flowing and careless writings. Wulfila was ordained bishop by Eusebius at the Council of Antioch in 341. In 369 Fritigern adopted Arianism, and with him his Visigoths, 'die vor und nach dem Übertritt ins römische Reich christlich wurden.'[1417] In 376, a *foedus* allowed Fritiger to settle down in Dacia Ripensis, Moesia Inferior, and Thracia with his Goths;[1418] he transmigrated to Pannonia after 378. At this time they were already Christian, 'an ihrer Bekehrung hatte vielleicht Wulfila noch unmittelbaren Anteil.'[1419] This fact offers an explanation to the Arianism of the converts, but does not explain either the Arianism of those who were converted later, or the fact that the Germans adhered to their Arian faith.

This Germanic conviction is not completely similar to actual Arian dogmas. It is true that they said *homoiusios*, but they did not emphasise Christ's finite, but rather his infinite, nature, so they should be called Docetists. Even apart from this, it is true that their heresy is based not on dogmatic, but rather political reasons. They did not enter theological debates. Their political motive was to express their independence from, if not their opposition to, the Roman emperor, especially after 381 when the emperor himself becomes a follower of Arian heresy.[1420] Although this exposition of Baetke's subject is very brief, it is enough to demonstrate that the Germans did not engage themselves in the propagation of the faith; so, the likelihood that they converted Pannonia's population to Arianism is not great. On the whole, Pannonia suffered rather from their restlessness, raids, or fights. Under such

[1411] Vide Historical Analysis - Aquincum. In his work previously referred to Mócsy expounds that in the province there was a layer of society that was objectively able to free slaves, and also used this ability.
[1412] See more examples in Germanen vol. 1988.
[1413] See ALFÖLDI 1938: 163 for more details, with the list of the sources.
[1414] NAGY T 1937: 136-137. The peculiar expression is the formal occurrence of *vivus* referring to the living Christ after his resurrection.
[1415] BAETKE 1962: 8.
[1416] BAETKE 1962: 9-14.
[1417] BAETKE 1962: 14.
[1418] MÓCSY 1974: 339.
[1419] BAETKE 1962: 15.
[1420] BAETKE 1962: 16-19. The article discusses further viewpoints of the above issues, together with their impacts on later times.

circumstances, it is hardly imaginable that they could spread the Gospel successfully among Pannonia's peoples.[1421] It was the Germanic presence that made life uncertain at the end of the 4th century and at the beginning of the 5th. As a territory, Savia suffered the most, where only Amantius, Bishop of Iovia, is said to have been able to remain in control.[1422] This is the period when the Romanised population and Romanised Christianity, to a certain extent, left the province.[1423] This is also the period when the Marcomann Queen, Fritigil, settles down in Pannonia Prima, which became Christian and was led by an orthodox bishop, ☞ Sopron. The first half of the 5th century was a period of adversity for Pannonia. The wild behaviour of the Germans resulted in destruction, and the presence of the Huns led to changes. In the distorting mirror of the sources, the Huns might be regarded as being as destructive as the Germans; but this would be a very superficial judgement.[1424] The Huns could fight hard, but their purpose was not destruction, but creating their own order. The Germans revolted against the Roman Empire and their purpose was disorganisation. The Huns wanted to establish a new order where normal life was possible. Research to date has ended Pannonia's history in the middle of the 5th century, as if its takeover by the Huns had meant a discontinuity of life. The objects and finds from the end of the 5th, and from the 6th centuries (4.I.g, 32.I.d, 32.I.e, 32.I.f and the ones referred to), and even those from the 7th century (22.I.a and the ones already referred to) prove that there was a Christian continuity, but that the Christian population did not have a Romanised, but rather a Hellenised, culture, and that they might just as well have arrived with Attila's multinational army, and might just as well have been given a warm reception by the descendants of the earlier oriental population.

2. Jewish Communities

The research has so far reckoned with the presence of a population of Jewish origin apart from the Syrians. This population did exist, but not in the expected areas. Their presence is clearly proved in Aquincum, including their synagogue (proseucha; 4.I.h), but the exact date of the building, and the duration of its use, have not yet been established. It was built in the 3rd century, at the latest, and it was presumably standing in the 5th century. The presence of Jews is made certain by their gravestones in Aquincum. Based on 4.I.i, we may also suspect Christians of Jewish origin in Aquincum. The Jewish tombstone found in Esztergom (14.I.a) leads to the conclusion that the Jewish population of Aquincum was moving towards the Danube Bend where they established a diaspora. Very carefully, I put forward the view that among the Jewish-origin population, the well-to-do and those who aspired after a career in public administration stayed in the towns, whereas the poor, or those who did not want to be officials, moved away from the centres.[1425] The research has supposed a Jewish population both in Savaria and Intercisa. In Savaria, two objects support this (62.II.b; 62.I.i); one of them is a clay lamp with the illustration of a menorah, the other one is a gem. The lamp is of unknown provenance, thus it has value in sketching the history of either Savaria or the province; the gem is not Roman, and its origin is uncertain. In Intercisa, the research pointed out the existence of a synagogue, but, unfortunately, this came from the misinterpretation of an inscription (12.II.c). This wrong interpretation caused other inscriptions to be regarded as Jewish (12.II.d-f), and now refuted by Heike Solin (1983, 1989). This means that we cannot assume a Jewish population in Intercisa, but that a considerable Syrian community did exist. The situation is the same in Brigetio, where István Tóth had presumed a Jewish community with a synagogue (1995), again based on completing and misinterpreting an inscription (52.II.c). His sole evidence, a Bar Kochba coin found here, was not satisfactory in proving the existence of a Jewish community. However, the presence of Christians and a considerable Syrian community is also verifiable in this town (52.I.a-e). In the research so far, we have not examined Pécs and its region (39.I.a) in terms of seeking the presence of Syrian, Jewish, or Christianised Jewish communities. Only Ferenc Fülep has written about some Jewish features that he suspected from amongst the available finds (1984). I suppose that, when the Carps were settled around Pécs, those who left the area and sought dwelling-places within the province moved to Tokod. Among them there might have also been Jewish people, which is reinforced by the anthropological examinations.

We may also suspect a Jewish community at Ságvár, where some finds (36.II.d-f) and certain phenomena in the cemetery written about by Alice Burger (1966) seem to be promising. I suppose that in the inner fortified settlement, one of the buildings that has been considered a horreum was actually the synagogue. The amulet found at Dombóvár (8.I.b) is not interpretable in this respect.

In Pannonia, the presence of people of Jewish origin is proved in the following places: Aquincum with the Danube Bend, and Pécs and its neighbourhood. It is

[1421] András ALFÖLDI does not consider Pannonia's Arianism a common phenomenon, in fact he writes about an Orthodox majority in the province (1938: 156).

[1422] EGGER 1962a: 57-67. The interpretation of an inscription of a sarcophagus gave a chance to verbalise different opinions. It is the epitaph of Amantius, Bishop of Iovia. He is the Amantius who took part in the Council of Aquileia in 381, and represented Niceanum. His epitaph mentions two chiefs (*geminis ducibus*) and two peoples (*bini populi*). Egger raised the possibility of the two chiefs being Alatheus and Saphrac, German chiefs, whereas the two peoples were their tribes. The Hungarian research partly accepted this opinion (MÓCSY 1974: 340-342); András ALFÖLDI rejected it (1938: 166).

[1423] The translations supply plenty of data: Anastasia's corpse is taken away from Sirmium, Demetrius' relics are taken to Thessaloníki; Quirinus' dead body is taken to Rome; ☞ Szombathely, from Cibalae the relics are taken to Rome.

[1424] OLAJOS T: A Kárpát-medence népei a honfoglalás előtt. In: Árpád előtt és után. Tanulmányok a magyarság és hazája korai történetéről - The Peoples in the Carpathian Basin before the Hungarian Conquest. In: Before and after Árpád. Treatises on the Early History of the Magyars and their Homeland (translaters). Ed. KRISTÓ Gy / MAKK F Szeged 1996: 13.

[1425] It was possible because Roman laws did not oblige the Jews to participate in public administration; it had to be their choice.

strongly presumable at Ságvár, and perhaps at Tokod. There is also a small chance at Pilismarót.

In contrast with the research so far, it is not verifiable in Intercisa, Savaria or Brigetio.

3. Bishops

The history of research shows that there are numerous unresolved questions and controversial points in connection with the bishops and episcopates in Pannonia. It is known that Quirinus was a bishop in Siscia, Amantius in Iovia, Eusebius in Cibalae, the Arian Valens in Mursa, Ireneus in Sirmium, Victorinus in Poetovio, and the Arian Ursacius in Singidunum. We cannot evidently localise bishoprics in Pannonia Prima or in Valeria. We have a list of bishops from Sirmium only. The most commonly accepted fact is that the list starts here with Ireneus.

The Council of Serdica (342-343) is a turning point in the organisation of the Church. The council settled its canons in connection with the affairs of bishops. These canons obliged the bishops to stay in their own town or city. If the bishop travelled to another town he had to return within three weeks. Canon XI is evidence of the fact that the bishops would travel.[1426] On such occasions, they had had the chance to settle in other towns.[1427] Thus, it would be fair to ask whether the same bishop had appeared in several towns in the earlier period, which means that the same name might occur in different places. The debate, therefore, on the bishoprics of Megasius, Migdonius, Gaius or Paulus seems to be pointless. They may well have been bishops in all the places that have emerged; only the chronological order is questionable. On the other hand, it is also true that we do not know the names of all the bishops who turned up in Pannonia; especially those who were not bishops, according to the organisation of the Church then taking shape in the 4th century, but were simply elders of Christian communities in the double hierarchy of the *episcopatus* and *diaconatus*.[1428]

Contrary to other views, I accept the authenticity of Epaenetus and Andronicus arriving in Sirmium in the fifth decade of the 1st century. Their presence is a guarantee of apostleship. On the other hand, it is also necessary, since we may suppose with good reason the appearance of Christianity in Pannonia, based on certain archaeological correlation; ☞ Aquincum or ☞ Dunaújváros. It is no accident that, in Sirmium, we cannot follow the list of bishops from the 1st, but only from the 4th century. It is due to the above-discussed fact, namely, that by this period, the organisation of the Church had not reached the point when the bishop performed his duties over a defined territory in his defined residence. We cannot, therefore, even expect such an early list. When the establishment evolved, the question was which places were defined as a *civitas*, and which was a simple village. The villages did not get bishops.[1429] In this respect, among Pannonia's Christian sites in Valeria and Pannonia Prima, only Savaria, Keszthely-Fenékpuszta, and Pécs (and also perhaps Kapospula-Alsóheténypuszta and Ságvár) met the requirements of a *civitas* in the organisation of the Church. Aquincum, however important its role in political history might have been, could hardly have been called a *civitas* in the organisation of the Church because of its small community. This refers to all the places in Valeria where Christianity may be traced, except Pécs.

Of those so far considered as Bishop of Scarbantia,[1430] Episcopus Vigilius, was certainly bishop of Bassiana I; ☞ Sopron.

The research has already discovered that the borders of dioceses do not follow the administrative borders. That is why we can mention Ursacius from Singidunum, or all Poetovio's bishops, among the bishops of Pannonia. However, it is difficult to say where the borders of the dioceses used to run; if we may talk about such borders at all. It is also possible that during the first three centuries an unwritten system of hierarchy, that is the hierarchy of bishops, evolved and determined the relations between the bishops of different towns. The end of the 4th, and the beginning of the 5th century have also to be considered central in this respect. In any case, when Arian Bishop Secundianus, in his appeal to Gratianus, referred to the fact that he was not obliged to appear at the Council of Aquileia because his territory had been detached,[1431] the emperor reproved him in his reply, and stated that the political detachment had not concerned the diocesan borders. He was obliged to attend at councils called by his bishop.

I do not wish to write in detail about organisational questions in Pannonia. I would rather submit a comprehensive table that has been compiled based on Tibor Nagy's similar table,[1432] with some alterations.

[1426] ERDŐ 1983: 310-311.

[1427] It is proved in the case of Amantius, who went to Iovia from Aquileia. EGGER 1962°: 57-67.

[1428] LIETZMANN 1958: 141-185.

[1429] EnchF No. 503, p. 308; ERDŐ 1983: 307, Serdica canon 6. Cf. n. 495.

[1430] BERG 1985: 85.

[1431] Noricum, Pannonia (together with Valeria and Savia) and Dalmatia used to belong to the Illyricum occidentale diocesis. The provinces Moesia Superior, Dacia and Dardania were, however, under the Illyricum orientale praefectura. Macedonia also belonged to the latter. The office of the prefecture was granted to Sirmium by Constantine the Great (pl GAMBER 1982b: 77), and was later transfered to Thessaloníki in the above mentioned year.

[1432] NAGY T 1939: 217. The dates appear in the table according to Tibor Nagy. These dates differ from the ones given by GAMS (1893). Cf. No. 73.

Sirmium		
Andronicus 50s	following Epaenetos	
Irenus †303	orthodox, martyred here	gnostic community
Domnus 325-330	present at the Council of Nicaea	exiled
Eutherius 330-343	Arian, until 337	
Photinus 343-345-351	leader of Photinian heresy	exiled, returns
Germinius 351-378	Arian, assistant of Ursacius and Valens	
Anemius 376-382	orthodox	
Cornelius before 392	orthodox	
Laurentius 401-417	orthodox	
Siscia		
Quirinus †304	orthodox, martyred in Savaria	translatio to Rome
Marcus 343	orthodox	
Constantius 381	orthodox	
Cibalae		
Eusebius †259	orthodox	
Mursa		
Valens 330-371?	Arian, Ursacius's mate	
Iovia		
Gaius? 359-370?	Arian	
Amantius 379-399	orthodox, comes from Aquileia	
Poetovio		
Victorinus †303	orthodox, martyred here, scholarly activity	large gnostic community
Aprianus 343	orthodox	
Marcus -353?	orthodox	
Agrippinus 363?-	orthodox	
Iulianus Valens 353?-375	Arian	
Singidumun		
Ursacius 330-371?	Arian, Valens's mate	
Secundianus 378-384	Arian	
Bassiana I		
Vigilius 6th century	orthodox, present at the Council of Grado	

BISHOP'S NAME	SUPPOSED LOCATION	DATE
Paulus, Arian	Bassiana I, Sopianae	359-370?
Ioannes	Siscia	around 530 / 532
Migdonius	Savaria	?
Megasius	Savaria	?
Gaius, Arian	Iovia, vide above, Savaria	359-370?

4. EDIFICES OF TEMPLES AND CHURCHES

Let me start this chapter by quoting Friedrich Wilhelm Deichmann: 'Die Kirche als christliches Gotteshaus geht aus einer Wandlung der Vorstellung vom Heiligtum hervor, die sich während der ersten Jahrhunderte nach Christi Geburt vollzog und im Laufe des vierten Jahrhunderts abgeschlossen war. … Die Kirche bedeutetet wohl im Kultbau den tiefsten Wandel seit jenen Urzeiten, als der Mensch die Gottesverehrung vom Zufälligten des Natur-Elementes oder des Natur-Ortes löste und an künstlich geschaffene, geweihte Stätten band. Diese wurden bald zum Tempel: ein Bezirk, ein Haus war göttliche Wohnstatt, Stätte dauernder Anwesenheit des Numen; hier steht das Götterbild oder Göttermal, als der Gott selbst.'[1433] The main subject of his article is not the evolution of the building of the temple or church, but rather the evolution of the idea that will later create the building. Its main issue is that, in antiquity, two forms of worship existed: one that represented its gods in pictures, and one that did not. The first was characterised by 'Bildkult', the

[1433] 1964: 52.

second by 'Mahl-Kult'. Egypt and ancient Greece belonged to the first; the Middle Eastern cultures, Asia Minor, the Arabians and the Israelites belonged to the second. For the Jewish who lived in the Diaspora, the synagogue did not represent a building, but rather the community. Where this community gathered for worship was not a temple, but a profane room. Very similarly, for the early Christians the *ecclesia* did not mean a building, but Christ's community, the community of the invited.[1434] Early Christianity held this idea up to the end of the 3rd century. 'Dem himmlischen Tempel - Gott - entspricht auf Erden die Gemeinde der Gläubigen. ... Ein solcher (sc. Tempel) wäre Götzenwerk gewesen. ... Das Urchristentum hatte keine sakrale Architektur.'[1435] Where they gathered for prayer, breaking bread and giving thanks, the room was a profane place, which we today call a 'house-church'. Deichmann interprets one of Clemens Alexandrinus' sites (Strom 7,5), and states that the word ἐκλησια was first used for a building around 200, but without the word losing its original and sole meaning. 'Dennoch blieb der Kirchenraum im dritten Jahrhundert offiziell profan.' In early Christianity the hierarchy of priests did not exist. In parallel with its evolution, the name *ecclesia* began to be used for the building as sacred room. In this process, the first decade of the 3rd century were a turning point. At that time, Eusebius of Caesarea consecrated a church in Tyrus, as the reflection of the holy Jerusalem. 'Dieses irdische Heiligtum entsprach also dem himmlischen und wird daher Tempel genannt ... Die Kirsche als Gottes Heiligtum leitete eine neue Epoche des Sakralbaus überhaupt ein.'[1436]

The private home, as profane room, offered a place for thanksgiving. In the beginning, it was not the whole house that served the liturgy, only a certain part was appointed for this purpose. These houses were transformed after a certain time, as exemplified and verified by the baptistery found in Dura-Europos.[1437] Dura's example in Pannonia should be considered very closely, as Dura's impact may be demonstrated in several places: in Aquincum, in frescos of six different buildings, cf. 4.I.g; at Kővágószőlős cf. 26.II.a; conceptually in Dunaújváros, cf. 12.I.c. Thus, we may well assume that Dura's influence was also expressed in the usage of house-churches. The existence of house-churches would solve a seeming contradiction: the fact that some grave goods refer to Christianity, whereas no churches have been found.[1438] This fact very clearly leads to the conclusion that the places of both the liturgy and the baptism were private houses, where the house and the room were profane, while the gathered community was God's living church, the living *ecclesia*, that is sacred. We have to presume such a house-church in Aquincum, both in the civil town, and in the canabae, cf. figs 56, 60. In the same way, there were presumably house-churches in Intercisa, where there are many Christian grave goods, and at Kékkút, where the research names them basilica I and II. According to the rules of liturgical history, we cannot say that the two buildings at Kékkút were basilicas, and, especially, we cannot use the marks I and II. They are, rather, buildings or parts of buildings that used to belong to the same community. One was the place remembering the foundation of the Eucharist, or thanksgiving (i. e. the place of liturgy), while the other was the place for preaching, or the agape. The part of the building called the 'hall church without an apse' served as a place for liturgy; while the rooms of the other building provided spaces for all the other activities of the living *ecclesia*. Dyggve, who used the term 'hall church without an apse', supposed the existence of a detached 'Priesterbank',[1439] referring to the antique custom that the altar could be walked around. Here, however, we cannot speak about an altar, only a table. At Sümeg (43.I.a), a part of the villa served as a house-church. In Pannonia, we can certainly expect the usage of a house-church in every place, or town, where Christians lived; in Arrabona, for instance, where there are no such buildings.

The outer features and criteria of a house-church. In Pannonia, the fundamental unit of the house-church is a rectangular room, which is divided into two by a wall, in parallel with its shorter side. Although other rooms may also belong to it, it is never built into the house completely. This two-room fundamental unit can always be distinguished from the rest of the rooms (as at Dura), although its relationship with the other rooms differs from place to place. The presence of an apse is not typical. We may say that the apse does not belong to the house-church. In Pannonia there are no baptisteries as separate buildings. In this respect, Dura is again an analogue, where the suitable place for the baptism is found within the house-church. This was also the case in Pannonia. Transforming the bathing pool of the house, they may have prepared the site for Christenings. We may perhaps find an example in Aquincum, with the bathing pool of the so-called large dwelling house.

The length of time a house-church was used. We may assume an early appearance of Christianity in Sirmium, and then in Aquincum (the middle, or end, of the 1st century). In this period, naturally, only a house-church might have existed, if we can talk about house-churches at all; it was possibly just a dwelling house where a room was used for assemblies. The term used in this early period is the Mahlgemeinde, rather than the Hausgemeinde, which evidently lasted until the end of the 2nd century; later we may use the term house-church. In Pannonia, the practice of erecting churches did not start after 311 or 313. Pannonia's population kept the tradition that said that

[1434] Deichmann here refers to Joh. 4, 21-25.
[1435] DEICHMANN 1964: 55.
[1436] DEICHMANN 1964: 58.
[1437] KRAELING 1967.
[1438] Besides the possibility of wooden churches, this can be very rarely found in archaeology.

[1439] DYGGVE 1952: 41, 45: 'Die innerste Voraussetzung für die Einrichtung des Presbyteriums muß die Anbringung loser Sitzplätze, die für die ältesten der Versammlung reserviert sind, um den transportablen Altartisch herum, sein. So war es in den ersten christlichen privaten Wohnungen, apsidenlosen Räumen wie unsere Saalkirchen. Diese freistehende Ordnung muß folglich alte Ahnen haben, darf aber auf keinen Fall als christliche Neuschöpfung angesehen werden.' Also see in KAHL 1980: 35-38.

God's church was the people, and the place of liturgy a profane room. The parts of buildings and finds put to the 5th century let us conclude that, by this time, the building of churches had started, but the usage of the house-church, the *domus ecclesiae*, had not diminished. This custom here is thought to exist because Pannonian Christianity was mainly an eastern population, whose pagan forefathers, in their homelands, had been the worshippers of the Mahl-Kult.

The building with an apse, as an early Christian basilica, has some significance beyond the 5th century in Pannonia. The sole site where such a basilica has been opened is Keszthely-Fenékpuszta (22.I.a); it can be dated to the 7th century. Buildings with an apse, because of their basilican system, are called 'basilica' in architecture. This name did not come from Christianity, however, but from its structure.[1440] Although the Pannonian research deals with the reconstruction carried out in forts and fortresses during Constantine and later, it has never been examined whether these alterations had any non-military purposes. The Christian finds of Leányfalu (27.I.a), Szentendre (48.I.a-b) and Dunaújváros (12.I.g), or even Aquincum's fortress and canabae, instruct us to conduct such an examination. There were surely Christians among the soldiers. During the persecutions of Diocletian's reign, soldiers were executed, but, on the other hand, Constantine needed to gain the sympathy and support of exactly these Christian soldiers.

Archaeology can facilitate the reconstruction, or alteration, of existing buildings, but it is difficult to define what kind of buildings were involved, and how the thing was carried out. This question has been elaborated by Vaes.[1441] Based on the examples and types of buildings highlighted by him, in Pannonia we have to take the horrea. We must not forget that the ground plans of horrea are not far from the plans of 'pagan' shrines, or even from the basilican style. Consequently, they achieved more space by means of minor alterations where necessary. The Tokod 'horreum' (Site 57) may be interpreted in this sense. Such alterations may also be traced at Pilismarót (33.I.a) and Tokod, and there is a need to consider such alterations at the horrea of the inner fortified settlements. In the case of the Keszthely-Fenékpuszta horreum (22.I.b), even the fact that it was ever a horreum is disputable. The cemetery next to it, which follows this construction, makes the function of this 'horreum' very doubtful, as László Barkóczy has defined it. At Ságvár, in the inner fortified settlement, the number of horrea raises suspicions.

A. Remains of the interiors of churches

The pieces found in Szombathely and defined as parts of a screen (51.II.a) proved later to belong to the imperial palace, since in Sirmium, where there were also imperial constructions, similar pieces were found. The carving found in Székesfehérvár (46.I.a), and earlier thought to be a screen, based on similar examples, was defined as a mullion or pillar. As Székesfehérvár is only its secondary location, I attempted to define its primary location. Very probably, it is from Aquincum. This type of mullion belongs to Byzantine architecture, or at least to the Hellenised territory. This phenomenon is not unique in Aquincum. The building with apses (4.I.g) may be classified here as well. These buildings (associated with further buildings), bearing their Romanised Hellenised features, are fastidious pieces and represented antiquity in the 5th century.

Capitals, pillars. This theme is rather neglected in Pannonia. They have been collected as elements of buildings,[1442] but as for their execution, type or dating there have been few attempts to study them. In this respect, Sylvia Palágyi's work[1443] is a significant development and it serves as a starting point here. It is rather hard work, especially if we would like to be very accurate concerning early Christian aspects. It is difficult because we do not have such a building - in contrast with other provinces - that can be defined as a standing, or partially standing, church or basilica. We would certainly be aware of the Christian relation of the pillars of such a building, and the criteria would be easier to define; we must rely on parallel examples (provided by Marušić's paper).[1444] The capitals come from the following places: Poreč, where the basilica was built in the first decade of the 5th century,[1445] and the area near Betika and Peroj. In Betika, the three-aisled basilica of Saint Andrew was built in the 5th century. The capitals found here were 'corinthiens avec les fouilles d'acanthe lisses en un rang sont typiquespour la moitié du Ve siecle ...'[1446] Comparing the examples with Sylvia Palágyi's typological groups, we may state that two types resemble the analogues from among Palágyi's groups. One of them is the 'geteilte Vollblattkepitelle', the other is the 'Kapitelle mit Eckblatt-Volutenverzierung'.[1447] Palágyi dates the former to the 4th century and the latter to the 3rd century. The capital with a parted leaf ornament was found in Kékkút (21.I.a-b-c), while spurs with volute scroll ornaments were found at Zánka and Kékkút, cf. figs. 123 and 124. As the two buildings at Kékkút (21.I.b-c) have probable Christian associations, and because the same may be assumed for the sites from where the parallels came, we may presume that these types of capitals appear in house-churches, churches and basilicas. The question whether they appear exclusively in such buildings, or perhaps other, profane buildings as well, remains open. The question is appropriate, because at Kékkút, although we always use the term basilica, there was only a house-church, and it was, as a room, profane.

We should mention the phenomenon of a capital type belonging to the Balaton area[1448] appearing in the medieval church of St Mary of Apollonia.[1449] The

[1440] LANGLOTZ/DEICHMANN 1950: 1225-1259. Especially the chapter written by A. Langlotz deals with the non-Christian basilica. Cf. Notes 1127-1128.

[1441] VAES 1980, 1986.

[1442] KISS Á 1987.

[1443] PALÁGYI 1996.

[1444] MARUŠIĆ 1978: 559-560, 571 fig. 9.

[1445] MARUŠIĆ 1978: 570.

[1446] MARUŠIĆ 1978: 572.

[1447] PALÁGYI 1996: 111 Abb. 6, 117 Abb. 9.

[1448] PALÁGYI 1966: 115-116 Abb. 8.

[1449] Albanien. Kulturdenkmäler eines unbekannten Landes aus 2200 Jahren. Photoausstellung. Hrsg. KOCH G. Photographieren von GLOC J /

Hungarian sites are Balatonakali-Ságipuszta, Vörösberény and Mencshely. The dating of the capital is dubious, 4th century or medieval. According to recent research, we may consider its type of illustration as ancient. Palágyi's describes the capital as follows: 'Characteristisch für diese Kopfkapitelle sind ihre Würfelform, der einfache Blattkranz, die unter den Köpfen wie eine Girlande zusammenlaufenden Voluten und die Kopfdarstellungen an der Stelle der Abakusblüten.'[1450] The similarity between Albanian and Pannonian capitals is so great that we have to presume the spread of motifs, or at least an influence.

Special attention should be given to those capitals ornamented with chip carving (the capital found near 4.I.h; the capital mentioned under 48.I.b). The origin of this technique was discussed by András Alföldy[1451] in a debate from the beginning of the last century. It was developed from carpentry, which can be proved by the fact that this type of ornament appears on stone-cuttings that feature gates or buildings. This ornamentation was imitated by the Germans. The illustration of carved triangles facing each other in pairs is a common motif. The nature of the decoration is that the motif is cut deep into the surface, which leads to an enhanced light-shadow effect. The patterns are generally geometrical motifs: triangles, semicircles, or a row of eggs, but they can be in the shape of a leaf, heart, or other designs; the main feature is the deep cutting and it is very common on metal or bone objects.[1452] András Alföldi deals with the carvings of the early Christian church of Breza in detail, 'deren Schäfte wie Holz sgedrechselt sind und wo die merkwürdigen Kapitelle zweifellos Holzwerk nachahmen.'[1453] Although it is true that this decoration was mainly imitated by the Germans, its origin is not connected to them. 'Diese Art Ornamentik wieder stammt aus der altskytisch-iranischen Tradition, die in den griechischen Goldschmiedewerkstätten am Nordgestade des Pontus sich bis in diese Epoche fortvererbte."[1454] The two capitals concerned well show this chip carving (Kerbschnitt) technique. Although Germanic finds along the River Danube already had these decorations by the 4th century, I think the pillars were made in the 5th century, and that they have no connection with the Germans. Based on parallels, the 46.I.a mullion appears 5th century. These parallels refer back to Greece, and it seems only proper to presume that the origin of the decoration of other building-elements should be sought in that direction.

Summary. In Pannonia, before the Constantinean changes, the necessary and most satisfactory setting for the liturgy was a two-part room in a building, or in a part of a building. The participants were, strictly speaking, the community of the house (Hausgemeinde). Alterations may have taken place to achieve such a space, but the purpose of these alterations was never to create a different image, but to serve practical reasons. That is why I would not reduce the question to the basilican style, whose first appearance as an early Christian basilica, in Pannonia, cannot be expected in the 4th or the 5th centuries (and very probably not even in the 6th century either). The builders of the first basilican-style church are open to question. The basilica at Keszthely-Fenékpuszta (22.I.a) gained its last form after alterations in the 7th century. The builders of the last phase are unknown. The alterations may generally be dated to after the Constantinean changes, but these alterations did not necessarily mean the dominance of the basilican style. The conclusion, however, that the possible early Christian building/buildings (i.e. the church) should be sought in accordance with the features of the construction (whenever it took place), seems correct. The alteration might have included the creation of a lavacrum for Christenings. The Constantinean era did not immediately result in the construction of new churches. The *extra muros* basilica at Pécs (32.I.g) was a new building, erected in the second half of the 5th century at the earliest, or at the beginning of the 6th century at the latest.

B. Building with apses

On the basis of F.W. Deichmann's summary, it is evident that the buildings with apses, such as the cella trichora, were also used before Christianity. However, it is indisputable that they also became popular among the Christians,[1455] and that they survived antiquity. The Óbuda building regarded as a cella trichora (4.II.a) is not actually a trichora. A larger building, perhaps a villa, had been reconstructed, and, in one of its later periods, the trichora probably belonged to it. The trichora remained a part of this building, which was, otherwise, not related to Christianity. The trichora of Pécs (32.II.1) has remained from the Middle Ages. It was not built before the 8th century. In Pannonia, trichoras from ancient times are recorded only in the south of Pannonia, or more correctly, Pannonia Secunda. Here, it is a memoria: the memoria of Demetrius.

The *septichora* (32.I.d.) is a building without parallels, whose ground plan I assumed symbolised the sigma table, and consequently I defined it as an oratory.

In Aquincum, Dalmatian examples help to describe the building with six apses (4.I.g) as a church. Nevertheless the accurate date of its construction and periods may only be determined by new excavations. At present we may only assert that its final period was not before the end of the 5th century , or the beginning of the 6th.

The pillar of Székesfehérvár (46.I.a.) also originally stood in a building with apses, or at least a building constructed centrally with apses (based on the analogous church). However, the place where church 46.I.a. stood cannot be identified at this time. I have assumed that its location should be searched for in Aquincum.

HAAG P / KOCH G. Marburg 1985: 48, Abb. 79.

[1450] PALÁGYI 1996: 116.

[1451] ALFÖLDI A 1929-1930: 17-18.

[1452] The formal and chronological classification of the motifs has recently been conducted by Margit NAGY, 1998. Her discourse shows that this technique was very generally used in the 5-6th centuries.

[1453] ALFÖLDI A 1929-1930: 19.

[1454] ALFÖLDI A 1929-1930: 24. Here Alföldi referred to M. Rostovtzeff's work, Iranians and Greeks in south Russia 1922.

[1455] DEICHMANN 1954: 942-948.

In the early Migration Period, the buildings with apses preserved and represented orthodoxy among the Arian Germans. They point towards the east, as I have already alluded to previously. We may say that Pannonia never accepted Arianism in general, although, in the reign of Constantius II, it contributed to the deabate (on the Arian side), or, rather, that in this period, only a few bishops represented Arianism, expressing their political loyalty.

In Pannonia, it is not the presence of orthodoxy and Arianism that results in division,[1456] but the different mysticisms of the West and the East within orthodoxy. The East is more mystical and symbolic in its thinking, while the West is less mystical, laying its emphasis, rather, on those matters concerning organisation, power and ethics.

5. SAINTS, MARTYRS AND THE VENERATION OF MARTYRS

As regards Pannonia, András Mócsy argued that the Diocletian persecutions did not harass the Roman Empire in the same measure.[1457] In Pannonia the only place where the edicts were taken seriously was Sirmium, the imperial centre of this quarter of the Empire as well as the centre of Pannonia Secunda. Although this region - due to its geographical situation - is not included in my topography, I have to allude to its martyrs. The *diaconus*, Saint Demetrius is an outstanding figure among the several martyrs of Sirmium because of the later veneration of his personality. I intend here to summarise his story and veneration, as well as the main points of debate arising from the studies of Vladislav Popović.[1458] The *Passio Sancti Demetrii Thessalonicensis* has remained in three versions from different periods. The first version serves as a basis for the second one. In the first version, Emperor Maximianus - during gladiatorial combats arranged by him - confined Demetrius in the heating chamber of the nearby *thermae*, where he was murdered. The congregation buried him there. His power to work miracles soon came to light, and the sick were healed at his grave. From among those restored to health, Leontios (*praefectus praetorio Illyrici*), had a real church built between the *thermae* and the stadium out of gratitude and respect for the saint. The second version draws on the earlier version but differs in one significant point, that Demetrius was a senator, a *primus exceptor*, who preached in the so-called bronze *stoa*. He was held in the heating rooms of the *thermae*, where he was then executed.

The problem concerning the origins of the Passio has induced a debate involving the different events. It would be most helpful to settle this debate as some put the origin of Demetrius' veneration to Saloniki, from where, they think, it moved to Sirmium, while others consider that his veneration originated from Sirmium, and only later spread to Saloniki. Those who favour the Saloniki opinion draw a parallel between the dates of the building of the Saloniki basilica with the beginning of the Saint's veneration. The basilica was erected in the middle of the 5th, or at the beginning of the 6th century *here*. According to Popović, the construction of the basilica bears no relation to the question of the origin of Demetrius' veneration. The basis of his opinion is that *Martyrologium Hieronymianum* does not include a saint called Demetrius in Saloniki. However, it is a well-known fact that Sirmium was attached to the eastern part of the Empire in 424-425.[1459]

Vladislav Popović definitely states that Demetrius' veneration started from Sirmium and spread to Thessaloníki from here. He argues that the corpse of the martyr has never been in the present Demetrius church in Saloniki, because, when Maurikios visited Saloniki, he complained that 'außer dem Grab der Heiligen Matrona nicht wußten, wo ihre Märtyrer bestattet worden seien.'[1460] He also refers to the Just. Novell. XI, 535,[1461] which says that the office of the Sirmium Prefecture with Prefect *Appenius* was taken to Thessaloníki in Attila's time, and that the bishop of Thessaloníki owed his respect not to his own authority but to the significance of the prefecture. All this happened in 442, the time when Demetrius' cult and veneration spread to Thessaloníki. He became one of the most venerated saints of the Eastern Orthodox Church.

Passio prima was made before the building of the Saint Demeter church in Sirmium; *Passio altera*, on the other hand, was made when Sirmium was attached to the eastern part, i.e. in 424 or 437.

The other great martyr of Sirmium is Bishop Ireneus,[1462] whose veneration also reached beyond Sirmium. Both the Orthodox and Latin Church celebrate the day of his martyrdom, the former on 23 August, the latter in March (he was martyred on 25 March, 304).[1463]

'Gardener' Synerotas was also highly venerated here. (Versions of his name include Sirenus, Serenus, Sinerius and Sinerus.) One account of his story says that he was martyred on 23 February,[1464] another on 22 February. Thanks to the study made by Tibor Nagy, we know that the latter date is the correct one. It was also Tibor Nagy who solved the problem of the name varriants. Synerotas became the commonly accepted name, deriving from an incorrect reading of the Breviarium Syriacum: σύν Ερωτος.[1465] Ruinart puts the martyrdom to 307,[1466] but

[1456] Concerning Pannonia, this division is valid only to the same extent as in the case of the whole empire. The division is not deeper here either. What is beyond all question is that the two bishops of Pannonia, Ursacius and Valens played a prominent part throughout.

[1457] MÓCSY A: Pannonia a késői császárkorban - Pannonia in the late period of the Empire. Budapest 1974: 157-164.

[1458] POPOVIĆ 1987: 95-139; JARAK 1996: 274-276.

[1459] According to other scholars, it happened in 437; the attachment is out of debate.

[1460] POPOVIĆ1987: 100.

[1461] Partially quoted at Sopron, cf. Note 980. A further passage: *Quum enim in antiquis temporibus Firmi praefectura fuerit constituta, ibique omne fuerit Illyrici fastigium tam in civilibus quam in episcopalibus causis, postea autem Attilanis temporibus eiusdem locis devastatis Appenius praefectus praetorio de Firmitana civitate in Thessalonicam profugus venerat, tunc ipsam praefecturam sacerdotalis honor secutus est, et Thessalonicensis episcopus non sua auctoritate, sed sub umbra praefecture meruit aliquam praerogativam.* CIC P. 3. p. 84. Ed. E. OSENBRÜGGEN [15]Lipsiae 1875.

[1462] Passio Sancti Irenei RUINART 1859 pp. 432-434.

[1463] Admunition in passionam S. Irenei episcopi. RUINART 1859 p. 432.

[1464] Admunition in acta S. Sereni Martyris. RUINART 1859 p. 516-517.

[1465] NAGY T 1939: 59. n. 49.

Tibor Nagy has it later.[1467] Synerotas' offence was that he had reproved a woman who was walking in his garden at night. He told her that her intentions must have been improper as no respectable woman would be walking in a garden at such an hour. The woman sued him, and, because she was the wife of one of Galerius' officials, gardener Synerotas was singled out for revenge. It is clear that he was not martyred during the persecutions of Diocletian, but of Galerius. One of Sirmium's cemeteries was formed around his grave.

Fortunatus and Donatus were also venerated locally. Their passions were written in the 5th century based on an unknown Pannonian tradition.[1468] The year of Agrippinus' and Secundus' martyrdom is unknown, but the date is 20 July. They are also martyrs of Sirmium, and presbyter Romulus also belongs here. We do not know the names of the seven women of Sirmium who were mentioned in Martyrologium Hieronymianum as the seven virgins killed, with Timotheus, on 15 May. Timotheus was thought by Tibor Nagy to be an saint from Antioch.[1469] The casket mount (36.I.a) that was found later drew attention to a Pannonian saint, about whom Endre Tóth wrote: 'Our saint can securely be identified with Timotheus, who was venerated in the Pannonian provinces and who is known from the Pannonian sources.'[1470] As well as the seven virgins mentioned above, we know of three other virgins, Basilla, Anastasia and Sabbatia[1471] as Sirmium martyrs.

The first Pannonian martyr was Eusebius, Bishop of Cibalae.[1472] Although Eusebius' file has not remained, we know of him indirectly. The Cibalaeian Pollio, as he was being tortured in Diocletian's reign, referred to Eusebius, whose example he was following, and who, during a previous persecution,[1473] had been martyred.[1474] Lector Pollio was executed on 28 April, 303. That is why András Mócsy calls Pollio and his fellows the second generation of Pannonian martyrs.[1475] Research puts this earlier persecution to the time of Valerianus (253-260).[1476] Others refer to the persecution of Traianus Decius, and they mention a martyr named Tiballus as well as Eusebius.[1477]

The Sancti Quattuor Coronati. The passion of the four crowned saints contains the story of five stone-carvers: Simpronianus, Claudius, Nicostratus, Castorius and Simplicius. Their execution took place in Pannonia, in the presence of Diocletian (and so dated by the Pannonian research to 8 November 293).[1478] Vulić says there are two different stories, but the first is the main one. In it, the fifth stone-carver, Simplicius, decides to join the original group of four.[1479] Both Vulić (1931) and Tibor Nagy (1939) note the chronological mistakes, the deficiencies of topographical data, and the unusualness of the legal proceedings in the passion. In addition, Tibor Nagy discovers the construction and compositional features of the gospels in the construction and composition of the passion.[1480] Although the authenticity of the passion of the Sancti Quattuor Coronati cannot be doubted, identification of the persons has not as yet taken place. In fact, this group of martyrs used to be venerated in Rome, where the names of Clemens, Sempronius, Nicostratus and Claudius appeared in the Via Labicana catacomb.[1481]

A short summary of their story: Diocletian ordered different statues and columns from a stone-quarry. Among the workers, Clemens, Sempronius, Nicostratus and Claudius began their work in the name of Christ, entreating Him to speed their work and make it more skilful and pleasing. When reports of their excellent work had circulated, Diocletian praised them and made them supervisors, making the other supervisors jealous. By this time, Simplicius had also joined the group. They performed their subsequent commissions accurately, except for one single statue, the statue of Aesculapius. This was an opportunity for their jealous colleagues to denounce the five stone-carvers to Diocletian as Christians, who charged Tribune Lampadius to examination them. In the meantime, the supervisors prompted the other stone-carvers to take the lives of the five Christians: they were thrown into the river in a leaden box, and so martyred.

Victorinus was Poetovios martyr bishop. Quirinus was Siscia's bishop, ☞ Szombathely. We do not know of any martyrs from Valeria and Pannonia Prima, although Christians also lived Christians in these two territories. This fact reinforces András Mócsy's idea that the edicts were not observed so strictly in these areas.

Only for the sake of completeness do we mention Saint Martin (☞ Szombathely) and Saint Anthony (☞ Tokod), who were not venerated because of their martyrdoms.

6. Events of history

Research so far has thought that the burial of pagan shrine-treasures, cult pictures or statues in the province was connected to the spread of Christianity; these were either destroyed by the Christians, or the pagans hid them from the Christians. 'In their sudden disappearance under

[1466] RUINART 1859 p. 517.

[1467] NAGY T 1939: 59.

[1468] NAGY T 1939: 59. Fortunatus' veneration later merged into the figure of the Aquileian Hermagoras-Fortunatus. NAGY T ibid.

[1469] NAGY T 1939: 61.

[1470] TÓTH E 1995: 141.

[1471] She is not unanimously accepted as a martyr of Sirmium, but is possibly identical with the Antiochian martyr, Sabbatius. NAGY T 1939: 60, n. 56.

[1472] The previous research doubted the fact that Eusebius was Pannonian; BALOGH 1932: 47, n. 6.

[1473] 'In qua superiori persecutione Eusebius ...' Passio Sancti Pollionis I. RUINART 1859 p. 435.

[1474] ... agonem suum implevit Martyr (sc. Pollio) intrepidus, laudans, benedicens, et glorificans Deum, qui eius venerabilem passionem, sed et sancti Episcopi eiusdem civitatis Eusebii, ante plurimos annos, eodem die vita (functi) martyrium prenovit ad caelestem gloriam: ... Passio Sancti Pollionis III. RUINART 1859 p. 436.

[1475] MÓCSY 1974: 325.

[1476] MÓCSY 1974: 325-326.

[1477] BRUNŠMID 1979: 115; JARAK 1996: 276-278.

[1478] MÓCSY 1974: 326.

[1479] VULIĆ 1931: 359-360.

[1480] NAGY T 1939: 63.

[1481] For the problem in detail , Tibor NAGY, 1939: 61-65.

Constantine, we must recognise political and religious reasons.'[1482]

In fact, it is a common opinion that Christians demolished the pagan shrines. However, it is contradicted by many archaeological observations.[1483] We certainly know of definite instances where Christians actually destroyed things, but these did not happen in Constantine I's reign, and we cannot generalise them. On the other hand, the protection of a shrine-treasure by means of burying is not destruction. Such issues prompted me to examine this phenomenon, so much linked with Constantine's reign from another aspect, that of pagan worshippers and the equipment of their shrines.

The wealth of the pagan temples belonged to the town and the state (the emperor). Constantine never abolished pagan shrines and temples by decree, or other means. He built two new temples in Constantinople, one to Rhea, and one to Tyche. Any kind of official and non-official destruction would have been against the 313 *constitutio* of Mediolanum.[1484] The burial of the treasures may have seemed to be a practical protection at a time when the emperor was continually short of money, partly because of his wars and partly because of his new constructions. Constantine I stayed in Pannonia several times, at some length, and the communities of the Pannonian pagan shrines may well have been afraid that the emperor might lay claim to the treasures, so they buried them. Thus, in the discoveries of such treasures, I do not see the destructive intentions of Christians, but the efforts of those communities associated with the temples to save their wealth.

In thc chronological order, we have to mention first the existence of Marcionites and Gnostic communities. Their presence is verifiable indirectly, through Victorinus' scholarly activity. Two of his works remained, the *de fabrica mundi*, and the *Apocalypsis*-commentary. We know the catalogue of his works from Hieronymus' list.[1485] On the basis of these works, Tibor Nagy has proved the existence of both Marcionites and Gnostics in Pannonia.[1486] Some examples serve to illustrate his argument. Victorinus mainly wrote commentaries to books of the Old Testament, and in his texts he always highlighted 'the inner unity and organic connection' of the two Testaments.[1487] It was needed against the Marcionites, who separated off the two Testaments. They denied the Old Testament, and selected the New. His exegetical method is similar to the one of Origenes, whom he joined in his argument. In his remaining works, he considered all the arguments that had been raised against the Gnostic schools, including that the Lord created the world first, and only then the angels, 'so that some people cannot claim in deceitful speeches that the angels helped Him with the Creation.'[1488] This argument is against the Gnostic teaching which says that the world was created by different demiurges. The *symbolum* text noted down by Victorinus is very significant. It was compiled in Pannonia, and verbalises the creed with peculiar phrases.[1489] Let me quote some parts of it in association with the *Symbolum Nicaenum.*[1490] The part of the *symbolum* may be found in a Pannonian inscription, cf. 51.I.d.

VICTORINUS	SYMBOLUM NICAENUM
Confitemur	Credimus
patrem omnipotentem	in unum Deum Patrem omnipotentem
et huius filium dominum nostrum Iesum Christum	Et in unum dominum nostrum Iesum Christum filium Dei,
ante originem seculi spiritaliter apud patrem genitum,	natum ex Patre unigenitum,
factum hominem	et homo factus est
et morte devicta	et resurrexit tertia die
in caelis cum corpore a patre receptum	et ascendit in coelos

Tibor Nagy, after having analysed Victorinus' *symbolum*, pointed out that it showed a Syro-Palestinian influence, or an influence from Asia Minor; but it was surely compiled in Pannonia.

If we analyse the different phrases of the symbolums on, then it turns out that there is a special one from Pannonia, although it comes from a later time.[1491] This special phrase is the *descendit ad inferna,* which appears for the first time in the formula at the Council of Sirmium in 358. This symbolum was composed by Ursacius and Valens. This formula was also under debate at later councils. Ultimately it remained longer, because even Martinus Bracarensis taught it in this manner, and afterwards it still remained in the symbolums.[1492] The addition may also be found at Rufinus.[1493] This special phrase cannot be ignored, however, and, from this point, it should be treated as a theological rather than a liturgical problem.

Constantine's reign made the life of Christians in Pannonia eventful in certain respects. At the Church

[1482] MÓCSY 1974: 323.

[1483] MIGOTTI 1992b.

[1484] LACTANTIUS de mort. 48, 2 '... ut daremus et Christianis et omnibus liberam potestatem sequendi religionem quam quisque voluisset, ... ' Ed. CREED J. L. Oxford 1984 p. 70.

[1485] de virilis illustribus 74. Ed. HERDINGII G. Lipsiae 1879 p. 48.

[1486] NAGY T 1937.

[1487] NAGY T 1937: 123.

[1488] Tibor Nagy's translation from the *de fabrica mundi* 4.

[1489] VICTORINUS Commentarii in Apocalypsin XI, 1. Ed. HAUSLEITER (CSEL 49) Vindobonae , Lipsiae 1916 p. 96, 5-13. The whole creed is as follows: *patrem confiteri omnipotentem, ut didicimus et huius filium dominum nostrum Jesum Christum ante originem saeculi spiritaliter apud patrem genitum, factum hominem et morte devicta in caelis cum corpore a patre receptum, sanctum dominum et pignus inmortalitatis, hunc per prophetas praedicatum, hunc per legem conscriptum hunc per manum dei et per verbum patris omnipotentis et conditorem orbis totius mundi, haec est arundo et mensura fidei, ut nemo adoret ad aram sanctam, nisi qui haec confitetur : dominum et Christum eius.*

[1490] The quotation is not complete; these are only extracts from the *articuli.* DENZINGER H: Enchiridion symbolorum definitionum et declarationum de rebus fidei et morum. Denuo compositum a Clemente BANNWART. Ed. UMBERG J. B. Friburgi Brisgoviae [21-23] 1937. No. 54.

[1491] BRATOŽ 1994: 46-47.

[1492] NAGY T 1937: 134-139.

[1493] Cf. BRATOŽ 1994. BRATOŽ also states that it is included in the *Symbolum Romanum*, though maybe not in the form composed above.

Council of Nicaea, Arius and his followers were exiled to Illyricum, where Ursacius and Valens also became his disciples.

In 326, Constantine was staying in Pannonia when Arius personally handed over his creed to him.[1494] There is an assumption that Arius met Presbyters Ursacius and Valens, who later, after their inclination to Arianism had circulated, were excluded from among the presbyters (between 326 and 327). In 327, Arius was restored, which meant the acceptance of the Arian Eusebius of Nicodemia. Around Eusebius of Nicodemia, by this time, there was a group called the Eusebians, and Ursacius and Valens also belonged here.[1495] Eusebius' activities brought about the conflict between the two groups, which resulted in the exile of the bishops. Bishop of Sirmium Domnus was also exiled,[1496] supposedly between 328 and 330.[1497]

Ursacius and Valens did not only contribute to matters concerning Pannonia, they were also involved in cases against Athanasius. These two bishops gained high respect as they were elected into Arian special committees. They appeared in Mareotis, where they had to look into the accusations against Athanasius. They disclosed their false records of evidence later, at the Council of Mediolanum, and in their appeal to the Pope.

The strength of the Arians was not broken by the death of Arius or Constantine I. Among his sons, Constantius II was an especially fanatic Arian, who would not shrink from violence to defend the Arian faith. Ursacius and Valens were his staunch adherents, and they were joined by a henchman, Germinius.

Councils between the Arians and the orthodox:

338 Constantinople	Athanasius dismissed	Ursacius and Valens also sign the document.
339-341 Antiocha	The council of Eusebians; the formulation of the Creed IV, brought to the west by Ursacius and Valens; it is handed over to the emperor Constans in Mediolanum.	The participants are known. The creed is free from Arian extremes.
342-343 Serdica	Bishops Ursacius and Valens are dismissed; they join the Orthodox.	Bishops Eutherius, Marcus and Aprianus are present. The Eusebians leave for Philippopolis.
345-347 Mediolanum	Constans is present. Ursacius and Valens request their acceptation to the Church. In their *libellus supplicationis*, they condemn Arius. The council excludes Photinus from the Church and dismisses him from his episcopate, but he does not obey.	Italian and Gallian bishops. They turn down Ursacius' and Valens' request; they have to turn to Rome. They do so and condemn Arius' dogmas, and they are accepted.
348 or 349 Sirmium	They try to make Photinus leave his office.	It is an unsuccessful attempt.
351 Sirmium	Photinus is summoned to give up his opinions. He requests a debate. When he loses the debate he is exiled to Galatia.	Ursacius and Valens are again Arian.
355 Mediolanum	Ursacius and Valens, in Constantius II's group, back the Arians. Valens hinders the signature of the Nicaene symbolum by force. The emperor decided, had Athanasius' sentence signed, and exiled three bishops who denied the signature.	The council is dissolved in disorder. Ursacius and Valens assisted in the execution of the sentence. Gallia opposes, counter-council.
355 Baeterrae, counter-council	Called by Bishop of Pictavium Hilarius. Ursacius and Valens unexpectedly arrive.	All the Gallian bishops who refuse to sign Athanasius' sentence are exiled.
355 outside the council	Constantius II exiles Pope Liberius to Beroea, Thrace; and the old Hosius to Sirmium.	The Pope did not sign Athanasius' sentence.
357 Sirmium	The Creed II of Sirmium was formulated by Ursacius and Valens. Constantius II is in Sirmium.	The creed was not accepted. The creed was anathematised in Gallia.
358 Sirmium	Polemics on the different creeds proposed by Basilius. Pope Liberius is here, too. The formulation of the Creed III of Sirmium. The emperor is asked to call a Church Council.	Bishops from Africa, Italy, Asia Minor and Pannonia are present. The creed was partially accepted.
359 Seleukia/Ariminium, Church Council	Called by Constantius II. Ursacius and Valens protest against the Church Council; they formulate a creed (the Creed IV of Sirmium). Later in the Thracian Nice, the Creed IV of Sirmium was altered.	This Arian creed is accepted with the exception of some bishops. Pope Liberius is not among the acceptors.
366 Singidunum	Management of local affairs. Ursacius, Valens, Gaius and Paulus are present.	Proposals for compromise.
378 Sirmium	The eastern Arians are condemned.	
381 Aquileia	Iulianus Valens is condemned.	The end of Pannonian Arianism.

[1494] NAGY T 1936: 38.
[1495] ibid. 41.
[1496] The list of the exiled was disclosed by Athanasius. This part is quoted by Tibor Nagy - NAGY T 1936: 42.
[1497] NAGY T 1936: 60.

Ursacius' and Valens' hypocritical orthodoxy, rough ambition or sycophantic loyalty, is the opposite of Athanasius' dogged endurance, or Hosius' or Hilarius' incorruptible endurance and faith. The unfolding events testify to the two groups of very real characters as drawn in the works of Bishop Victorinus: 'As well as the strong, simple, and uneducated believers, of firm faith, and who are in opposition to the world, we find the compromising and false Christians, inclined to false doctrines, but who are proud of their knowledge.'[1498] Bishop Victorinus' observation remains true in the retrospect of almost 2,000 years.

[1498] NAGY T 1937: 127.

Museum abbreviations

BBM	Balassa Bálint Múzeum, Esztergom
BM	Balatoni Múzeum, Keszthely
BTM	Budapest Történeti Múzeum (Aquincum Múzeum), Budapest
FM	Ferenczy Múzeum, Szentendre
GM	Göcseji Múzeum, Zalaegerszeg
HM	Hansági Múzeum, Mosonmagyaróvár
IKM	Szent István Király Múzeum, Székesfehérvár
IM	Intercisa Múzeum, Dunaújváros
JPM	JanusPannonius Múzeum, Pécs
KJM	Katona József Múzeum, Kecskemét
KoJM	Koszta József Múzeum, Szentes
LDM	Laczkó Dezső Múzeum, Veszprém
LFM	Liszt Ferenc Múzeum, Sopron
MNM	Magyar Nemzeti Múzeum, Budapest
RRM	Rippl-Rónai Múzeum, Kaposvár
SM	Savaria Múzeum, Szombathely
TGyM	Thury György Múzeum, Nagykanizsa
WMM	Wosinszky Mór Múzeum, Szekszárd
XJM	Xantus János Múzeum, Győr

Abbreviations

ActaAntHung	Acta Antiqua Academiae Scientiarum Hungaricae
ActaAntSzeged	Acta universitatis Szegediensis. Acta Antiqua et Archaeologica. Szeged
ActaArchHung	Acta Archaeologica Academiae Scientiarum Hungaricae
ActaDebrecen	Acta Classica Universitatis Scientiarum Debrecensis
AnnÉp	L'Année Épigraphique
ANRW	Aufstieg und Niedergang der römischen Welt
AntTan	Antik Tanulmányok
AqNo	Aquincum Nostrum
ArchÉrt	Archaeológiai Értesítő
ArchHung	Archaeologia Hungarica
ArchJug	Archaeologia Jugoslavica
ArchKözl	Archaeológiai Közlemények
Aregia	Alba Regia
AV	Arheološki Vestnik
BÁMÉ	A Szekszárdi Béri Balogh Ádám Múzeum Évkönyve
BCH	Bulletin de Correspondance Hellénique
BH	Burgendländische Heimatblätter
BpR	Budapest Régiségei
BSAA	Boletin de Seminario de Estudios de Arte y Arqueologia
BullBAHongr	Bulletin de Musée Hongrois des Beaux-Arts. A Szépművészeti Múzeum Közleményei
CarJb	Carnuntum Jahrbuch.
CIL	Corpus Inscriptionum Latinarum
CommArchHung	Communicationes Archaeologicae Hungariae
CSIR	Corpus Signorum Imperii Romani
DACL	Dictionnaire d' Archéologique Chrétienne et de Liturgie, Paris 1924-
DissPann	Dissertationes Pannonicae
DOP	Dumbarton Oaks Papers
EDH	Epigraphischen Datenbank Heidelberg
EML	Egyházművészeti Lap
EPRO	Études Préliminaires aux Religions Orientales dans l'Emprie Romain.
FolArch	Folia Archaeologica
GID	Glasnik istoriskog društva u Novom Sadu
GZMS	Glasnik Zemaljskog Muzeja. Sarajevo
HTR	Harvard Theological Revue
IIL	Instrumenta Inscripta Latina, Pécs
ILCV	Inscriptiones Latinae Christianae Veteres. Ed. E. DIEHL Berlin. 1925-1928. J. MOREAU [2]1961.
ILS	Inscriptiones Latinae Selectae. Ed. H. DESSAU Berolini apud Weidmannos. 1892-1916
JbAC	Jahrbuch für Antike und Christentum.
JÖAI	Jahreshefte des Österreichischen Archäologischen Instituts, Wien
JPMÉ	Janus Pannonius Múzeum Évkönyve
JRGZ	Jahrbuch des Römisch-Germanischen Zentralmuseums, Mainz
KölnJb	Kölner Jahrbuch für Vor- und Frühgeschichte, Berlin
LIMC	Lexicon Iconographicum Mythologiae Classicae, Zürich, München.
LThK	Lexikon für Theologie und Kirche
MEFRA	Mélanges de l'École Française de Rome
MEV	Magyar Egyháztörténeti Vázlatok
MittArchInst	Mitteilungen des Archaeologischen Instituts der Ungarischen Akademie der Wissenschaften, 1986-tól Antaeus
MKÉ	Múzeumi és Könyvtári Értesítő
NHQ	The New Hungarian Quarterly
NumKözl	Numizmatikai Közlemények
PBMÉ	Pécs – Baranyamegyei Múzeumi Egyesület Értesítője
PKG	Propyläen Kunstgeschichte

RAC	Reallexikon für Antike und Christentum
RCRFA	Rei Cretariae Romanae Fautorum Acta
RégFüz	Régészeti Füzetek
RLiÖ	Der römische Limes in Österreich
RÖ	Römisches Österreich, Jahresschrift der Österreichischen Gesellschaft für Archäologie
RömMitt	Römische Mitteilungen
RömQuart	Römische Quartalschrift
SMK	Somogy Megyei Múzeumi Közlemények
SSz	Soproni Szemle
SzSz	Székesfehérvári Szemle
TIR	Tabula Imperii Romani
TVMK	A Tapolcai Városi Múzeum Közleményei
VAMZ	Vjesnik Arheološkog muzeja u Zagrebu
VEAB Értesítő	A Magyar Tudományos Akadémia Veszprémi Akadémiai Bizottságának Történelmi Szakbizottsága Értesítője
VHAD	Vjesnik Hrvatskoga arheološkog društva.
VMMK	A Veszprém Megyei Múzeumok Közleményei
VSz	Vasi Szemle
WMMÉ	A Wosinsky Mór Múzeum Évkönyve
ZM	Zalai Múzeum
ZPE	Zeitschrift für Papyrologie und Epigraphik

Bibliography

ABBIANI M G 1969	Lucerne fittili paleocristiane nell'Italia settentrionale. Bologna.
ADAM A 1985	Grundriß Liturgie. Freiburg, Basel, Wien.
ÁDÁM I 1882	Sümegi földalatti épületmaradványok. ArchÉrt II 14-28.
ADAMESTEANU D 1966	Sopianae. Enciclopedia dell'arte antica. Roma VII 408.
ADAMIK T 1983	Quisquis eris post me (CE 578). ArchÉrt CX 3-9.
Age of Spirituality	Late Antique and early Christian Art, third to seventh century. Catalogue of the exhibiton at The Metropolitan Museum of Art, November 19, 1977, through February 12, 1978. Ed. Kurt WEITZMANN. The Metropolitan Museum of Art, New York 1979.
ALBEKER M 1978	Későrómai temető Dombóváron. ArchÉrt CV 66-77.
ALEKSOVA B 1982-1983	The Old Episcopal basilica at Stobi. ArchJug XXII-XXIII 50-62.
ALFÖLDI A 1920-1922	Újabb irodalom Pannonia keresztény archaeológiájához. ArchÉrt XXXIX 127-129.
ALFÖLDI A 1926	Materialien zur Klassifizierung der gleichzeitigen Nachahmungen von römischen Münzen aus Ungarn und den Nachbarländern. NumKözl XXV 37-48.
ALFÖLDI A 1929-1930	Materialien zur Klassifizierung der gleichzeitigen Nachahmungen von römischen Münzen aus Ungarn und den Nachbarländern. NumKözl XXVIII-XXIX 10-25.
ALFÖLDI A 1931-1932	A pannoniai őskereszténységnek néhány numizmatikai vonatkozású emléke. NumKözl XXX-XXXI 1-8.
ALFÖLDI A 1932	Leletek a hun korszakból és ethnikai szétválasztásuk. (Funde aus der Hunnenzeit und ihre ethnische Sonderung) ArchHung IX, Budapest.
ALFÖLDI A 1937	A festival of Isis in Rome under the Christian emperors of the 4th century. Budapest, DissPann Ser II Fasc. 7.
ALFÖLDI A 1938	A kereszténység nyomai Pannoniában a népvándorlás korában. SZI I. 151-170.
ALFÖLDI A 1939	Epigraphica II. ArchÉrt LII 101-103.
ALFÖLDI A 1940	Epigraphica III. ArchÉrt III. Ser. I 195-235.
ALFÖLDI A 1942	Latin keresztény felirat Erdélyből a Kr. u. IV. századból. (Eine lateinische christliche Inschrift aus Siebenbürgen) ArchÉrt III 252-258.
ALFÖLDY G 1958	Collegium-Organisationen in Intercisa. ActaAntHung VI 177-198.
ALFÖLDY G 1959	Liber pater oltára Scarbantiából. SSz XIII 158-162.
ALFÖLDY G 1990	Revidierte Inschriften aus der Gegend des Plattensees. Specimina Nova 85-108.
ALFÖLDY G 1992	Birodalmak tündöklése és bukása: a római történelem tanulsága. Doctores Philosophiae honoris causa Universitatis de Iano Pannonio nominatae. Pécs.
ALFÖLDY G 1997	Die großen Götter von Gorsium. ZPE CXV 225-241.
Alt-ägina	Alt-ägina, Hrsg WALTER H 1 Bd. 2T. Mainz am Rhein 1975.
APPEL G 1909	De Romanorum praecationibus. Gissae [Religionsgeschichte Versuche und Vorarbeiten VII. Bd 2. Heft].
Archaeology	The Archaeology of Roman Pannonia. Ed. RADAN G. Budapest, Lexington 1980
Atlas 1959	van der MEER F. / MOHRMANN CH.: Atlas of the early Christian world. [Nelson].
Atti 1965	Atti del VI Congresso Internazionale di Archeologia Cristiana. Roma 1965.
Atti 1969	Akten des VII. Internazionalen Kongress für christliche Archäologie. (Trier 5-11 September 1965). Cittŕ del Vaticano, Berlin 1969.
Atti 1978	Atti del IX Congresso Internazionale di Archeologia Cristiana. Roma Vol. I–II.
Atti 1984	Actes du Xe Congres International d'Archeologie Chretienne. (Thessalonique 28 Septembre - 4 Octobre 1980). Cittŕ del Vaticano. Vol I-II.
Atti 1998	Acta XIII Congressus Internationalis Archaeologiae Christianae.

(Split-Poreč 25. 9 - 1. 10 1994) 1998. Arheološki Muzej, Split - Pontificio Istituto di Archeologia Cristiana.

Ausstellung 1969 — Römer in Rumänien. Ausstellung des Römisch-Germanischen Museums Köln und des Historischen Museums Cluj. Köln 12 Februar bis 18 Mai 1969.

Ausstellung 1986 — Römisch-frühchristliche Denkmäler in Pécs (Ungarn) Würzburg.

BAETKE W 1962 — Die Aufnahme des Christentums durch die Germanen. Ein Beitrag zur Frage der Germanisierung des Christentums. Sonderasugabe. Darmstadt

BAKAY K 1975 — Kaposvár földjének őstörténete a legrégibb időktől a honfoglalásig. In: Kaposvár. Várostörténeti tanulmányok. Kaposvár.

BALICS L 1901 — A kereszténység története hazánk mai területén a magyarok letelepedéséig. Budapest.

BALLA L/KÁDÁR Z 1958 — Savaria. Budapest.

BALOGH A 1927 — A római Pannonia kereszténységéről. Katolikus Szemle XLIV 193-204.

BALOGH A 1932 — Pannonia őskereszténysége. 1. Történeti rész. [Szent István Könyvek 102.] Budapest

BALOGH A 1934 — Néhány adat Esztergom városának és vármegyének római koráról. Esztergom Évlapjai VII 41-52.

BALOGH A 1936 — Pécs kereszténysége szent Mór előtt. Szent Mór emlékkönyv. Pécs 241-251.

BÁNKI ZS 1965 — A basilica feltárása. Aregia IV-V 213-215.

BARB A A 1953 — Diva matrix. Journal of the Warburg Institute XVI 193-238.

BARB A A 1953a — Die römischen Ausgrabungen von Donnerskirchen und das älteste Denkmal christlichen Kults in Österreich. BH XV 97-118.

BARB A A 1959 — Ein rätselhafte Emailscheibe aus Carnuntum. CarJb Beiheft V 7-16, Taf I-IV.

BARB A A 1964 — Krippe, Tisch und Grab. In: MULLUS 17-27.

BARIŠIĆ F 1955 — Vizantiski Singidunum. Zbornik radova Vizantološkog Instituta III 1-14.

BARKÓCZI L 1951 — Brigetio. DissPann II,22 Budapest 1944-1951.

BARKÓCZI L 1954 — Tábor és lakótelep. In: Intercisa I. ArchHung XXXIII. Budapest 1-42.

BARKÓCZI L 1957-1958 — Hadrianus-kori éremlelet Brigetióból. NumKözl LVI-LVII 1957-58 17-19.

BARKÓCZI L 1960 — Későrómai temető Pilismaróton. (Ein spätrömisches Gräberfeld in Pilismarót) FolArch XII 111-132.

BARKÓCZI L 1964 — The Population of Pannonia from Marcus Aurelius to Diocletian. ActaArchHung XVI 257-356.

BARKÓCZI L 1965 — New Data on the History of Late Roman Brigetio. ActaAntHung XIII 215-257.

BARKÓCZI L 1968 — A 6th Century Cemetery from Keszthely-Fenékpuszta. ActaArchHung XX 275-311.

BARKÓCZI L 1973 — Beiträge zur Steinbearbeitung in Pannonien am Ende des 3. und zu Beginn des 4. Jahrhunderts. FolArch XXIV 67-112.

BARKÓCZI L 1981 — Kelche aus Pannonien mit Fadenauflage und Gravierung. Acta ArchHung XXXIII 35-70.

BARKÓCZI L 1982 — A keletpannoniai sírsztélék ábrázolásainak délkeleti és keleti kapcsolatai. (Die südöstlichen und orientalischen Bezihungen der Darstellungen an den ostpannonischen Grabstelen) ArchÉrt CIX 18-50.

BARKÓCZI 1983 — Die südöstlichen und orientalischen Beziehungen der Darstellungen auf den ostpannonischen Grabstelen. MittArchInst XII-XIII 1982-1983 123-151, 333-369.

BARKÓCZI L 1984 — Az intercisai mitológikus és állatjelenetes domborművek időrendje és felhasználása. (Chronologie und Anwendung von Reliefs mit mythologischen Darstellungen und Tierszenen aus Intercisa) ArchÉrt CXI 170-193.

BARKÓCZI L 1986 — A 3. század első feléből származó vésett díszű üvegek Pannoniában.

(Geschliffene Gläser aus der ersten Hälfte des 3. Jhs in Pannonien) ArchÉrt CXIII 166-189.

BARKÓCZI L 1988 Pannonische Glasfunde in Ungarn. Budapest.

BARKÓCZI L 1992 Frührömische glasierte Keramik in Ungarn. In: Glasierte keramik in Pannonien. Székesfehérvár 7-35.

BARKÓCZI L /LENGYEL I 1971 Das Gräberfeld von Keszthely-Fenékpuszta aus dem 6. Jahrhundert und die frühmittelalterlichen Bevölkerungsverhältnisse am Plattensee. Mit einem Beitrag von Imre LENGYEL. JRGZ XVIII 179-199.

BARKÓCZI L/SALAMON Á 1984 Tendenzen der strukturellen und organisatorischen Änderungen pannonischer Siedlugen im 5. Jahrhundert. Aregia XXI 1984 147-187.

BARTA P 1989 Agrippa H. C. mágikus quadrátjai. Egyiptológiai füzetek III. Budapest.

BARTON P F 1975 Frühzeit des Christentums in Österreich und Südostmitteleuropa bis 788. Wien-Köln-Graz.

BAUS K 1962 Von der Urgemeinde zur frühchristlichen Großkirche. [Handbuch der Kirchengeschichte I. hrsg von JEDIN H.] Freiburg, Basel, Wien.

BÉL M 1737 Notitia Hungariae nova historica. Tom. 3. Viennae.

BELLA L 1890 A sopronmegyei múzeum néhány érdekesebb tárgyáról. ArchÉrt X 362-364.

BELLA L 1891 A sopronmegyei múzeum néhány érdekesebb tárgya. A sopronmegyei régészeti Társulat második évkönyve. 45-46.

BELLA L 1895 BELLA Lajos: Római borostyánkő leletek Sopronban. ArchÉrt XV 392-400.

BELLA L 1911 Római leletekről Sopronban. ArchÉrt N.S. XXXI 271-275.

BELLA L 1914 Torzrajzos római párkányos tégla. MKÉ VIII 1914 63-64.

BENYIK 1992 BENYIK György: A páli és a jánosi hagyomány. Szeged.

BERG H 1985 Bischöfe und Bischofssitze im Ostalpen- und Donauraum vom 4. bis zum 8. Jahrhundert. In: Die Bayern und Ihre Nachbarn I. Hrsg. WOLFRAM H/ SCHWARCZ A. Wien 61-108.

BERWIG D 1970 Mark Aurel und die Christen. München.

BEYER H W /KARPP H 1954 Bischof. (B. Christlich) RAC 2 (Stuttgart) 400-407.

BEZECZKY T 1991 A tribunus laticlaviusok háza az aquincumi 2-3. sz-i legiotáborban. Amphorák. (Das Haus der tribuni laticlavii aus dem Legionslager von 2-3. Jh. in Aquincum. Amphoren) BpR XXVIII 133-134.

BIANCHI BANDINELLI R 1971 Rom. Das Ende der Antike. Die römische Kunst in der Zeit von Septimius Severus bis Theodosius I. München.

BILKEI I 1979 Die griechischen Inschriften des römischen Ungarns. Aregia XVII 23-48.

BILKEI I 1990 A Keszthelyi Balatoni Múzeum római kőtára. (Die römische Steinsammlung im Balaton-Museum von Keszthely) Zalai Múzeum II 117-145.

BÍRÓ M 1974 Roman villas in Pannonia. ActaArchHung XXVI 23-57.

BLAISE A 1954 Dictionnaire latin-français des auteurs chrétiens. Revue spécialement pour le vocabulaire théologique par H. CHIRAT. Turnhout.

BLAISE A 1975 Lexicon Latinitatis medii aevi. Corpus Christianorum Continuatio Mediaevalis. Turnhout.

BLAZQUEZ J M 1983 La pintura helenistica de Qusayr 'Amra II. Archivo Español de Arqveologia LVI 169-196.

BOJÁR I 1967-1968 Le pilier de pierre de Székesfehérvár. Aregia VIII-IX 43-53.

BÓNA I 1960 VI. századi germán temető Hegykőn I. (Gräberfeld aus dem VI. Jahrundert in Hegykő I.) SSz XIV 233-241.

BÓNA I 1961 VI. századi germán temető Hegykőn II. (Germanisches Gräberfeld aus dem 6. Jh. in Hegykő II.) SSz XV 131-140.

BÓNA I 1963 VI. századi germán temető Hegykőn III. SSz XVII 136-144.

BÓNIS É 1991 Glasierte Keramik der Spätrömerzeit aus Tokod. ActaArchHung XLIII 87-150.

BONNET CH 1985 Dévelopment urbain et topographie chrétienne de Genčve. CRAI avril-juin, Paris 323-338.

BONNET CH 1986 Les premiers édifices chrétiens d'Augusta Praetoria (Aoste, Italie). CRAI juillet-october, Paris 477-496.

BONNET CH 1986a Geneva in early Christian times. Geneva.

BONNET CH 1987 The Archeological Site of the Cathedral of Saint Peter (Saint-Pierre), Geneva. World Archaeology XVIII 330-340.

BONNET CH 1988 Chronique des découvertes archéologiques dans le canton de Genčve en 1986 et 1987. Geneva XXXVI 37-56.

BONNET CH 1991 Le groupe épiscopal de Genève. Archäologie der Schweiz XIV 221-228.

BONNET/PERINETTI L'età della Cristianizzazione. BONNET Charles: Introduction, PERINETTI R.: Augusta Praetoria. Le necropoli, le sepolture e i riti funerari cristiani. Separatum. – Aosta, progetto per una storia della città. s. a. 97-163.

BOROS I 1899 Régiségek a pápai református főiskola gyűjteményében. ArchÉrt XIX 148-150.

BOŠKOVIĆ et al. 1974 BOŠKOVIĆ D. / DUVAL N. / GROS P. / POPOVIĆ V.: Recherces archéologique à Sirmium. Campagne Franco-Yougoslavie de 1973. MEFRA LXXXVI 597-656.

BOTTE B 1963 Riten und liturgische Familien, Handbuch I. 16-59.

BÖRZSÖNYI A 1904 Római leletekről Győr város területén. ArchÉrt XXIV 251-254.

BRATOŽ R 1994 Der Einfluß Aquileias auf den Alpenraum und das Alpenvorland. (Von den Anfängen bis um 700). In: Christentum 29-61.

BRATOŽ R 1996 Christianisierung des Nordadria- und Westbalkanraumes im 4. Jahrhundert. In: Westillyricum und Nordostitalien in der spätrömischen Zeit. Situla 34.299-366.

BRENDEL O J 1977 Symbolism of the sphere. A contribution to the history of earlier Greek philosophy. [EPRO 67] Leiden.

BRENK B 1978 Zu den Grundrisstypen der frühesten Synagogen Palestins. Atti Vol. I. 539-550.

BRILLANT R 1966 Simboli e Attributi 2. Arte Romana. Enciclopedia dell' arte antica. Roma VII 308-311.

BRUNŠMID J 1908-1909 Kameni spomenici Hrvatskoga narodnoga muzeja u Zagrebu. VHAD N.S. X 44-276.

BRUNŠMID J 1910-1911 Kameni spomenici Hrvatskoga narodnoga muzeja u Zagrebu. VHAD N.S. XI 61-144.

BRUNŠMID J 1979 Colonia Aurelia Cibalae. Vinkovci in alter Zeit. (Kommentare von Stojan DIMITRIJEVIĆ). Corolla memoriae Josepho Brunšmid dicata. Vinkovci 107-129.

BT 1942 Budapest Története I, Budapest az ókorban. 1-2. I/1: TOMPA Ferenc, ALFÖLDI András, NAGY Lajos. I/2: ALFÖLDI András, LÁSZLÓ Gyula, NAGY Lajos, NAGY Tibor. Budapest 1942.

Budapest 1986 Das römische Budapest. Neue Ausgrabungen und Funde in Aquincum. Hrsg. POLENZ H. Münster, Lengerich.

BULLINGER E W 1974 The witness of the stars. Michigan rep. 1974[4].

BUOCZ T 1962 Adatok Savaria topográfiájához. (Angaben zur Topographie von Savaria) ArchÉrt LXXXIX 181-187.

BUOCZ T 1967 Savaria topográfiája. Szombathely.

BUOCZ 1971-1972 Savaria úthálózatának kronológiája a Járdányi Paulovics Romkertben. Savaria V-VI 231-241.

BURGER A 1966 The late Roman cemetery at Ságvár. ActaArchHung XVIII 99-234.

BURGER A 1973 Rómaikori szarkofág a pécsi Aranyhegyen. (Roman Sarcophagus at Aranyhegy in Pécs) ArchÉrt C 42-49.

BURGER A 1979 Das spätrömische Gräberfeld von Somogyszil. Budapest.

BURGER A 1984 Későrómai sírok Aquincumban BpR XXV 65-102.

BURGER A 1985-1986 The Roman villa and mausoleum at Kővágószőlős, near Pécs (Sopianae). Excavations 1977-1982. JPMÉ XXX-XXXI 1985-86 65-228.

BURGER A 1991 Ungarn. CSIR Bd VII: Die Skulpturen des Stadtgebietes von Sopianae und des Gebietes zwischen der Drau und der Limesstrecke Lussonium – Altinum. Budapest.

BUSCHHAUSEN H 1971 Die spätrömischen Metallscrinia und frühchristlichen Reliquiare. I.

Teil, Katalog. Wiener Byzantinistische Studien Band IX. Wien.

CALZA G/BECATTI G 1977 — Ostia. Roma.

CAMBI N 1975 — La figure du Christ sur les monuments paleochretiens de Dalmatie. Disputationes Salonitanae 1970. Split 51-68.

CAMBI N 1978 — Unpublished Excavations and Finds of Early Christian Period in Yugoslavia. Atti II 141-156.

Catalogue 1994 — From the Invincible Sun to the Sun of Justice. Early Christianity in Continental Croatia (Catalogue of Exhibition) Arheološki Muzej, Zagreb 22 September - 4 November 1994. Ed. DEMO Ž.

CBTEL — Cyclopaedia of Biblical, Theological, and Ecclesiastical Literature. By J. M'CLINTOCK / J. STRONG. Vol I-II A-D. New York 1969.

CERMANOVIĆ A 1965 — Die dekorierten Sarkophage in den römischen Jugoslawien. ArchJug VI 89-103.

CETEDOC I. — CETEDOC Library of Christian Latin Texts. Turnhout 1991.

CHRISTENSEN T 1981 — Christus oder Jupiter. Göttingen.

Christentum 1994 — Das Christentum im bairischen Raum von den Anfängen bis ins 11 Jahrhundert. [Passauer Historische-Forschungen 8] hrsg. von BOSHOF E. / WOLFF H. Wien.

CIC — Corpus Iuris Civilis P.3 Cod.Just. Ed. E. OSENBRÜGGEN Lipsiae [15]1875.

CZOBOR B 1879 — Kiadatlan őskeresztény műemlékek a Magyar Nemzeti Múzeumban. ArchÉrt XIII 169-180.

CZOBOR B 1880 — A pécsi őskeresztény cubiculum. EML I 327-339.

CZOBOR 1883 — CZOBOR Béla: Rövid közlemények. EML IV 242-248.

DAREGGI G 1991 — Generi e suiluppo della tipologia del loggiato imperiale nelle raffigurazioni degli edifici circensi. MEFRA CIII 71-89.

DE ROSSI G B 1871 — Epigrafe d'un sacro donario. Bulletino di archaeologia cristiana II 65-70, tav. V-VI.

DEICHMANN F W 1954 — Cella, Cella trichora. RAC 2 (Stuttgart) 942-944-954.

DEICHMANN F W 1964 — Vom Tempel zur Kirche. In: Mullus 52-59.

DEICHMANN F W 1983 — Einführung in die christliche Archäologie. Darmstadt.

DINKLER-VON SCHUBERT E 1980 — Arca und scrinium. JbAC XXIII 141-157.

DOBÓ Á 1940 — Publicum portorium Illyrici. DissPann II, 16. Budapest

DOBÓ Á 1968 — Die Verwaltung der römischen Provinz Pannonien von Augustus bis Diocletianus. Budapest.

DOBSTAL 1914 — Ein Bronzemonogramm Christi aus Emona. RömQuart XXVIII 187-194.

DOMBAY J 1957 — Későrómai temetők Baranyában. (Spätrömische Friedhöfe in Komitat Baranya) JPMÉ 181-330.

DU BOURGUET P 1965 — Die frühchristliche Malerei. [Epochen der Kunst 3]. Gütersloh.

DU CANGE D C 1954 — Glossarium mediae et infimae Latinitatis. Rep. Graz.

DUFRÈN M 1731 — Rudimenta historica. Op. 4. Tyrnaviae.

DUVAL N / CINTAS J 1976 — Le martyrium de Cincari et les martyria triconques et tétraconques en Afrique. MEFRA LXXXVIII 853-856, 857-870, 871-927.

DUVAL N 1978 — Édifices de culte des origines à l'epoque Constantienne. Atti Vol I. 513-537.

DYGGVE E 1935 — Das Mausoleum in Pécs. (Ein christliches Heroon aus Pannonia Inferior). Pannonia 1 62-77.

DYGGVE E 1952 — Über die freistehende Klerusbank (Bema). In: Festschrift für Rudolf EGGER. Beiträge zur älteren europäischen Kulturgeschichte. Bd.I. Klagenfurt 41-52.

EGGER R 1962a — Historisch-epigraphisches Studien in Venezien. RAFC I. 45-69.

EGGER R 1962b — Das Mausoleum von Marusinac und seine Herkunft RAFC I. 181-188.

EGGER R 1962c — Die Begräbnisstätte des Kaisers Konstantin. RAFC I. 1-21.

EGGER R 1963a — Aus dem Leben der donauländischen Wehrbauern. RAFC II. 51-68.

EGGER R 1963b — EGGER Rudolf: Die ecclesia secundae Raetiae. RAFC II. 69-85.

ELBERN V H 1979 — Altar implements and liturgical objects. Age of Spirituality 592-598.

ELIADE M 1979 — Images et symboles. Essais sur le symbisme magico-religieux. Paris

ELIADE M 1983	Geschichte der religiösen Ideen. Vol. III/1. Von Mohammed bis zum Beginn der Neuzeit. Freiburg, Basel, Wien.
ELIADE M 1993	Az örök visszatérés mítosza (Le mythe de l'éternel retour). Budapest.
Emlékkonferencia 1999	Pannoniai kutatások. A Soproni Sándor emlékkonferencia előadásai Bölcske 1998. Wosinsky Mór Múzeum kiadványa. Szekszárd.
EnchF	Enchiridion Fontium Historiae Ecclesiasticae Antiquae, quod in usum scholarum collegit KIRCH Conradus. Editio quinta, aucta et emendata quam curavit UEDING Leo. Friburgi Brisgoviae 1941.
Encyclopedia	Encyclopedia of Archaeological Excavations in the Holy Land. Ed. Michael Avi-Yonah / Ephraim Stern. Oxford, Jerusalem, USA, Canada 1977 Vol. I. 1975 Vol. II. 1976, Vol. III. 1977, Vol IV. 1978.
Erdély 1988	Erdély története. I. A kezdetektől 1606-ig. Szerk. MAKKAI L / MÓCSY A Budapest.
ERDÉLYI G 1974	A római kőfaragás és kőszobrászat Magyarországon. Budapest.
ERDÉLYI G/FÜLEP F 1954	Kőemlék-katalógus. Intercisa I. ArchHung XXXIII. Budapest 233-276.
ERDÉLYI I 1982-1983	A fenékpusztai régészeti kutatások rövid története. Zalai Gyűjtemény XVIII 59-67.
ERDÉLYI I/SÁGI K 1978-1979	Ergebnisse der Ausgrabungen von Keszthely-Fenékpuszta 1976-1677. MittArchInst VIII-IX 151-156.
ERDŐ P 1983	Az ókeresztény kor egyházfegyelme. Budapest.
ESCHBACH H 1978	Pompej. Leipzig 1978.
FARLATUS D 1751	Illyricum sacrum. Liber primus: Ecclesia Salonitana ab eius exordio usque ad saeculum quartum aerae Christianae. Venetiis.
FERJANČIĆ B 1969	Syrmium in the byzantine period. Sirmium 33-58.
FETTICH N 1920-1922	Adalékok Savaria amphitheatrumához. Az Országos Magyar Régészeti Társulat Évkönyve. I 58-64.
FÉVRIER P.-A 1978	Le culte des morts dans les communautés chrétiennes durant le IIIe siècle. Atti Vol I 211-274.
FINÁLY G 1907	Castra ad Herculem. ArchÉrt XXVII 45-57.
FINK R O 1971	Roman Military Records on Papyrus. Ann Arbor, Mich.
FINLEY M I 1979	Atlas der klassischen Archäologie. München.
FISCHER T 1988	Zur römischen Offizierensrüstung im 3. Jahrhundert n. Chr. Bayerische Vorgeschichtsblätter LIII 167-190.
FITZ J 1958	A Fejér megyébe hurcolt római kövek kérdéséhez. Székesfehérvár. (Zur Frage der ins Komitat Fejér verschleppten Steindenkmaeler aus der Römerzeit)
FITZ J 1959	Mikor került a cohors I milliaria Hemesenorum Intercisába? (Wann kam die cohors I milliaria Hemesenorum noch Intercisa?) ArchÉrt LXXXVI 139-144.
FITZ J 1961	Gorsium. NHQ II (4) 147-156.
FITZ J 1962	Gorsium. RE Suppl IX 73-75.
FITZ J 1962a	Gorsium. Das Altertum VIII 155-173.
FITZ J 1962b	Intercisa. RE Suppl. IX 84-104.
FITZ J 1961-1962	Floriana. Jelentés a csákvári római település ásatásáról, 1960. Aregia II-III 153-158.
FITZ J 1967-1968a	Gorsium. Ötödik jelentés a táci rómiakori település feltárásáról, 1965-1966. Aregia VIII-IX 215-231. [et BÁNKI Zs.]
FITZ J 1967-1968b	Les rapport locaux du pilier de pierre de Székesfehérvár. Aregia VIII-IX 55-56.
FITZ J 1970	A concilium provinciae Pannonia Inferiorban. Aregia XI 152-153.
FITZ J 1972	Feltárások. (Ausgrabungen) In: Kutatások Gorsiumban 1972-ben. (Forschungen in Gorsium im Jahre 1972) Aregia XIII 195-203, 229-233. [et BÁNKI Zs.]
FITZ J 1972a	Les Syriens à Intercisa. Latomus 112. Bruxelles.
FITZ J 1972b	Forschungen in Gorsium 1962-1970. Das Altertum XVIII 119-129.
FITZ J 1972c	The excavations in Gorsium. ActaArchHung XXIV 3-52.
FITZ J 1975	Feltárások. (Ausgrabungen) In: Kutatások Gorsiumban 1973-ban. (Forschungen in Gorsium im Jahre 1973) Aregia XIV 289-292, 318-

322. [et LÁNYI V. / BÁNKI Zs.]

FITZ J 1976 Gorsium-Herculia im 4. Jh. ActaAntHung XXIV 383-389.

FITZ J 1976a Gorsium, Later Herculia. The Princeton Encyclopedia of Classical Sites, Princeton 361-362.

FITZ J 1976b Gorsium - Herculia. Székesfehérvár.

FITZ J 1978 Feltárások. (Ausgrabungen) In: Kutatások Gorsiumban 1975-ben. (Forschungen in Gorsium im Jahre 1975) Aregia XVI 169-174, 236-240. [et LÁNYI V. / BÁNKI Zs.]

FITZ J 1979 Ausgrabungen. In: Forschungen in Gorsium im Jahre 1976. Aregia 17 1979 191-196. [et SZÉKELY Gy. V. / BÁNKI Zs.]

FITZ J 1980 A római temetők éremanyaga és a pénzforgalom. (Das Münzmaterial der römischen Gräberfelder und der Münzverkehr) NumKözl 78-79 1979-1980 23-40.

FITZ J 1980a Die Innenbefestigungen der Provinz Valeria. Aregia XVIII 53-60.

FITZ J 1980b Ausgrabungen. In: Forschungen in Gorsium in Jahren 1977-1978. Aregia XVIII 157-166. [et LÁNYI V. / BÁNKI Zs.]

FITZ J 1981 Ausgrabungen. In: Forschungen in Gorsium im Jahre 1979. Aregia XIX 201-211. [et LÁNYI V. / BÁNKI Zs.]

FITZ J 1983 Grabungen. In: Forschungen in Gorsium in Jahre 1980. Aregia XX 201-204. [et LÁNYI V. / BÁNKI Zs.]

FITZ J 1984 Notes. Aregia XXI 279-281.

FITZ J 1985a Ausgrabungen. In: Forschungen in Gorsium in den Jahren 1981/82. Aregia XXII 109-113. [et FÜLÖP Gy. / LÁNYI V. / BÁNKI Zs.]

FITZ J 1985b Concilia provinciae in Pannonien. In: Römishce Geschichte, altertumskunde und Epigraphik Festschrift für A. Betz zur Vollendung seines 80. Lebensjahres. Hrsg. WEBER E / DOBESCH G. Wien 257-265.

FITZ J 1986 Geschichtliche Probleme des Forums von Gorsium. Oikumene V 329-369.

FITZ J 1987 Grabungen. In: Forschungen in Gorsium in den Jahren 1983/82. Aregia XXIII 179-185. [et JUNGBERT B. / LÁNYI V. / BÁNKI Zs.]

FITZ J 1990 Grabungen. In. Forschungen in Gorsium in den Jahren 1985/86. Aregia XXIV 93-97. [et LÁNYI V. / BÁNKI Zs.]

FITZ J 1991-1992 A császári IV-V. századi temető érmei. (Münzfund im Gräberfeld von Csákvár aus dem 4-5. Jh.) NumKözl XC-XCI 61-103.

FITZ J 1993 Templum provinciae in Gorsium? BpR XXX 161-176.

FITZ J 1993a Die Verwaltung Pannoniens in der Römerzeit I. Budapest.

FITZ J 1993b Die Verwaltung Pannoniens in der Römerzeit II. Budapest.

FITZ J 1994 Die Verwaltung Pannoniens in der Römerzeit III. Budapest.

FITZ J 1995 Die Verwaltung Pannoniens in der Römerzeit IV. Budapest.

FITZ 1997 Area sacra des niederpannonischen Kaiserkultes in Gorsium. Aregia 26 1997 61-73.

FONTAINE J 1986 Martin v. Tours. LThK VII 118-119.

FRÖHLICH R 1890 Római föliratok Alsó- és Felső-Pannoniából. ArchÉrt X 227-236.

FRÖHLICH R 1891 Római feliratok Alsó- és Felső Pannoniából ArchÉrt XI 224-238.

FREMERSDORF F 1958 Das römische Buntglas. Köln.

FREMERSDORF F 1967 Die römische Gläser mit Schliff, Bemalung und Goldauflagen aus Köln. Köln.

FÜLEP F 1954 Epigráfia. Intercisa I. ArchHung XXXIII. Budapest 194-229.

FÜLEP F 1959a Neuere Ausgrabungen in der cella trichora von Pécs (Fünfkirchen) ActaArchHung XI 399-417.

FÜLEP F 1959b Újabb ásatások a pécsi cella trichorában JPMÉ 75-90.

FÜLEP F 1961a Adatok a pécsi székesegyház alatt húzódó csatorna kérdéséhez. (Angaben zur Problematik des Kanals unterhalb des Domes zu Pécs-Fünfkirchen) JPMÉ 59-67.

FÜLEP F 1961b Archäologische Forschungen im Jahre 1960 Pécs, Komitatsbibliothek, Geisler Eta u. 8. ArchÉrt LXXXVIII.

FÜLEP F 1962a Újabb kutatások a pécsi későrómai temetőben. (Neuere Forschungen im spätrömischen Gräberfeld von Pécs-Fünfkirchen) ArchÉrt

LXXXIX 23-46.
FÜLEP F 1962b Pécs-Gyárváros, Buzsáki Imre út. RégFüz I 15 Budapest.
FÜLEP F 1963 Pécs rómaikor emlékei. Pécs.
FÜLEP F 1966 New Remarks on the Question of the Jewish Synagogue at Intercisa. ActaArchHung XVIII 93-98.
FÜLEP F 1967a Ókeresztény üvegedények a Magyar Nemzeti Múzeumban. AntTan XIV 239-248.
FÜLEP F 1967b The Stratigraphy of Sopianae as Reflected by Pottery. RCRFA IX 23-28.
FÜLEP F 1968 Early Christian Gold Glasses in the Hungarian National Museum. ActaAntHung XVI 401-412.
FÜLEP F 1969a Scavi archeologici a Sopianae. XVI Corso di cultura sull'arte Ravennate e Bizantina. 151-163.
FÜLEP F 1969b Nuove indicazioni per la storia del cristianesimo in Pannonia. XVI Corso di cultura sull'arte Ravennate e Bizantina. 165-178.
FÜLEP F 1969c Későrómai temető Pécs - Geisler Eta u. 8. sz. alatt. (Early Christian - *sic*- [Late Roman] Cemetery at Pécs, no 8 Geisler Eta Street) ArchÉrt XCVI 3-42.
FÜLEP F 1970 Roman Relics in Pécs (Sopianae). NHQ XI (39) 175-179.
FÜLEP F 1971 Neuere Forschungen in der ausgemalten, frühchristlichen Grabkammer Nr. II von Pécs. JPMÉ XVI 91-103.
FÜLEP F 1973a The Survival of an Early Christian Chapel at Pécs. The NHQ XIV (51) 191-196.
FÜLEP F 1973b Archäologische Forschungen im Jahre 1972. No. 44: Pécs, Jókai u. 10, Rákóczi út 1, Ecke Szalai András u. – Jókai u. ArchÉrt C 265.
FÜLEP F 1974 Neuere Ausgrabungen in der Römerstadt Sopianae (Pécs) RégFüz II 16 Budapest.
FÜLEP F 1977a A pécsi ókeresztény mauzóleum ásatása. (Excavation of the Early Christian Mausoleum of Pécs) ArchÉrt CIV 246-257.
FÜLEP F 1977b Roman Cemeteries on the Territory of Pécs (Sopianae). Fontes ArchHung Budapest.
FÜLEP F 1980 Pannonian Research in Hungary. in: Archaeology 34-56.
FÜLEP F 1984 Sopianae. The History of Pécs during the Roman Era, and the Problem of the Continuity of the Late Roman Population. Budapest.
FÜLEP F 1987 A pécsi későrómai ókeresztény mauzóleum feltárásáról. (The Excavation of the Late Roman - Old Christian Mausoleum in Pécs) JPMÉ XXXII 31-44.
FÜLEP F /BACHMAN Z /PINTÉR A 1988 Sopianae – Pécs ókeresztény emlékei. Budapest.
FÜLEP F /BURGER A 1974 Neuere Ausgrabungen in der Römerstadt Sopianae (Pécs) RégFüz Ser. II. No. 16. Budapest.
FÜLEP F /DUMA GY 1972a Adatok a római kori emlékek továbbéléséhez. AntTan XIX 64-80.
FÜLEP F /DUMA GY 1972b Examinations of the Wall Paintings in the Cella Trichora of Pécs. FolArch XXIII 195-213.
FÜLEP F/ FETTER A 1969–1970 Neuere Forschungen in der ausgemalten frühchristlichen Grabkammer Nr. II. von Pécs. JPMÉ XVI 91-103.
FÜLEP F /TÖRÖK Gy 1977 Ungarn. In: Propyläen Kunstgeschichte. Supplementband I. Spätantike und frühes Christentum. Hrsg. B. BENK. Frankfurt am Main, Berlin, Wien 310-316.
FÜLÖP GY 1978 Forschungsbericht über das römische Strassennetz bei Gorsium. Aregia XVI 281-285.
GABLER D 1964 Az import terra szigilláták forgalma Pannoniában. (Angaben zur Verbreitung der Sigillaten in Pannonien) ArchÉrt XCI 94-110.
GABLER D 1964a Adatok az itáliai barbotinos szigilláták kérdéséhez. (Contributions to the Question of Italian Terra Sigillata Decorated en Barbotine) Arrabona VI 5-17.
GABLER D 1967 Pannoniai terra sigillata-feliratok. AntTan XIV 59-66.
GABLER D 1972 Sigillaten auf dem Gebiet des Palatiums von Gorsium. Aregia XIII 9-67.

GABLER D 1988 — Spätantike Sigillaten in Pannonien. CarJb. 9-40.

GABLER D 1991 — A tribunus laticlaviusok háza az aquincumi 2-3. sz-i legiotáborban. Terra sigillaták. (Das Haus der tribuni laticlavii aus dem Legionslager von 2-3. Jh. in Aquincum. Terra Sigillata) BpR XXVIII 135-140.

GABLER D /KOCZTUR É 1976 — Terra sigillata Depotfund aus Gorsium. Aregia XV 65-88.

GAMBER K 1968 — Domus Ecclesiae. Die ältesten Kirchenbauten Aquilejas sowie im Alpen- und Donaugebiet bis zum Beginn des 5. Jh. Liturgiegeschichtlich untersucht. Regensburg.

GAMBER K 1970 — Churrätische Saalkirchen mit Dreiapsiden-Chor liturgiegeschichtlich untersucht. RömQuart LXV 98-126.

GAMBER K 1971 — Der Ordo Romanus IV, ein Dokument der ravennatischen Liturgie des 8. Jh. RömQaurt 66 1971 154-170.

GAMBER 1972 — GAMBER Klaus: Conversi ad Dominum - Die Hinwendung von Priester und Volk nach Osten bei der Meßfeier im 4. Und 5. Jahrhundert. RömQuart LXVII 49-64.

GAMBER K 1976 — Liturgie und Kirchenbau. Studien zur Geschichte der Meßfeier uénd des Gotteshauses in der Frühzeit. Regensburg 1976.

GAMBER K 1981 — Sancta Sanctorum. Studien zur liturgischen Ausstattung der Kirche, vor allem des Altarraums. Regensburg.

GAMBER K 1982a — Sarmannia. Studien zum Christentum in Bayern und Österreich während der Römerzeit. Regensburg.

GAMBER K 1982b — Die lateinischen liturgischen Quellen Illyriens vom 4. bis zum 6. Jahrhundert. Sirmium IV 77-84.

GAMBER K 1987 — Das Geheimnis der sieben Sterne. Zur Symbolik der Apokalypse. Regensburg.

GAMS B 1893 — Series episcoporum ecclesiae catholicae quotquot innotuerunt a beato Petro apostolico. Ratisbonae.

GARBSCH J 1982 — Terras sigillata. Ein Weltreich im Spiegel seines Luxusgeschirrs. Katalog der Ausstellung. München.

GÁSPÁR D 1971 — Spätrömishce Kästchenbeschläge in Pannonien. ActaAntSzeged XV/1-2

GÁSPÁR D 1986 — Römische Kästchen aus Pannonien. MittArchInst 15/I-II.

GÁSPÁR D 1989 — Lituus vagy baculus vagy…? FolArch XL 103-111.

GÁSPÁR D 1989-1990 — Episcopi Pannoniae. RÖ XVII-XVIII 99-105.

GÁSPÁR D 1993a — When did Chi–rho become a Christian Symbol? Bronces y religion romana. Madrid 207-212.

GÁSPÁR D 1993b — Gondolatok a pannóniai ókereszténységről. MEV V (1-2) 5-22.

GÁSPÁR D 1994 — 1 Kor 13, 12 és a görög filozófia egyetlen szempont szerint. Biblikus konferencia 1994. Szeged 1995 163-172.

GÁSPÁR D 1995 — Christliche Forschungen in Pannonien seit der Tätigkeit von István Járdányi Paulovics. ActaDebrecen XXX 111-120.

GÁSPÁR D 1997 — Brick Inscription, CIL III 11467. XI Congresso Internazionale di Epigrafia Greca e Latina. Preliminary Publication 567-569.

GASSIOT-TALABOT G 1966 — Römische und frühchristliche Malerei. Lausanne.

GÉFIN GY 1992 — A romkert feltárása 1938-1943. (Die Aufdeckung des Ruinengartens) A szöveget gondozta és jegyzetekkel ellátta KISS G. / SOSZTARICS O. / VÍZVÁRI Zs. Szombathely.

GERKE F 1939 — Die Zeitbestimmung der Passionssarkophage. ArchÉrt LII 1-130.

GERKE F 1952 — Die Wandmalereien der neugefundenen Grabkammer in Pécs (Fünfkirchen). Neue Beiträge zur Kunstgeschichte des 1. Jahrtausends. Erster Halbband. Spätantike und Byzanz. Baden-Baden 115-157.

GERKE F 1954 — Die Wandmalereien der Petrus-Paulus-Katakombe in Pécs (Südungarn). Neue Beiträge zur Kunstgeschichte des 1. Jahrtausends. Zweiter Halbband. Frühmittelalterliche Kunst. Baden-Baden 147-199.

GERMAIN S 1969 — Les mosaiques de Timgad. Études descriptive et analytique. Paris.

Germanen 1988 — Germanen, Hunnen und Awaren. Schätze der Völkerwanderungszeit. Die Archäologie des 5. und 6. Jahrhunderts an der mittleren Donau und des örtlich-merowingische Reihengräberkreis. Frankfurt am Main.

GNIRS A 1919 — Zur Frage der christlichen Kultauflagen aus der ersten Hälfte des vierten Jahrhunderts. ÖstJh XIX-XX Beibl 165-206.

GOSZTONYI GY 1939 — A pécsi Szent Péter székesegyház eredete. Pécs.

GOSZTONYI GY 1940a — A pécsi hétkarélyos ókeresztény temető épülete. (Ein altchristliches Gebäude mit 7 Apsiden in Pécs) ArchÉrt I 56-61.

GOSZTONYI GY 1940b — A pécsi hétkaréjos ókeresztény temetői épület építészeti jelentősége. A Magyar Mérnök- és Építészet Egylet Közlönye LXXIV (október 13.) 207-210.

GOSZTONYI GY 1941 — A pécsi ókeresztény temető. Katolikus Szemle, február 41-45.

GOSZTONYI GY 1942 — A pécsi II. sz. ókeresztény festett sírkamra és sírkápolna. (Das bemalte altchristliche Grabkammer und Kapelle No II) ArchÉrt III 196-206.

GOSZTONYI GY 1943 — A pécsi ókeresztény temető. Pécs.

GOSZTONYI GY 1944-1945 — A fenékpusztai I. számú bazilika kérdése. ArchÉrt V-VI 251-265.

GÖMÖRI J 1984 — Scarbantia fazekastelepe és a város melletti római kori téglaégető kemencék. VEAB Értesítő 111-135.

GÖMÖRI J 1985 — Scarbantia foruma. (Il foro di Scarbantia) Sopron.

GÖMÖRI J 1986 — Grabungen auf dem Forum von Scarbantia (1979-1982). ActaArchHung XXXVIII 343-396.

GÖMÖRI J 1992 — Hírek. Iparrégészeti és archeometriai Tájékoztató X 5.

GRABAR A 1946 — Martyrium. Recherches sur le culte des reliques et l'art chrétien antique. I. Vol. Architecture, II. Vol Iconographie. Paris.

Gräberstraßen 1987 — Römische Gräberstraßen. Selbstdarstellung - Status - Standard. Kolloquium in München vom 28. bis 30. Oktober 1985. Hrsg. Henner von HESBERG, Paul ZANKER. München.

GRAF A 1936 — Übersicht der antiken Geographie von Pannonien. DissPann I. 5. Budapest.

GRAMATOPOL M 1982 — Dacia antiqua. Bucureşti.

GRBIĆ M 1936 — Architektura u Basijani (Sremski Petrovci). GID IX 19-31.

GRBIĆ M 1937 — Architektura u Basijani kod Donjich Petrovaca u Sremu. GID X 1-7.

GROENEWEGEN H A /ASHMOLE B — Art of the Ancient World. Painting, Pottery, Sculpture, Architecture from Egypt, Mesopotamia, Crete, Geece, and Rome. Englewood Cliffs, New York.

GROLLER M von 1911 — Archaeologiai ásatások Sopronmegyében. ArchÉrt N.S. XXXI 64-74.

GUARDUCCI M 1978 — Dal gioco letterale alla crittografia mistica. ANRW II/16 1736-1773.

GUYON J 1974 — La vente des tombes à travers d'épigraphie de la Rome chrétienne (III-VIIe siècle): le role des fossores, mansionarii, praepositi et prêtre. MEFRA LXXXVI 549-596.

GUYON J 1975 — Un étui-reliquaire retrouvé dans les fouilles de Sirmium. MEFRA LXXXVII 419-424.

GUZSIK T 1988 — Szakrális építészeti terek funkcióelemzése. I. (Keleti liturgiák és építészetük) Budapest.

GUZSIK T 1994 — Középkori építészettörténeti ábraanyag. I. Róma és a keresztény Kelet szakrális építészete. Budapest.

GYÁRFÁS I T 1889 — Pannónia őskeresztény emlékei. Régészeti tanulmány. Budapest.

HAJNÓCZI GY 1985-1986 — The Conceptual and Actual Reconstruction of the Villa and the Burial Vault at Kővágószőlős JPMÉ XXX-XXXI 229-236.

HAJNÓCZI GY 1987 — Pannonia római romjai. (Römische Ruinen in Pannonien) Budapest.

HAMMERSCHMIDT E 1986 — Antiochia am Orontes. LThK 1 648-650.

HAMPEL J 1880 — Jelentések, gyűjtemények és társulatok. ArchÉrt XIV 165-174.

HAMPEL J 1886 — A bronzkor emlékei Magyarországon. I-III. Budapest, 1. Képesatlasz 1886, 2. Leletek statisztikája 1892, 3. Áttekintő értékelés 1896. Alterthümer der Bronzezeit in Ungarn I-III. Budapest 1887.

HAMPEL J 1894 — Keresztény emlékek a régibb középkorból. ArchÉrt XIV 23–53.

HAMPEL J 1903a — Lovas istenségek dunavidéki antik emlékeken. ArchÉrt XXIII 305-365.

HAMPEL J 1903b — A M. N. Múzeumi régiségosztály gyarapodása. ArchÉrt XXIII 425-447.

HAMPEL J 1906 — Intercisa emlékei. ArchÉrt XXVI 221-274.

HAMPEL J 1907 — A pannoniai síremlékek áttekintő osztályozása. ArchÉrt XXVII 285-341.

HAUCK A [8]1954	Kirchengeschichte Deutschlands. Berlin, Leipzig. Bds I-V.
HAUGH B 1902	A szekszárdi múzeum újabb szerzeményei. ArchÉrt XXII 364-370.
HAYES J W 1972	Late Roman pottery. London.
HEKLER A 1910	Római ezüst övdíszek. ArchÉrt XXX 242-249.
HENSZLMANN I 1869	Pécsnek középkori régiségei. Monumenta Hungariae Archaeologica. I/1. Pest.
HENSZLMANN I 1873a	Pécsnek régiségei. Monumenta Hungariae Archaeologica II/1. Pest.
HENSZLMANN I 1873b	Die altchristliche Grabkammer in Fünfkirchen. Mitteilungen der k. k. Central–Comission zur Erforschung und Erhaltung der Baudenkmale, Wien XVIII 57-83.
HENSZLMANN I 1876	Magyarország ókeresztény, román és átmeneti stylű műemlékeinek rövid ismertetése. Budapest.
HESBERG H von /ZANKER P 1987	Einleitung. Gräberstraßen 9-20.
HESBERG H von 1987	Planung und Ausgestaltung der Nekropolen Roms im 2. Jh. n. Chr. Gräberstraßen 43-60.
HEUSSI K 1958	Kompendium der Kirchengeschichte. Berlin.
HOFFMANN D 1969	Das spätrömische Bewegungsheer und die Notitia Dignitatum. Epigraphische Studien Bd 7/I-II. Düsseldorf.
HOFMANN H 1978	Satorquadrat. RE Suppl. XV 479-565.
HORVÁTH A 1997	A pannoniai ókeresztény sírfeliratok néhány tanulsága. Századok CXXXI 105-118.
HORVÁTH E 1985-1986	Wooden Remains from the Roman Villa at Kővágószőlős. JPMÉ XXX-XXXI 237-241.
HÖRING E 1950	Die geographischen Namen des antiken Pannonien. Inaugural Dissertation. Heidelberg.
H.-TÓTH E(lvira) 1969	Kora-bizánci üvegpohár egy szabadszállási magányos sírból. (Early Byzantine Glasscup in a solitary Grave at Szabadszállás) Kecskemét.
H.-TÓTH E(lvira) 1971	Ein spätantiker Glasbecherfund aus Szabadszállás. ActaArchHung. XXIII 115-138.
H.-TÓTH E(lvira) 1990	Négy évtized régészeti kutatásai Bács-Kiskun megyében (1949-1989) Cumania XII 81-233.
HÜBNER W 1983	Zodiacus Christianus. Jüdishc-christliche Adaptationen des Tierkreises von der Antike bis zur Gegenwart. Königstein/Ts.
Itinerarium	Pannonia Hungarica Antiqua [Itinerarium Hungaricum I] Edd. HAJNÓCZI Gy / MEZŐS T / NAGY M / VISY Zs. Budapest 1995.
IVÁNYI D 1935	Die pannonischen Lampen. Eine Typologische Übersicht. Budapest. DissPann. Ser. 2. No. 2.
JARAK M 1994	The History of Early Christian Communities in Continental Croatia. Catalogue 17-39, 155-179.
JARAK M 1996	*Martyres Pannoniae* – The Chronological Position of the Pannonian Martyrs in the Course of Diocletian's Persecution'. In: Westillyricum und Nordostitalien in der spätrömischen Zeit. Situla 34. 263-289.
JARAK M 1998	"Pannonian" Benedictine Possessions in the 6th Century. Atti II 779-786.
JUNGMANN J A 1948	Missarum Sollemnia. I-II. Wien.
JUNGMANN J A 1977	A szentmise. Eisenstadt.
KABA M 1964	Az aquincumi cella trichora újabb helyreállítása. (Das erneute Wiederherstellung der cella trichora von Aquincum) BpR XXI 337-343.
KABA M 1991	Thermae maiores legionis II adiutricis. Budapest.
KÁDÁR Z 1938-1939	Pannonia ókeresztény emlékeinek ikonográfiája. Regnum 3-58.
KÁDÁR Z 1939	Pannonia ókeresztény emlékeinek ikonográfiája. Budapest.
KÁDÁR Z 1940-1941	A triumphus-eszme a pécsi ókeresztény héroon egyik freskóján. Regnum 65-69.
KÁDÁR Z 1942	Pannónia keresztény emlékeinek kapcsolatai az ókor és a középkor triumfális művészetéhez. Emlékkönyv Gerevich Tibor születésének hatvanadik fordulójára. Budapest 1-15.
KÁDÁR Z 1962	Bemerkungen über byzantinische Amulette und magische Formeln.

ActaAntHung X 403-411.

KÁDÁR Z 1965 — Ricerche sull'arte e sull'archeologia cristiano-romana in Ungheria dal 1954 al 1962. Atti del VI congresso internazionale di archeologia cristiana, Ravenna 23-30 settembre 1962. Roma 413-421.

KÁDÁR Z 1968a — A ságvári későrómai szinkretisztikus ábrázolású ládikaveret. (La rapprezentatione sincretistiche dello serigno tardoromano di Ságvár) ArchÉrt XCV 90-92.

KÁDÁR 1968b — Ikonographische und religionsgeschichtliche Bemerkungen zu den Zierscheiben von Rakamaz. FolArch XIX 105-112.

KÁDÁR Z 1969 — Lineamenti dell'arte della Pannonia nell'epoca dell' antichità tarda e paleocristiana. Corsi di cultura sull'arte ravennate e bizantina. Ravenna XVI 179-201.

KÁDÁR Z 1972 — Die kleinasiatisch–syrischen Kulte zur Römerzeit in Ungarn. EPRO 2. Leiden.

KÁDÁR Z /HORVÁTH I A /GÉFIN Gy 1981 — Szombathely. Budapest.

KÁDÁR Z /NÉMETH GY /TOMPOS E 1987 — A Hagia Szophia. Budapest.

KAHL H-D 1980 — Zwischen Aquilejä und Salzburg. Beobachtungen und Thesen zur Frage romanischen Restchristentums im nachvölkerwanderungszeitlichen Binnen-Noricum (7.-8. Jh.). Die Völker an der mittleren und unteren Donau im 5. und 6. Jahrhundert. Hrsg. WOLFRAM H. / DAIM F. DenkschrÖsterrAkdWiss CXLV 33-81.

KAISER A /PÓCZY K 1985 — Budapest római öröksége. Budapest

KANAEL E 1961 — Die Kunst der antiken Synagoge. München, Frankfurt am Main.

KANTOROWITZ E H 1960 — On the Golden Marriage Belt and the Marriage Rings of the Dumbarton Oaks Collection. DOP XIV 1-16.

KARP H 1954 — Bischof, christlich. RAC 2 Stuttgart 400-407.

KAŠIĆ S 1975 — Rano hriščanstvo na tlu Jugoslavije. Magistarski rad.

Katalog 1998 — MIGOTTI B/ ŠLAUS M / DUKAT Z/ PERINIĆ L, Accede ad certissiam. The archeological site of Štrbinci in the Roman and early Christian periods. Zagreb.

KATANCSICH M P 1826 — Istri adcolorum geographia vetus e monumentis epigraphicis, marmoribus, nummis, tabellis eruta et commentariis illustrata. Vol. 1-2. Budae.

KAUFMANN C M 1922 — Handbuch der christlichen Archäologie. Paderborn.

KÉRDŐ K 1998 — Die Erforschungen des claudischen Auxiliarkastells und seiner Umgebung in Budapest- Víziváros als typisches Beispiel der stadtarchäologischen Forschung. AqNo. II 246-263.

KERÉNYI A 1941 — Dáciai személynevek. DissPann. Budapest.

KIRÁLY I 1979 — Árpádházi Szent Margit és a Sziget. Budapest.

KISS Á 1973 — Roman mosaics in Hungary. Budapest.

KISS Á 1987 — Pannonische Architekturelemente und Ornamentik in Ungarn. Budapest.

KISS G / TÓTH E 1993 — A szombathelyi Szent Márton templom régészeti kutatása 1984-1992 (Archäologische Untersuchung der St.-Martinskirche in Szombathely 1984-1992) CommArchHung 175-199.

KITZINGER E 1960 — A Marble Relief of the Theodosian Period. DOP XIV 17-42.

KISS ZS 1973 — Les ampoules de St. Ménas découvertes à Kom el-Dikka (Alexandrie) en 1969. Études et Travaux VII 137-154.

KLEINBAUER W E 1984 — Remark on the Building Historiy of the Acheiroporetos Church at Thessaloniky. Atti II 241-257.

KNIFIC T/SAGADIN M 1991 — Arheologija o prvih stoletjih krščanstva na Slovenskem. (The Archaeology of the First Centuries of Christianity in Slovenia.) Carta sine litteris. Ljubljana.

KOCH G/ SICHTERMANN H 1982 — Römische Sarkophage. München.

KOCZTUR É 1972 — Ausgrabungen im südlichen Stadtviertel von Gorsium (Tác-

Margittelep). Aregia XIII 69-148.

KOCSIS L 1989 Inschriften aus dem Mithra-Heiligtum des Hauses des Tribunus Laticlavius im Legionslager von Aquincum aus dem 2-3. Jahrhundert. ActaArchHung XLI 81-92.

KOCSIS L 1990 Zur Periodisierung des Hauses des tribunus laticlavius im Legionslager von Aquincum. RLiÖ XXXVI 709-713.

KOCSIS 1991a A glass bowl with engraved ornament from the Mithraeum in the legionary fort at Aquincum (Budapest). Journal of Glass Studies XXXIII 29-31.

KOCSIS L 1991b A tribunus laticlaviusok háza az aquincumi 2-3. sz-i legiotáborban. (Das Haus der tribuni laticlavii aus dem Legionslager von 2-3. Jh. in Aquincum.) BpR XXVIII 117-121, 141-142.

KOLLER J 1804 Prolegomena in historiam episcopatus Quinqueecclesiarum. Posonii.

KOLNÍK T 1984 Römische und germanische Kunst in der Slowakei. Bratislava.

KOROŠEC P 1980 Zwei altchristliche Kandelaber aus Rogoznica in Ptuj. AV XXXI 55-61.

KOVÁTS V 1983 The Marcus Aurelius Bronze Head of Dunaszekcső. NHQ XXIV (89) 181-184.

KOVÁTS V 1984 KOVÁTS V.: Mark Aurel Porträt aus Lugio. Aregia XXI 89-92.

KOVRIG I 1937 A császárkori fibulák főformái Pannoniában. (Die Haupttypen der kaiserzeitlichen Fibeln in Pannonien) Budapest DissPann Ser. II. No. 4.

KRAELING C H 1956 The Synagogue. [The Excavations at Dura–Europos. Final Report VIII/1. Ed. BELLINGER A R /BROWN F E /PERKINS A /WELLES C B] New Haven, London, Oxford.

KRAELING C H 1967 The Christian Building. With a contribution by C. BRADFORD WALLES. [The Excavations at Dura–Europos. Final Report VIII/2. Ed. BELLINGER A R /BROWN F E /PERKINS A /WELLES C B] New Haven, New York.

KRAUSZ S 1904 Jelvényes ókori zsidó sírkövek. ArchÉrt XXIV 170-172.

KUBINYI Á 1857a Szekszárdi régiségek. Pest.

KUBINYI Á 1857b Szekszarder Alterthümer. Pest

KUZSINSZKY B 1895 Pannonia és Dacia. A magyar nemzet története, szerk. SZILÁGYI S I. Budapest LVII-CCLII.

KUZSINSZKY B 1908 Magyarországon talált római kőemlékek a vidéki múzeumokban s egyéb helyeken. MKÉ XI 74-112.

KUZSINSZKY B 1920 A Balaton környékének archaeologiája. Budapest.

KUZSINSZKY B 1934 Aquincum, Ausgrabungen und Funde. Führer mit einer topographischen und geschichtlichen Einleitung. Budapest 1934.

KUZSINSZKY 1937 KUZSINSZKY Bálint: Az aquincumi múzeum római kőemlékeinek ötödik sorozata. (Die römischen Steindenkmäler des Museums von Aquincum) BpR XII 61-152, 302-307.

LAKATOS P 1993 Fontosabb bátai emlékek az ókorból. In: Báta évszázadai. Emlékkönyv a bátai apátság alapításának 900 éves évfordulójára. Szerk. KÁPOLNÁS M. Báta. 7-10.

LANGLOTZ E / DEICHMANN F W 1950 LANGLOTZ E. / DEICHMANN F. W.: Basilika. RAC I 1225-1259.

LASSUS J 1968 Frühchristliche und byzantinische Welt. Gütersloh.

LECLERCQ H 1907 Manuel d'archéologie chrétienne depuis les origines jusqu'au VIIIe siècle. Tome I. Paris.

LIETZMANN H 1958 Kleine Schriften I. Hrsg. ALAND K Berlin.

LimesAalen 1986 13. Internationaler Limeskongreß Aalen 1983. Forschungen und Berichte zur Vor- und Frühgeschichte in Baden–Württemberg 20. Stuttgart.

LINIGER H 1976 Einige Lösungen der Schalenstein- und Felsbildprobleme (Schluss). Basler Beiträge zu den Felsbildproblemen. Heft X 1-32.

LJUBIĆ S 1876 Inscriptiones quae Zagrabiae in Museo Nationali. Zagrabiae.

LONOVICS J 1865 Népszerű egyházi archaeologia 3. Bécs.

LORTZ J 1962 Geschichte der Kirche in ideengeschichtlicher Betrachtung. Bd I.

	Altertum und Mittelalter. Münster.
LŐRINCZ B 1978	Pannonische Stempelziegel II. Limes-Strecke Vetus Salina - Intercisa. DissArch II/7. Budapest 1978.
LŐRINCZ B 1980	Pannonische Ziegelstempel III. Limes-Strecke Ad Flexum – Ad Mures. DissArch Ser II. No. 9. Budapest.
LŐVEI P 1998	Építészet a Kárpát-medencében a honfoglalás előtt. In: Magyarország építészetének története. Ed. SISA J / WIEBENSON D. Budapest 13-19.
MADARASSY O 1991a	Die bemalte Kultwand im Mithräum des Legionslagers von Aquincum. KölnJb XXIV 207-211.
MADARASSY O 1991b	A tribunus laticlaviusok háza az aquincumi 2-3. sz-i legiotáborban. A Mithraeum falfestménye. (Das Haus der tribuni laticlavii aus dem Legionslager von 2-3. Jh. in Aquincum. Wandgemälde des Mithrasheiligtums) BpR XXVIII 123-126.
MAHLER E 1905	Újabb római feliratos emlékek Dunapenteléről. ArchÉrt XXV 222-234.
MANO-ZISI Đ 1978	The problems of the old Christian mosaics in Yugoslavia. Materijali (Bitola) XVIII 16-18. [Ranokrišćanski mozaici u Jugoslaviji]
MARASOVIĆ 1984	Regionalizam u ranosrednjovjekovnoj arhitekturi Dalmacije. Starohrvatska prosvjeta, Split XIV 135-157.
MARIN E 1988	Starokršćanska Salona. Zagreb.
MÁRITY E 1993	Kronológiai problémák és városszerkezeti sajátságok Aquincum polgárvárosában. BpR XXX 137-148.
MAROSI A 1934	A Tác-fövenypusztai ásatás SzSz IV 53-71.
MAROSI A 1936	Fejérmegyei őskeresztény emlékek és a múzeum Krisztus–monogram köve. SzSz 1936 105-108.
MARSHALL F H 1968	Catalogue of the finger rings. London, British Museum. Oxford.
MARSIGLI L 1726	Danubius Pannonico-Mysicus observationibus geographicis, astronomicis, hydrographicis, historicis, physicis perlustratus et in sex tomos digestus. Hagae Amsterdam.
MARTIMORT A-G 1963	Die Grundelemente der Liturgie, Aufbau und Gesetze der liturgischen Feier. Handbuch I 61-200.
MARUCCHI O 1912	Handbuch der christlichen Archäologie. Einsiedeln, Waldshut, Cöln.
MARUŠIĆ B 1978	La Christianisme et le Paganisme sur le sol de l'Istrie aux IVe et V[e] siecles. AV XXIX 549-572.
MEDGYES M 1977	Savaria településszerkezete és városfala az újabb kutatások tükrében. Lapok Szombathely történetéböl. 52 szám, Panniculus ser. C. no. 70.
MEDGYES M 1998	The Settlement Structure and Town Walls of Savaria as reflected in the latest excavations. AqNo II 87-91.
MERRIFIELD R 1987	The Archaeology of Ritual and Magic. London.
MESZLÉNYI A 1977	Szent Márton püspök életútja (317-397). A 200 éves szombathelyi egyházmegye emlékkönyve 1777-1977. Szombathely 95-126.
METZGER CH 1975	Trois lustres cultuels en mètal. Revue du Louvre et des Musées de France 319-322.
MIGOTTI B 1990	Ranokršćanska topografija na Području između Krke i Centine. Zagreb.
MIGOTTI B 1991	Dekorativna ranokršćanska plastika Jaderskog i Salonitanskog područja. Temeljne osobine i međusobne razlike (Decorative Stone Sculpture of the Jader and Salona Regions: Basic Characteristics and Individual differences) Diadora XIII 291-312.
MIGOTTI B 1991-92	Dalmacija na razmeđi istoka i zapada u svjetlu međusobnog odnosa jaderske i salonitanske ranokršćanske crkve (Dalmatia on the Watershed between East and West in the Light of the Relationship between the Jadertine and the Salonitan Early Christian Church). VAMZ XXIV-XXV163-182.
MIGOTTI B 1992	Zusatz zut Datierung der außerstädtischen frühchristlichen Architektur des breiteren salonitanischen Bereiches. AV XLIII 111-133.
MIGOTTI B 1992a	Ranokršćanska biskupija Scradona-Skradin (The Early Christian Episcopal See of Scardona-Skardin). In: Prilozi Institua za Arheologiju

u Zagrebu vol. 9 101-112.

MIGOTTI B 1992b — Antičko-srednjovjekovni sakralni kontinuitet na području Dalmacije (Kontinuiät der Widmung der Sakralbauten von der Antike bis zum Mittelalter in Dalmatien) Opuscula Archaeologica XVI 225-249.

MIGOTTI B 1994 — Arheološka građa iz ranokrščanskog razdobilja u kontinentalnoj Hrvatskoj - The Archaeological Material of the Early Christian Period in Continental Croatia. Catalogue 41-67, 187-209.

MIGOTTI B 1997 — Evidence for Christianity in Roman southern Pannonia (Northern Croatia). A Catalogue of Finds and Sites. BAR International Series 684.

MIKL-CURK I 1978 — Poetovio in der Spätantike. AV XXIX 405-411.

MILOŠEVIĆ P 1971 — Earlier Archaeological Activity in Sirmium. Sirmium II 3-12.

MILOŠEVIĆ P / PRICA R 1979 — Kroz vekove Sirmiuma. Sremska Mitrovica.

MIRABELLA ROBERTI M 1953 — Considerazioni sulle aule teotodoriane di Aquileia. 209–244. Studi Aquileiesi offerti il 7 ottobre 1953 a Giovanni Brusin nel suo 70. compleanno. Aquileia.

MIRKOVIĆ L 1956 — La nécropole paléochrétienne de Niš. ArchJug II 85-100.

MIRKOVIĆ M 1971 — Sirmium - its history from the I century A.D. to 582 A.D. Sirmium I 5-60.

MÓCSY A 1955 — MÓCSY András: A Százhalombatta-dunafüredi római tábor és település. (Roman Camp and Settlement at Százhalombatta) ArchÉrt LXXXII 59-69.

MÓCSY A 1956 — Die Entwicklung der Sklavenwirtschaft in Pannonien zur Zeit des Prinzipates. ActaAntHung IV 221-250.

MÓCSY A 1962 — Pannonia. RE IX Suppl. 515-776.

MÓCSY A 1968 — Bacensis civitas. RE XI. Suppl. 346.

MÓCSY A 1970 — Murocincta. Adriatica praehistorica et antiqua. Zagreb 583-586.

MÓCSY A 1971 — Zur frühesten Besatzungsperiode in Pannonien. ActaArchHung XXIII 41-46.

MÓCSY A 1973 — Pannonia-Forschung 1976-1972. ActaArchHung XXV 375-403.

MÓCSY A 1974 — Pannonia and Upper Moesia. A History of the Middle Danube Provinces of the Roman Empire. London, Boston.

MÓCSY A 1977 — Pannonia-Forschung 1973-1976. ActaArchHung XXIX 373-401.

MÓCSY A et al. 1983 — Nomenclatur provinciarum Europae Latinarum et Galliae Cisalpinae cum indice inverso

MOELLER W O 1973 — The Mithraic origin and meanings of the Rotas-Sator square. Leiden.

MRÁV ZS 1996 — Castellum contra Tautantum. Egy későrómai erőd azonosításának problémájához. Savaria 22/3 1992-1995 (1996) 11-19.

MRT — Magyar Régészeti Topográfia
1. BAKAY K / KALICZ N / SÁGI K: Veszprém megye, a keszthelyi és tapolcai járás. Szerk. SÁGI K. Budapest 1966.
2. ÉRI I. / KELEMEN M. / NÉMETH P. / TORMA I.: Veszprém megye, a veszprémi járás. Szerk. ÉRI I. Budapest 1969.
3. BAKAY K. / KALICZ N. / SÁGI K: Veszprém megye, a devecseri és sümegi járás. Szerk. BAKAY K. Budapest 1970.
4. DAX M. / ÉRI I. / MITHAY Sándor / PALÁGYI S. / TORMA I.: Veszprém megye, a pápai és zirci járás. Szerk. TORMA I. Budapest 1972.
5. HORVÁTH I. / H KELEMEN M. / TORMA I.: Komárom megye, az esztergomi és dorogi járás. Szerk. TORMA I. Budapest 1979.
7. DINNYÉS I. / KŐVÁRI K. / LOVAG ZS./ TETTAMANTI S. / TOPÁL J. / TORMA I.: Pest megye régészeti topográfiája, A budai és szentendrei járás. (XIII/l). Budapest 1986.

Mullus — Festschrift Theodor Klauser. Jahrbuch für Antike und Christentum Ergänzungsband 1 1964 (Münster Westfalen)

MÜLLER R 1978 — V. századi bronzműves műhely maradványai Keszthely – Fenékpusztáról. (Reste einer Bronzewerkstätte aus dem 5. Jh. im Keszthely–Fenékpuszta) ArchÉrt CV 11-29.

MÜLLER R 1979 — A Keszthely – fenékpusztai erőd északi kapujának feltárása 1971-ben.

(Die archäologische Erschliessung des Nord-Tores der Befestigung von Keszthely-Fenékpuszta im Jahr 1971) VMMK XIV 123-156.

MÜLLER R 1987 — Megjegyzések Fenékpuszta történetéhez. (Bemerkungen zur Geschichte von Fenékpuszta) ZM I 105-122.

MÜLLER R 1988 — Die spätrömische Festung Valcum am Plattensee. In: Germanen 270-273.

NÁDORFI G 1992 — Glasierte Keramik in den spätrömischen Gräberfeldern Pannoniens. In: Glasierte Keramik in Pannonien. Ausstellung 1992. Székesfehérvár. 45-51.

NAGY L 1926 — Die römisch-pannonische dekorative Malerei. RömMitt XLI 79-131.

NAGY L 1927 — A pannoniai-római díszítő rendeltetésű stukkó párkányok. ArchÉrt XLI 114-132.

NAGY L 1929 — A dunaszekcsői keresztény-római pohár. Theologiai Szemle V 158-163.

NAGY L 1930a — A brigetiói vas diatretum. ArchÉrt XLIV 111-123.

NAGY L 1930b — A tatai ókeresztény aranygyűrű. ArchÉrt XLIV 243-246.

NAGY L 1931a — Keresztény római emlékek Magyarország területéről. ArchÉrt XLV 29-42.

NAGY L 1931b — Az óbudai cella trichora a Raktár utcában. Budapest.

NAGY L 1934 — I ricordi cristiano-romani trovati recentemente in Ungheria. Atti del III. Congresso Internazionale di Archeologia Cristiana. Ravenna, 25-30. settembre 1932. Roma 293-310.

NAGY L 1937a — Ácsszerszámokból álló tömeglelet az aquincumi múzeumban. BpR XII 153-178.

NAGY L 1937b — Aquincumi vonatkozású kiadatlan feliratos kőemlékek Szentendréről. ArchÉrt L 85-115.

NAGY L 1938 — Pannonia sacra. SZI I. Budapest 29-148.

NAGY L 1939 — A szír és kisázsiai vontakozású emlékek a Duna középfolyása mentében. ArchÉrt LII 115-147.

NAGY L 1940 — Az aquincumi ókereszténység újabb emlékei. ArchÉrt III 246-256.

NAGY L 1950 — Üvegserleg hálós díszítés utánzatával a budakalászi őrtoronyból. BpR XV 535-540.

NAGY M(argit) 1998 — Ornamenta Avarica I. Az avar kori ornamentika geometrikus elemei. A Móra Ferenc Múzeum Évkönyve IV 377-459.

NAGY Mi(hály) 1988 — Über einen Monopodiumtyp. Versuch zur Umwertung der sog. pannonischen frühchristlichen Reliquienaltäre. FolArch XXXIX 135-158.

NAGY T 1936 — Az eusebianus reakció és a pannóniai ariánizmus kezdetei. Regnum 32-60.

NAGY T 1937 — Orthodoxia és eretnekség a pannoniai őskereszténységben. Történetírás I 111-142.

NAGY T 1939 — Die Geschichte des Christentums in Pannonien bis zu dem Zusammenbruch des römischen Grenzschutzes. Budapest DissPann Ser.II. No.12.

NAGY T 1939a — NAGY Tibor: Dissertationes Pannonicae ex Instituto Numismatico et Archaeologico Universitatis de Petro Pázmány Nominatae Budapestinensis provenientes. Századok LXXXIII 196-207.

NAGY T 1941 — A pannóniai ókeresztény egyházközségi bazilikák. Szépművészet II 1940-1941 144-148.

NAGY T 1943 — A Fővárosi Régészeti és Ásatási Intézet jelentése az 1938-1942. évek között végzett kutatásairól. BpR XIII 361-399.

NAGY T 1944-45 — Az aquincumi kereszténység egy eddig félreismert emléke. [Tanulmányok Pannonia kereszténységének történetéhez II]. (Un monument méconnu jusqu'ŕ présent du Christianisme d'Aquincum) ArchÉrt 266-273.

NAGY T 1945 — Die Lehre des Photeinos in den theologischen Streitigkeiten am Anfang des V. Jahrhunderts. (Tanulmányok Pannonia kereszténységének történetéhez I.) Oriens Antiquus I.

NAGY T 1946 — A Quirinus-reliquiák translatiójának időpontja Pannonia prima késői római sorsának tükrében. (Tanulmányok a pannoniai egyháztörténethez IV.) Regnum VI 1944-46 244-257.

NAGY T 1949 — Ókeresztény gyűrű Magyaregregyről. [Tanulmányok Pannónia kereszténységéhez III.] (Une bague paléochrétienne de Magyaregregy. Études sur le christianisme de la Pannonie III) ArchÉrt LXXVI 80-82.

NAGY T 1955 — Megjegyzések Intercisa történetének új feldolgozásához. ArchÉrt LXXXII 232-244.

NAGY T 1958 — Das Mithras-relief von Paks. ActaAntHung VI 407-431.

NAGY T 1961a — Nuove scoperta dell'antica cultura in Pannonia. Atti del settimo congresso internazionale di Archeologia Classica. Roma Vol. III 135-142.

NAGY T 1961b — Das pannonischen Christentum und die barbarischen Völkerschaften im IV. Jahrhundert. In: Bericht über den V. internationalen Kongress für Vor- und Frühgeschichte, Hamburg 1958. Berlin 591-596.

NAGY T 1964 — Perióduskutatások az aquincumi polgárváros területén. (Erforschung der Perioden im Zentral-Gebiete der Zivilstadt von Aquincum) BpR XXI 9-54.

NAGY T 1965 — Quelques aspects de la romanisation dans la Pannonie Orientale. Congrès international d'archéologie classique (Paris 1960) Paris 375-381 pl. 86-88.

NAGY T 1966 — R. Noll, Eugippius. Das Leben des heiligen Severin. Berlin 1963. Rec. ArchÉrt XCIII 311-313.

NAGY T 1967 — The frontier of Pannonia as reflected by recent research. Separatum: Roman Frontier Studies 145-150.

NAGY T 1971 — Kőfaragás és szobrászat Aquincumban. BpR XXII 103-160.

NAGY T 1976 — Aquincum. Stadt und Lager im 4. Jh. ActaAntHung XXIV 369-382.

NAGY T 1978 — Die bekannten Ehrennamen der syrischen Kohorte von Ulcisia Castra. Studia in honorem Veselini Beševljev. Sofia 208-216.

NAGY T 1983 — Die Gens Marcomannorum in Pannonia Prima. MittArchInst XII-XIII 1982-83 113-121.

NÉMETH M 1971 — Vezető az Aquincumi Múzeum kőtárában. Budapest.

NÉMETH M 1990 — Forschungen im Alenkastell von Aquincum. RLiÖ XXXVI 1990 675-681.

NÉMETH M 1991a — Stand der Forschungen in den Militärlagern von Aquincum. CarnuntumJb 81-87.

NÉMETH M 1991b — Aquincum. Feltárások az 1981-1988 közötti időszakban. (Aquincum. Erforschungen in den Jahren 1981-1988) BpR XXVIII 91-105.

NÉMETH M/KÉRDŐ K 1986 — Zur Frage der Besatzung von Aquincum im 1. Jahrhundert. LimesAalen 384-388.

NÉMETHY L 1886 — Voltak-e keresztények Aquincumban? Család és Iskola III (16) 241-243.

NIKOLAJEVIĆ I 1978 — Les ensevelissements dans les églises paléochrétiennes sur le territoire de la Serbie. AV XXIX 678-693.

NOETHLICHS K L 1983 — Die historische Entwicklung vom 1-5 Jh. n. Chr. In: Spätantike 1-8.

NOLL R 1954 — Frühes Christentum in Österreich von den Anfangen bis um 600 nach Chr. Wien.

NOLL R 1968 — NOLL Rudolf: Zwei unscheinbare Kleinfunde aus Emona. AV XIX 79-88.

NOLL R 1974-1975 — Neuere Funde und Forschungen zum frühen Christentum in Österreich 1954-1974. Mitteilungen der österreichischen Arbeitsgemeinschaft für Ur- und Frühgeschichte Bd. 25 Wien 195-216.

NOLL R 1981 — Literatur zur Vita sancti Severini aus den Jahren 1975-1980. Anzeiger der phil.-hist. Klasse der Österreichischen Akademie der Wissenschafaten 118 So. 12. 196-221.

NOVÁKI GY 1964 — Adatok a soproni belváros rétegeinek kronológiájához. SSz XVIII 236-243.

Onomasticon 1994 — Onomasticon provinciarum Europae Latinarum. Vol. I. Aba-Bysanvs. Ed. LŐRINCZ B/ REDŐ F. Budapest.

OROSZLÁN Z 1920-22 — MNM Régiségtárának ásatása Dunapentelén 1922 július havában.

ArchÉrt XXXIX 93-96.

ORTVAY T 1875 A Magyar Nemzeti Múzeum érem és régiség-gyűjteménye 1874-ben. ArchKözl X 81-135.

OTTOMÁNYI K 1991 Késő római kerámia a leányfalui őrtoronyból. (Die Keramik von Burgus in Leányfalu) Studia Comitatensia XXII 5-144.

PALÁGYI S 1996 Architektonische Gliederungs- und Dekorationselemente aus der Gegend des Balaton. Akten des 3. internationalen Kolloquiums über Probleme des provinzialrömischen Kunstschaffens. Köln, Bonn 105-119.

PALOL P de 1964 Nuevos bronces liturgicos Hispanovisigodos. BSAA XXX 311-318.

Pannonia 1990 Pannonia régészeti kézikönyve. Szerk. MÓCSY A/ FITZ J. Budapest 1990.

PARISI PRESICCE C 1990 Il Marco Aurelio in Campidoglio. Milano.

PAROVIĆ-PEŠIKAN M 1973 Excavations of a late Roman villa at Sirmium. Sirmium III 1-44.

PARRAGI GY 1976 Karéjos épület a Kiscelli utcában. BpR XXIV 177-181, figs 57-59, 179-199.

PAŠKVALIN V 1959 Zwei spätantike gewölbe-Kammer-Gräber, zugleich eine Übersicht solcher auf dem bosnisch-herzegowinischen Territorium. GZMS 149-162.

PAŠKVALIN V 1974 Antičko staklo spodrucja Bosne i Hercegovine. AV XXV 109-138.

PATSCH K 1893 Zur Verwaltung des illyrischen Zolles. RömMitt VIII 192-200.

PAULOVICS I 1927 Dunapentelei római telep. – Die römische Ansiedlung von Dunapentele (Intercisa). ArchHung II. Budapest.

PAULOVICS I 1938a Vezető a Magyar Nemzeti Múzeumban. Budapest.

PAULOVICS I 1938b Szent Quirinus savariai bazilikájának feltárása. VSz 5 138-152.

PAULOVICS I 1943a Savaria - Szombathely topográfiája. Az 1938-41. évi kutatások eredményei. Szombathely.

PAULOVICS I 1943b Lapidarium Savariense. Római kőemlékek új felállítása a szombathelyi múzeumban. Szombathely.

PAULOVICS I 1944 A szombathelyi Szent Márton-egyháznak savariai Szent Márton születéshelyének rómaikori eredete. Szombathely .

PAUR I 1859 Huszonöt kiadatlan pannoniai latin felirat. ArchKözl I 217-228.

PEIRCE H /TYLER R 1926 L'art Byzantin. London.

PEKÁRY T 1955 Későrómai sírok Fenékpusztán. ArchÉrt LXXXII 19-29.

PELEKANIDIS S 1977 Studien zur frühchristlichen und byzantinischen Archäologie. Thessaloniki.

PERKINS A 1973 The art of Dura–Europos. Oxford.

PÉTERFI ZS 1993 A Bátaszék-Kövesd pusztai későrómai temető. (Der spätrömische Friedhof von Bátaszék-Kövesd puszta) WMMÉ XVIII 47-128 I-XL. tábla.

PLESNIČAR-GEC L 1976 Frescos from Emona. ArchJug XVII 29-36.

PLESNIČAR-GEC L 1983 Old Christian cemetery in Emona. Ljubljana.

PLESNIČAR-GEC L 1990 Emona nel IV secolo problemi di collegamento con Milano e l'area Padana. Felix Temporis Reparatio 219-226.

PLOUG G 1985 The Graeco-Roman town. Nationalmuseets Skrifter Storre Beretninger IX Hama, Fouilles et recherches de la fondation Carlsberg 1931-1938. III. 1. København.

PÓCZY K 1955 Római épületek Óbudán a Kiscelli u. 10. sz. alatt. (Römische Gebäude von Óbuda, Kiscelli-Strasse Nr. 10) BpR XVI 41-87.

PÓCZY K 1964 Aquincum a IV. században. BpR XXI 55-77.

PÓCZY K 1967 PÓCZY Klára: Scarbantia városfalának korhatározása. (La datation de l'enceinte de Scarbantia) ArchÉrt XCIV 137-154.

PÓCZY K 1971a Aquincum polgárvárosa. Vezető. Budapest.

PÓCZY K 1971b Die Anfänge der Urbanisation in Scarbantia. ActaArchHung XXII 93-110.

PÓCZY K 1976a Pannoniai városok. Budapest.

PÓCZY K 1976b Städtbau in Pannonia. In: Kultur und Natur in Niederösterreich. Bd 1. Neue Forschungen in Carnuntum. Wien 95-113.

PÓCZY K 1977 Scarbantia. A római kori Sopron. Budapest.

PÓCZY 1980 — Pannonian cities. In: Archaeology. Budapest 239-274.

PÓCZY 1983 — Aquincum polgárvárosa. Budapest.

PÓCZY 1984 — Későcsászárkori falfestmények Aquincumban. (Spätkaiserzeitliche Wandmalereien in Aquincum) BpR XXV 53-63.

PÓCZY K 1989 — A zsidók Pannoniában, in: Magyarországi zsinagógák. Szerk. GERŐ László. Budapest. 13-18.

PÓCZY K 1991 — Egy aquincumi pénzverde nyomában. BpR XXVIII 45-56.

PÓCZY K 1994 — Aquincum. The Roman town in Budapest. Budapest.

PÓCZY K /ZSIDI P — Aquincum. Római kori ásatások Budapest 1970-1985. Budapest, s. a.

POLASCHEK E 1937 — Noricum RE XVII 971-1048.

POPOVIĆ V 1971 — A survey of the topography and urban organization of Sirmium in the late Empire. Sirmium I 119-133.

POPOVIĆ V 1987 — Die süddanubischen Provinzen in der Spätantike vom Ende des 4. bis zur Mitte des 5. Jahrhunderts. Die Völker Südosteuropas im 6. bis 8. Jahrhundert. Hrsg. Bernhard HÄNSEL. München, Berlin [Südosteuropa Jahrbuch Bd 17] 95-139.

PULSZKY F 1897 — Magyarország archaeologiája. Budapest Vol. I-II.

PURCELL N 1987 — Tomb and Suburb. Gräberstaßen 25-41.

PUSZTAI R 1973 — A mosonmagyaróvári Hansági Múzeum története, gyűjteményei, kiállításai. Mosonmagyaróvár.

R.-ALFÖLDI 1956 — Beiträge zur Frage der Cistophori Kaisers Hadrian. FolArch VIII 89-95.

RADAN G 1973 — Comments on the History of Jews in Pannonia. ActaArchHung XXV 265-278.

RADNÓTI A 1939 — Római kutatás Ságváron. ArchÉrt LII 148-164.

RADNÓTI A 1941 — Római tábor és feliratos kövek Környéről. Laureae Aquincenses. Budapest II. DissPann Ser 2. No. 11. 77-105.

RADNÓTI A 1954 — Pannoniai városok élete a korai feudalizmusban. MTA II. Osztályának Közleményei V 489-508.

RADNÓTI A 1955 — Buda régészeti emlékei. In: Budapest Műemlékei II. [Magyarország műemléki topográfiája IV.] Budapest 13-35.

RADNÓTI A 1956 — Sopron és környéke régészeti emlékei. In: Sopron és környéke műemlékei. Budapest 13-38.

RADNÓTI A 1957 — Möbel und Kästchenbeschläge, Schlösser und Schlüssel. In: Intercisa II (Dunapentele). Geschichte der Stadt in der Römerzeit. Budapest ArchHung 36.

RAFC — Römische Antike und frühes Christentum. Ausgewählte Schriften von Rudolf Egger. Band I (1962) Band II (1963) Klagenfurt

RB 1986 — Das römische Budapest. Neue Ausgrabungen und Funde in Aquincum. Hrsg. POLENZ H. Münster/Westf.

R-D — RÓMER F. / DESJARDINS E.: Acta nova Musei Nationalis Hungariae. Tom I. Budapest 1873.

RE — Paulys Realencyclopedie der classischen Altertumswissenschaft, neu begonnen von WISSOWA G., fortgeführt von KROLL H. / ZIEGLER K. etc.

RIU — Die römischen Inschriften Ungarns
1. BARKÓCZI L / MÓCSY A Budapest 1972.
2. BARKÓCZI L / MÓCSY A Budapest 1976.
3. BARKÓCZI L / SOPRONI S Budapest 1981.
4. BURGER A / FÜLEP F Bonn, Budapest 1984.
5. FITZ J Bonn, Budapest 1991.

RIVET A L F 1969 — The Roman villa in Britain. London.

RODZIEWICZ M 1976 — Alexandrie I. La céramique romaine tardive d'Alexandrie. Varsovie.

RÓMER F 1864 — A Buda környékén felásott római sírokról ArchKözl IV 101-124.

RÓMER F 1864a — Pannonia újabb kiadatlan római feliratai. ArchKözl IV 45-64.

RÓMER F 1868 — Magyar régészeti krónika. ArchKözl VII 181-198.

RÓMER F 1875 — Kiadatlan római felriatok. ArchKözl X 24-57.

RÓMER F 1886 — Az újabb időben Ó-szőnyben kiásott római régiségek. ArchKözl VI 151-163.

ROSTOWTZEFF M 1938 — Dura-Europos and Its Art. Oxford 1938.

Römer 1969 — Römer in Rumänien. Ausstellung des Römisch-Germanischen Museums Köln und des Historischen Museums Cluj. 12. Februar – 18. Mai 1969. Kunsthalle Köln.

RUINART 1859 — Acta Martyrum P. Theodorici RUINART opera ac studio collecta, selecta atque illustrata accedunt praeterea in hoc editione Acta SS. Firmii et Rustici ex optimis codicibus Veronensibus. Ratisbonae.

RUPP J 1868 — Buda-Pest és környékének helyrajzi története. Pesten.

SÁGI K 1951 — La colonie romaine de Fenékpuszta et la Zone intérieure de forteresses. ActaAntHung I 87-90.

SÁGI K 1954 — Temetők. Intercisa I. Budapest ArchHung XXXXIII 43-100.

SÁGI K 1960 — Die spätrömische Bevölkerung der Umgebung von Keszthely. ActaArchHung XII 187-256.

SÁGI K 1961 — Die zweite altchristliche Basilika von Fenékpuszta. ActaAntHung IX 397-460. Anhang: FÜZES M.: Pflanzenreste in der zweiten altchristlichen Basilika bei Fenékpuszta 461-470.

SÁGI K 1968a — A Balaton szerepe Fenékpuszta, Keszthely és Zalavár IV-IX. századi történetének alakulásában. AntTan XV 15-46.

SÁGI K 1968b — Darstellung des altchristlichen Kreuzes auf einem römischen Ziegel. ActaAntHung XVI 391-400.

SÁGI K 1970 — Das Problem der pannonischen Romanisation im Spiegel der völkerwanderungzeitlichen Geschichte von Fenékpuszta. ActaAntHung XVIII 147-196.

SÁGI K 1972 — Ókeresztény bazilikának vélt villa rustica hitelesítő ásatása. (Kontrollgrabung einer für frühchristliche Basilika gehaltenen Villa rustica in Kékkút) VMMK XI 121-137.

SÁGI K 1978 — Keszthely és környéke. Veszprém.

SÁGI K 1981 — Das römische Gräberfeld von Keszthely-Dobogó. Budapest.

SÁGI K 1989 — Adatok a fenékpusztai erőd történetéhez. (Über die Geschichte der Festung in Fenékpuszta) TVMK I 261-317.

SÁGI K 1991 — Egy VI. századi keszthelyi temető és mondanivalója a "Keszthely kultúra" etnikumának szempontjából. (Keszthelyer Friedhof aus dem 6. Jahrhundert und dessen Bedeutung bezüglich des Ethnikums "Keszthely Kultur") TVMK II 113-141.

SÁGI K /FÜZES FRECH' M 1967 — Régészeti és archeobotanikai adatok a pannoniai kontinuitás kérdéséhez. Agrártörténeti Szemle IX 79-98.

SALAMON Á/ BARKÓCZI L 1970 — Bestattungen von Csákvár aus dem Ende des 4 und dem Anfang des 5. Jahrhunderts. Aregia XI 35-80.

SALAMON Á/ BARKÓCZI L 1973 — Archäologische Angaben zur spätrömischen Geschichte des pannonischen Limes. Gräberfeld von Intersica I. MittArchInst IV 73-95.

SALAMON Á /BARKÓCZI L 1982 — Pannonien in nachvalentinischer Zeit (376-476). Severin 147-178.

SALAMON Á /ERDÉLYI I 1971 — Das völkerwanderungszeitliche Gräberfeld von Környe. Budapest.

SANIE S 1968 — Deus Aeternus et Theos Hypsistos en Dacie romaine. Hommages à Maarten J. Vermaseren. [EPRO 68.] Vol. III. 1092-1115.

SARNOWSKI T 1978 — Les représentation de villas sur les mosaiques Africaines tardives. Wroclaw, Warszawa, Kraków, Gdansk.

ŠAŠEL J 1974 — Siscia RE Suppl. XIV 702-741.

SAUSER E 1966 — Früchristliche Kunst. Sinnbild und Glaubensaussage. Innsbruck, Wien, München.

Schamanen — Schamanen. Mittler zwischen Menschen und Geistern. Begleitband zur Ausstellung im Kultur- und Stadthistorischen Museum Duisburg 14. April bis 30 Juni 1991. Duisburg 1991.

SCHEIBER [A] 1955 — Jews at Intercisa in Pannonia. The Jewish Quarterly Review XLV (January) 189-197.

SCHEIBER S 1960 — Magyarországi zsidó feliratok. Budapest.

SCHEIBER [A] 1983 — Jewish Inscriptions in Hungary from the 3rd Century to 1686. Budapest, Leiden.

SCHMAUS M 1986 — Bischof. LThK II 491-497.

SCHOBER A 1914 — Römischer Friedhof in Au am Leithaberg. ÖstJh 17 Beibl 203-256.

SCHOBER A 1923 — Die römischen Grabsteine von Noricum und Pannonien. Wien.

SCHOENVISNER [St] 1791 — Antiquitatum et historiae Sabariensis ab origine usque ad praesens tempus libri novem studi et opera. Pestini.

SELLYE I 1990 — Ringfibeln mit Ansatz aus Pannonien. Savaria XIX 17-102.

SESTON W 1947 — Constantine as a 'Bishop'. Journal of Roman Studies XXXVII 128-131.

Severin, Katalog — Severin. Zwischen Römerzeit und Völkerwanderung. Ausstellung im Stadtmuseum Enns. Linz 1982.

SHELTON K J 1981 — The Esquiline Treasure. London.

SIMONI K 1988 — Ein Silberlöffel aus Sisak. VAMZ XXI 79-86.

SOLIN H 1983 — Juden und Syrer im westlichen Teil der römischen Welt. Eine ethnisch-demographische Studie mit besonderer Berückscihtigung der sprachlichen Zustände. ANRW II, 29 587-789.

SOLIN H 1989 — Zu pannonischen Juden. ActaArchHung XLI 233-236.

SOPRONI S 1974a — Die spätrömische Festung von Jovia. Actes du IXe Congrès International d'Études sur les Frontières Romaines, Mamaia 1972. Bucureşti, Köln, Wien 181-186.

SOPRONI S 1974b — Beiträge zur Frage der Liste von Valeria der Notitia Dignitatum. ActaArchHung XXVI 59-70.

SOPRONI S 1975 — Előzetes jelentés az alsóhetényi későrómai erőd feltárásáról. (Vorläufiger Bericht über die Ausgrabungen der spätrömischen Festung von Alsóhetény – Iovia) SMK II 173-182.

SOPRONI S 1978 — Der spätrömische Limes zwischen Esztergom und Szentendre. Budapest.

SOPRONI S 1985 — Die letzten Jahrzehnte des pannonischen Limes. München.

SOPRONI S 1994 — Ein römischer Villenbesitzer aus Aquincum. Balácai Közlemények III 312-320.

Spätantike 1983 — Spätantike und frühes Christentum. Ausstellung im Liebieghaus Museum alter Plastik. Frankfurt am Main. Frankfurt am Main.

SQUARCIAPINO M F 1965 — La sinagoga di Ostia. Atti 299-315.

STADLER J E/HEIM F J 1858 — Vollständiges Heiligen-Lexikon der Lebensgeschichten aller Heiligen ... in alphabetischer Ordnung. Bd I Augsburg

STADLER P 1988 — Die Bevölkerungsstrukturen nach Eugippius und dem archäologischen Quellen. In: Germanen 297-310.

STAMBAUGH J E 1978 — The Function of Roman Temples. ANRW II. 16. 554-608.

Steindenkmäler 1971 — BALLA L / P.-BUOCZ T / KÁDÁR Z / MÓCSY A / SZENTLÉLEKY T: Das römische Steindenkmäler von Savaria. Budapest.

STUTZINGER D 1983 — Die kirchliche Selbstverständnis im ersten christlichen Jahrhundert. In: Spätantike 9-19.

SUPKA G 1913 — Frühchristliche Kästchenbeschläge aus Ungarn. RömQuart 162-191

[SZALÁGYI] 1784 — [SZALÁGYI] SALAGIUS Stephanus: De statu ecclesiae Pannoniae. Libri VII: De antiquo ecclesiae Pannoniae patriarcha. Quinque Ecclesiis 1777-1784.

SZÁNTÓ K [2]1987 — A katolikus egyház története I. Budapest.

SZENTLÉLEKY T 1960 — A szombathelyi Isis-szentély. Budapest.

SZENTLÉLEKY 1965 — Das Isis-Heiligtum von Szombathely. Szombathely.

SZEREDY J [3]1883 — Egyházjog különös tekintettel a Magyar Szent Korona területének egyházi viszonyaira, valamint a keleti és protestans egyházakra. Pécs.

SZI — Emlékkönyv Szent István király halálának kilencszázadik évfordulójára. Szerk. SERÉDI J. I-III. Budapest 1938.

SZILÁGYI J 1954 — Ein Ziegelstein mit Zauberformel aus dem Palast des Statthalters in Aquincum. ActaAntHung II 305-310.

SZILÁGYI J 1956 — Aquincum. Budapest.

SZILÁGYI J 1968 — Aquincum. RE Suppl. XI 61-131.

SZIRMAI K 1972 — A római eredetmítosz újabb ábrázolása Pannoniában. AntTan XIX 259-263.

SZIRMAI K 1973a — Die Bildtypen auf den Denkmälern des Kaiserkults in Aquincum und ihr Ursprung. ActaDebrecen IX 83-89.

SZIRMAI K 1973b — Újabb adat a pannoniai Liber Pater és Libera kultuszhoz. AntTan XX 191-196.

SZIRMAI K 1975 — A Vihar utcai sírlelet. (Der Grabfund aus der Vihar-Straße) ArchÉrt CII 77-83.

SZIRMAI K 1980 — Kutatások az aquincumi II-III. századi legióstábor praetentúrájának északi részén 1973-1977-ben. (Forschungen im nördlichen Teil der Praetentura des Legionslagers im 2-3 Jh. zu Aquincum 1973-1977) ArchÉrt CVII 187-200.

SZIRMAI K 1984a — Az aquincumi 2-3. sz.-i légióstábor praetentúrája. BpR XXV 135-179.

SZIRMAI K 1984b — Újabb adatok az aquincumi légióstábor falfestészetéhez. BpR XXV 247-253.

SZIRMAI K 1985a — Későrómai sírok az aquincumi canabae északi részén. (Spätrömische Gräber im nördlichen Teil der Aquincumer Canabae) ArchÉrt CXII 221-239.

SZIRMAI K 1985b — Kutatások a II-III. századi aquincumi legiostábor keleti védműveinél, 1974-1980. (Forschungen bei den östlichen Wehrbauten des Aquincumer Legionslagers aus dem 2-3 Jh.) CommArchHung 49-68.

SZIRMAI K 1986a — Neue Keramik mit bemalter Tierfigur aus Aquincum. CommArchHung 49-54.

SZIRMAI K 1986b — Das Straßennetz des Legionslagers von Aquincum im 2. und 3. Jahrhundert. LimesAalen 426-428.

SZIRMAI K 1986a — Neue Keramik mit bemalter Tierfigur aus Aquincum. CommArchHung 49-54.

SZIRMAI K 1988a — Stratigrafiai megfigyelések az aquincumi későrómai erőd északi körzetében. (Startigraphische Beobachtungen im Bereich der spätrömischen Festung von Aquincum) CommArchHung 49-70.

SZIRMAI K 1988b — Monumental Bronze Fragments from Aquincum. In: Griechische und römische Statuetten und Großbronzen. Akten der 9. Tagung über antike Bronze in Wien. Wien. 144-147.

SZIRMAI K 1989 — 49. Barrack-blocks in the praetentura of the legionary fortress in Aquincum (1987-1988). Roman Frontier Studies. 259-262.

SZIRMAI K 1991a — Új női sírszobor Aquincumban. (Neue weibliche Grabstatue in Aquincum) BpR XXVII 99-103.

SZIRMAI K 1991b — Újabb adatok az aquincumi 2-3. századi legiotábor principiájának nyugati traktusához. (Neuere Angaben zum westlichen Trakt der Principia in Aquincum aus dem 2-3 Jahrhuéndert) BpR XXVIII 107-116.

SZIRMAI K 1991c — Neue Wandgemälde auf dem Gebiet der Lagerstadt von Aquincum. KölnJb XXIV 203-206.

SZŐKE B M 1998 — A korai középkor hagyatékai a Dunántúlon. Ars Hungarica II 257-319.

SZŐNYI E 1981 — Kutatások Mursellában. Beszámoló az Árpás–Dombiföldön 1975-80 között végzett feltárásokról. (Ausgrabungen in Mursella) CommArchHung I 87-119.

SZŐNYI E /TOMKA P 1985 — Győrszentiván – Söprűhegy: későrómai és kora Árpád–kori temető. CommArchHung 97-123.

SZŐNYI O 1906 — A pécsi püspöki múzeum kőtára. Pécs.

SZŐNYI O 1907 — A pécsi ókeresztény sírkamra. Budapest.

SZŐNYI O 1911 — Római leletekről Pécsett. ArchÉrt XXXI 371-373.

SZŐNYI O 1913 — A Pécsi Városi Múzeum gyarapodása Horváth Antal hagyatékából. MKÉ VII 66-67.

SZŐNYI O 1913a — A pécsi őskeresztény cubiculum vízmentesítése és a régi puskaporos-torony restaurálása. PBMÉ VI 84-85.

SZŐNYI O 1913b — A pécsi ókeresztény temető. I. közlemény. Religio LXXII (30) 577-579.

SZŐNYI O 1913c — A pécsi ókeresztény temető. II. közlemény. Religio LXXII (31) 596-597.

SZŐNYI O 1926 — Ásatások a pécsi székesegyház környékén 1922-ben. Az Országos Magyar Régészeti Társulat Évkönyve II 1923-26 172-195.

SZŐNYI O 1929 — A pécsi ókeresztény temető sírkamrái és kápolnája. Magyar Művészet V 537-544.

SZÖRÉNY[I] L 1734 — Vindiciae Syrmienses, seu descriptio Syrmii una cum suo episcopatu, aliisque eo spectantibus ex variis authoribus depromta. Posonii.

TAU S / NICU M 1985 — Ein Beschrifteter Glasbecher aus der Nekropole von Barcea-Tecuci (4. Jahrhundert u.Z.) Dacia XXIX 1-2.

TESTINI P 1966 — Le catacombe e gli antichi cimiteri cristiani in Roma. Bologna.

TESTINI P 21980 — Archeologia Cristiana. Puglia.

TESTINI P 1985 — Note per servire allo studio del complesso paleocristiano di S. Felice a cimitile (Nola). MEFRA XCVII 329-371.

Thesaurus — Thesaurus Linguae Latinae

THOMAS E 1954 — Későrómai márvány asztallap Csopakról. FolArch VI 74-85

THOMAS E 1955 — Bruchstück einer frühchristlichen Marmortischplatte mit Reliefverzierung aus Csopak. ActaAntHung III 261-282.

THOMAS E 1964 — Römische Villen in Pannonien. Budapest.

THOMAS E 1973-74 — Arius-Darstellung, eine römerzeitliche Ziegelritzzeichnung aus Kisdorog in Pannonien. BÁMÉ IV-V 77-116.

THOMAS E 1974 — Martyres Pannoniae. FolArch XXV 139-146.

THOMAS E 1975-1976 — Savaria Christiana. (németül). Savaria IX-X 105-160.

THOMAS E 1977a — "Eins ist der Gott!" Über jüdische Denkmäler der Römerzeit in Pannonien. Pannonia (Eisenstadt) V 3-4. 21-25.

THOMAS E 1977b — Savaria Christiana. A 200 éves szombathelyi egyházmegye emlékkönyve 1777-1977. Szombathely 35-93.

THOMAS E 1978a — Frühchristliche Kontinuität in westpannonischen Raum (zur Quirinus– und Martinsfrage). Forschungsberichte zur Ur- und Frühgeschichte. Nr. 10 Wien 97-102.

THOMAS E 1978b — Pannonische Reliquienaltäre AV XXIX 573-587.

THOMAS E 1978c — Martyres Pannoniae. Atti Vol. I. 615-635.

THOMAS E 1980 — Religion. In: Archaeology 177-206.

THOMAS E 1982a — Spätantike Metallfunde in Pannonia. In: Metallkunst von der Spätantike bis zum ausgehenden Mittelalter. Wissenschaftliche Konferenz anläßlich der Ausstellung "Spätantike und frühbyzantinische Silbergefäße aus Köpenik". 20 und 21 März 1979. Hrsg v. A. EFFENBERGER. Berlin 56-76.

THOMAS E 1982b — Das frühe Christentum in Pannonien im Lichte der archäologischen Funde. Severin. 255-293.

THOMAS E 1983 — Eine kleine Bronzebüste des Kaisers Justinianus in Pannonien. RÖ XI-XII 1983-1984 305-314.

THOMAS E 1986a — Spätantike und frühchristliche Charakterköpfe aus Pannonien. Klio LXVIII 501-541.

THOMAS E 1986b — Spätantike und frühbyzantinische Silbergegenstände im mittleren Donaugebiet, innerhalb und außerhalb der Grenzen des Römerreich. In: Argenterie romaine et byzantine, ed. DUVAL N / BARATTE Fr. Paris 135-145.

THOMAS E 1988 — Die Romanität Pannoniens im 5. und 6. Jahrhundert. In: Germanen 284-294.

TIMON S 1735 — Imago antiquae Hungariae. Tyrnaviae.

Tokod 1981 — Die spätrömische Festung und das Gräberfeld von Tokod. Hrsg. MÓCSY A Budapest.

TOMASCHEK 1897 — Bassiana. RE III 34-52.

TOMKA P 1980 — Das germanische Gräberfeld aus dem 6. Jh. in Fertőszentmiklós. ActaArchHung XXXII 5-30.

TOPÁL J 1981 — The Southern Cemetery of Matrica (Százhalombatta-Dunafüred). Budapest.

TOPÁL J 1985 — Torony modellek a szentendrei római villából. Studia Comitatensia XVII 303-325.

TOPÁL J 1994 — Die römische Villa von Szentendre. Balácai Közlemények III 321-335.

TÓTH E 1971-1972 — A római lakosság-kontinuitás kérdése a Nyugat-Dunántúlon. (Die Frage der Kontinuität der römischen Bevölkerung Westtransdanubia)

	Savaria V-VI 231-241.
TÓTH E 1972a	A késöantik császári paloták alaprajzi típusának kialakulásához. DissArch II.1. 37-62.
TÓTH E 1972b	Adatok Savaria-Szombathely és környéke történeti földrajzához. VSz 239-262.
TÓTH E 1972c	Figürlich verzierte loculus-Platte aus Savaria. FolArch XXIII 59-67.
TÓTH E 1973	Late antique imperial palace in Savaria (The question of the so-called Quirinus basilica) ActaArchHung XXV 117-137.
TÓTH E 1974a	Korabizánci kőfaragvány Felsődörgicséről. (An Early Byzantine Stone Sculpture from Felsődörgicse) FolArch XXV 161-177.
TÓTH E 1974b	Vigilius episcopus Scaravaciensis. ActaArchHung XXVI 269-275.
TÓTH E 1975a	Későantik császári palota Savariaban. VSz XXV 420-440.
TÓTH E 1975b	Későrómaikori belső pannoniai erődök kérdéséhez. (Előzetes jelentés az 1971-1974. évi ságvári ásatásokról) SMK II 183-189.
TÓTH E 1976	Pannonia történetének problémái. AntTan XXIII 114-126.
TÓTH E 1977a	Frühbyzantinisches Lampenhängeglied. FolArch XXVIII 143-156.
TÓTH E 1977b	A Savaria-Bassiana útszakasz. (Römische Strassenstrecke zwischen Savaria und Bassiana) ArchÉrt CIV 65-75.
TÓTH E 1977c	Vigilius episcopus Scaravaciensis. SSz XXXI 320-326.
TÓTH E 1978	TÓTH Endre: Cirpi. Pannonia kora-Flavius-kori hadtörténetéhez. AntTan XXV 181-191
TÓTH E 1978-1979	Von den spätrömischen Prunksaalen. Schild von Steier XV-XVI 189-195.
TÓTH E 1979	Römische Gold- und Silbergegenstände mit Inschriften im Ungarischen Nationalmuseum.FolArch XXX 157-184.
TÓTH E 1980a	Römische Metallgegenstände mit Inschriften im Ungarischen Nationalmuseum: instrumenta domestica. FolArch XXXI 131-156.
TÓTH E 1980b	Zur Geschichte des Nordpannonischen Raumes im 5. und 6. Jahrhundert. Die Völker an der mittleren und unteren Donau im fünften und sechsten Jahrhundert. Symposion 24-27. Okt. 1978. Wien 93-100.
TÓTH E 1982	Tetrarchia-kori Iovia–Herculia névadás Pannoniában? (Tetrarchiezeitliche Namensgebung von Jovia–Herculia?) ArchÉrt CIX 55-72.
TÓTH E 1985	Zur Chronologie der militärischen Bautätigkeiten des 4. Jh. in Pannonien. MittArchInst XIV 121-136.
TÓTH E 1986a	Late Roman Burials in Aquincum. NHQ XXVII (104) 116-117.
TÓTH E 1986b	Ásatások Alsóhetényben 1981-86. Dombóvári Krónika 3. Dombóvár.
TÓTH E 1987	Bemerkungen zur Kontinuität der römischen Provinzial-Bevölkerung in Transdanubien (Nordpannonien). In: Die Völker Südosteuropas im 6. bis 8. Jahrhundert. Hrsg von B. HANSEL. Berlin 251-264.
TÓTH E 1988a	Ásatások Alsóhetényben. Jovia az utánpótlási központ. Élet és Tudomány 16. sz. 483-485.
TÓTH E 1988b	Az alsóhetényi 4. századi erőd és temető kutatása, 1981-1986. Eredmények és vitás kérdések. (Vorbericht über die Ausgrabung der Festung und des Gräberfeldes von Alsóhetény 1981-1986. Ergebnisse und umstrittene Fragen) ArchÉrt CXIV-CXV 1987-88 22-61.
TÓTH E 1988c	Sopianae: római város és temető. Baranya. Történelmi és honismereti folyóirat I 23-34.
TÓTH E 1989	Liturgic brooch from Alsóhetény (Iovia). Studia Aegyptiaca XII 385-394.
TÓTH E 1989a	Provincia Valeria Media. ActaArchHung XLI 197-226.
TÓTH E 1989b	Die spätrömische Festtung von Jovia und ihr Gräberfeld. Antike Welt XX (I.) 31-39.
TÓTH E 1989c	TÓTH Endre: Templum provinciae in Tác? Specimina Nova 1989:1 43-58.
TÓTH E 1990a	A 4-8. századi pannóniai kereszténység forrásairól és a leletek forrásértékéről. MEV II (2) 17-33.
TÓTH E 1990b	Et lux perpetua luceat ei. RÖ XVII-XVIII 1989-1990 262-279.

TÓTH E 1991 Templum provinciae Tácon? (Templum provinciae in Tác?) TVMK II 97-112.

TÓTH E 1991a Adatok Valeria tartomány kereszténységének történetéhez. In: Szent István és Európa. Budapest 119-139.

TÓTH E 1991b Pannonia Christiana. A kereszténység kezdete a Kárpát-medencében. Életünk XXIX (8) 740-747.

TÓTH E 1991c A Quincque Basilicae–Quincque Ecclesiae helynevek lokalizálásához és értelmezéséhez. (Zur Lokalisierung und Auslegung der Ortsnamen Quinque Basilicae–Quincque Ecclesiae) JPMÉ XXXVI 101-108.

TÓTH E 1994a Későrómai sír Tihanyból - Spätrömische Grab aus Tihany. FolArch XLIII 127-166.

TÓTH E 1994b Das Christentum in Pannonien bis zum 7. Jahrhundert nach den archäologischen Zeugnissen. In: Christentum 241-272.

TÓTH E 1994-1995 A kismákfai ezüst tálka. (Das Silberschüsselchen von Kismákfa) CommArchHung 117-151.

TÓTH E 1995 Ókeresztény ládikaveretek Ságvárról. (Altchristliche Kästchenbeschläge aus Ságvár) FolArch XLIV 107-150.

TÓTH E / ZÁGORHIDI CZIGÁNY B 1994 Források Savaria-Szombathely történetéhez. A római kortól 1526-ig. Szombathely Acta Savariensia 9.

TÓTH I 1994 A római kori keleti vallások pannoniai történetének kérdései (I. A kisázsiai és iráni eredetű kultuszok). Kandidátusi értekezés tézisei. Pécs.

TÓTH I 1995 Jüdische Gemeinde in dem römischen Pannonien. Speicmina Nova XI 179-186.

TÖRÖK Gy 1942 Rómaikori sírkamrák Pécs felső-sétaterén. ArchÉrt 207-215.

TÖRÖK J 1991 TÖRÖK J.: A magyar föld szentjei. Budapest.

TÖRÖK L 1988 Late Antique Nubia. History and Archaeology of the Southern Neighbour of Egypt in the 4th-6th C. A.D. Budapest .

UBL H-.J 1982 Frühchristliches Österreich. In: Severin 295-336.

UBL H.-J 1994 Die Christianisierung von Noricum Ripense bis zum 7, Jahrhzbndert nach den archäologischen Zeugnissen. In: Christentum 130-151.

UTTENWEILER J 1934 Martin, hl. 3. Bischof v. Tours. LThK VI 984-987.

VAES J 1980 Alcune considerazioni archeologiche sulla cristianizzazione de edifici antichi a Roma. Acta Archaeologica Lovaniensia XIX 49-65.

VAES J 1986 Christliche Wiederverwendung antiker Bauten: Ein Forschungsbericht. Ancient Society XV-XVII 1984-86 305-443.

[B.-] VÁGÓ E 1970 Ausgrabungen in Intercisa (1957-1969) Aregia XI 109-119.

[B.-] VÁGÓ E /BÓNA I 1976 Die Gräberfelder von Intercisa. Der spätrömische Südostfriedhof. Budapest.

VALEVA J 1998 Les tombeaux ornés de croix et chrismes peints. Atti III 761-786.

VANYÓ L 1995 Katekézis, költészet és ikonográfia a 4. században. Budapest.

VÁRADY L 1961 Későrómai hadügyek és társadalmi alapjaik. A Római Birodalom utolsó évszázada (376-476). Budapest.

VÁRADY L 1984 Epochenwechsel um 476. Budapest.

VERSNEL H. S. 1981 Religious Mentality in Ancient Prayer. In: Faith Hope and Worship. Ed. H. S. VERSNEL. Leiden.

VETTERS H 1969 Zum 'Episcopus in Castellis'. Anzeiger. Österreichische Akademie der Wissenschaften. Philosophisch-historische Klasse CVI 75-93.

VISY ZS 1977 Intercisa. A rómaikori Dunaújváros. (Dunaújváros in the Roman Period) Kecskemét.

VISY ZS 1978 Pannóniai limes-szakaszok légifényképen. (Pannonische Limesabschnitte auf Luftaufnahmen) ArchÉrt. CIII 235-259.

VISY ZS 1980 Római jelzőtornyok és limes-út Intercisa térségében. (Römische Wahttürme und die Limesstrasse im Raum fon Intercisa) ArchÉrt CVII 166-175.

VISY ZS 1981 Pannonische Limesstrecken in Ungarn auf Luftaufnahmen. Antike Welt XII 39-52.

VISY ZS 1989 A római limes Magyarországon. (Der pannonische Limes in Ungarn) Budapest.

VOPEL H 1899 Die altchristlichen Goldgläser. Ein Beitrag zur altchristlichen Kunst- und Kulturgeschichte. Freiburg, Leipzig, Tübingen.

VÖRÖS I 1985-1986 The animal remains from the Roman imperial period of Kővágószőlős. JPMÉ XXX-XXXI 243-244.

VÖRÖS 1991 VÖRÖS I.: A tribunus laticlaviusok háza az aquincumi 2-3. sz–i legiotáborban. A mithraeum állatcsont-leletei. (Das Haus der tribuni laticlavii aus dem Legionslager von 2-3. Jh. in Aquincum. Das Tierknochenmaterial des Mithräums) BpR XXVIII 127-132.

VULIĆ N 1909 Antike Denkmäler in Serbien. JÖAI XII Beibl. 147-204.

VULIĆ N 1931 Фрушкогорски мученици. Glasnik istoriskog društva u novom sadu IV 359-373.

WEINSTOCK St 1934 Templum. RE V A1 480-485.

WELLNER I 1962 Terra Mater és Liber Pater közös oltára Aquincumban. (Gemeinsamer Altar der Terra Mater und des Liber Pater in Aquincum) BpR XX 299-301.

WELLNER I 1965 Aesculapius és Hygieiát ábrázoló gemma lenyomatával díszített edény Aquincumból. (Ein mit Gemmenabdruck des Aesculapius und der Hygieia verziertes Gefäss aus Aquincum) ArchÉrt XCII 42-45.

WELLNER I 1970 A magyarországi római kori épületek belső díszítő művészete. Építés-Építéstudomány II 327-400.

WELLNER I 1976 Mit Astralsymbolen verziertes Altarsteinbruchstück aus Aquincum. Neue Forschungen im Carnuntum I 114-120.

WELLNER 1990 WELLNER István: Die Militärlager Aquincums in spätrömischer Zeit und im Mittelalter. (Aquincum und das Nibelungenlied). RLiÖ XXVI 715-721.

WESSEL K 1946 Römische Frauenfrisuren. Jahrbuch d. DAI 1946-1947 62-75.

WESSETZKY V 1974 Données relatives à l'interpretation des représentations Égyptiennes de poissons. BullBAHongr XLII 3-12, 107-110.

WHITE M 1997 Synagogue and Society in Imperial Ostia: Archaeological and Epigraphic Evidence. HTR XC:1 23-58.

WILKEN R L 1984 The Cristians as the Romans Saw Them. New Haven, London.

WILPERT J 1897 Die Malereien der Sacramentskapellen in der Katakombe des hl. Callistus. Freiburg im Breisgau 1897.

WILPERT J 1916 Die römischen Mosaiken und Malereien der kirchlichen Bauten von VI-XIII. Jahrhunderten. Freiburg i. Br.

WOLFF H 1994 Die Kontinuität der Kirchenorganisation in Raetia und Noricum bis on die Schwelle des 7. Jahrhunderts. In: Christentum 1-28.

WOSINSKY M 1896 Tolna vármegye története az őskortól a honfoglalásig. Vol I-II. Budapest.

ZEILLER J 1918 Les origines chrétiennes dans les provinces danubiennes de l'Empire Romain. Paris.

ZEILLER J 1932 Bassiana. In: Dictionnaire d'histoire et de géographie ecclésiastiques. 6. Ed. BAUDRILLART A. / De MEYER A. / Van CAUWENBERGH Ét. Paris 1272-1273.

ZSIDI P 1994 Die römische Villa von Kaszás-dűlő-Csikós utca und die Frage der Villen im Territorium des Municipiums von Aquincum. Balácai Közlemények III 292-311.

INDEX

Deities and Persons

Places

Subjects and Depictions

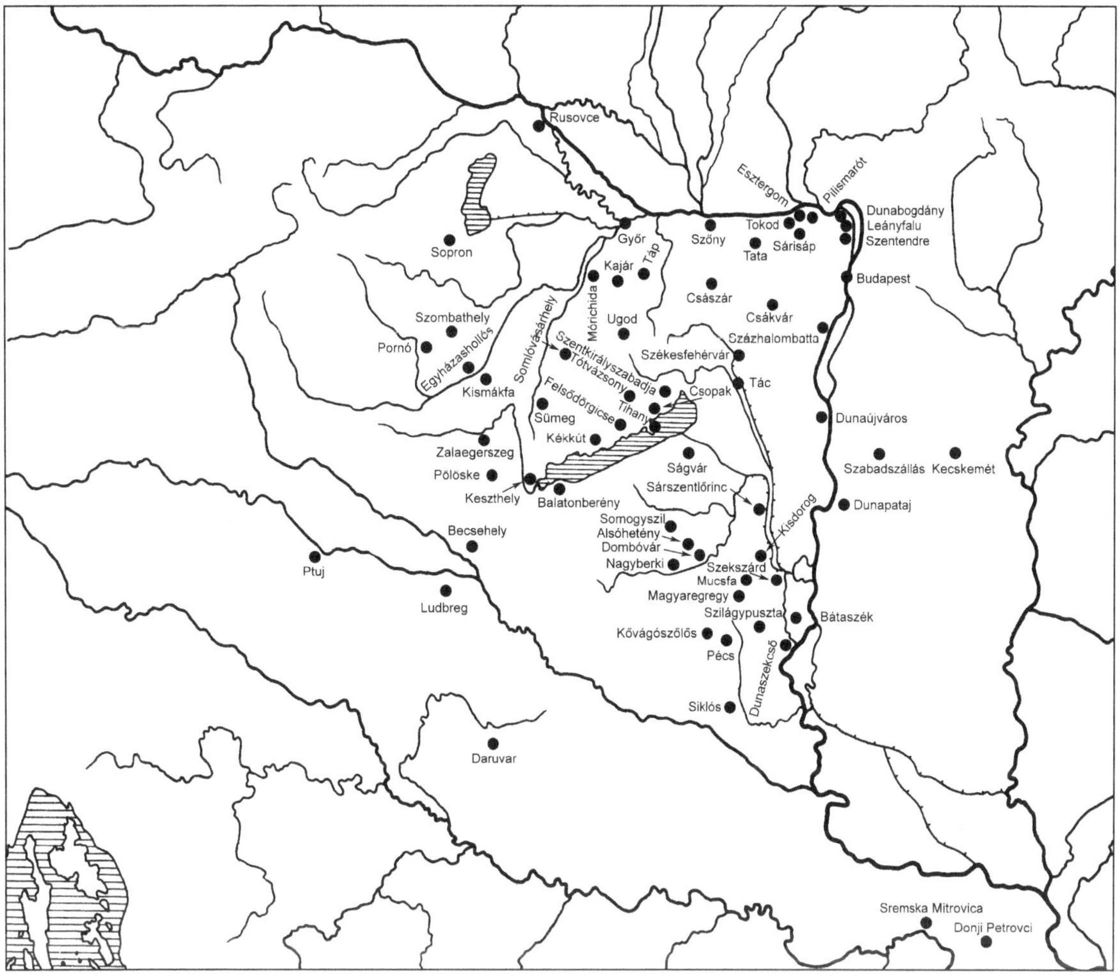

1. The sites of the finds that have so far been considered as early Christian.

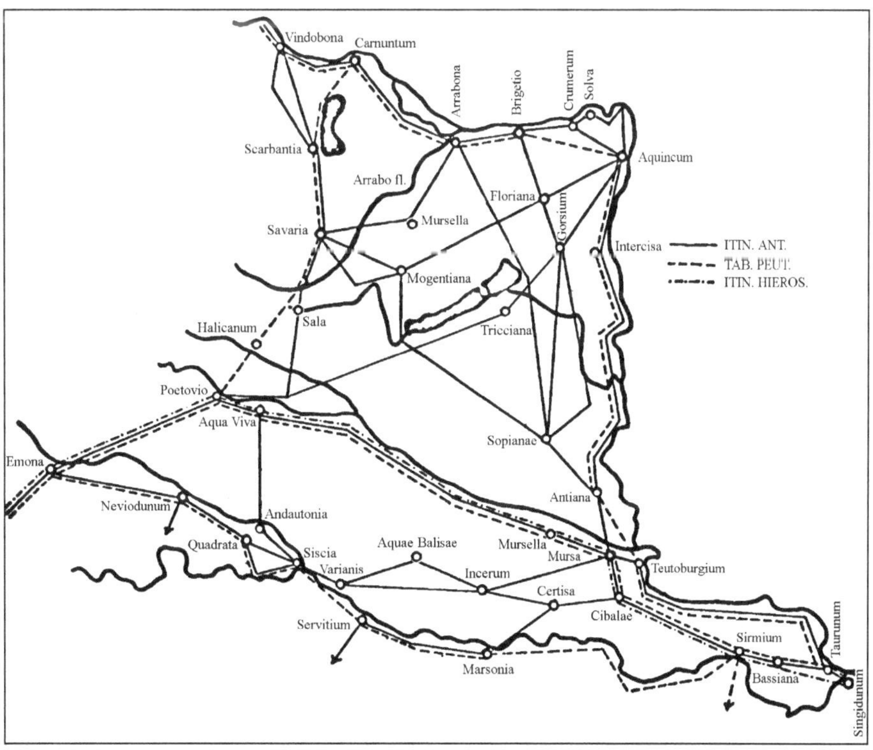

2. Road system in Pannonia. MÓCSY 1962, compiled based on KÉRDŐ 1998.

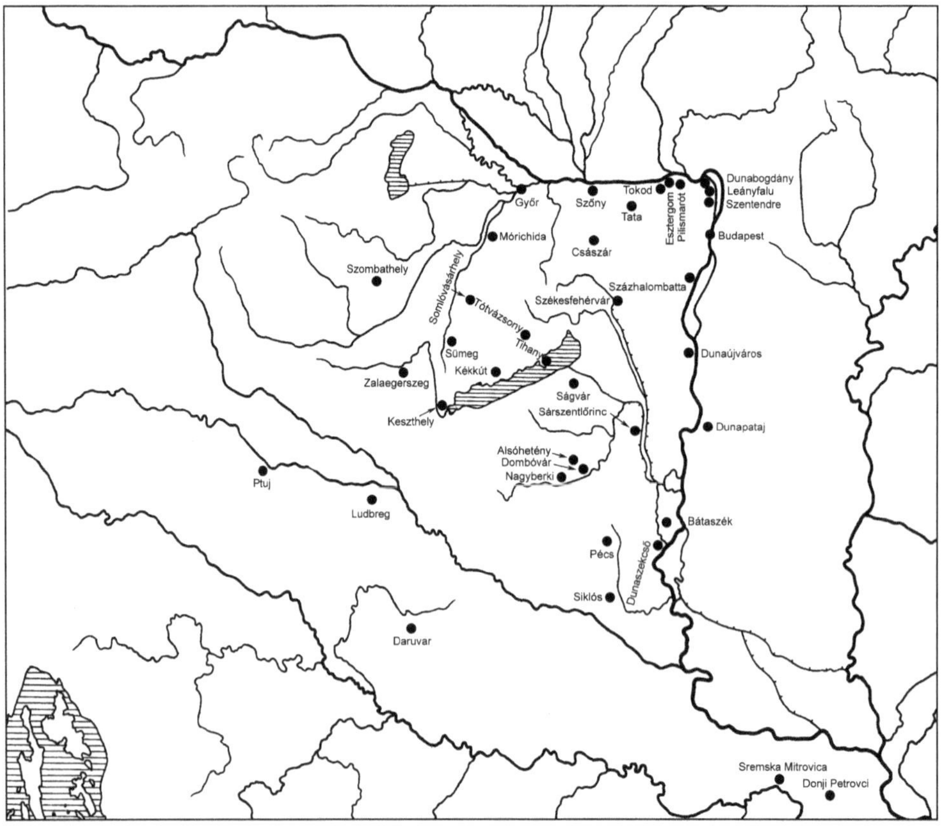

3. Provenance of early Christian sites and finds.

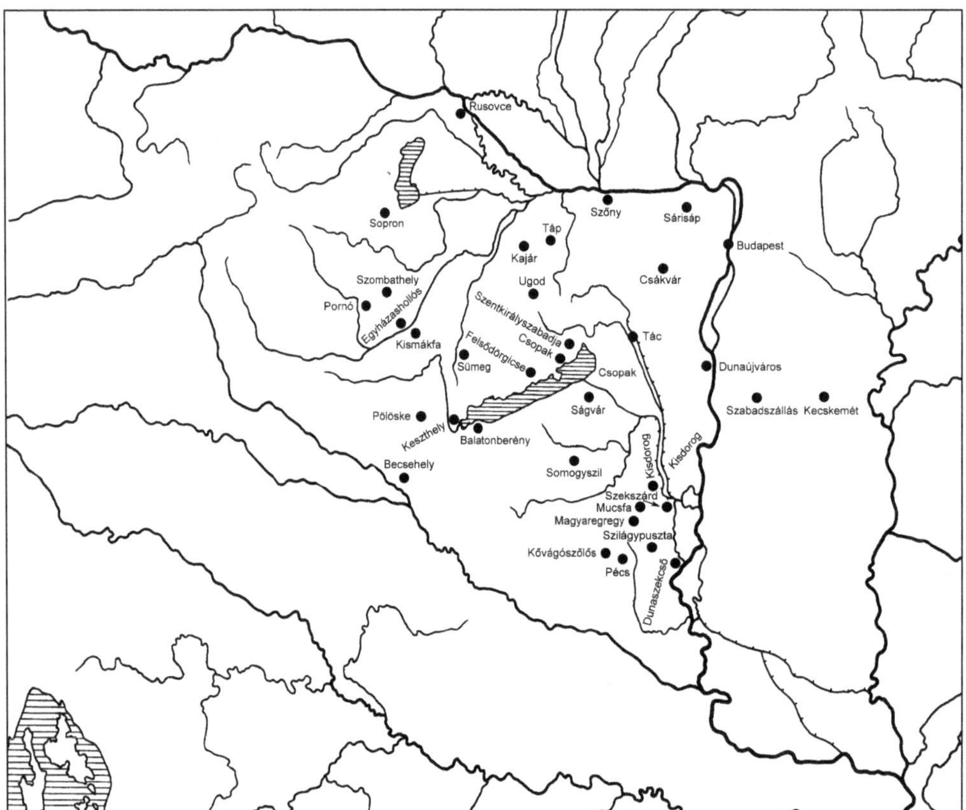

4. Provenance of sites and finds considered as early Christian, but actually not connectable to early Christianity. (Here the provenances of the dubious objects and buildings are also listed.)

5. Balatonberény, 1.II.a. roof tile. Based on SÁGI 1968b.

6. Bátaszék-Kövesdpuszta, 2.I.a. fibula. Based on PÉTERFI 1993.

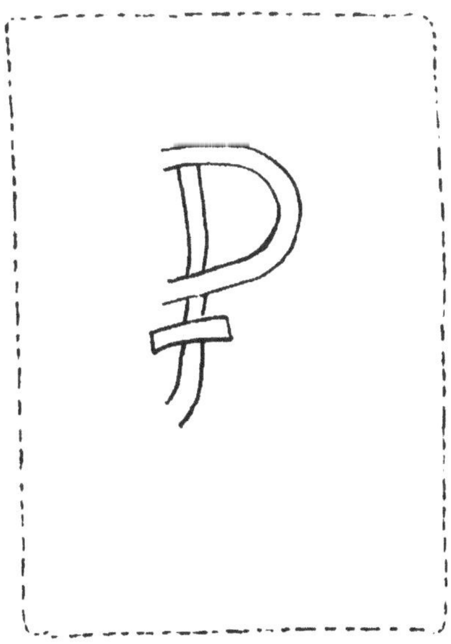

7. Becsehely, 3.II.a. roof tile. Based on THOMAS 1975-1976.

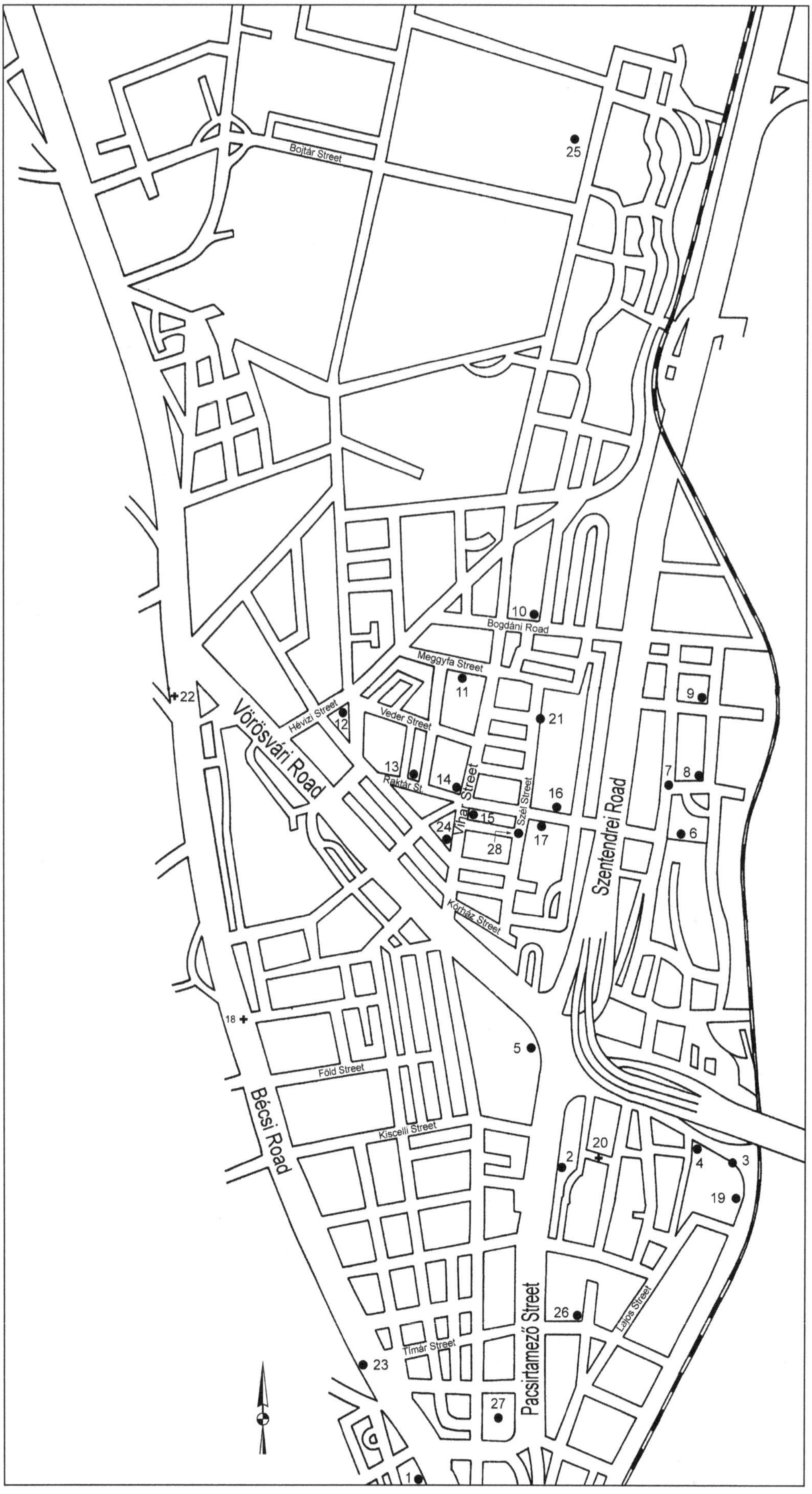

8. Budapest III. Aquincum. Sites of the environs of the legionary fortress. Sites of the early Christian objects and building are indicated with a cross. The map was produced according to the map published in 1992 by the Kartográfiai Vállalat. 1. beginning of the canabae's cemetery, 2. the Museum of Táborváros, 3. find-spot of the 4.I.a.pottery fragment, 4. building with an apse and a hall, 5. house of the *tribunus laticlavius,* 6. third horreum, 7. late graves, 8. remains of buildings, 9. fragments of frescos, mosaics and stuccoes from a building, 10. cemetery, 11. 'Hercules villa', 12. so-called *cella trichora*, 13. graves covered with stone slabs, 14. graves without goods, 15. settlement and road, 16. terrazzo floor, 17. road and workshop, 18. site of an early Christian burial chamber, 19. *ala* fort, 20. site of the 4.I.g. building with conches, 21. site of a iron find hoard, 22. site of the 4.I.e. glass vessel, 23. site of a grave vault, 24. cemetery at Vihar Street, 25. Kaszás dűlő– site of the villa at Csikós Street, 26. building erected in the second half of the 2nd century, 27. amphitheatre at Nagyszombat Street, 28. market-hall.

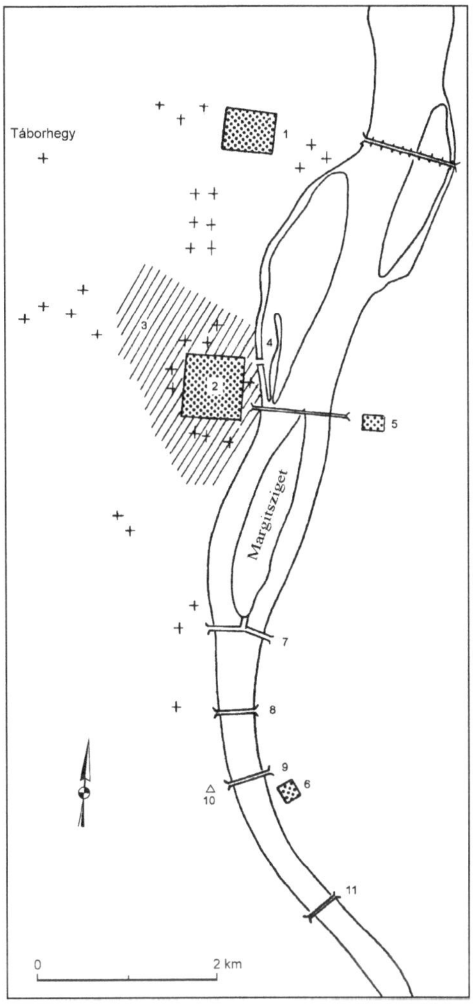

a

9. **a**: Map of sites along the Danube river. Compiled on basis of earlier disclosures from Aquincum and MRÁV 1996. 1. *municipium*, 2. legionary fortress, 3. canabae, 4. site of the 4.II.d roof tile - procuratorial palace, 5. Contra Aquincum, 6. Castellum contra montem Teutani, 7. Margit Bridge, 8. Chain Bridge, 9. Elisabeth Bridge, 10. Mons Teutani (Gellért Hill), 11. Szabadság híd (Bridge Freedom), **b**: Budapest's sites, No. 12 indicates the Kamaraerdő.

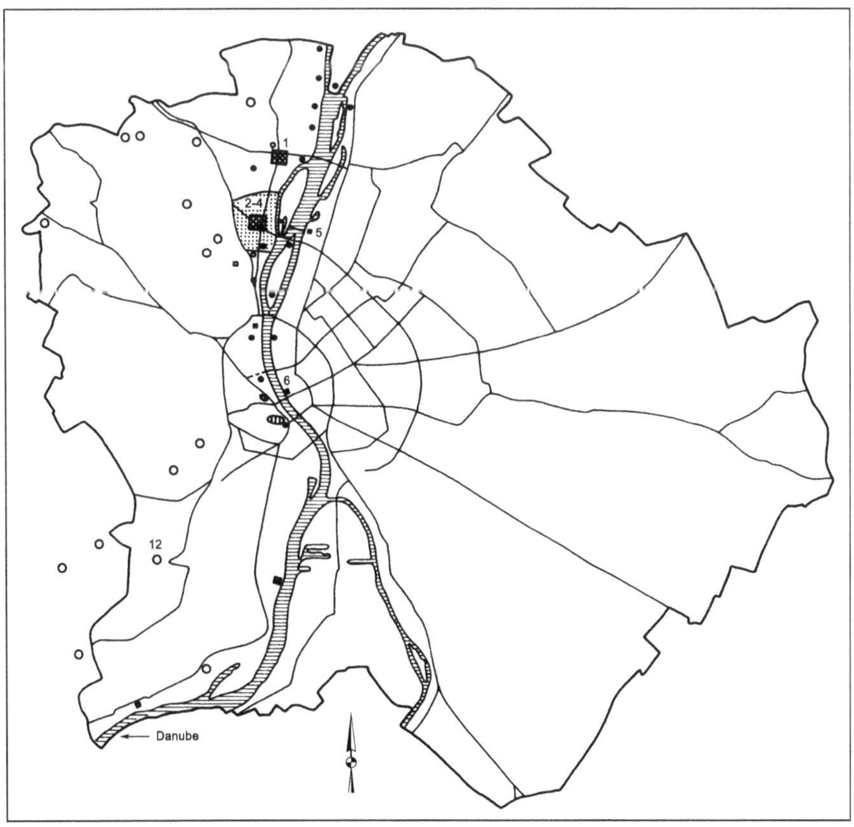

b

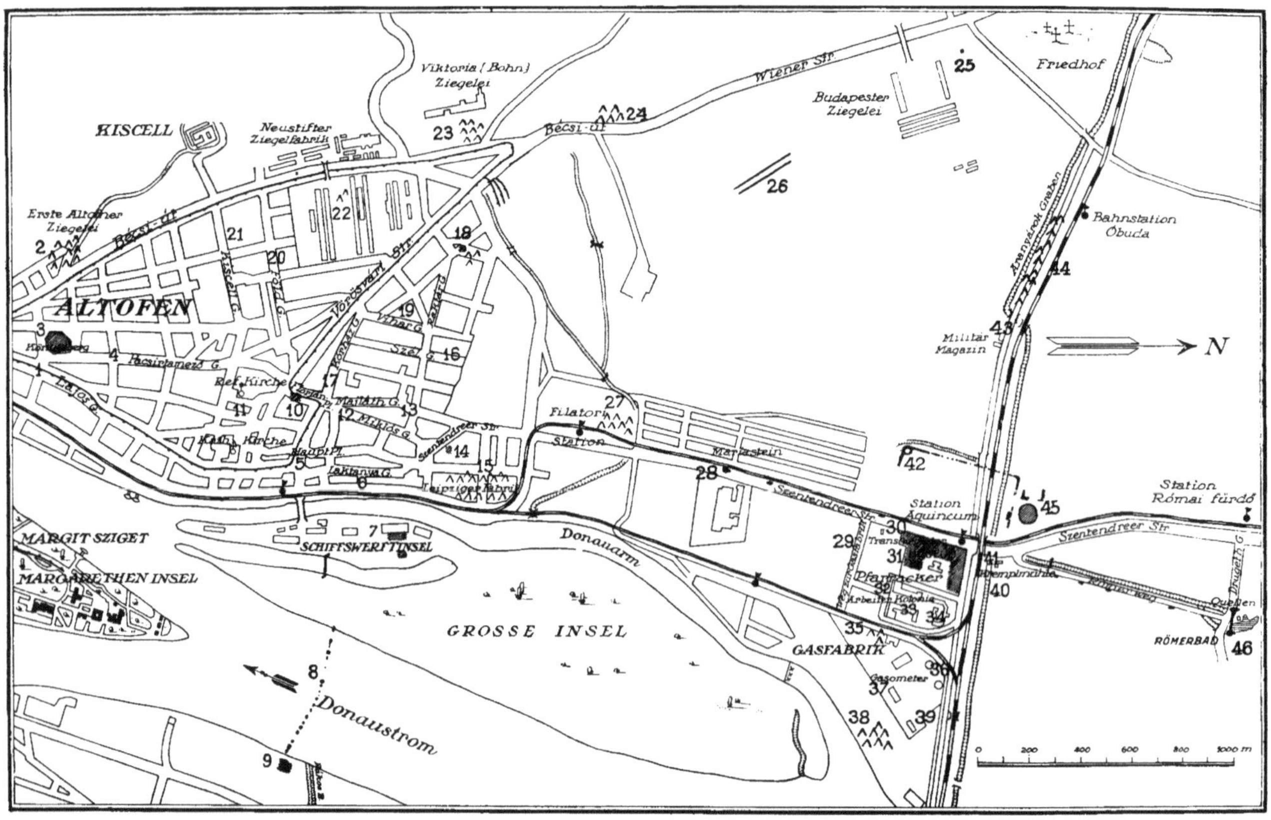

10. Environs of Aquincum in 1934. Site 4.II.c. Building in the cemetery No 38; map based on KUZSINSZKY 1934.

11. Aquincum, pottery fragment, 4.I.a. Drawing by the author.

12. Aquincum, marble fragment, 4.I.b

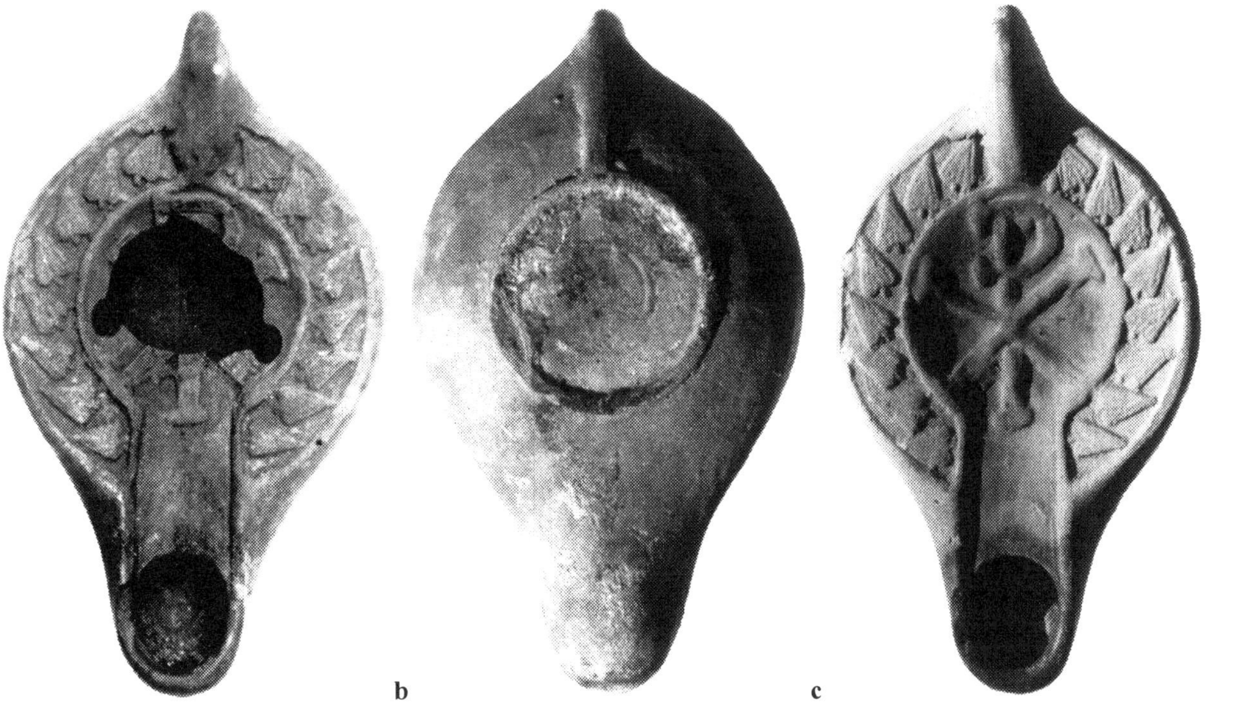

b c

a

13. Aquincum, clay lamp, 4.I.c. **a**: before restoration based on NAGY L 1931b, **b**: after restoration, the bottom of the clay lamp, **c**: restored. Photo: VIGH László.

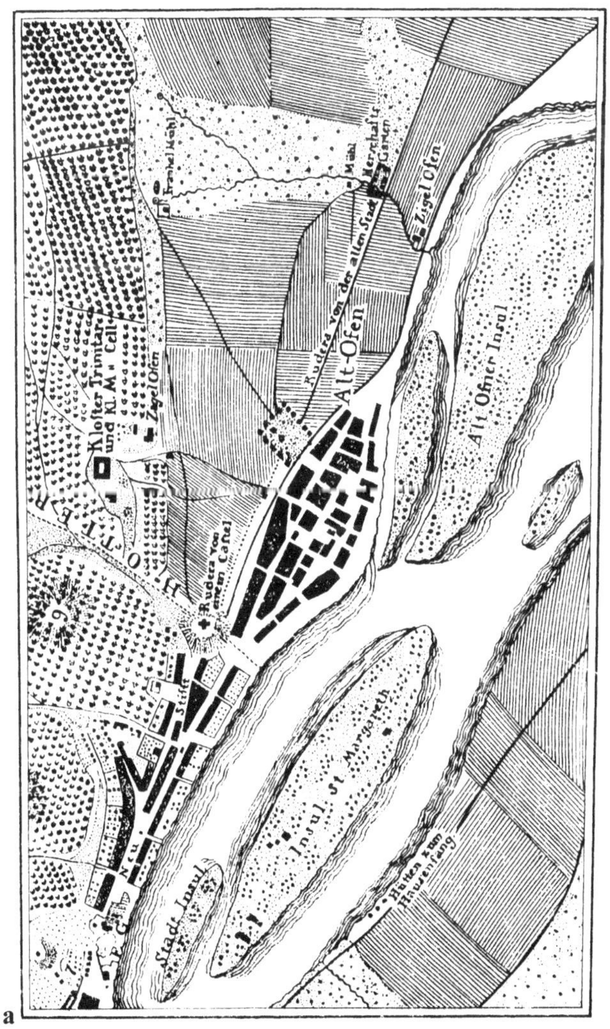

a

14. Fürdő Island in Aquincum, site 4.I.d.
a: map-section from 1756, KAISER / PÓCZY 1985 quotes Matthey's map-section. **b**: based on RUPP 1868.

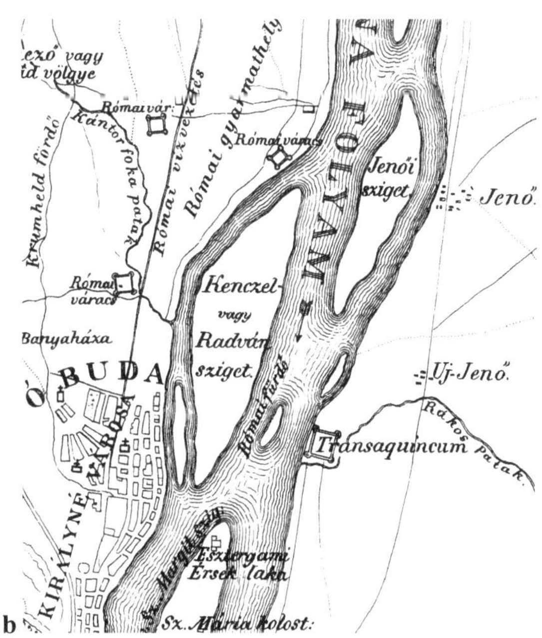

b

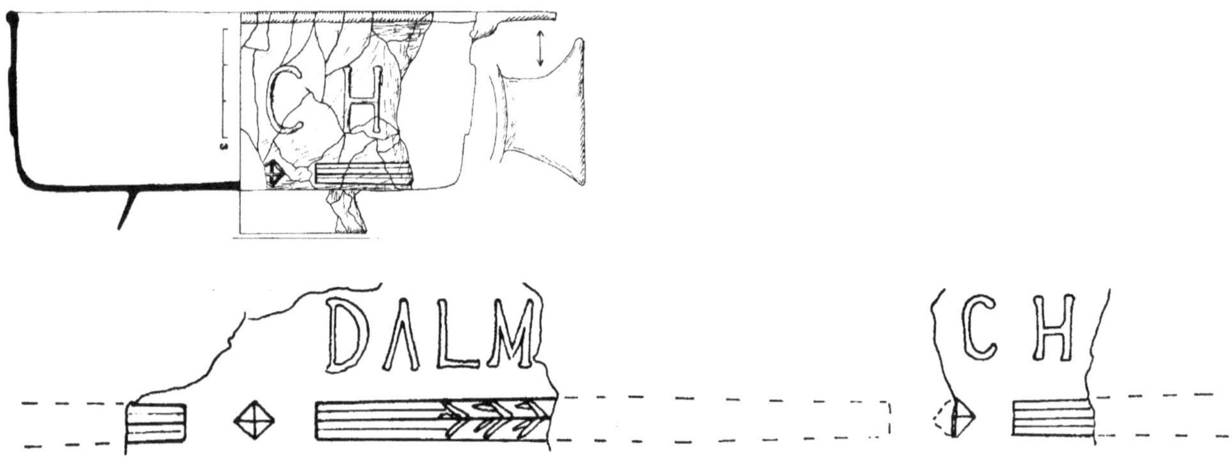

15. Aquincum, glass vessel 4.I.e. Drawing by Dr Judit Topál.

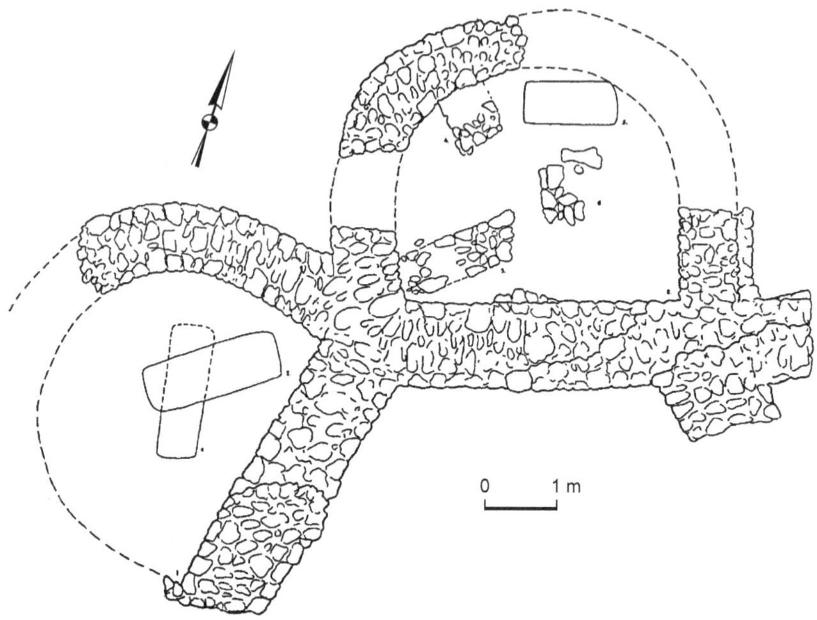

16. Aquincum, building with apses, 4.I.g, drawing of the excavated part based on PARRAGI 1976.

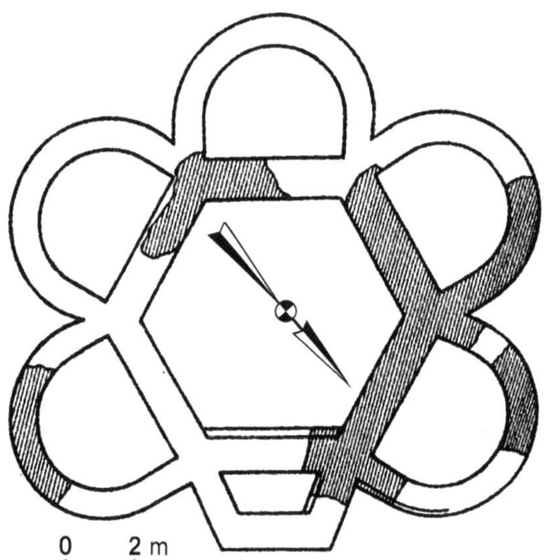

17. Ground plan of the building with apses based on TÓTH E 1994b.

18a-b. Excavation photos. of the building with apses (PARRAGI 1976).

19. Carving from the building with apses (PARRAGI 1976).

20a-f. Stucco and fresco fragments from the building with apses (PARRAGI 1976).

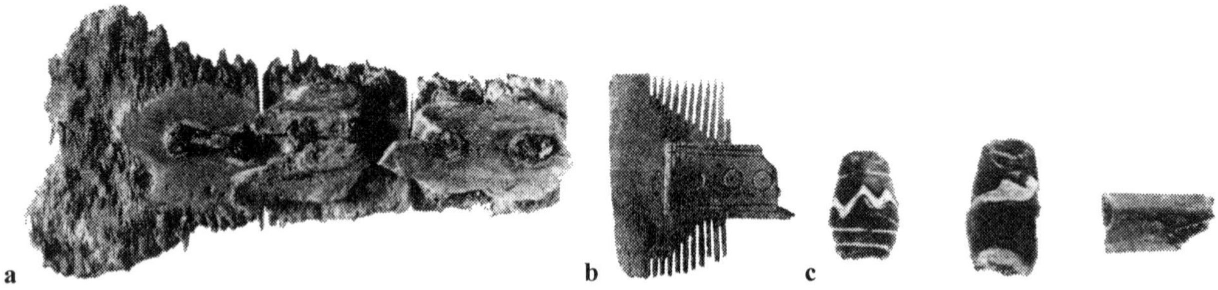

21a-c. Finds from the graves excavated in the building with apses (PARRAGI 1976).

22. Buildings with six apses from Dalmatia (MARASOVIĆ 1984).

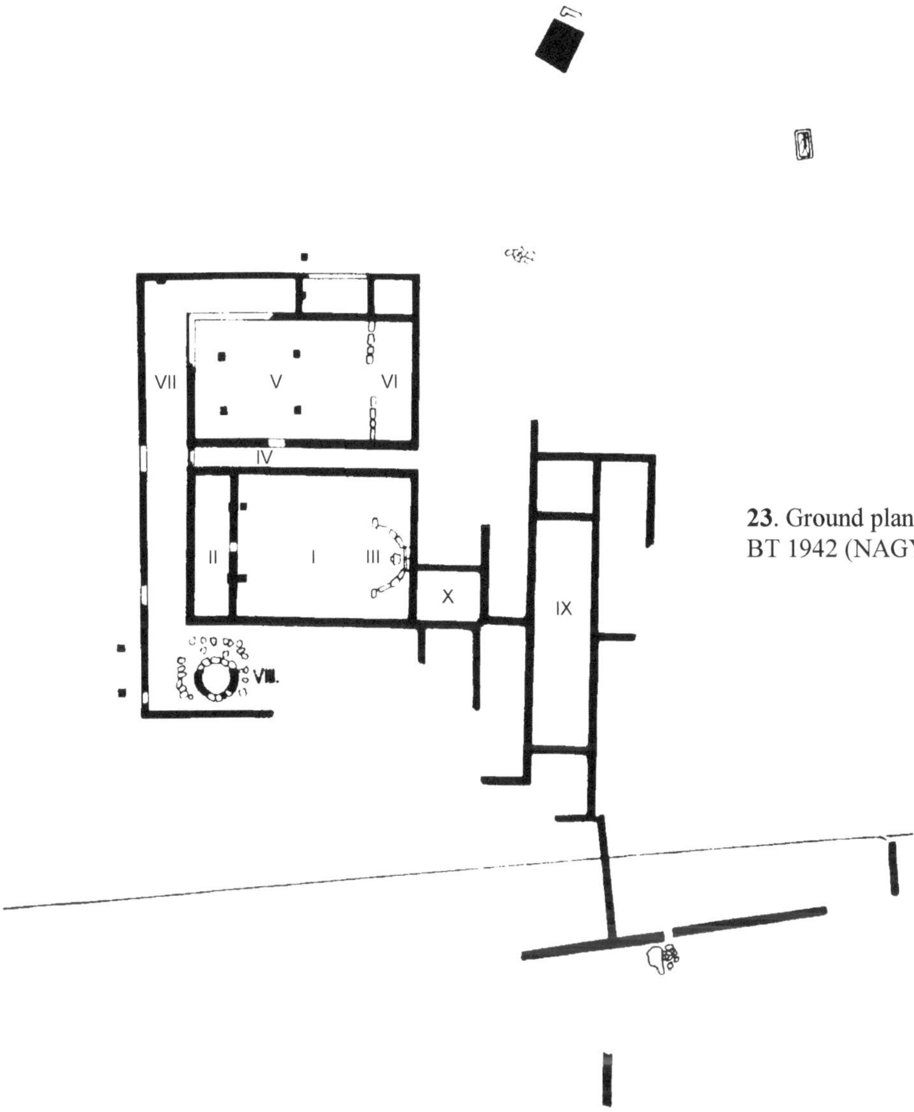

23. Ground plan of 4.I.h. Based on BT 1942 (NAGY L).

24. Marble slab from the 4.I.h.'doublechurch'. Based on NAGY L 1940.

25. Capital from the 4.I.h. building. NAGY L 1940 XLI/2 thinks this capital was found in the so-called double church.

26. Capital from the 4.I.h. building. NAGY L 1940 250 thinks this capital was found in the so-called double church.

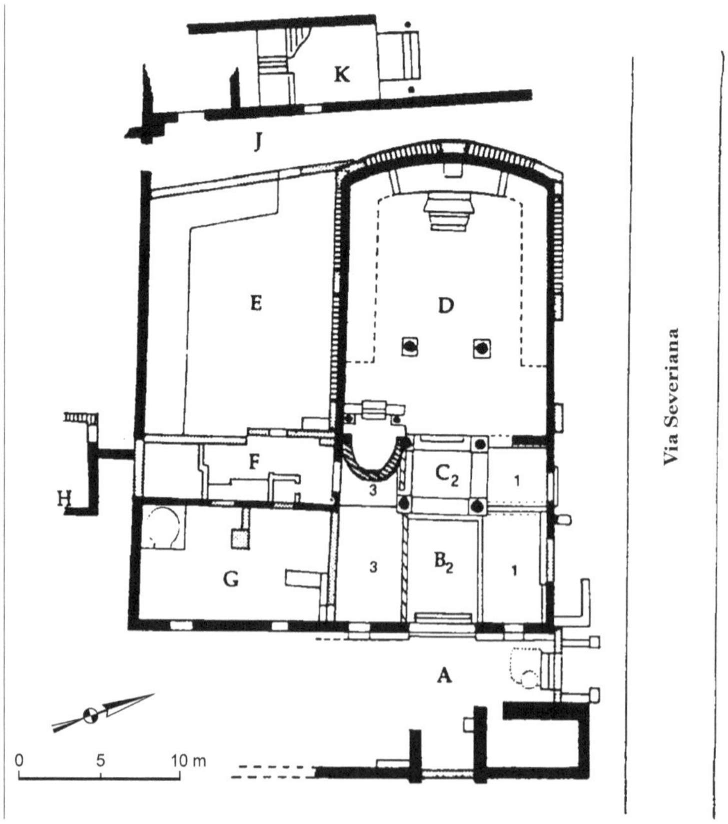

27. Ground plan of the Ostia synagogue (WHITE 1997).

29. Detail of the 4.I.i. tombstone, the picture panel.

28. Aquincum, Jewish tombstone of Anastasius, 4.I.i. (RADAN 1973).

30. Aquincum, Jewish tombstone of Ael. Silvanus, 4.I.j. Based on SCHEIBER 1983.

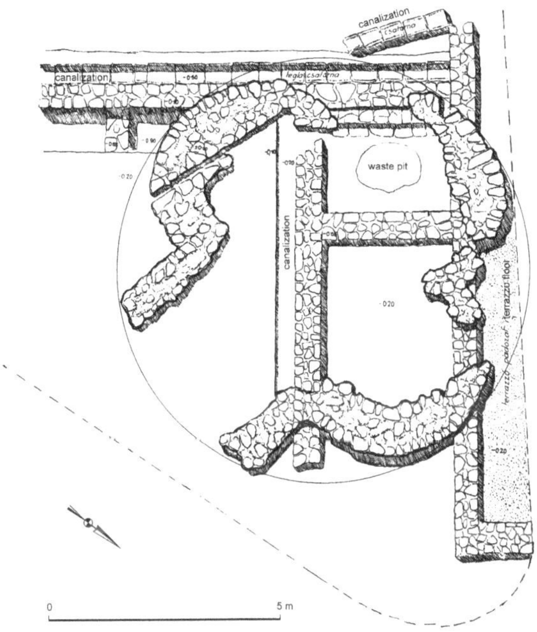

31. Ground plan of 4.II.a. as it was found. Based on NAGY L 1931b.

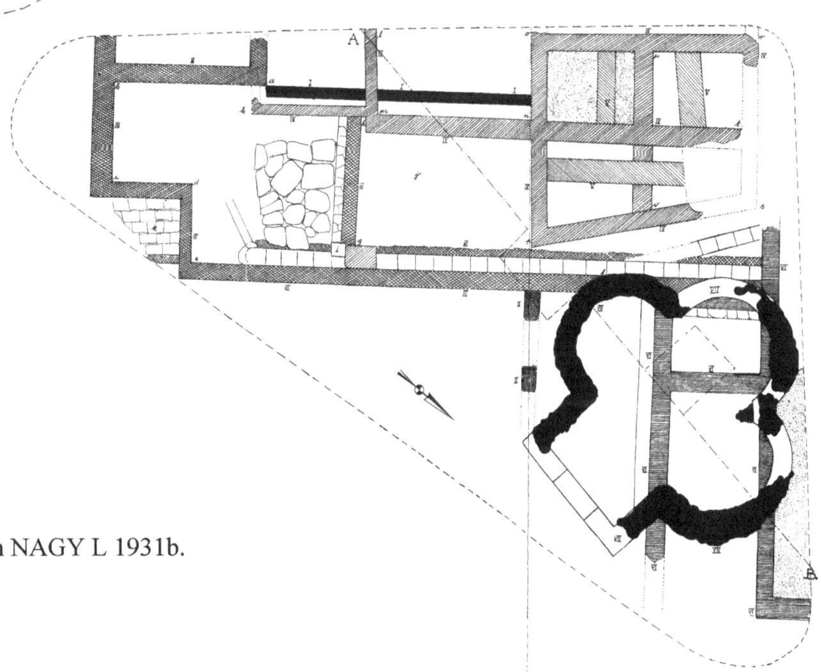

32. 4.II.a. Excavation map. Based on NAGY L 1931b.

33. Aquincum, so-called cella trichora (4.II.a), finds of the 3rd period:
a. pharos, **b**. clay object with the shape of a human head. Based on NAGY L 1931b.

34. 4.II.a. excavation photo of the southern apse, with the drain. Based on NAGY L 1931b.

35. 4.II.a. excavation photo from the eastern (entrance) side. In the background: the collected stones piled up. Based on NAGY L 1931b.

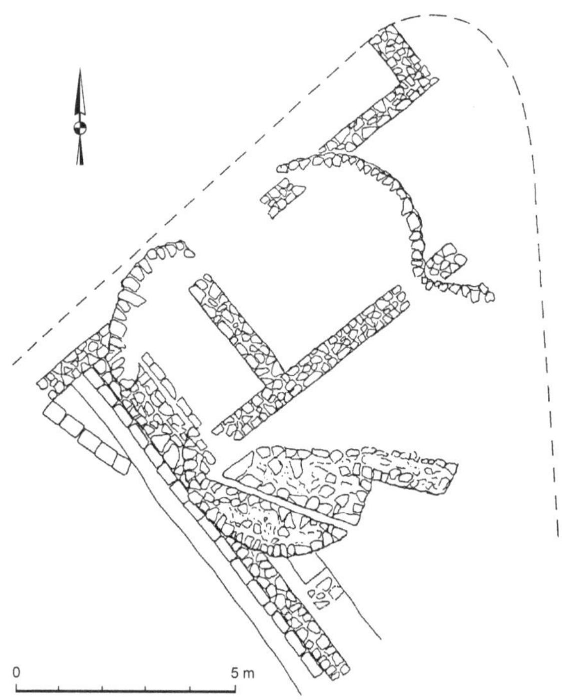

36. 4.II.a. An attempt to reconstruct the apses. Drawn by DUKAY Bernadett.

37. 4.II.a. Excavation photo of the whole area from the south. Based on NAGY L 1931b.

38. 4.II.a. Excavation photo from the northwest. Based on NAGY L 1931b.

39. 4.II.a. Excavation photo with the remnants of the western apse. Based on NAGY L 1931b.

40. Aquincum, Vihar Street, so-called chapel, 4.II.b. Threshold of 'The middle-nave'. Excavation photo. Based on NAGY L 1940.

41. Part of 4.II.b, cemetery behind the shrine. Excavation photo. Based on NAGY L 1940.

42. Part of 4.II.b, 'crypt', or cellar. Excavation photo. Based on NAGY L 1940.

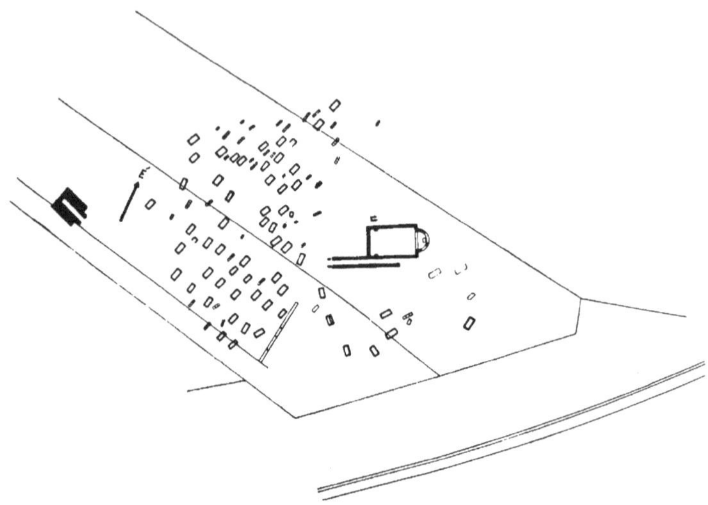

43. Aquincum, late Roman cemetery at Gázgyári housing estate, so-called cemetery chapel, 4.II.c. Based on NAGY L 1940.

45. Aquincum, tombstone, 4.II.e. Based on NAGY L 1938.

44. Aquincum, roof-tile, 4.II.d. Based on SZILÁGYI 1954.

47. Aquincum, detail of the 4.II.f. tombstone.

46. Aquincum, tombstone of Fl. Calvena, 4.II.f. Based on NAGY T 1944-1945.

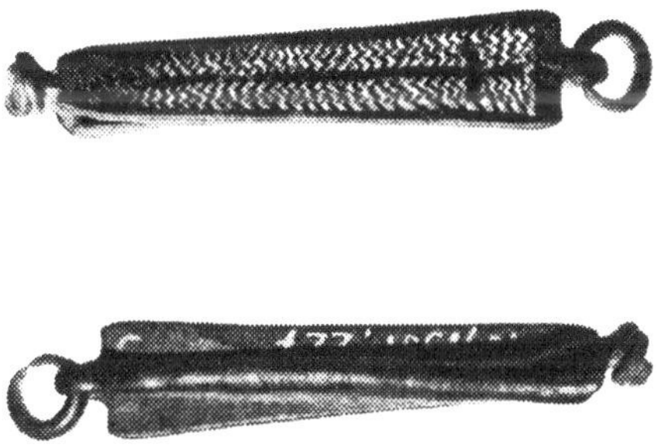

48. Budapest II. fibula, 4.II.g. Based on NAGY L 1831b.

49. Budapest II. Finds from grave 4.II.g. Based on NAGY L 1931b.

50. Budapest II. double grave 4.II.g. Based on NAGY L 1931b.

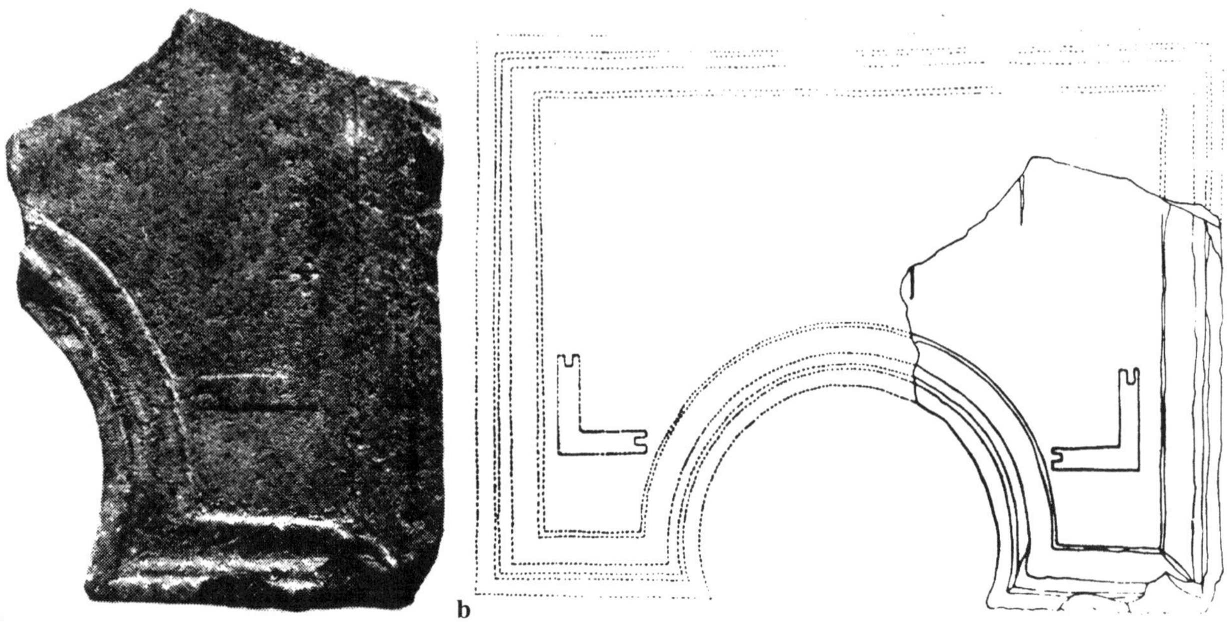

a b

51a, b. Aquincum, so-called altar slab and its completion, 4.II.h. Based on THOMAS 1978b.

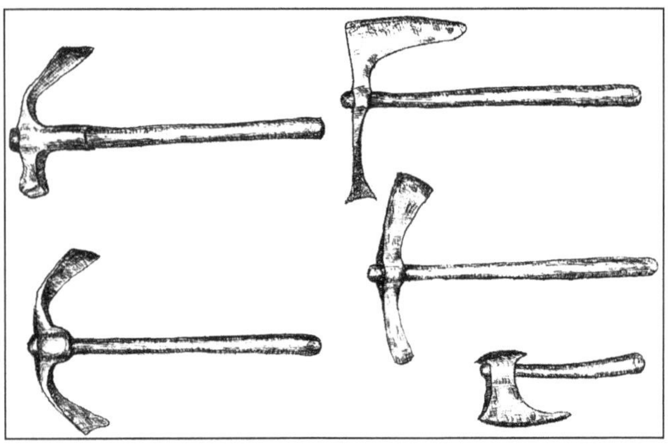

52. Aquincum, closed tool find from 23 Szél Street. Based on NAGY L 1937a.

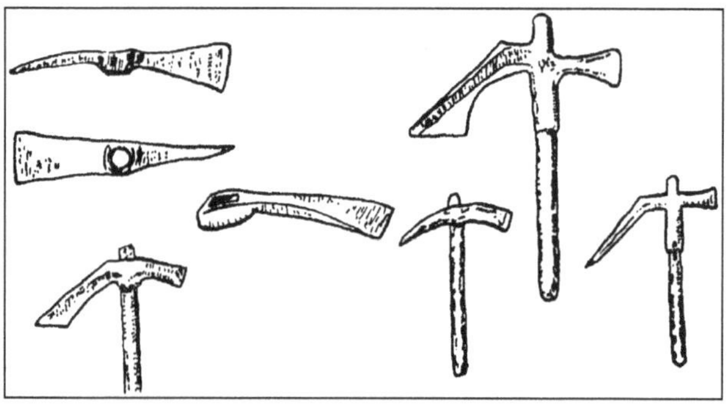

53. Fossors' tools. Based on TESTINI 1966.

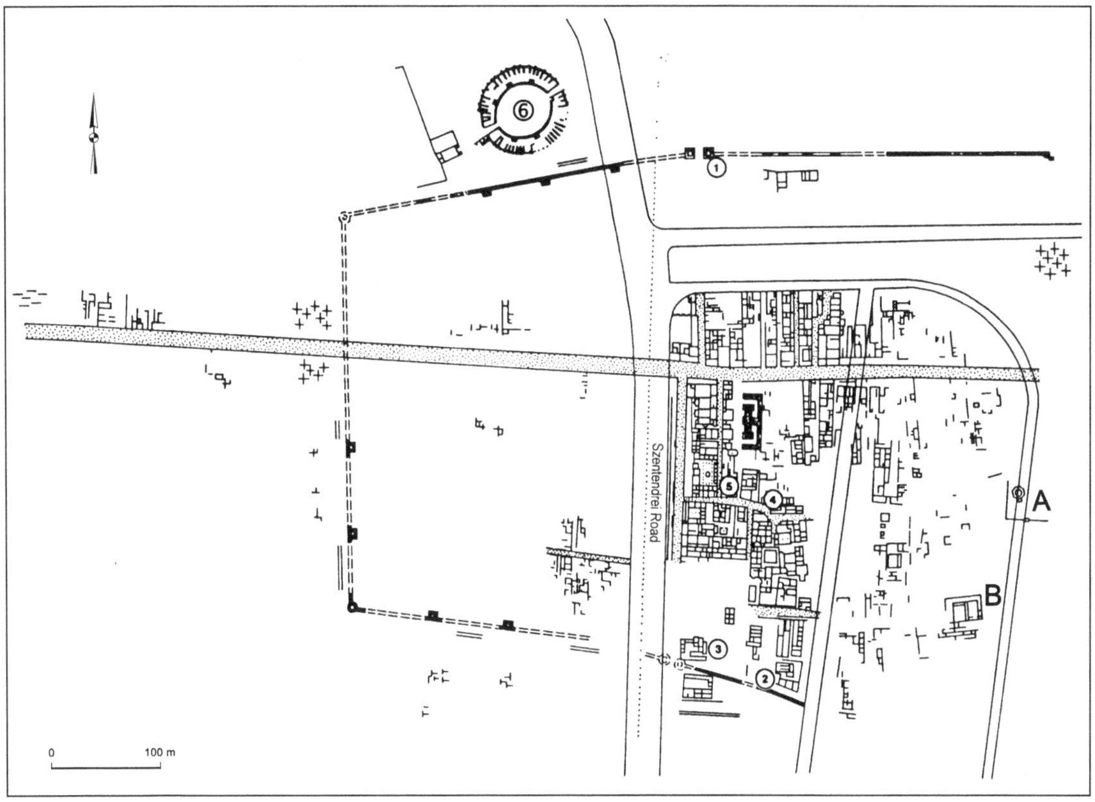

54. Aquincum, general plan of the civic town, based on NÉMETH 1991b. 1. Pók Street and the northern city wall, 2. Symphorus mithraeum, 3. Fire Brigade, 4. dwelling-house next to M. Ant. Victorinus mithraeum, 5. E-F, C-G streets, 6. amphitheatre of the civic town. **A**. Celtic shrine, **B**. synagogue.

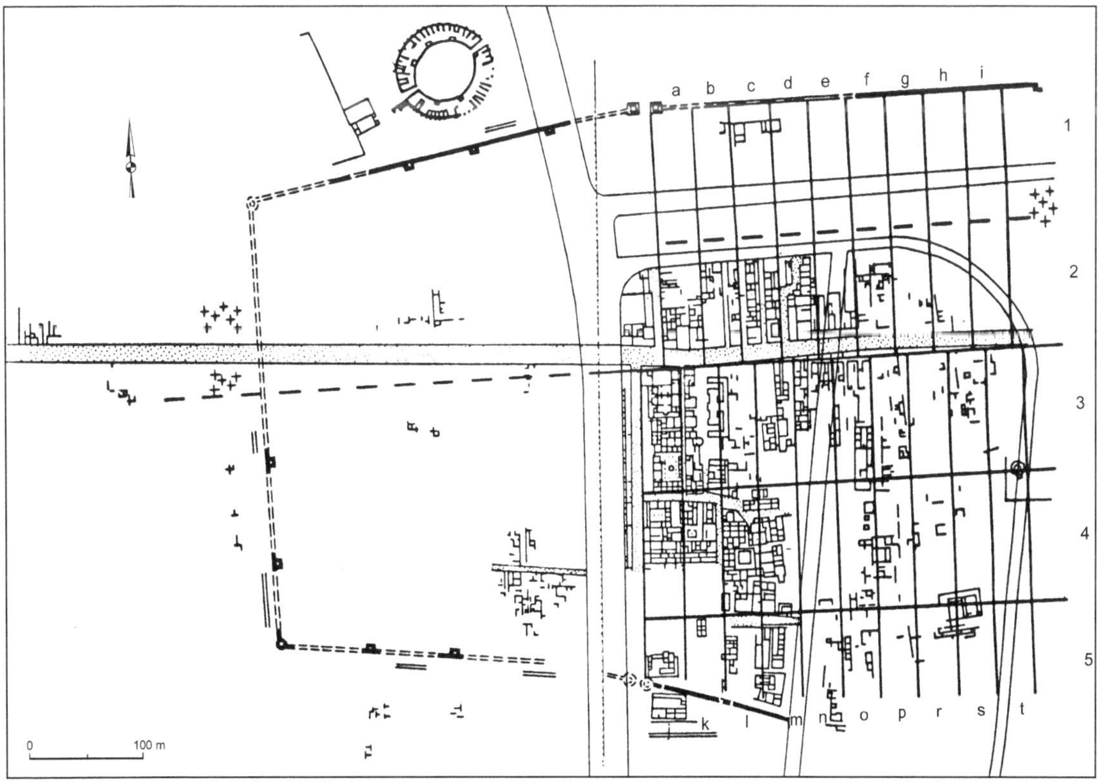

55. Aquincum, civic town, centuriatio based on MÁRITY 1993. A unit is composed by the a-b-c *insulae* north of the E-W main road, while the j-k-l-m *insulae* compose a unit to the south.

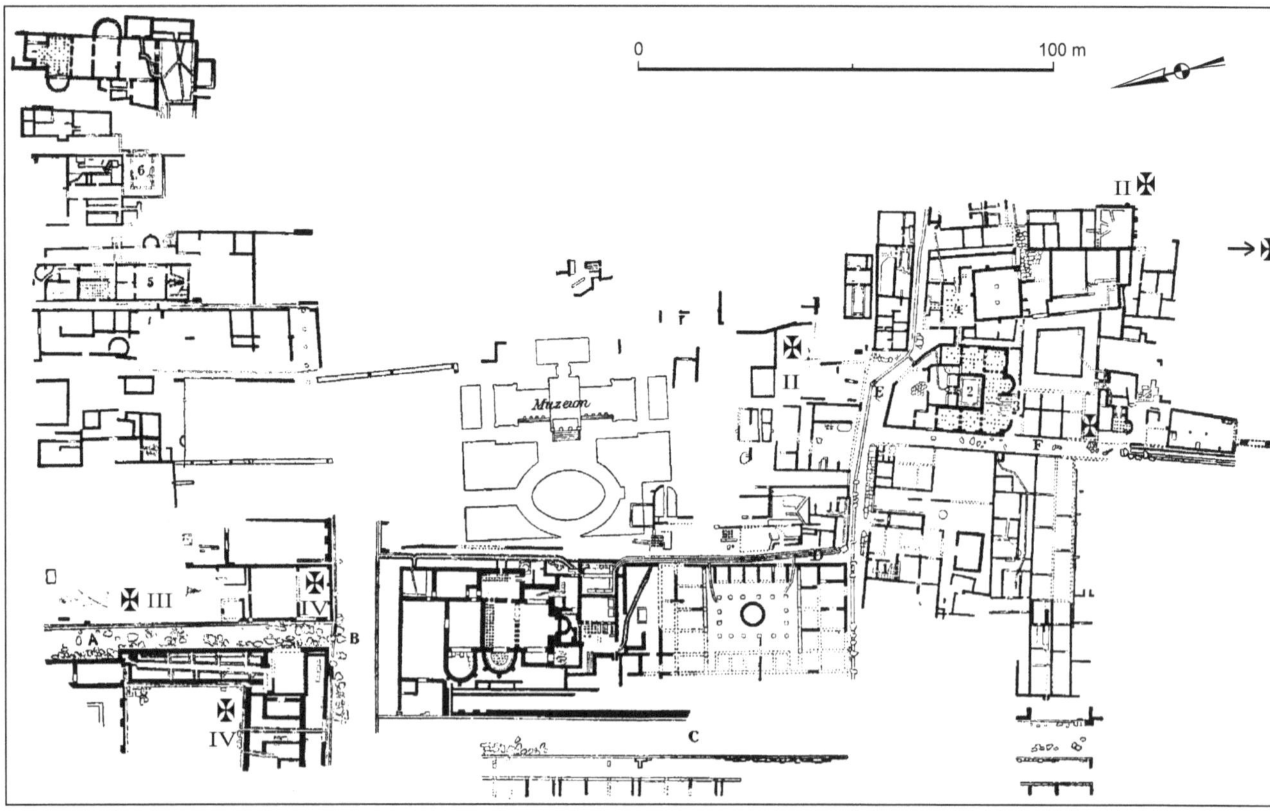

56. Aquincum, civic town, the hypothetical place of the Christians. The cross shows the hypothetical place, and the numbers besides the crosses mean the centuries.

57. Aquincum, civic town, M. Ant. Victorinus mithraeum. Excavation photo. Based on Budapest Topográfiája II 454.

58. Detail of the M. Ant. Victorinus mithraeum.

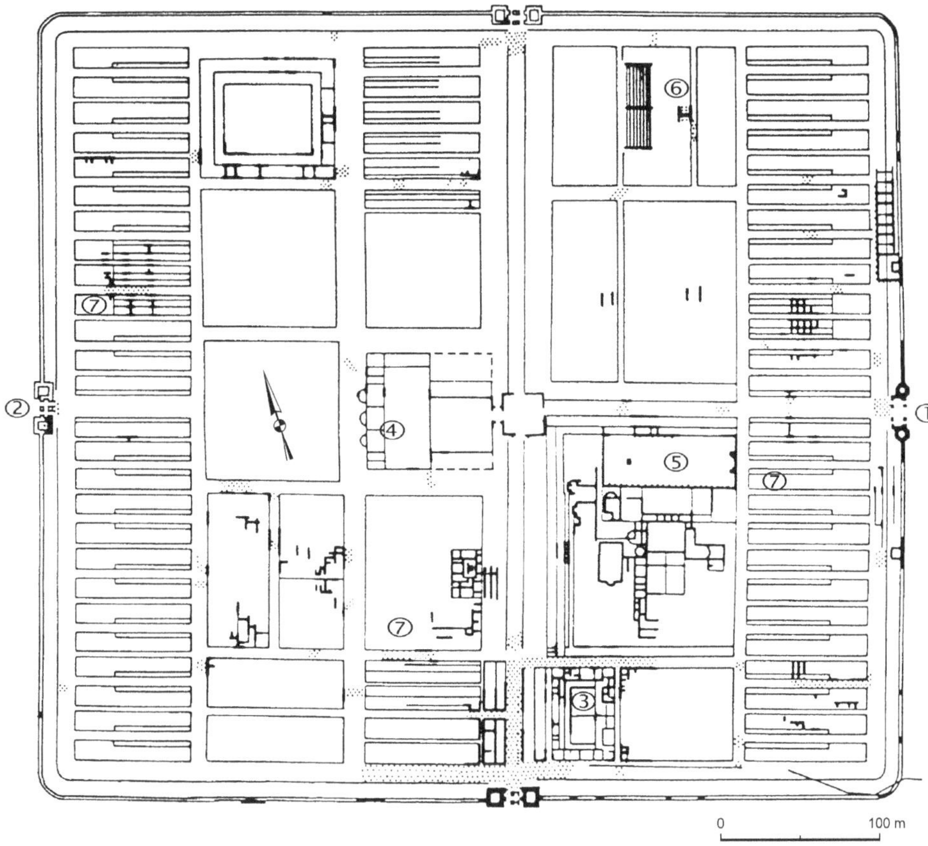

59. Aquincum, ground plan of the legionary fortress: 1. porta praetoria, 2. porta decumana, 3. the dwelling-house *tribuni laticlavii*, 4. sanctuary of standards of the principia, 5. thermae maiores, 6. horreum, 7. barracks (cf. fig. 60).

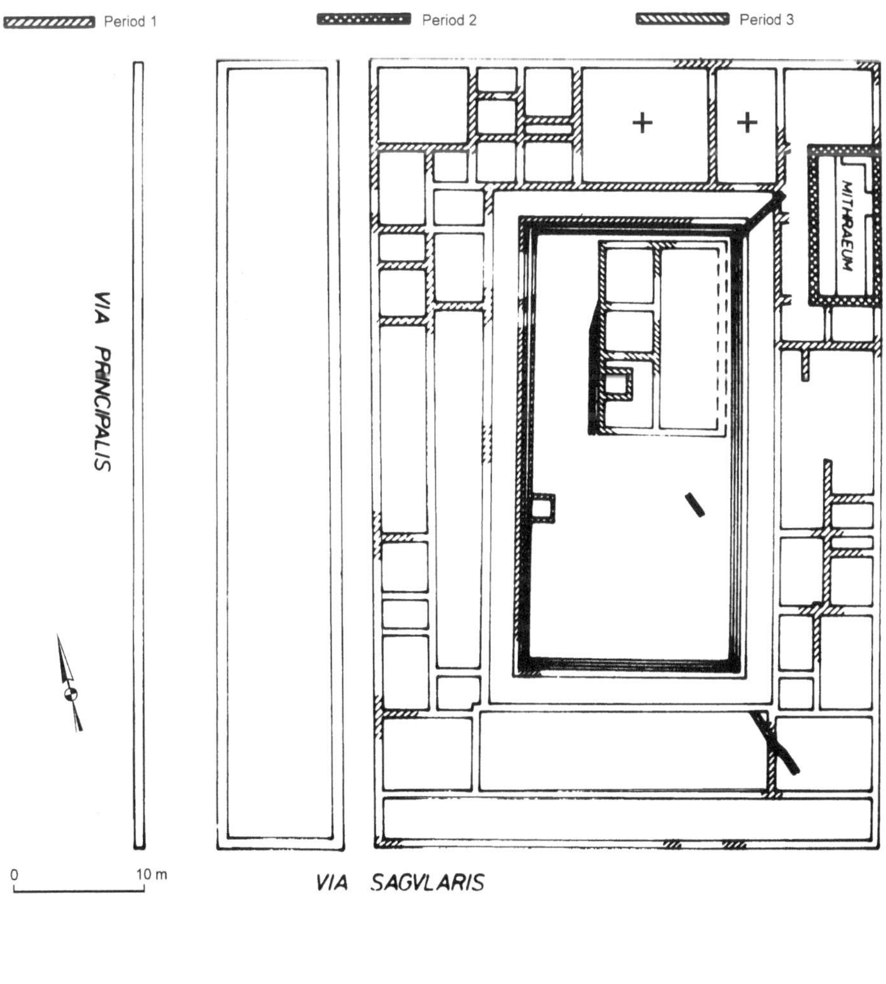

60. Aquincum, legionary fortress, dwelling-house of *tribuni laticlavii* with the mithraeum, Based on KOCSIS 1990. The places marked with a cross are supposed to have been in Christian use.

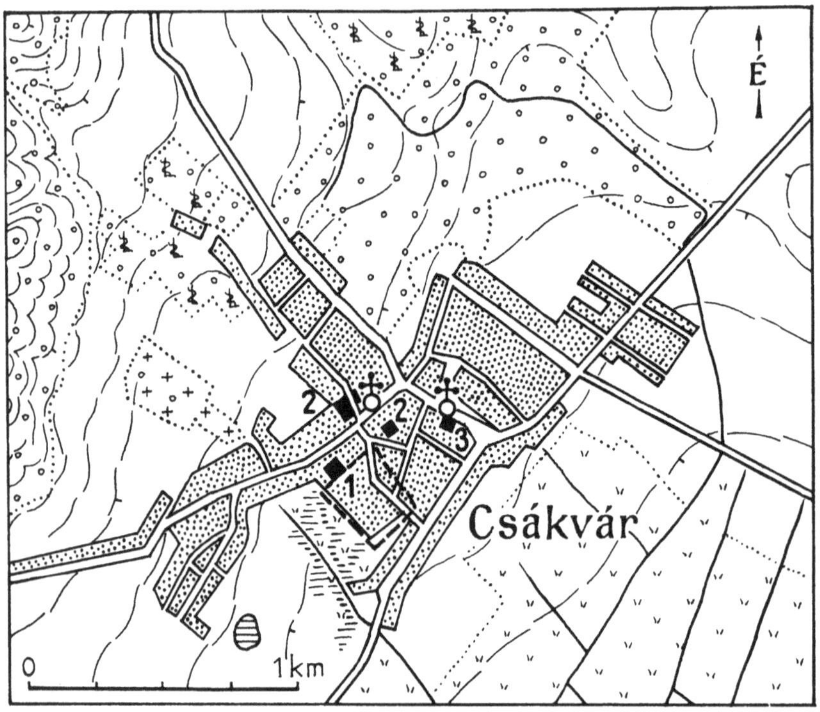

61. Csákvár, general plan of the excavations. Based on SALAMON / BARKÓCZI 1970. 1: excavation in 1937, 2: excavations between 1929 and 1960, 3: excavation in 1962.

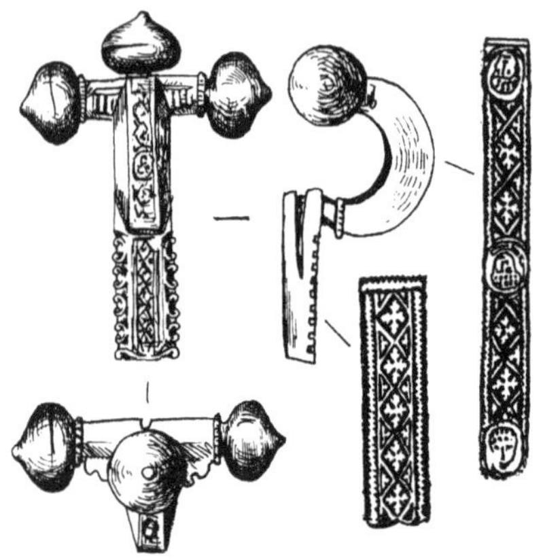

62. Csákvár, fibula, 5.II.a. Based on SALAMON/BARKÓCZI 1970.

63. Császár, casket mount, 6.I.a. Photo: SUGÁR Lajos.

64. Detail of 6.I.a.

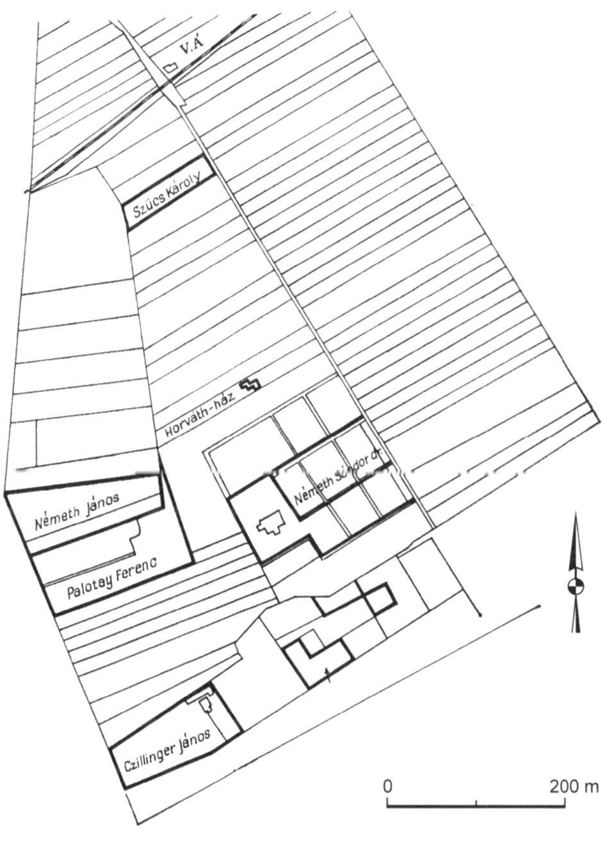

65. Csopak, general plan of the sites. Based on MRT 2, 17/4.

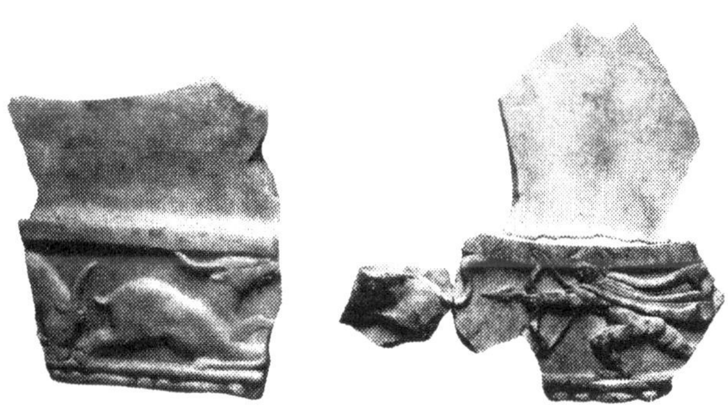

66. Csopak, marble fragments of sigma-shaped table, 7.II.a. Based on THOMAS 1954.

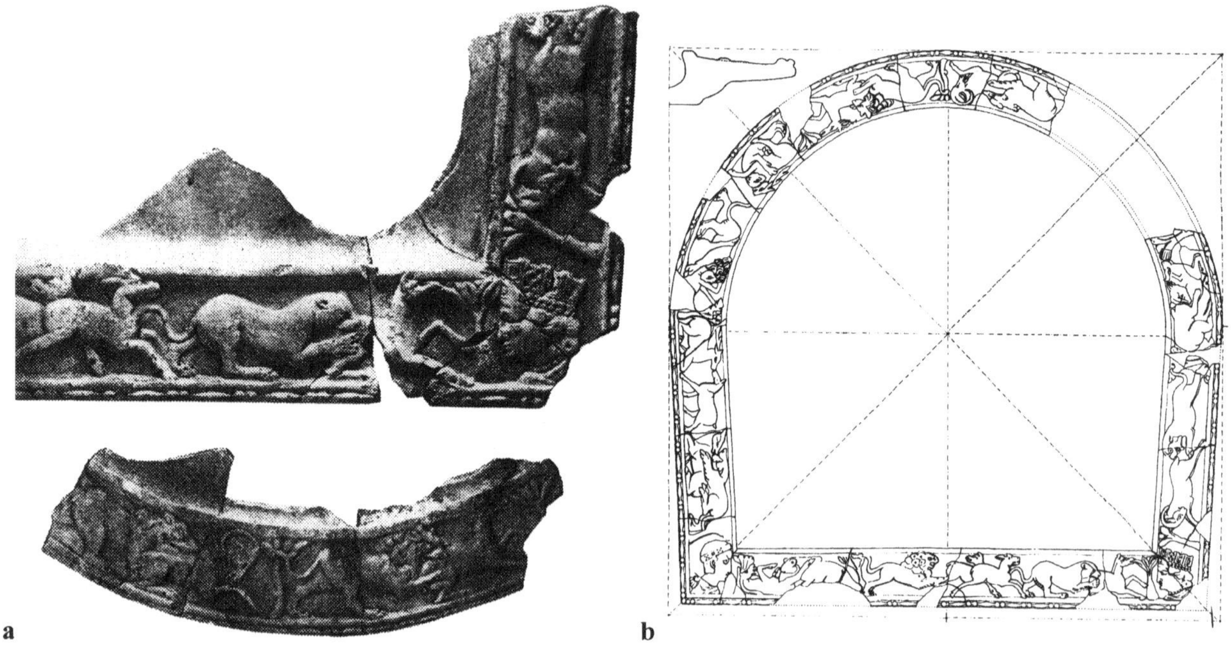

67. Hama, marble pieces of sigma-shaped table (**a**) and its reconstruction (**b**). Based on PLOUG 1985.

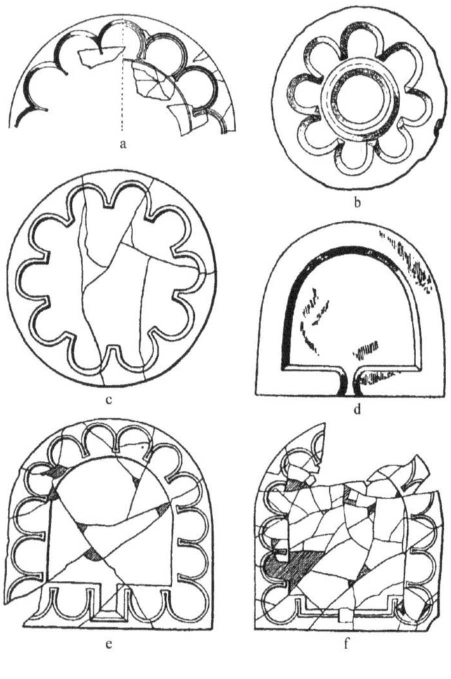

68. Sigma-shaped tables: a: Donnerskirchen, b: Lechaion, c: Delos, d: Al-Mu'allakah, e: Ephesos, f: Aquileia. Based on GAMBER 1968.

69. Illustration of sigma-shaped table. Chludov-Psalter, based on GAMBER 1968.

70. Dombóvár, roof tile, 8.I.a. Based on PÉTERFI 1993.

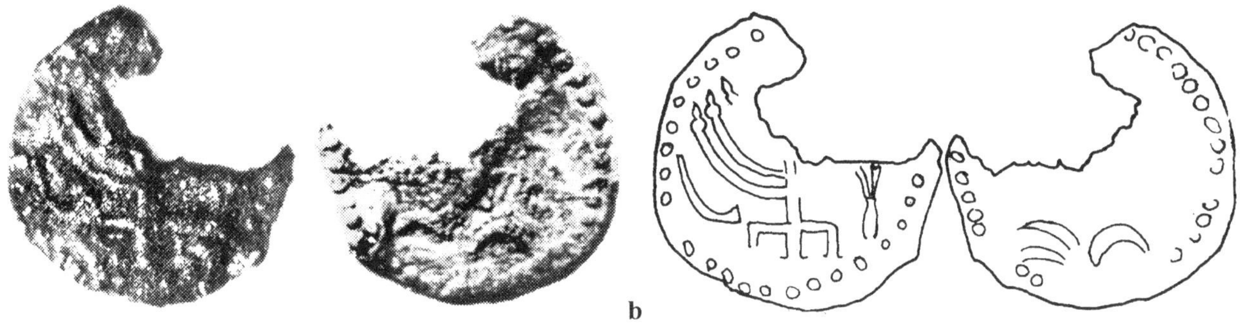

a b

71. Dombóvár, amulet, 8.I.b. **a:** photo, based on SCHEIBER 1983. **b:** drawing, based on ALBEKER 1978.

72. Dunapataj, finger ring, 10.I.a. Based on NAGY L 1938.

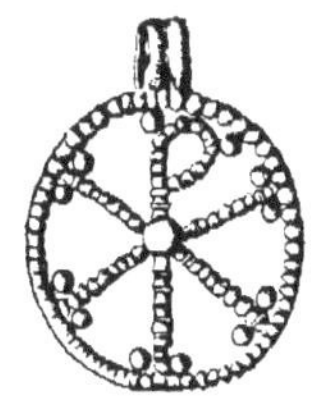

73. Dunaszekcső, pendant, 11.I.a. Based on HAUG 1902.

74. Dunaszekcső, fondo d'oro, 11.I.a. Based on FÜLEP 1968.

75. Dunaújváros, general plan of the excavation: 1. fort, northern cemetery and the northern part of the canabae, 2. canabae south of the fort, 3. hypothetical extension of the southern cemetery - here is the find-spot of the 12.II.a. Based on BARKÓCZI 1954 Appendix.

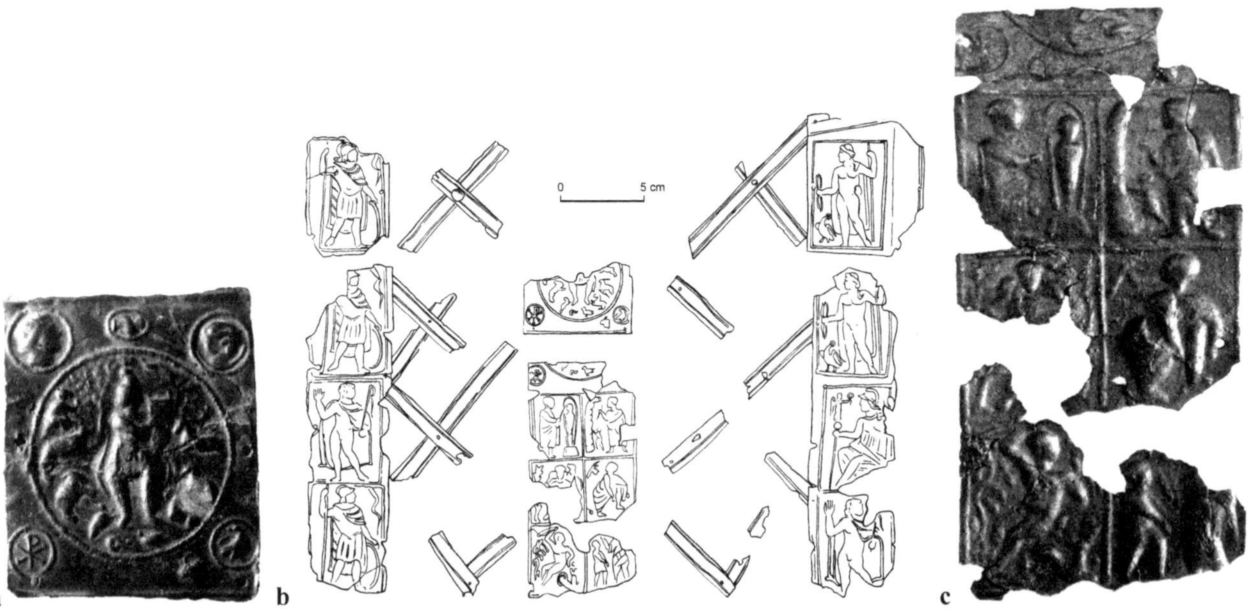

76. Dunaújváros, casket mount, 12.I.a. **a:** mount with an Orpheus scene, **b:** complete face of the mount, **c:** mount with biblical scenes. Photo: KÁDAS Tibor.

77. Dunaújváros, lid, 12.I.b. Photo: KÁDAS Tibor.

78. Dunaújváros, casket mount, 12.I.c. Photo: KÁDAS Tibor.

79. Dunaújváros, casket mount, 12.I.d. Based on BUSCHHAUSEN 1971.

80. Dunaújváros, casket mount, 12.I.e. Based on BUSCHHAUSEN 1971.

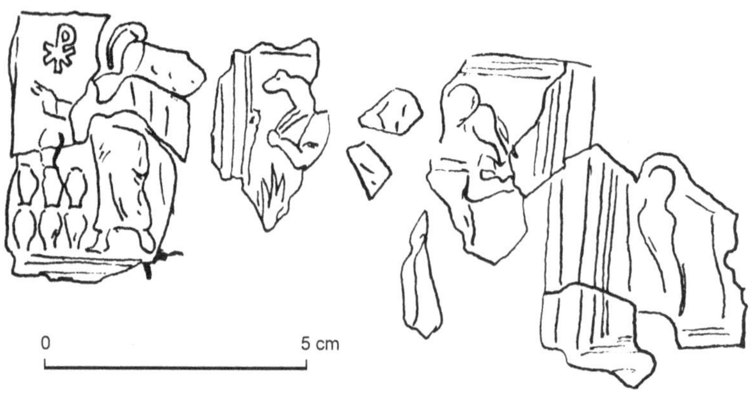

81. Dunaújváros, casket mount, 12.I.f. Drawing by author.

82. Dunaújváros, casket mount, 12.I.g. Based on BUSCHHAUSEN 1971.

83. Rome, fondo d'oro. Based on TESTINI 1966.

84. Dunaújváros, casket mount, 12.I.h. Based on NAGY L 1931b.

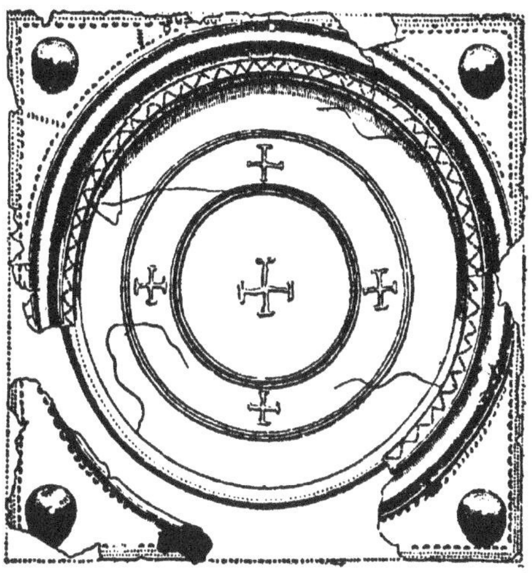

85. Dunaújváros, silver mirror, 12.I.i. Based on RADNÓTI 1957.

86. Dunaújváros, glass vessel, 12.I.j. Based on VISY 1977.

87. Dunaújváros, standard lamp, 12.I.k. Based on THOMAS 1988.

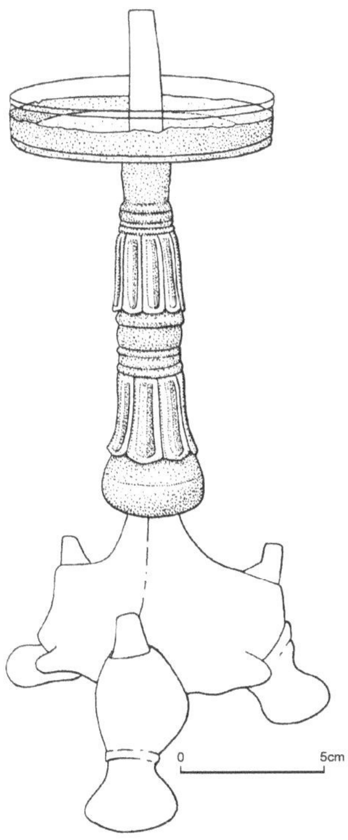

88. Dunaújváros, reconstruction of standard lamp. Drawing by ŐSI Sándor.

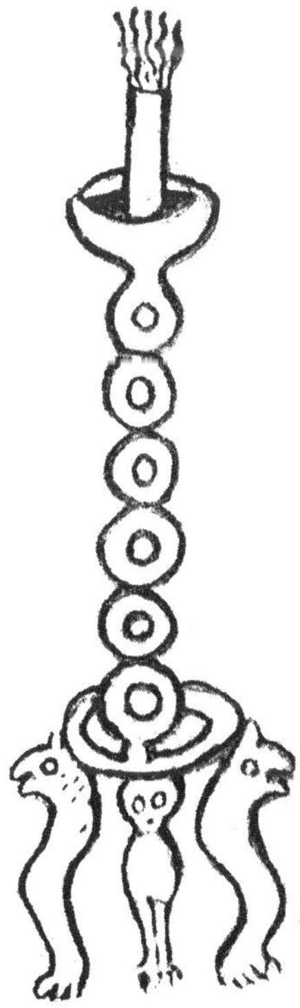

89. Illustration of standard lamp on a stone relief, The Lateran Museum. Based on TESTINI 1966.

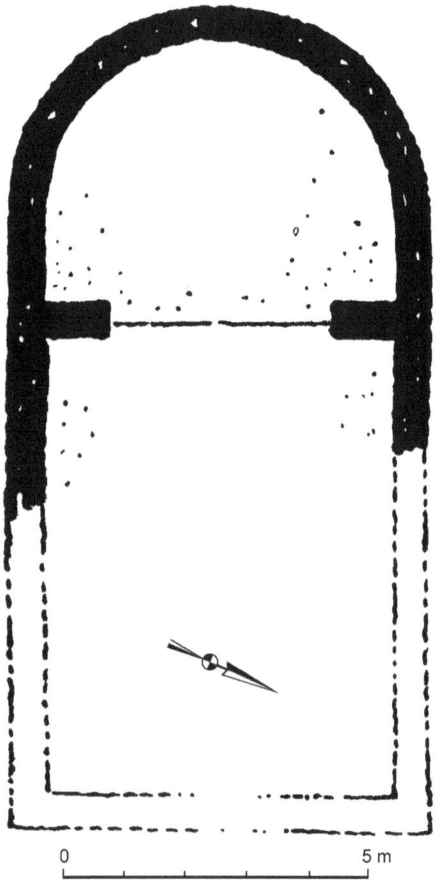

90. Dunaújváros, so-called early Christian church, 12.II.a. Based on VISY 1977.

91. Dunaújváros, fondo d'oro, 12.II.b. Based on FÜLEP 1968.

92. Dunaújváros, votive plaque, 12.II.c. Based on FÜLEP 1966.

93. Detail of 12.II.c. Based on FÜLEP 1966.

94. Detail of 12.II.c.

95. Dunaújváros, tombstone, 12.II.d. Based on SCHEIBER 1983.

96. Dunaújváros, sarcophagus of M. Aur. Sallumas, 12.II.e. Based on SCHEIBER 1983.

97. Dunaújváros, tombstone of Aurelia Baracha, 12.II.f. Based on SCHEIBER 1983.

98. Inscription field of tombstone 12.II.f.

99. Egyházashollós, marble screen, 13.II.a. Based on THOMAS 1975-1976.

a b

100a-b. Esztergom-Bánomi dűlő, Jewish tombstone of Kassia, 14.I.a. Based on SCHEIBER 1983, and RADAN 1973.

101. Felsődörgicse, carved stone plaque, 16.II.a. Based on TÓTH E 1980b.

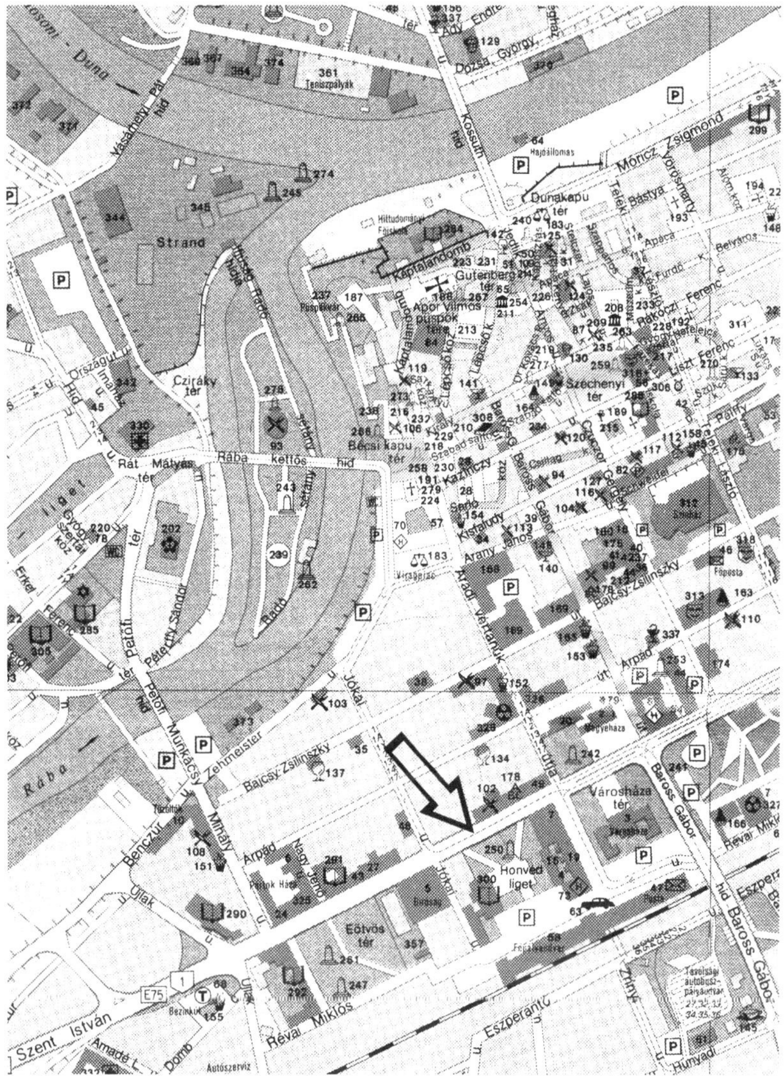

102. Győr Honvéd Liget (grove). Section of a tourist map.

103. Győr lamp, 17.I.b. Based on IVÁNYI 1935.

104. Kajár, inscribed roof tile, 18.II.a. Photo: GAJZÁGÓ Jolán.

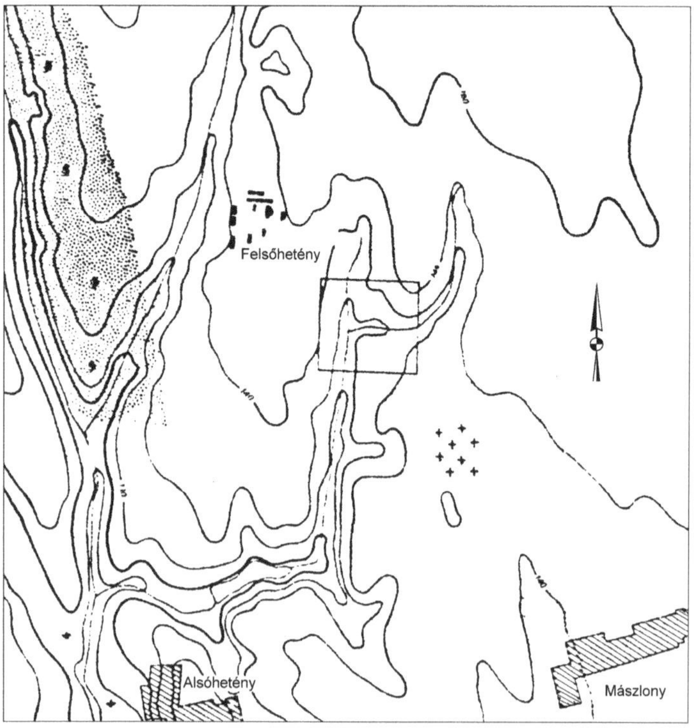

105. Kapospula-Alsóheténypuszta, site plan. Based on TÓTH E 1988b.

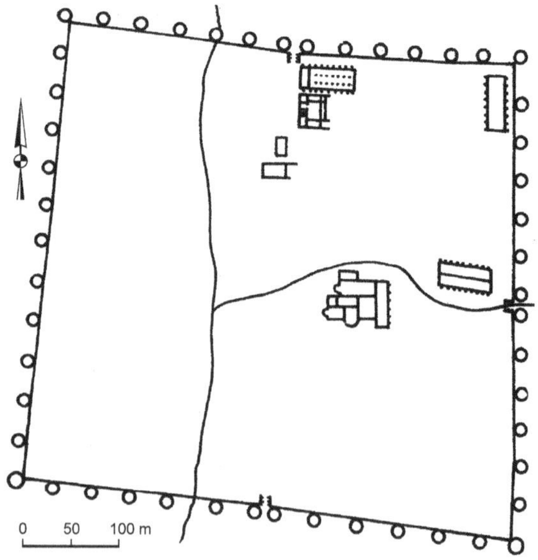

106. Kapospula-Alsóheténypuszta, ground plan of the fortified settlement. Based on TÓTH E 1988b.

107. Kapospula-Alsóheténypuszta, double grave. Based on TÓTH E 1989b.

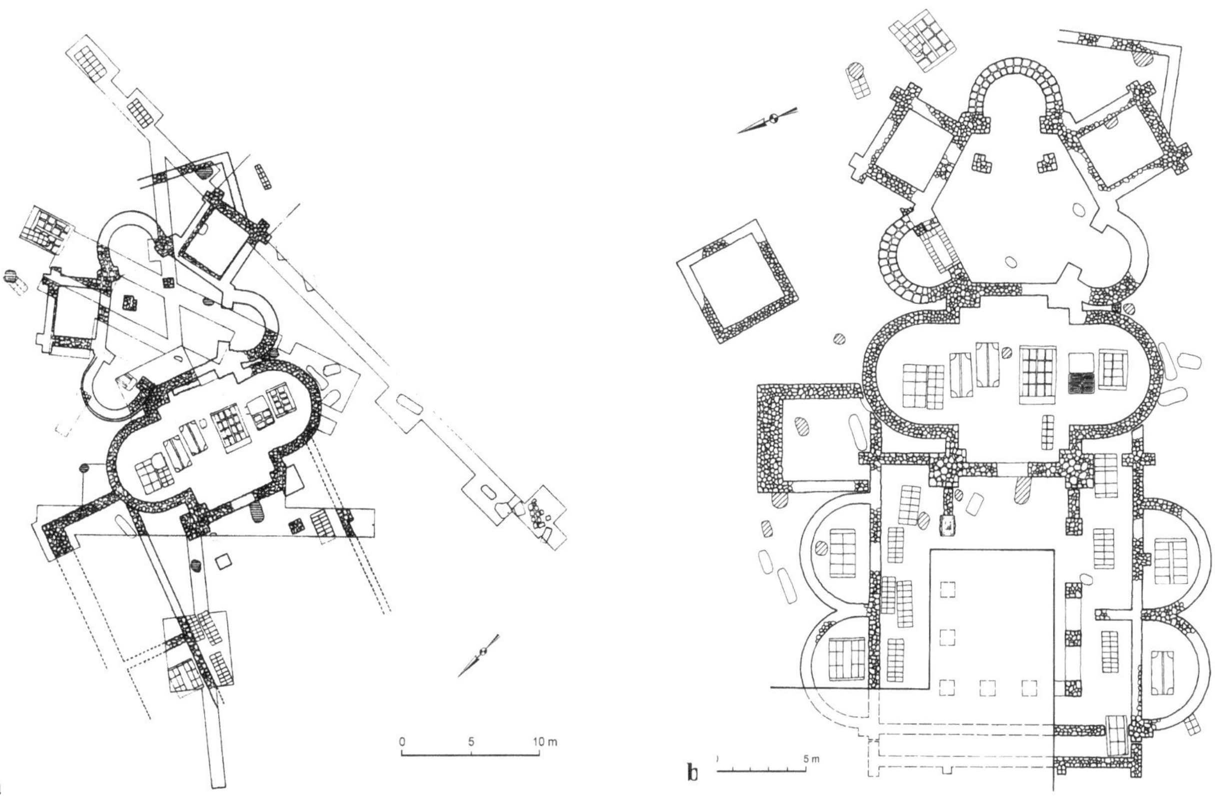

a b

108a-b. Kapospula-Alsóheténypuszta, ground plan of the mausoleum. Based on TÓTH E 1989b.

109. Kapospula-Alsóheténypuszta, mausoleum, excavation photo of one of the apses with a grave inside. Based on TÓTH E 1989b.

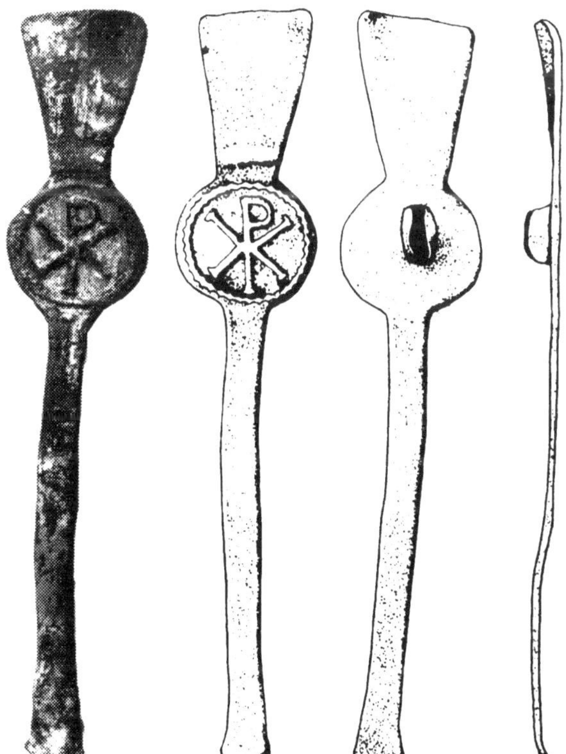

a b

110. Kapospula-Alsóheténypuszta, fibula? 19.I.a. **a:** photo: Dr TÓTH Endre **b:** drawing. Based on TÓTH E 1988b.

111. Kapospula-Alsóheténypuszta, suspension for lamp, 19.I.b. Based on TÓTH E 1990b.

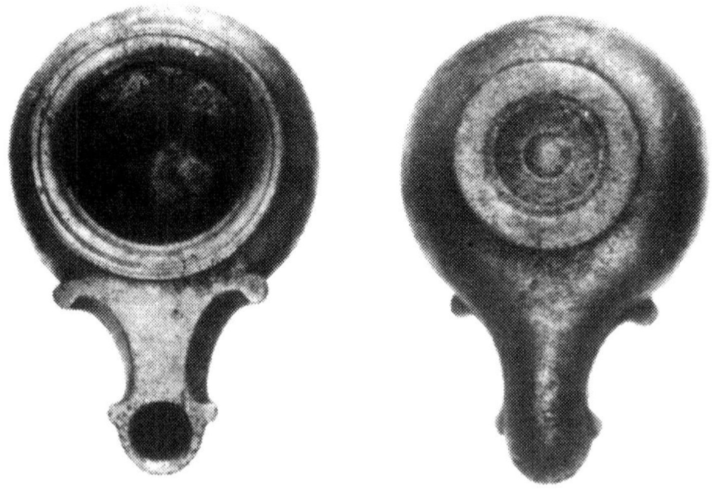

112. Kapospula-Alsóheténypuszta, lamp, 19.I.b. Based on TÓTH E 1990b.

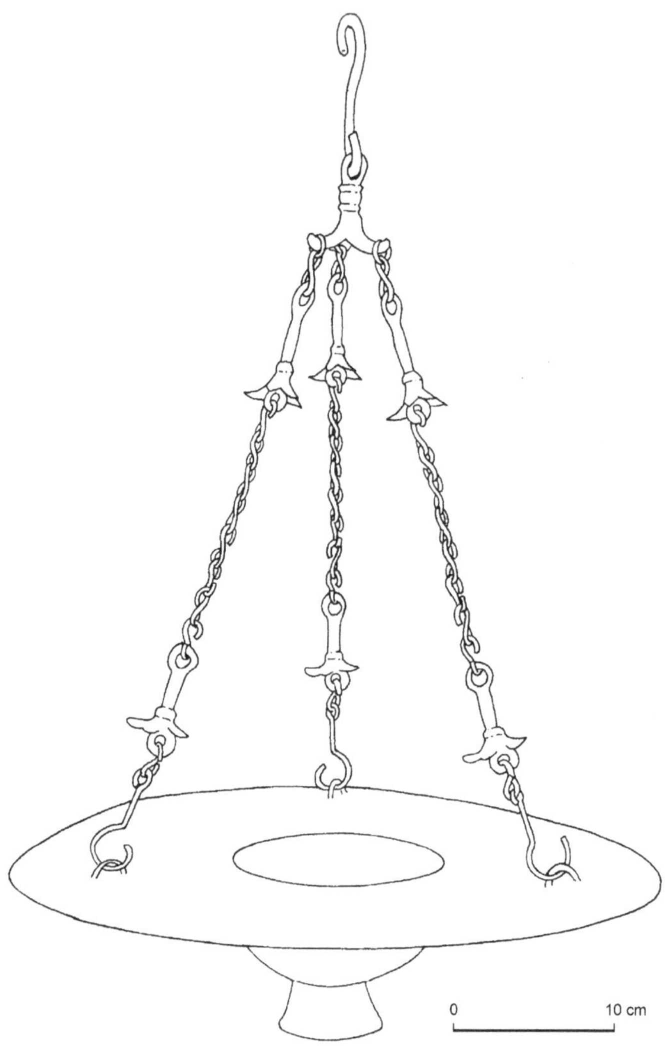

113. Reconstruction of 19.I.b. Drawing by ŐSI Sándor.

114. Kecskemét, finger ring, 20.II.a. Based on NAGY L 1938.

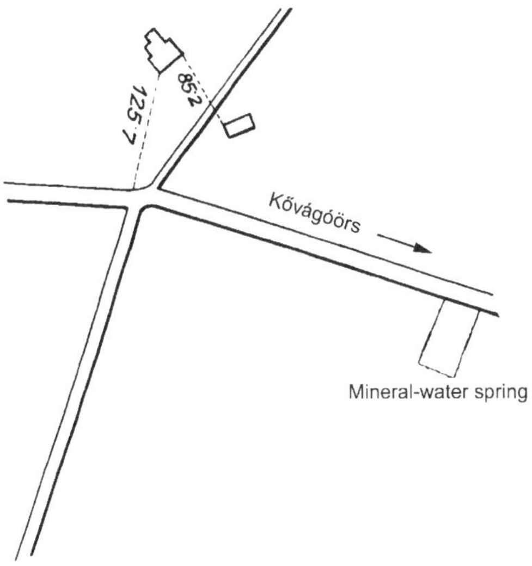

115. Kékkút, general plan. Based on KUZSINSZKY 1920.

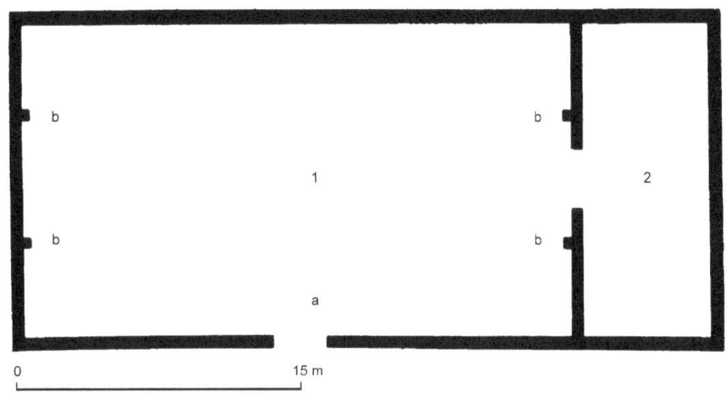

116. Kékkút, ground plan of basilica I, 21.I.b. Based on KUZSINSZKY 1920.

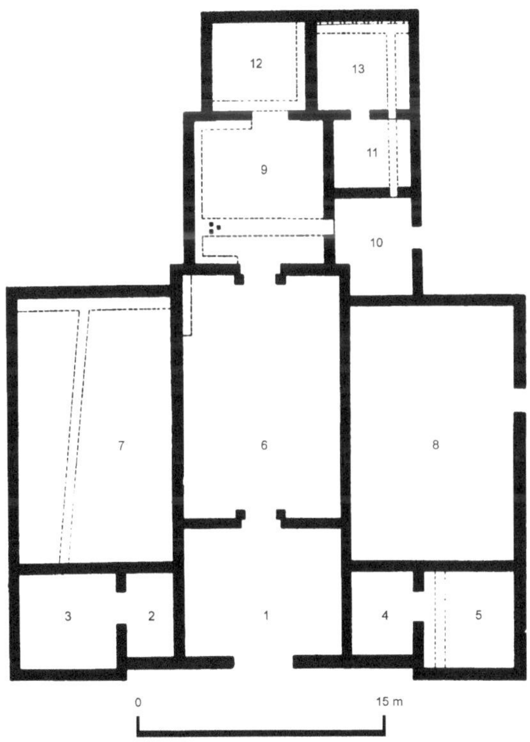

117. Kékkút, basilica II, 21.I.c. Based on KUZSINSZKY 1930.

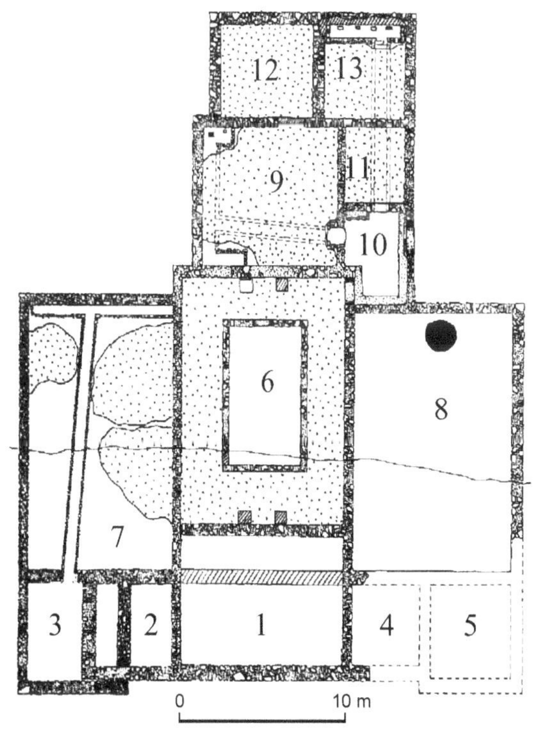

118. Kékkút, excavation sketch of basilica II (21.I.c). Based on SÁGI 1972.

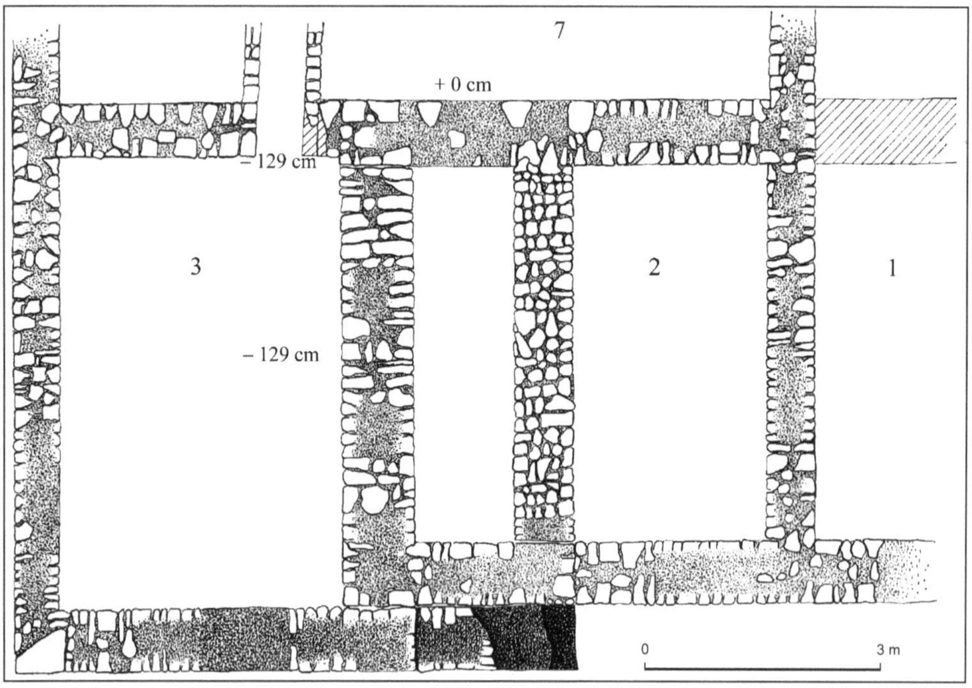

119. Kékkút, excavation sketch of rooms 2-3 in basilica II. Based on SÁGI 1972.

120. Kékkút, openwork clay brick, 21.I.a. Based on KUZSINSZKY 1920.

121. Kékkút, details of columns from basilica I. Based on KUZSINSZKY 1920.

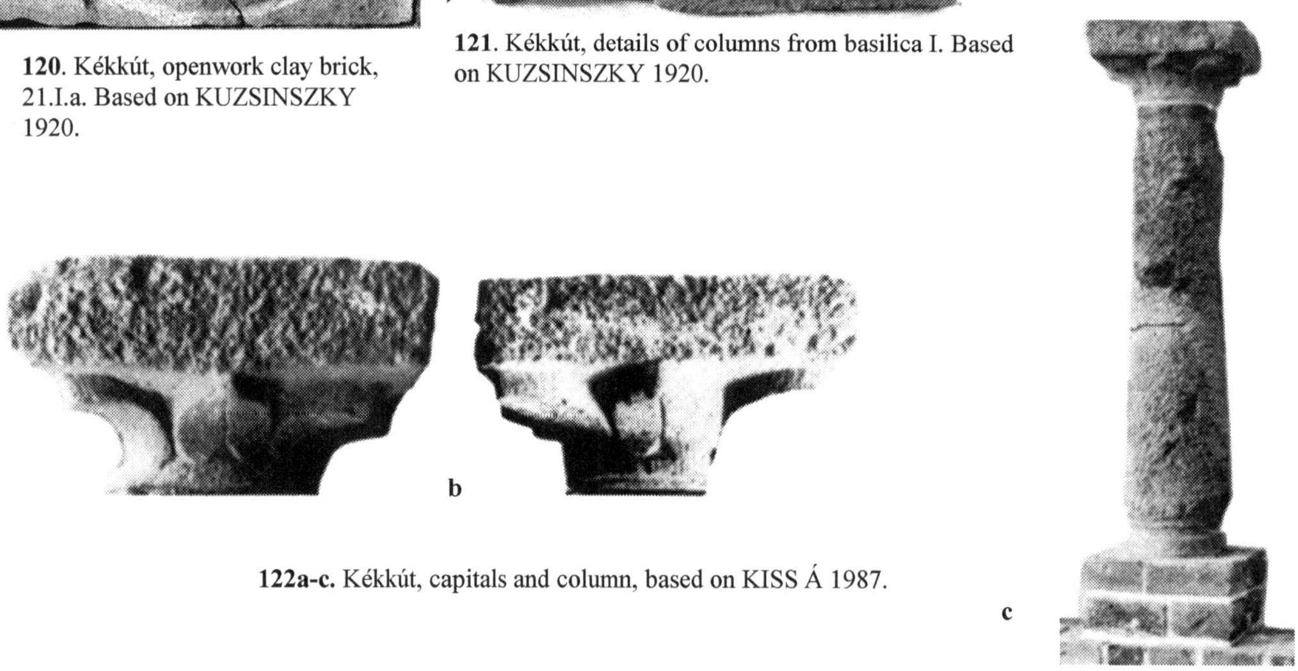

a

b

c

122a-c. Kékkút, capitals and column, based on KISS Á 1987.

123. Kékkút, capital, based on KISS Á 1987.

124. Kékkút? (Zánka) capitals, Based on KISS Á 1987.

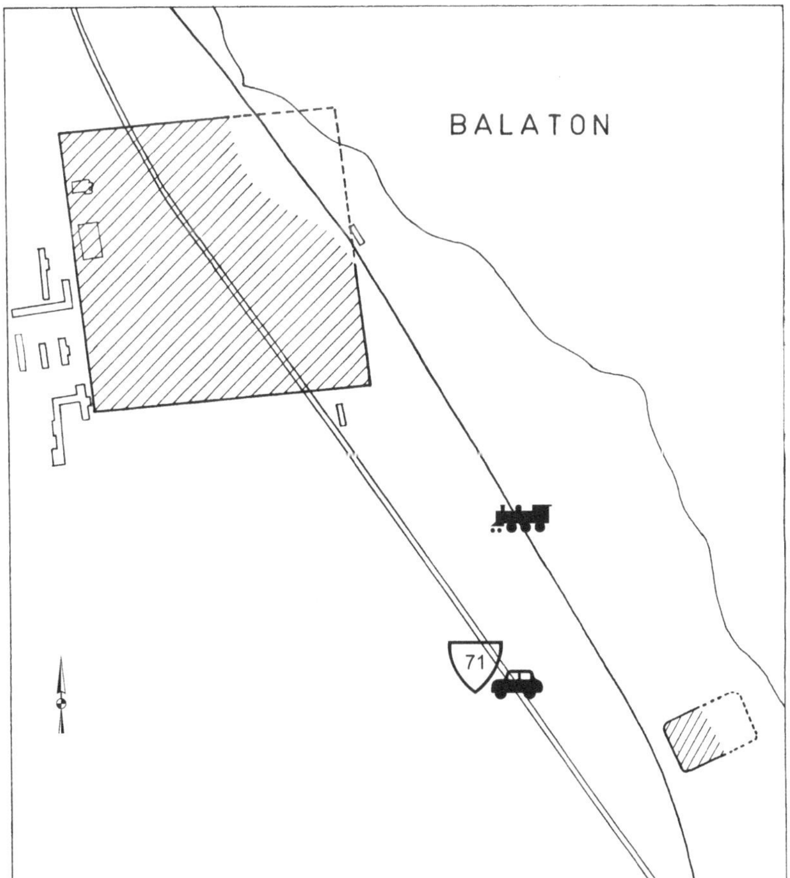

125. Keszthely-Fenékpuszta, site map. Based on MÜLLER 1987.

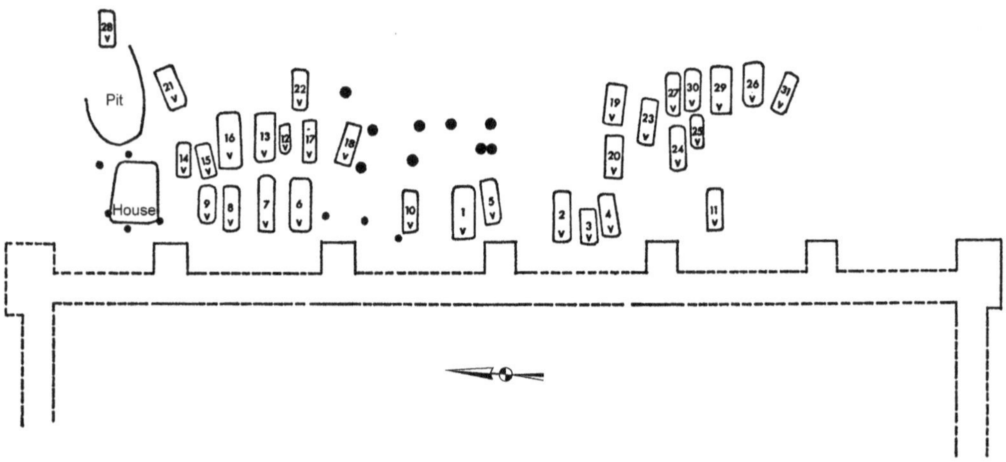

126. Keszthely-Fenékpuszta, cemetery with 31 graves at the eastern side of the horreum. Based on BARKÓCZI/LENGYEL 1971.

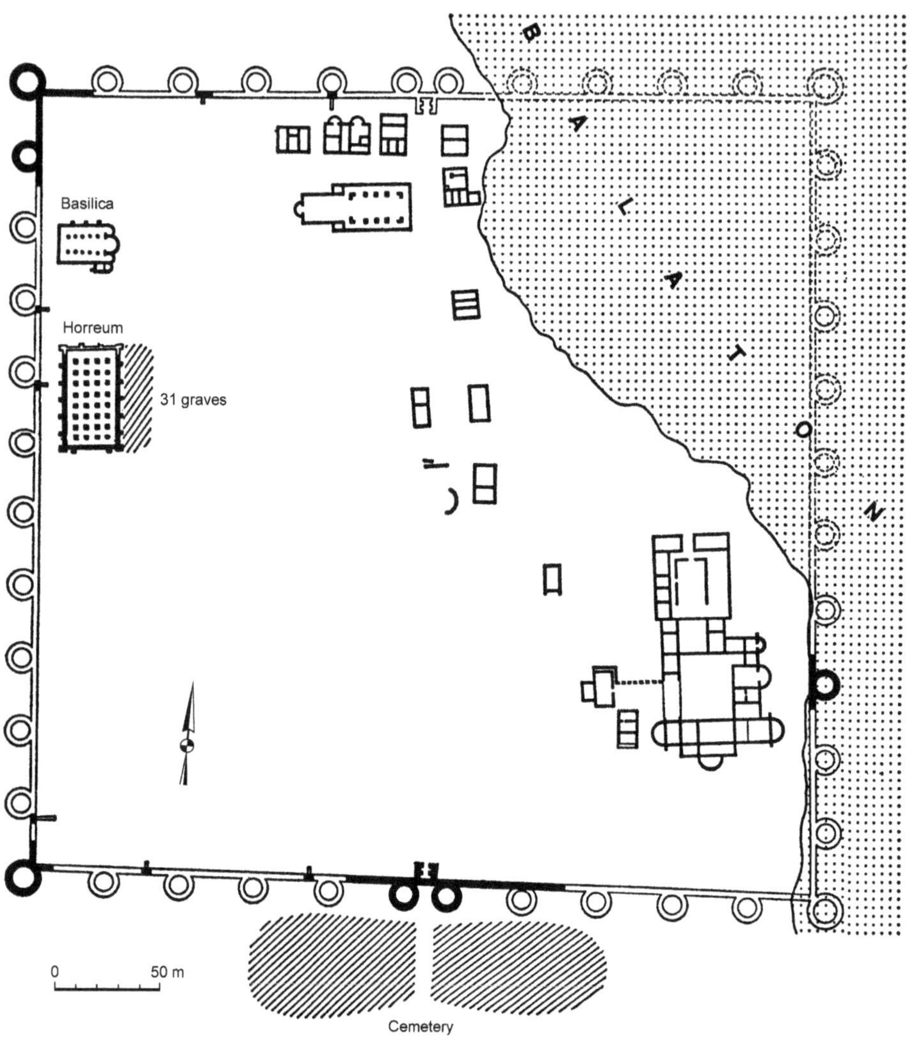

127. Keszthely-Fenékpuszta, ground plan of inner fortified settlement. Based on BARKÓCZI/LENGYEL 1971.

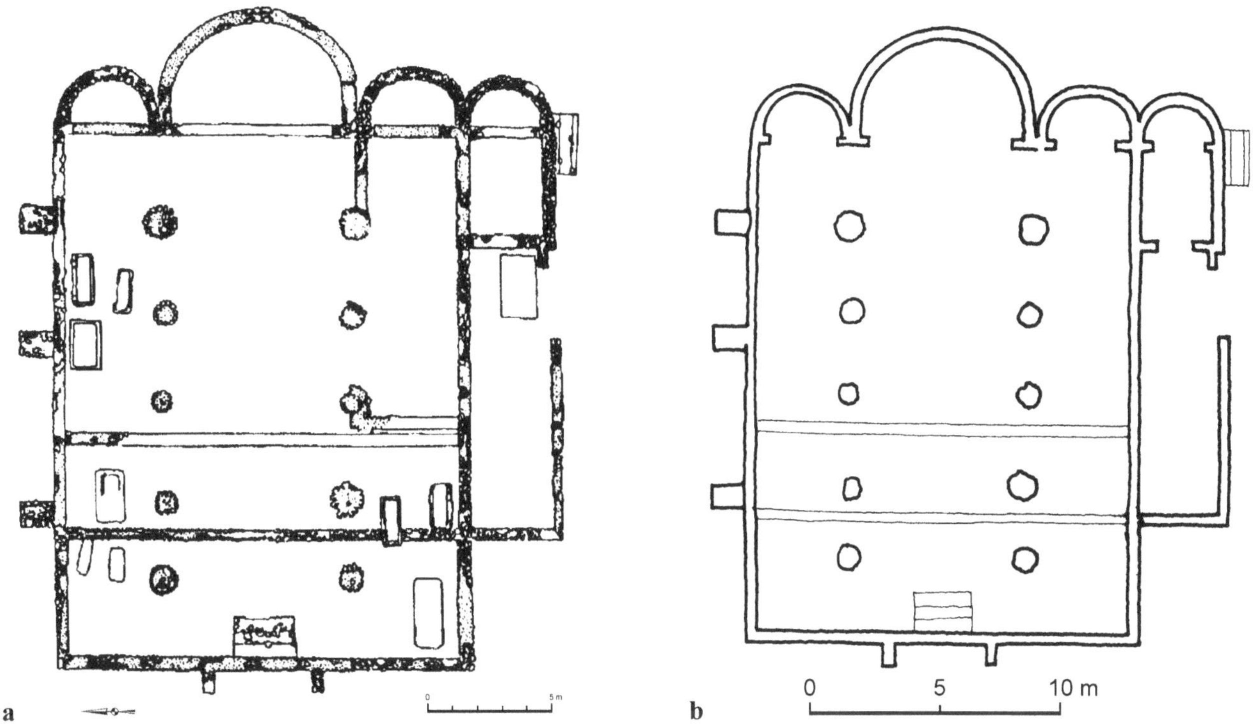

128a-b. Keszthely-Fenékpuszta, basilica, 22.I.a. Based on SÁGI 1961, and LŐVEI 1998.

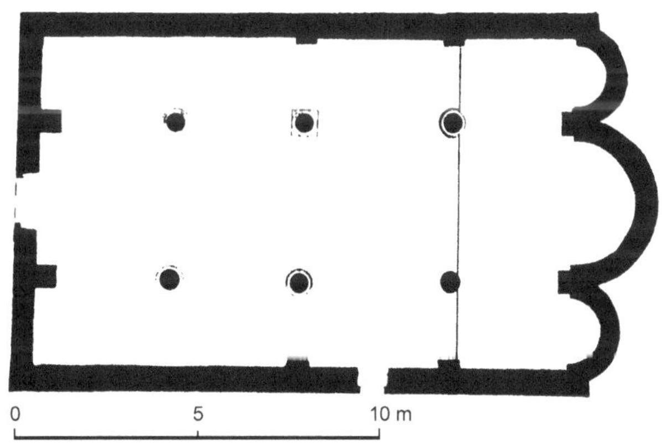

130. Saint Eufemia's Church, Dalmatia. Based on MIGOTTI 1990.

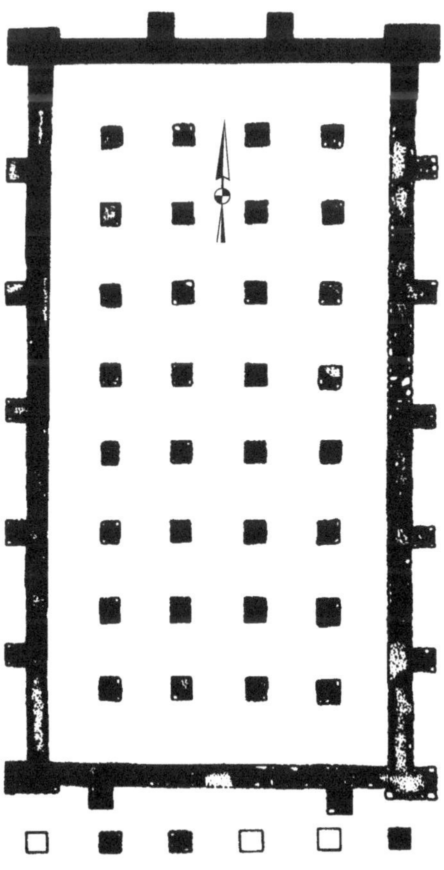

129. Keszthely-Fenékpuszta, horreum, 22.I.b. Based on SÁGI 1989.

131. Keszthely-Fenékpuszta, photo of the excavation of the basilica. The apses. Based on SÁGI 1961.

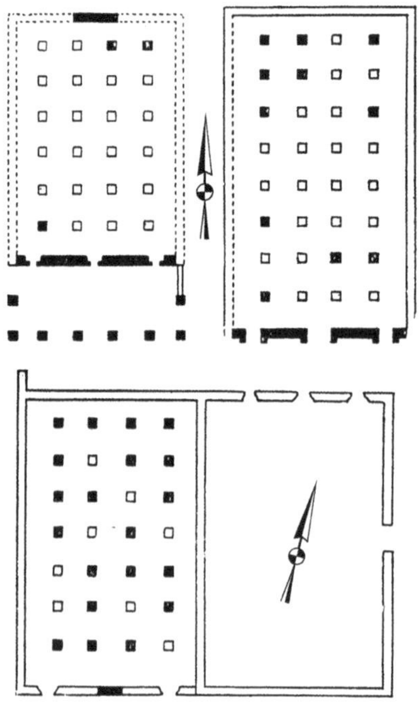

132. Ground plans of synagogues. Based on LECRERCQ 1907.

134. Keszthely-Fenékpuszta, silver spoon, 22.I.d. Based on Severin Katalog.

133. Keszthely-Fenékpuszta, omega, suspended by chain, 22.I.c. Based on Severin Katalog.

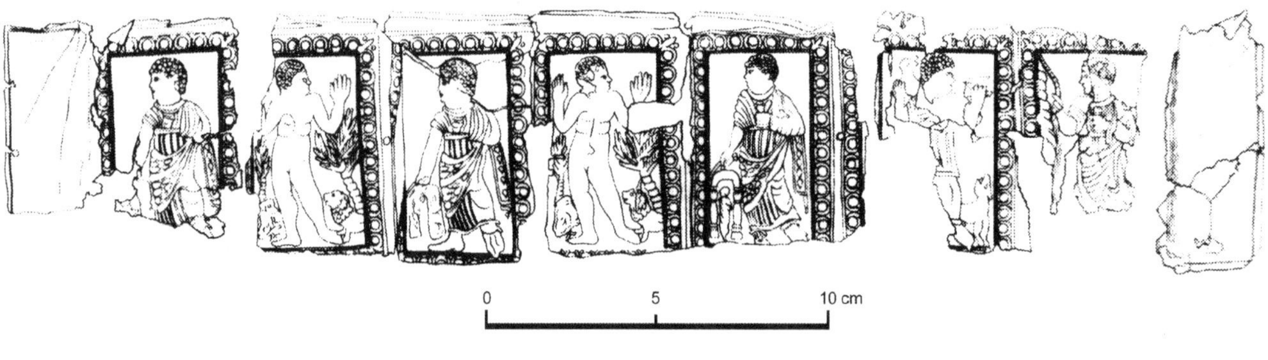

135. Keszthely-Fenékpuszta, casket mount with biblical scenes, 22.I.f. Drawing by SZATHMÁRY Ida.

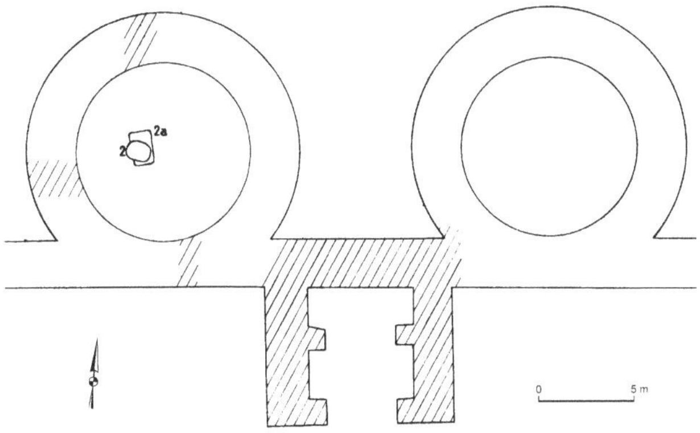

136. Site and environments of 22.I.f. Based on MÜLLER 1978.

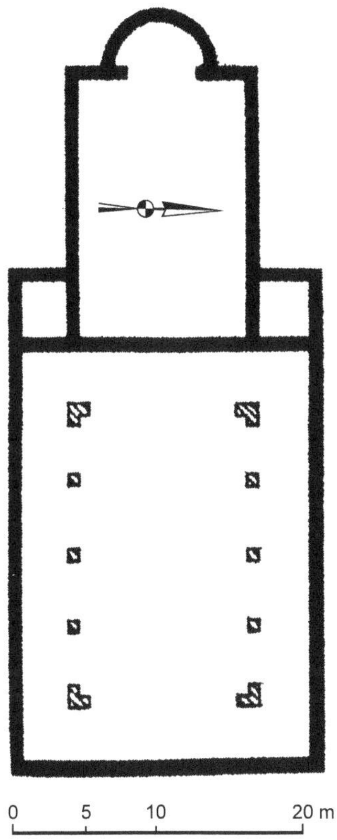

137. Keszthely-Fenékpuszta, so-called basilica No. I, 22.II.a. Based on THOMAS 1980.

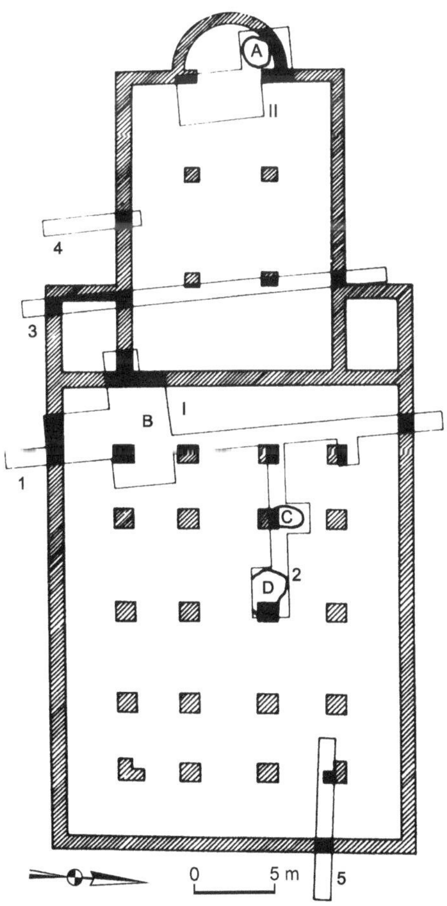

138. Excavation sketch of 22.II.a. Based on SÁGI 1989.

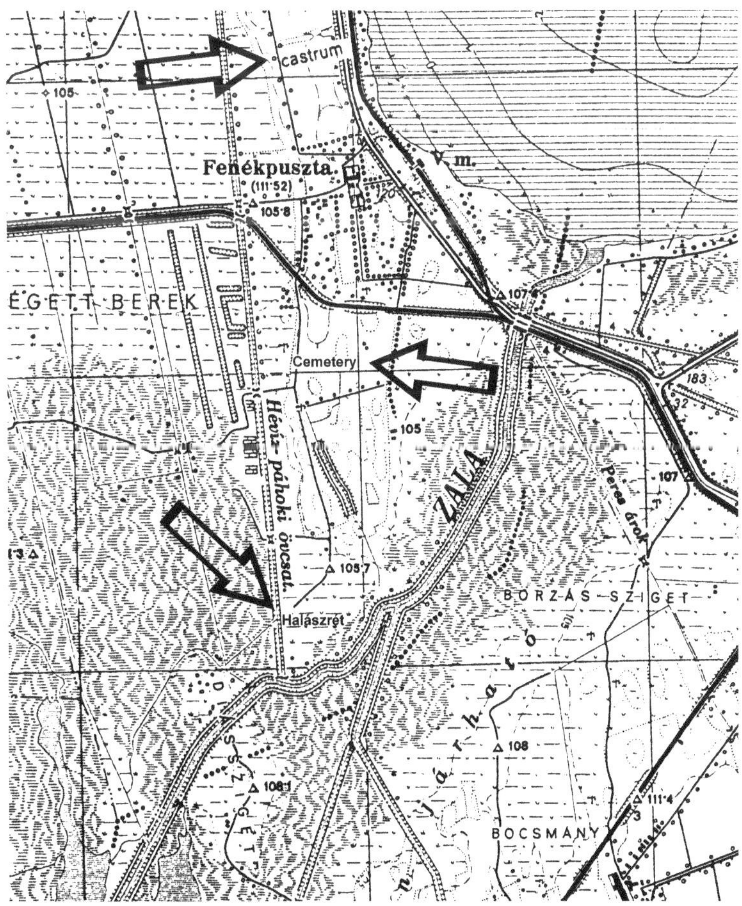

139. Keszthely-Fenékpuszta, Halászrét, general plan.

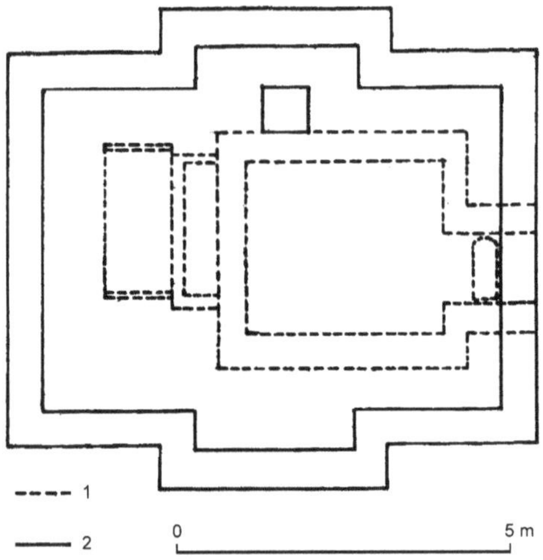

140. Keszthely-Fenékpuszta, Halászrét, ground plan of burial chamber, 22.II.b. Based on SÁGI 1960. 1:walls below the surface, 2: walls above the surface.

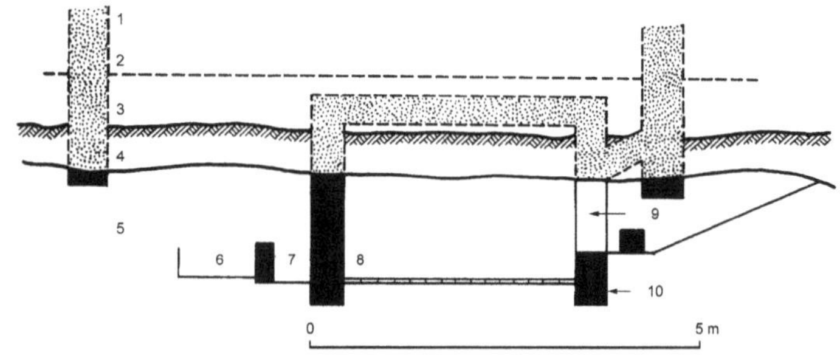

141. Sectional drawing of 22.II.b. Based on SÁGI 1960. 1: walls, took out, 2: supposed original surface, 3: surface to-day, 4: perturbed level, 5: unperturbed level, 6: grave 2, 7: grave 1, 8: grave vault, 9: entrance, 10: walls.

142. Keszthely-Dobogó, bottom stamp of pottery vessel, 23.II.a. Based on SÁGI 1968b

143. Kisdorog, incised roof tile, 24.II.a. **a:** photo, **b:** drawing. Based on THOMAS 1980.

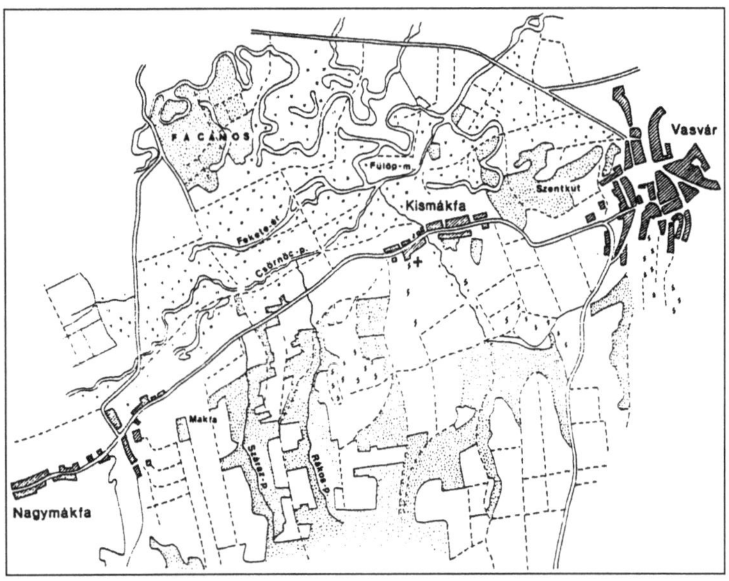

144. Kismákfa, general plan. The cross marks the site. Based on TÓTH E 1994-1995.

145. Kismákfa, so-called paten, 25.II.a. Based on THOMAS 1977b.

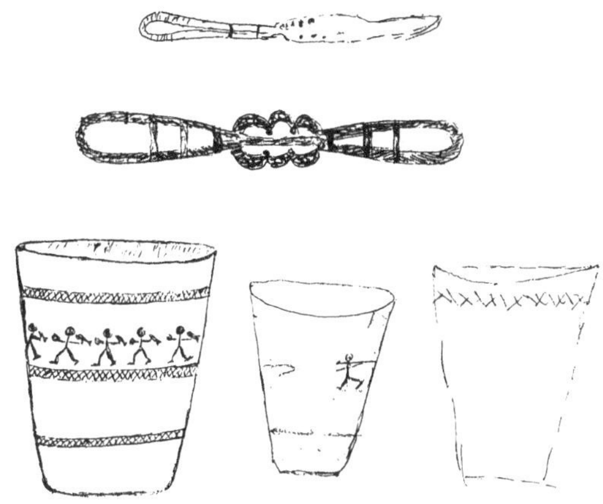

146. Kismákfa, chalice, buckle? (or brooch). Children drawings, based on THOMAS 1977b.

147. Besançon, *mensa*. Based on BARB 1964.

148. Shaman plate. Collection of the Ethnographical Museum.

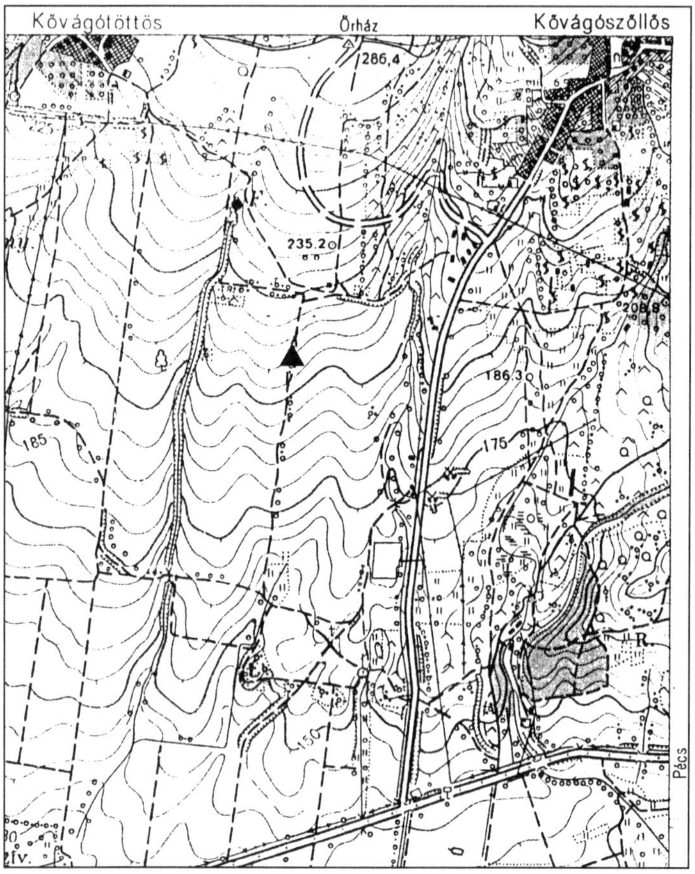

149. Kővágószőlős, general plan. The triangle shows the site. Based on a 19th century map.

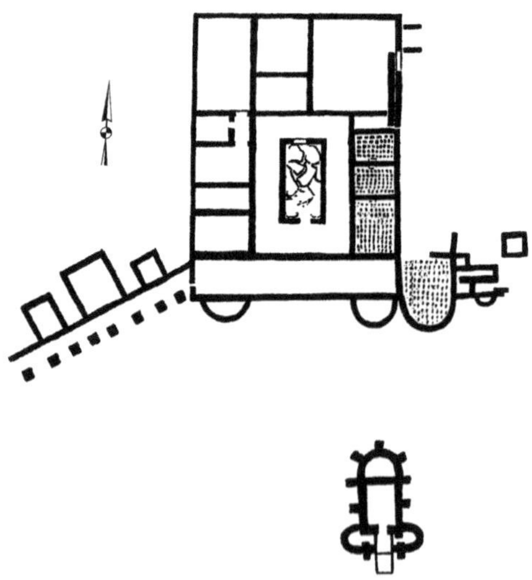

150. Kővágószőlős, ground plan of a villa and the so-called mausoleum. Based on BURGER 1985-1986.

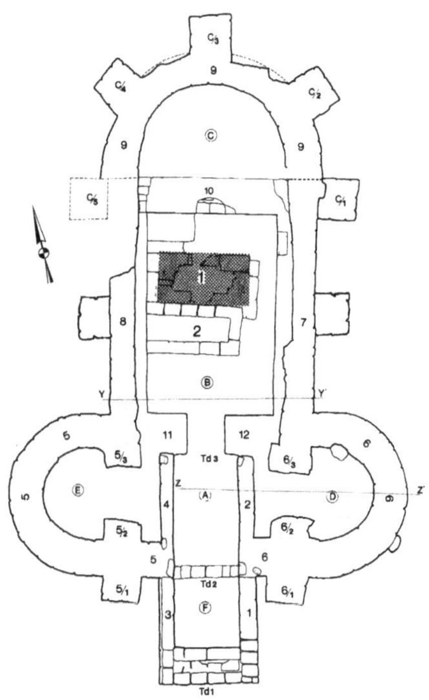

151. Kővágószőlős, ground plan of a burial chamber and the so-called mausoleum, 26.II.a. Based on BURGER 1985-1986.

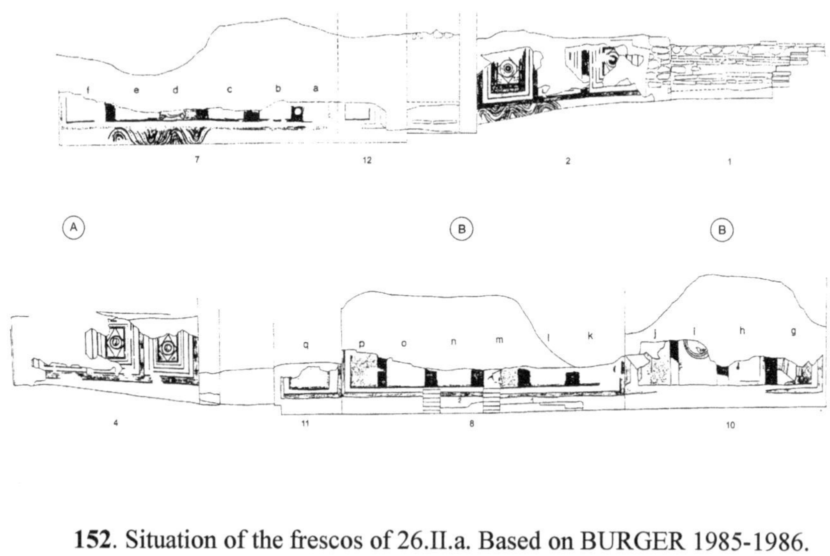

152. Situation of the frescos of 26.II.a. Based on BURGER 1985-1986.

153. Entrance and the two side-apses of 26.II.a, from the south. Based on BURGER 1985-1986.

154. Entrance corridor of 26.II.a. Based on BURGER 1985-1986.

155. Western wall of 26.II.a. at section *B*. Based on BURGER 1985-1986.

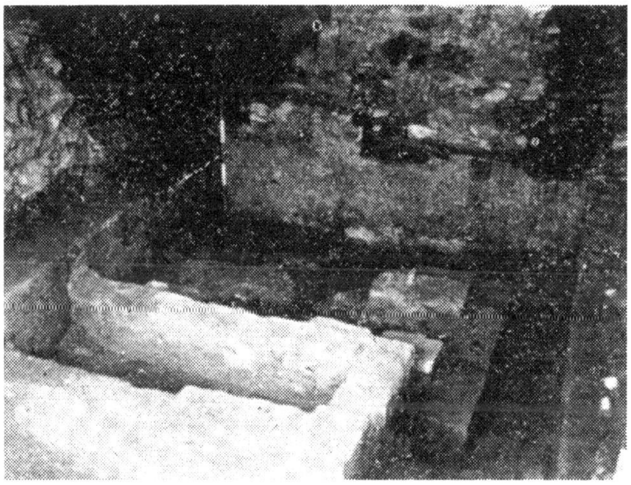

156. Grave of 26.II.a. Based on BURGER 1985-1986.

157. 26.II.a. from the inside, from the north. Based on BURGER 1985-1986.

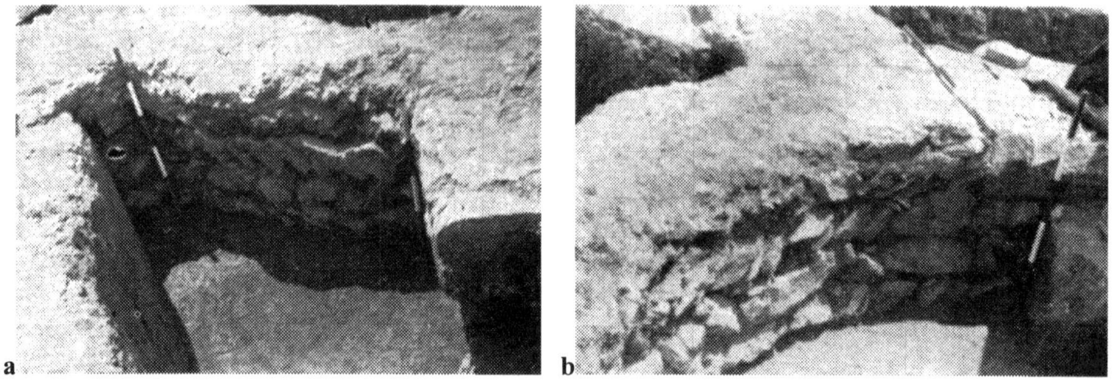

158. Buttress of 26.II.a, mark *C/3* on the ground-plan. **a:** the buttress and the arch from the outside, **b:** arch from the inside, where the walling is easily observable. Based on BURGER 1985-1986.

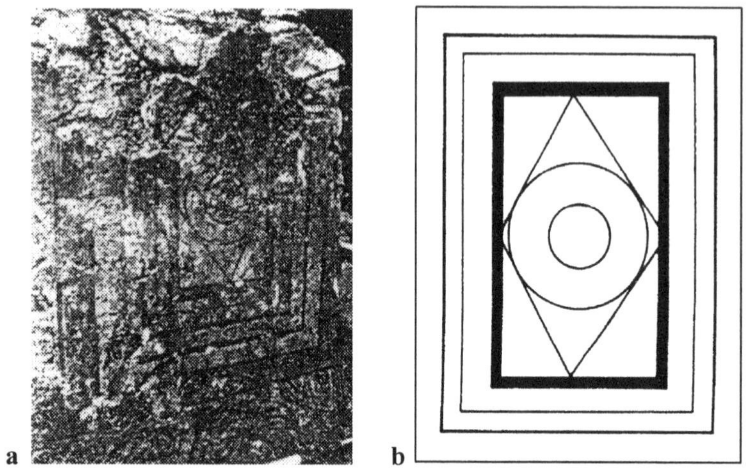

159. Geometric pattern of the fresco in the corridor of 26.II.a. **a:** photo, based on BURGER 1985-1986. **b:** drawing by DUKAY Bernadett.

160. Dura-Europos, fresco of the Torah shrine. Based on KRAELING 1956.

a

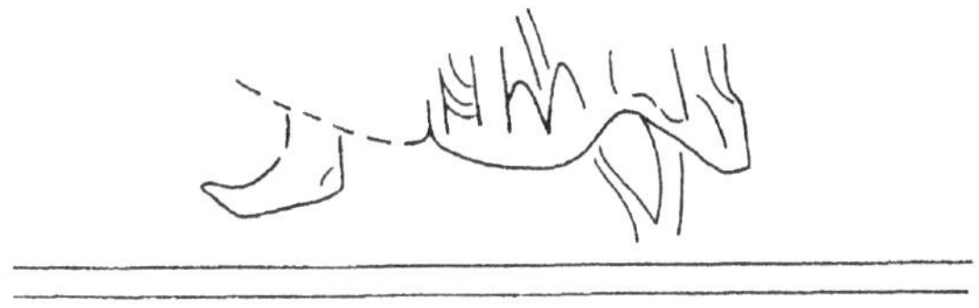

b

161. Part of the frescos of 26.II.a: standing figures. **a:** photo based on BURGER 1985-1986. **b:** drawing by DUKAY Bernadett.

a

b

162a, b. Dura-Europos, synagogue, and house-church; part of a fresco. Based on KRAELING 1956 and 1967.

a

b

163. Band below the frescos of 26.II.a. **a:** photo based on BURGER 1985-1986. **b:** drawing by DUKAY Bernadett.

164. Dura-Europos, band bDlow thDfresco. Based on KRAELING 1967.

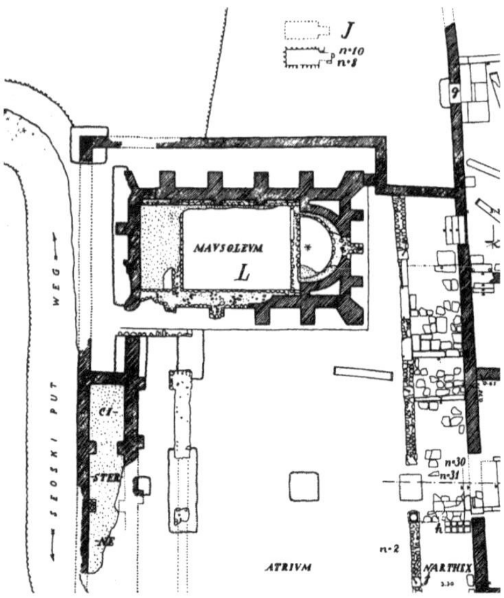

165. Marusinac, Anasthasius mausoleum. Based on MARIN 1988.

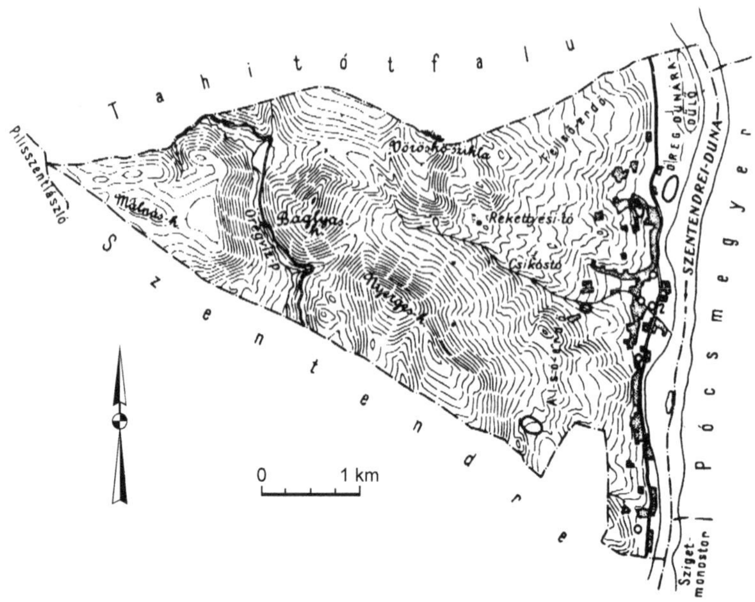

166. Leányfalu, general plan, based on MRT 7, 11. 11/2 is the provenance of the Roman watchtower.

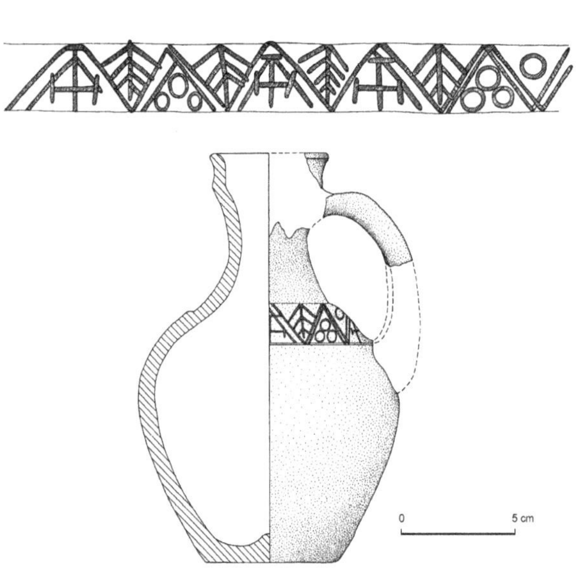

167. Leányfalu, jug, 27.I.a. drawn by Dr NAGY Margit.

168. Magyaregregy, finger ring, 28.II.a. **a:** photo, **b:** drawing. Based on Severin Katalog.

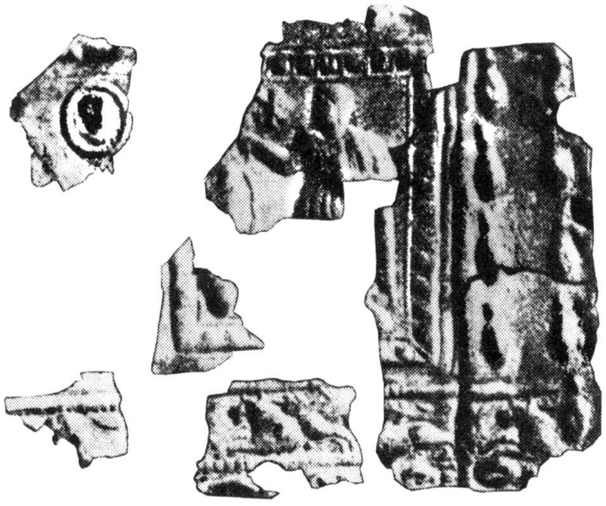

169. Mórichida-Kisárpás, casket mount, 29.I.a. Photo: SUGAR Lajos.

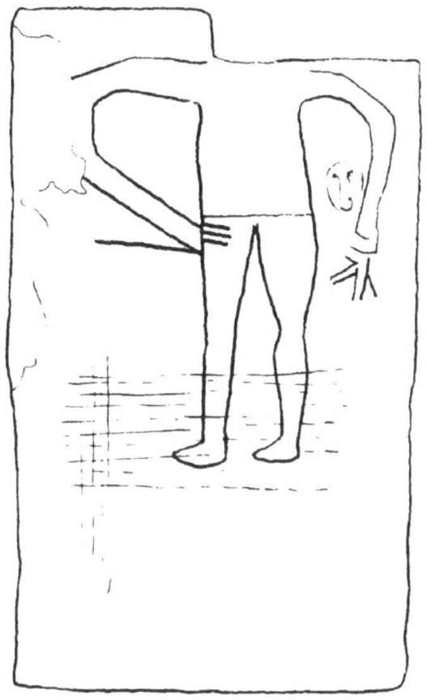

170. Mucsfa, roof tile, 30.II.a. Based on THOMAS 1982b.

171. Nagyberki-Szalacska, bronze lamp, 31.I.a. Based on IIL.

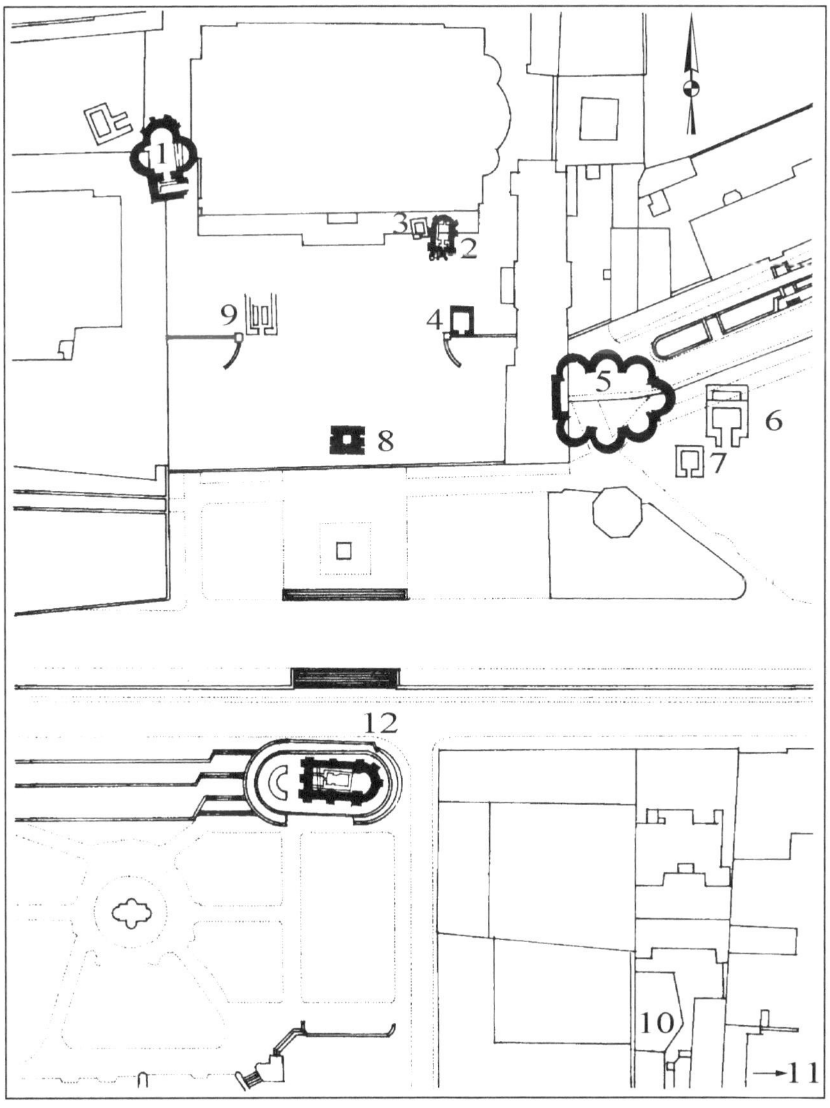

172a

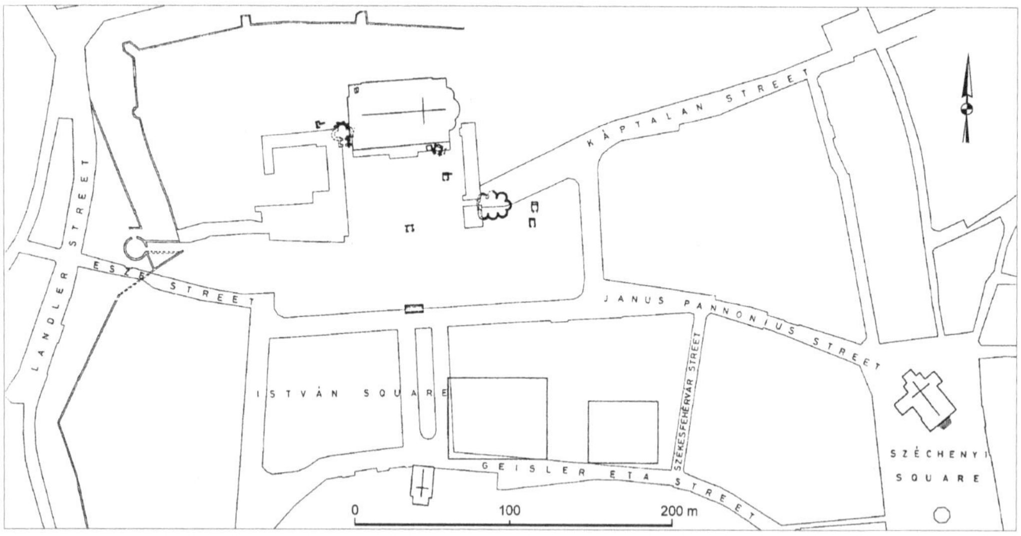

172. Pécs, general plan. Based on FÜLEP 1984. **a:** map of the graves found around the cathedral, **b:** map of the graves inside the present centre of town.

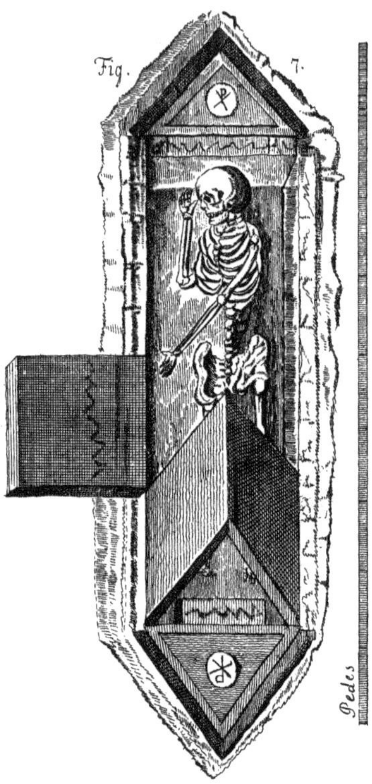

173. Pécs, one of the 13 graves, 32.I.a. Based on KOLLER 1804.

174. Glass finds of 32.I.a. Based on KOLLER 1804.

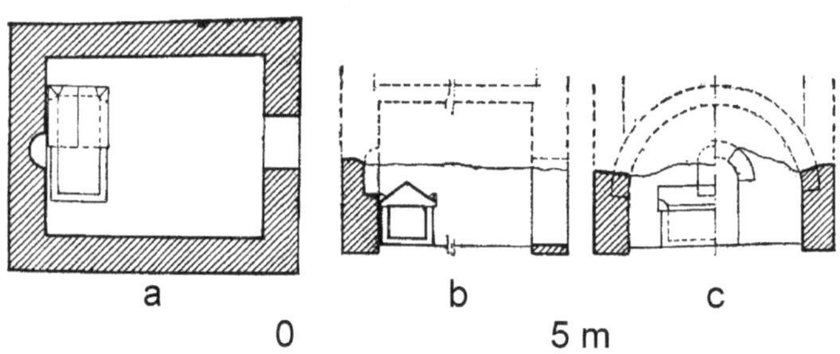

175. Pécs, tomb III, 32.II.a. Based on GOSZTONYI 1943.

176. Entrance of 32.II.a. Based on FÜLEP 1984.

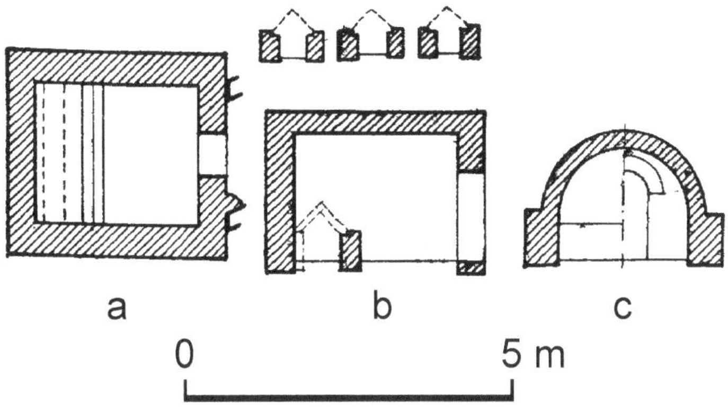

177. Pécs, tomb IV, 32.II.b. Based on GOSZTONYI 1943.

178. Entrance of 32.II.b. from the outside (south). Based on FÜLEP 1984.

179. Niche in the western wall, 32.II.b. Based on FÜLEP 1984.

180. Northern wall of 32.II.b. Based on FÜLEP 1984.

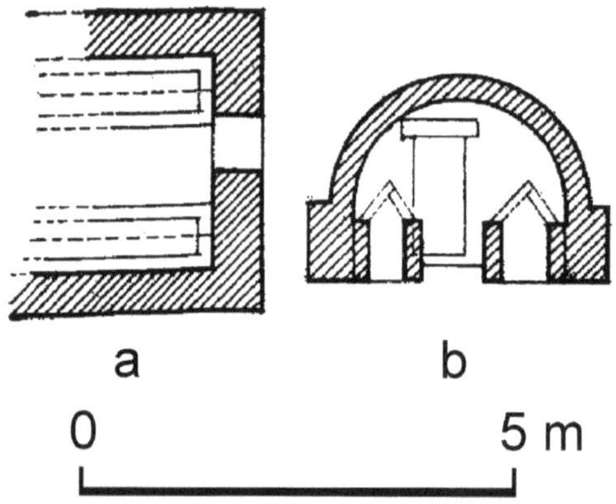

181. Pécs, tomb VI, 32.II.d. Based on GOSZTONYI 1943.

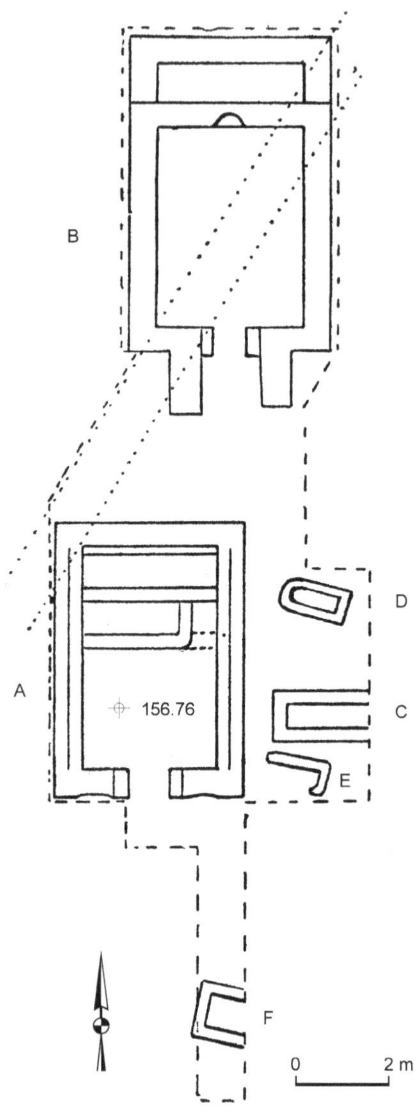

182. Pécs, general plan of the excavation of tombs VIII-IX. 32.II.f-g. Based on TÖRÖK Gy. 1942.

183. 32.II.f. from the entrance. Based on TÖRÖK Gy 1942.

184. 32.II.f. from the north. Based on TÖRÖK Gy 1942.

185. Northern wall of 32.II.f. with the place of burial. Based on TÖRÖK Gy 1942.

186. Southern wall of 32.II.g. with the entrance. Based on TÖRÖK Gy 1942.

187. Northern side of 32.II.g. Based on TÖRÖK Gy 1942.

188. Bronze buckle and pottery fragment found on the floor of 32.II.f. Based on TÖRÖK Gy 1942.

189. Fresco fragments found above 32.II.f-g. Based on TÖRÖK Gy 1942.

a  b 

c 

190. Pécs, tomb X, **a:** pieces of mosaic floor found near 32.II.h. **b:** reconstruction. Based on FÜLEP 1984.

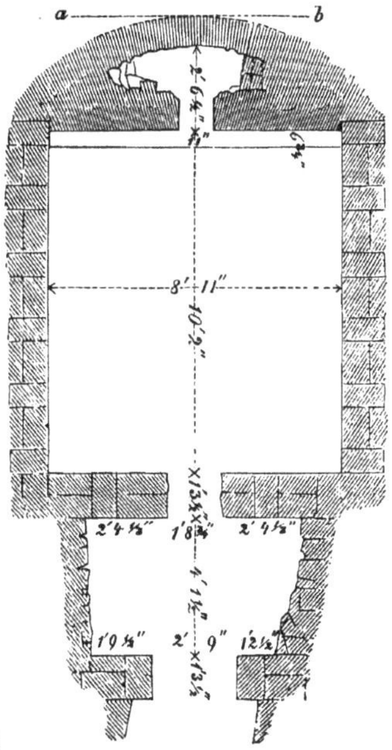
a

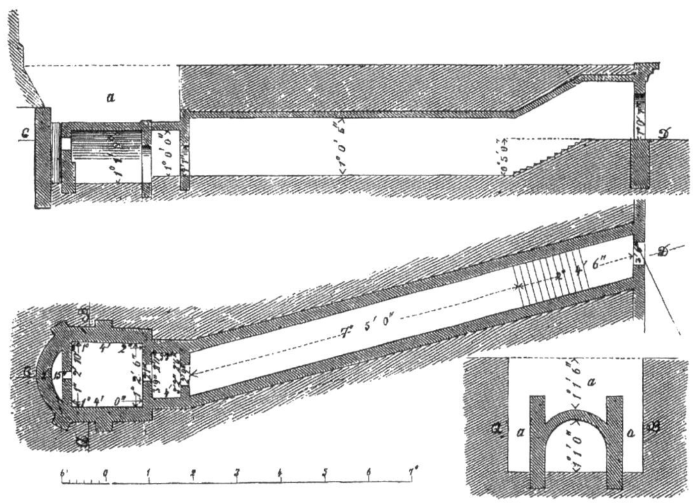
b

191. Pécs, painted tomb I, 32.I.b. **a:** ground plan**, b:** section, based on HENSZLMANN 1876. The measurements can be found here.

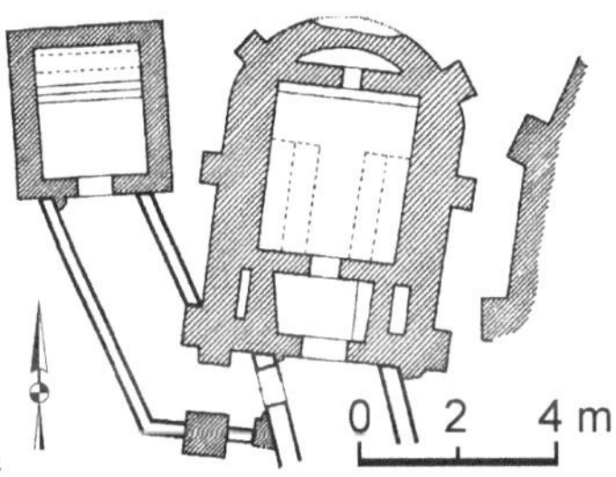
a

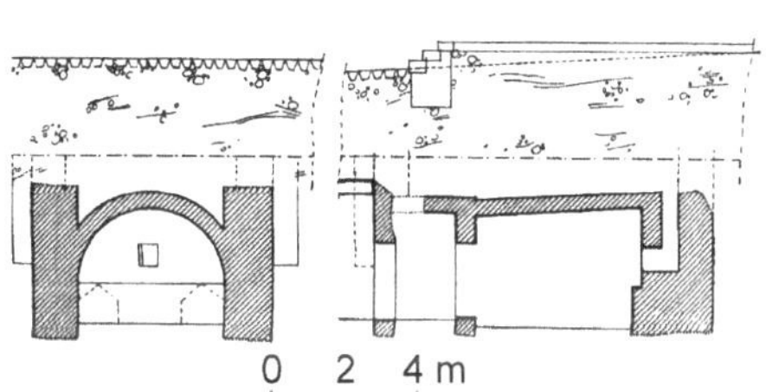
b

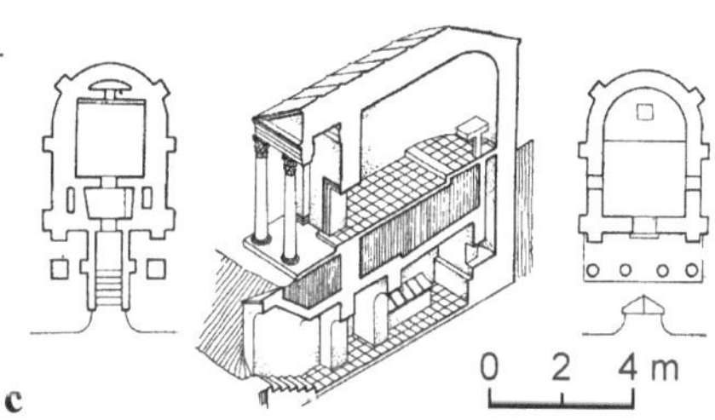
c

192. Pécs, painted tomb I, 32.I.b. **a:** ground plan, together with the ground plan of the 32.I.c. **b:** sectional, **c:** reconstruction based on GOSZTONYI 1943.

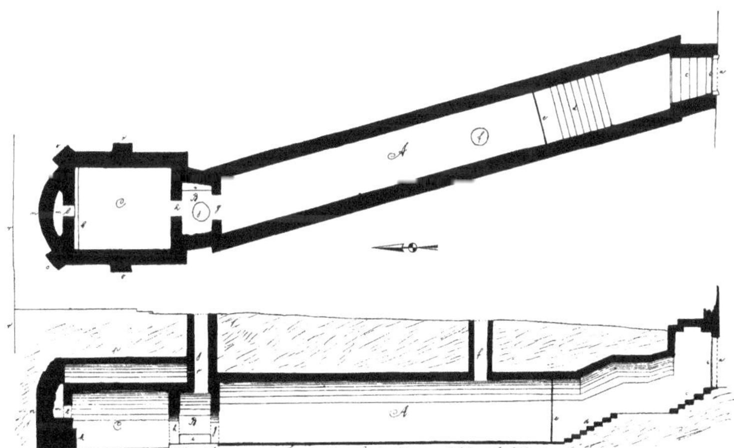
a

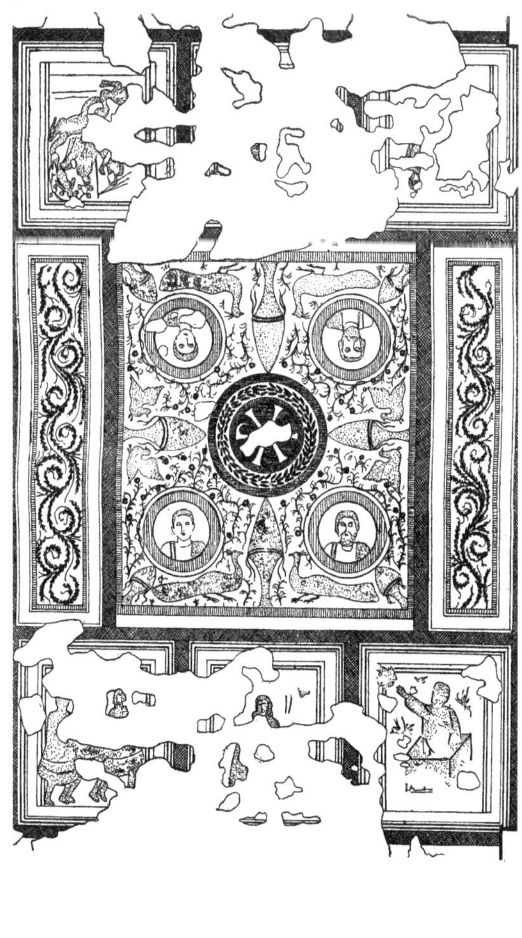
b

193. Pécs, painted tomb I, 32.I.b. **a:** Section, based on SZŐNYI O 1907. The description of the tomb follows the marks of this drawing. **b:** drawing of frescos, based on GOSZTONYI s.a.

194. Fresco on the entrance (southern) wall of 32.I.b. Based on FÜLEP 1984.

195. Fresco on the northern wall of 32.I.b. Based on SZŐNYI O 1907.

196. Frescos at the junction of northern and eastern walls of 32.I.b. Based on SZŐNYI O 1907.

197. Frescos at the junction of western and southern walls of 32.I.b. Based on SZŐNY O 1907.

198. Ceiling fresco of 32.I.b: portrait in medallion. Based on FÜLEP 1984.

199. Western wall of 32.I.b. Based on FÜLEP 1984.

a

b

200. 'Mary's' scene in the fresco of 32.I.b. **a:** based on FÜLEP 1984. **b:** based on BALOGH 1936.

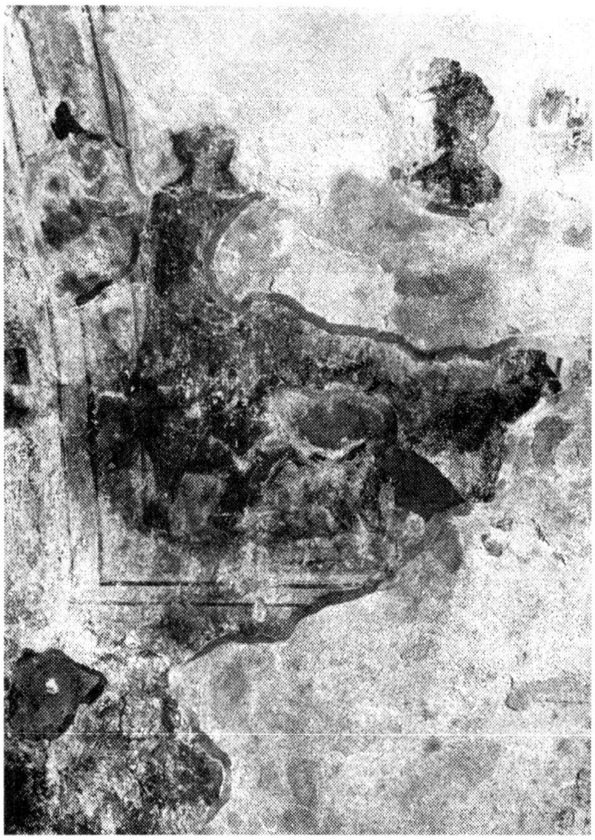

201. Western wall of the 32.I.b. Three youths in the furnace. Based on FÜLEP 1984.

202. Western wall of the 32.I.b. Noah in the Ark. Based on FÜLEP 1984.

203. Eastern wall of the 32.I.b. 'Eve'. Based on FÜLEP 1984. **a:** full-length figure, **b:** head.

204. Part of the fresco of 32.I.b.: Jonah's story. Based on FÜLEP 1984.

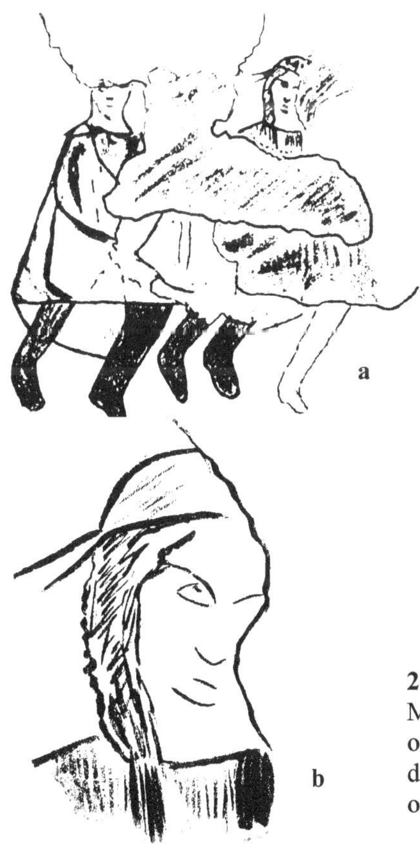

205. Scene of 'the Magi' of 32.I.b. Based on SZŐNYI O 1907. **a:** drawing, **b:** head of one of 'the Magi'.

206. Drawing of a portrait on the southeast part of the ceiling of 32.I.b. Based on SZŐNYI O 1907.

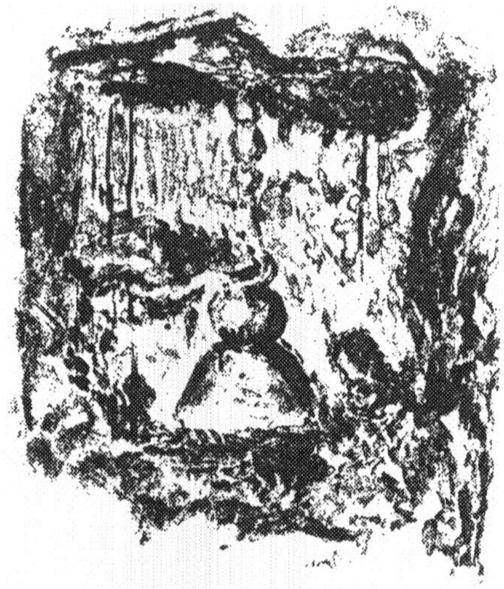

207. Fresco fragment in the antechamber of 32.I.b. Candelabrum in the southeast corner. Based on SZŐNYI O 1907.

208. Rome, Domitilla catacomb, Martyr Petronilla leads Veneranda. Based on TESTINI 1966.

209. Rome, part of the fresco in the hypogeum. Based on TESTINI 1966.

210a-b. Or ha-Ner, fresco: two portraits. Based on Encyclopedia 1978 (TSRAFRIR)

211. Or ha-Ner, fresco: series of portraits in medallions. Based on Encyclopedia 1978 (TSRAFRIR)

212. *Venus pudica*, bronze statue, H: 25 cm (Cf. Note 721.).

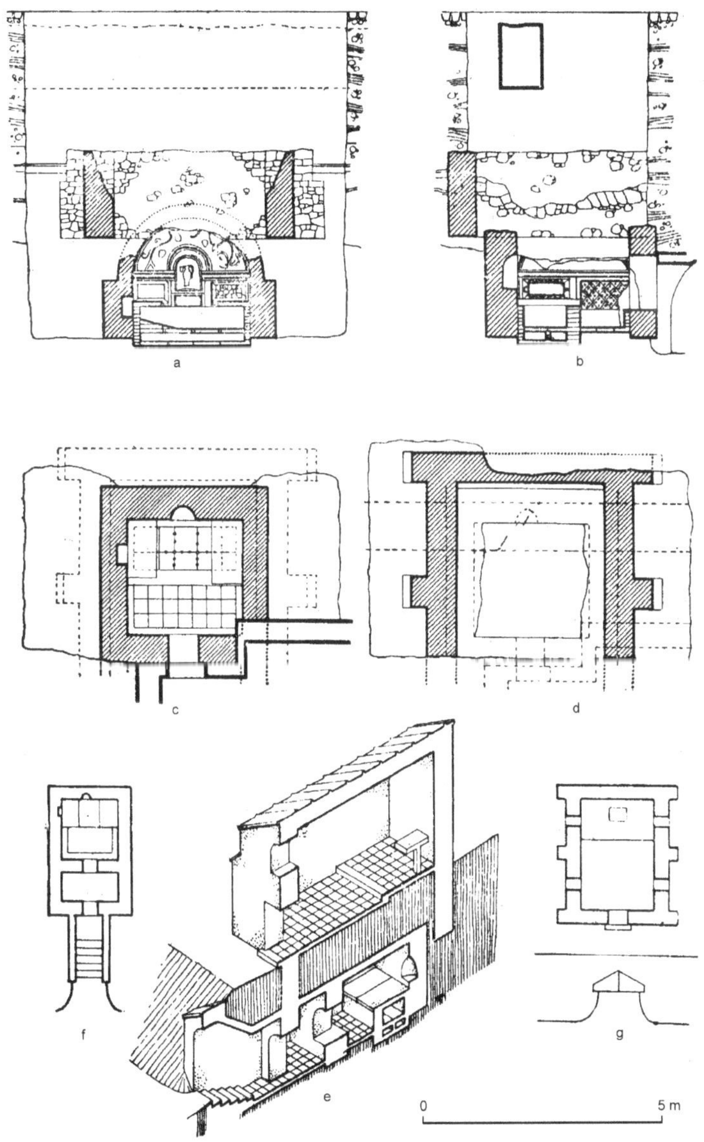

213. Pécs, painted tomb II, 32.I.c. Based on GOSZTONYI 1943. a: Cross-section oriented E-W, b: cross-section oriented N-S, c: ground plan of the tomb. The superstructure is only indicated nearby, d: ground plan of the superstructure, the tomb inside is only indicated. e, f, g: reconstruction of the superstructure.

214. Photo of the grave in 32.I.c. Based on FÜLEP 1984.

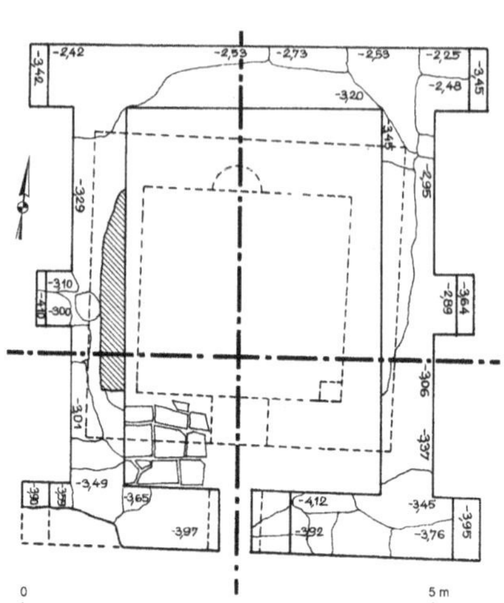

215. Measurements of 32.I.c. indicated on a drawing; ground plan. Based on FÜLEP/FETTER 1969-1970.

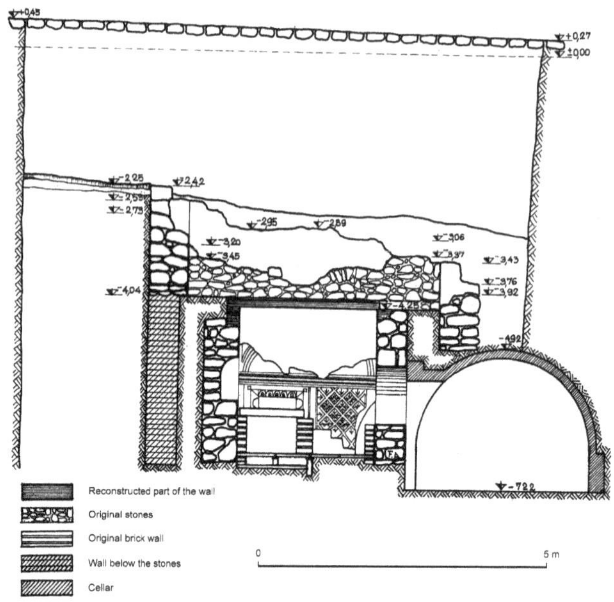

216. Measurements of 32.I.c. indicated on a drawing; section of the eastern wall. Based on FÜLEP/FETTER 1969-1970.

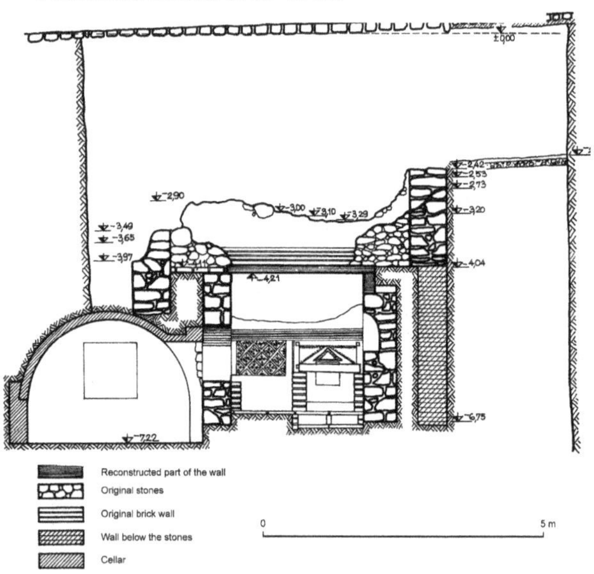

217. Measurements of 32.I.c. indicated on a drawing; section of the western wall. Based on FÜLEP/FETTER 1969-1970.

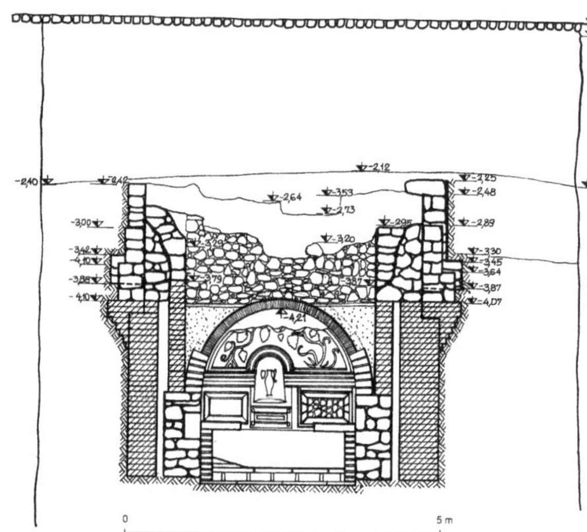

218. Measurements of 32.I.c. indicated on a drawing; section of the northern wall. Based on FÜLEP/FETTER 1969-1970.

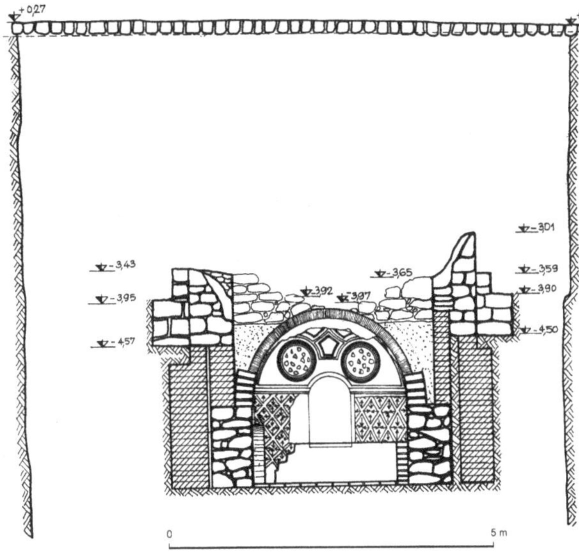

219. Measurements of 32.I.c. indicated on a drawing; section of the southern wall. Based on FÜLEP/FETTER 1969-1970.

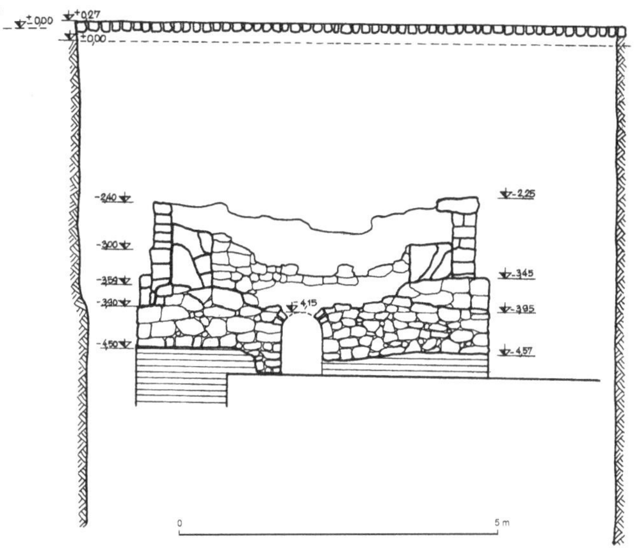

220. Measurements of 32.I.c. indicated on a drawing; picture of the southern outside wall. Based on FÜLEP/FETTER 1969-1970.

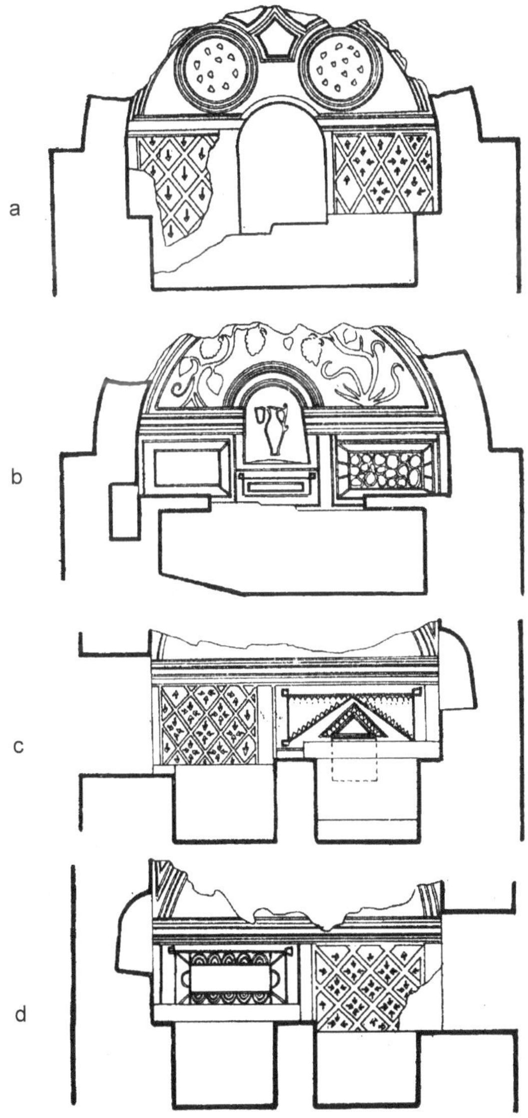

221. Drawing of the frescos of 32.I.c. Based on GOSZTONYI 1942. **a:** entrance (southern) side, **b:** northern side, **c:** western side, **d:** eastern side.

222. Fresco on the northern wall of 32.I.c. Based on FÜLEP 1984.

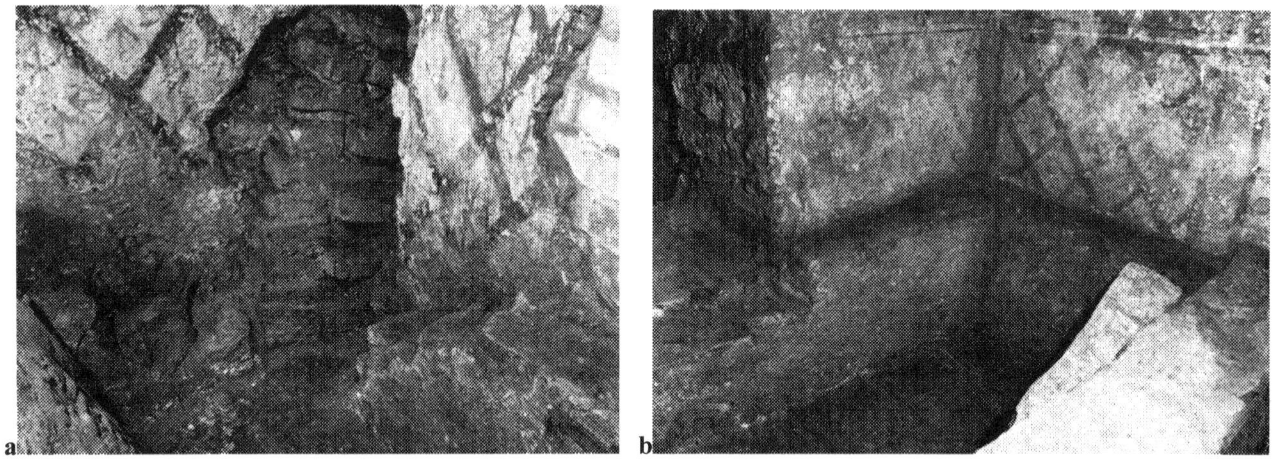

223a, b. Entrance and foundation of 32.I.c. Based on FÜLEP 1984.

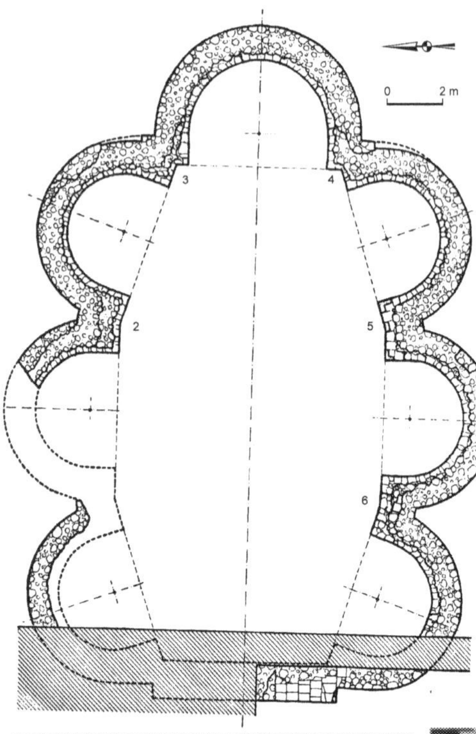

225. Excavation photos. of 32.I.d. Eastern apses. Based on GOSZTONYI 1940a.

224. Pécs, cella septichora, ground plan of 32.I.d. Based on GOSZTONYI 1940a.

226. Excavation photos. of 32.I.d. **a:** northern junction of the apses (*opus mixtum*). **b-c:** further wall sections with the beam-holes.

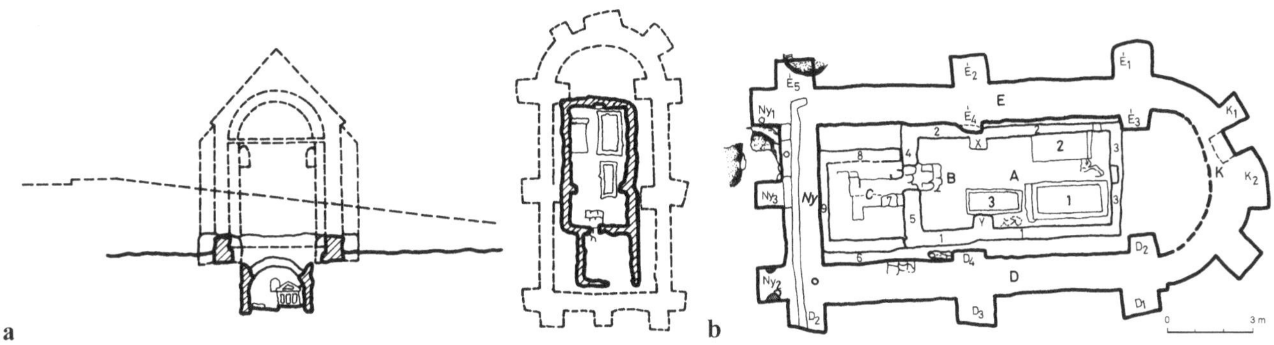

227. Pécs, so-called mausoleum, 32.I.e. Based on FÜLEP/BACHMAN/PINTÉR 1988. **a:** ground plan and section, **b:** ground plan in detail.

228. Excavation photos of 32.I.e. Based on FÜLEP 1977a. **a:** Western wall of the burial chamber with the entrance, **b:** eastern end wall and collapsed vault.

229. Fresco part of 32.I.e, the Fall. Based on FÜLEP/BACHMAN/PINTÉR 1988.

230. Frescos on the northern wall of 32.I.e.
Based on FÜLEP/BACHMAN/PINTÉR 1988.

231. Fresco on the eastern wall of 32.I.e.
Based on FÜLEP/BACHMAN/PINTÉR 1988.

232. Fresco on the southern wall of 32.I.e.
Based on FÜLEP/BACHMAN/PINTÉR 1988.

233. Sarcophagi of 32.I.e. Based on FÜLEP/BACHMAN/PINTÉR 1988.

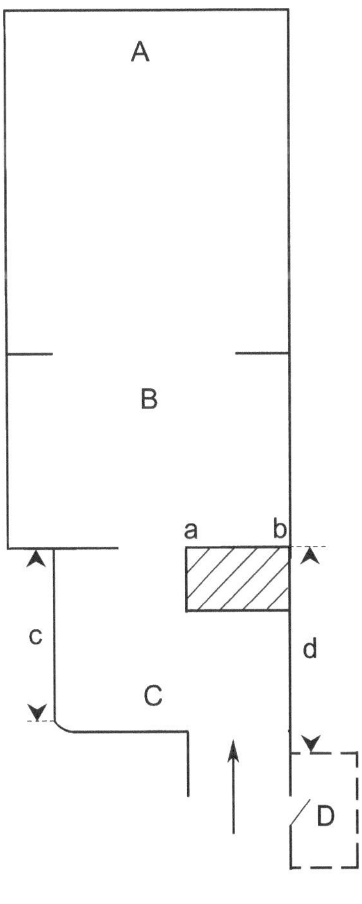

234. Author's sketch of the ground plan of 32.I.e.

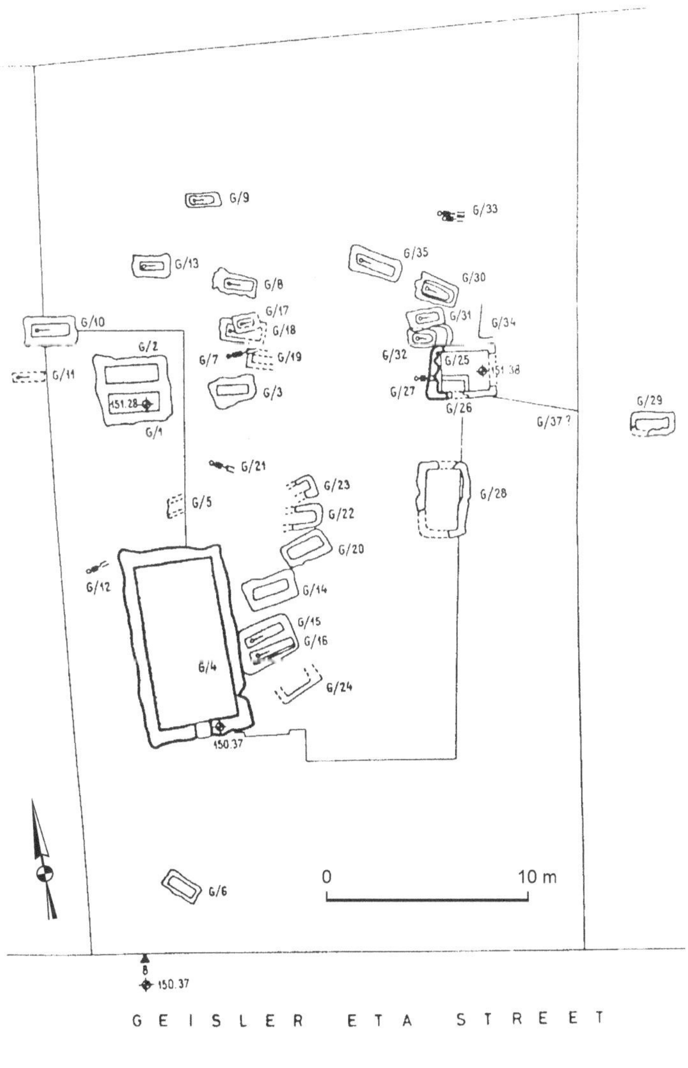

235. Pécs, general plan of the G/4 tomb, 32.I.f. Based on FÜLEP 1984.

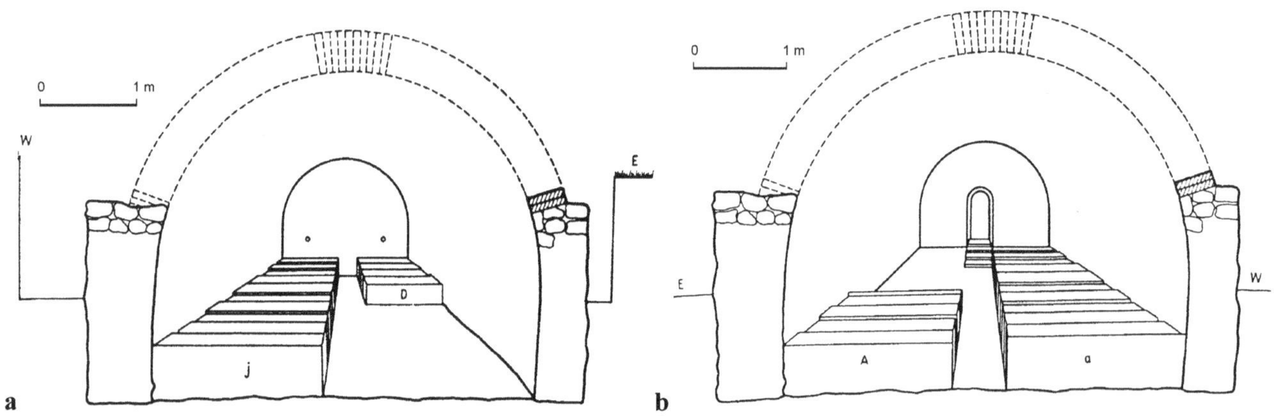

236. Reconstruction of 32.I.f. Based on FÜLEP 1984. **a:** southern side from the north, **b:** northern side from the south.

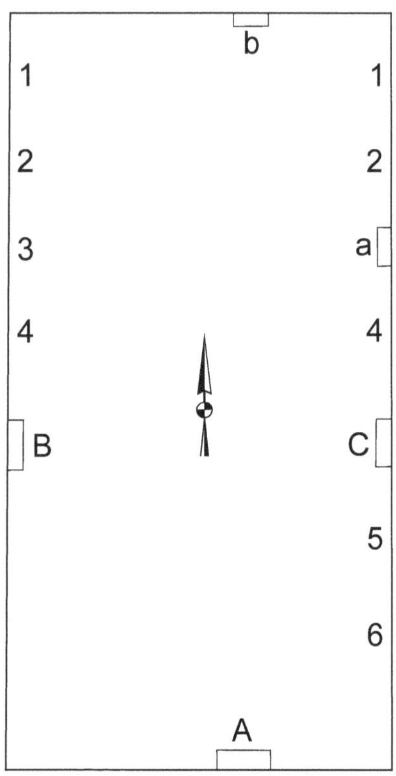

237. The author's sketch for the description of the phenomena of 32.I.f.

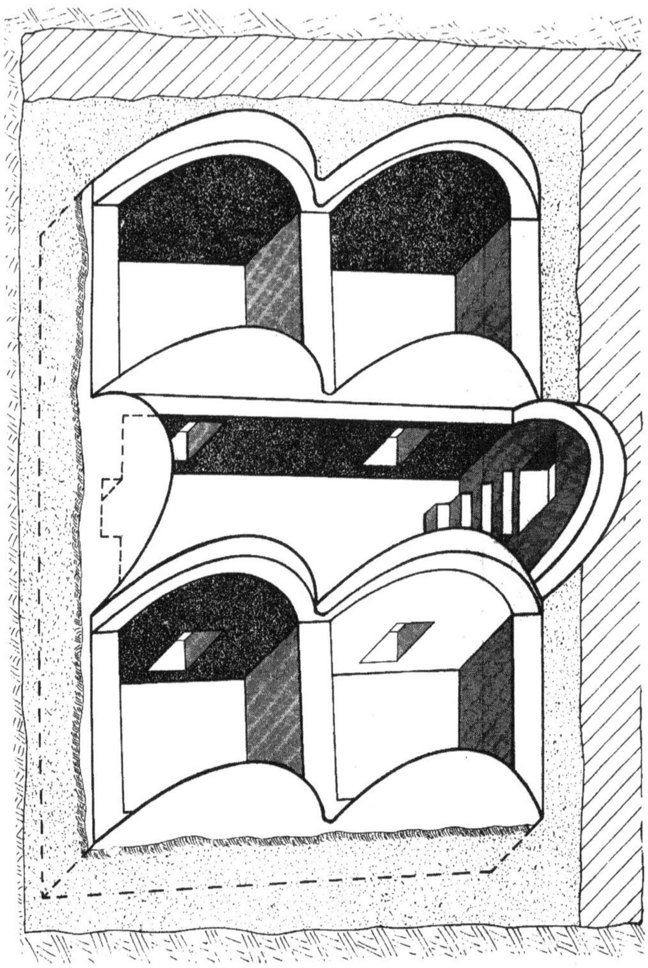

238. Or ha-Ner, isometric reconstruction of the burial chamber. Based on Encyclopedia 1978 (TSAFRIR)

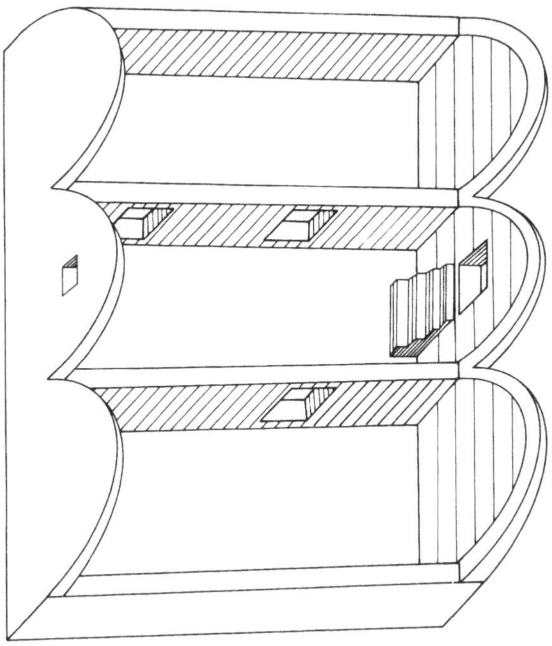

239. The reconstruction of 32.I.f. planed by author, drawn by DUKAI Bernadett.

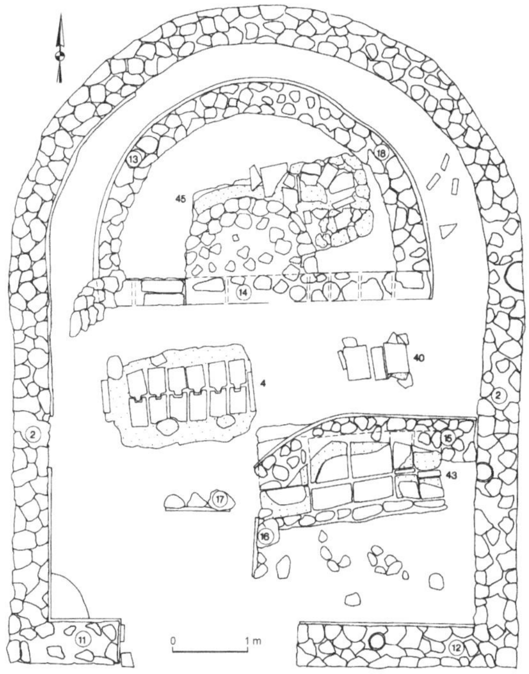

240. Pécs, ground plan of tomb XIII, 32.I.g. Based on FÜLEP 1984.

241. Excavation photos. of 32.I.g. View from the south. Based on FÜLEP 1984.

242. Excavation photos. of 32.I.g. View from the north. Based on FÜLEP 1984.

243. Excavation photos. of 32.I.g. The apse with the so-called altar and with grave L45. Based on FÜLEP 1984.

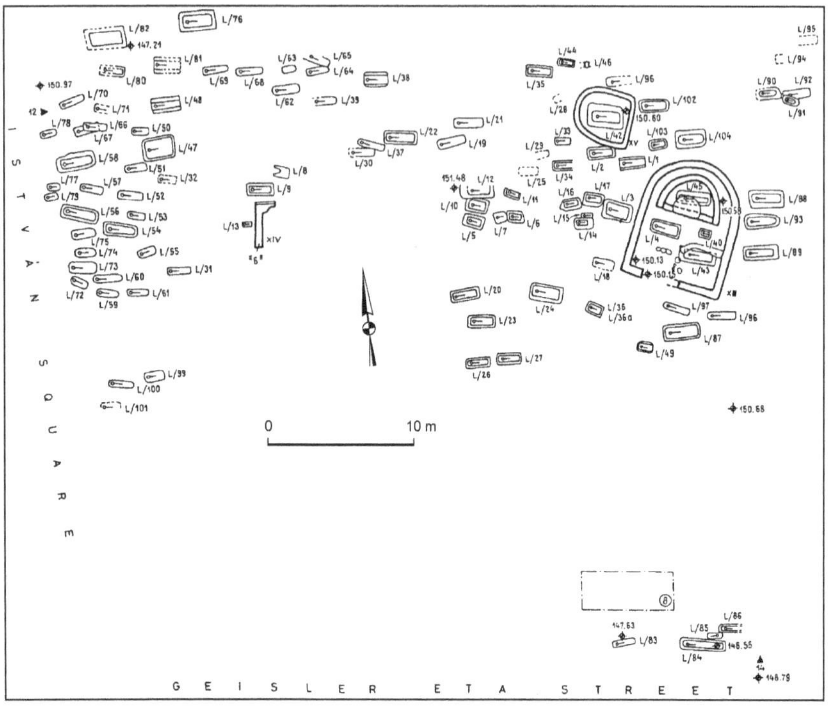

244. Map of the cemetery between Saint Stephen Square and 8 Geisler Eta Street. FÜLEP 1984 Suppl. VII.

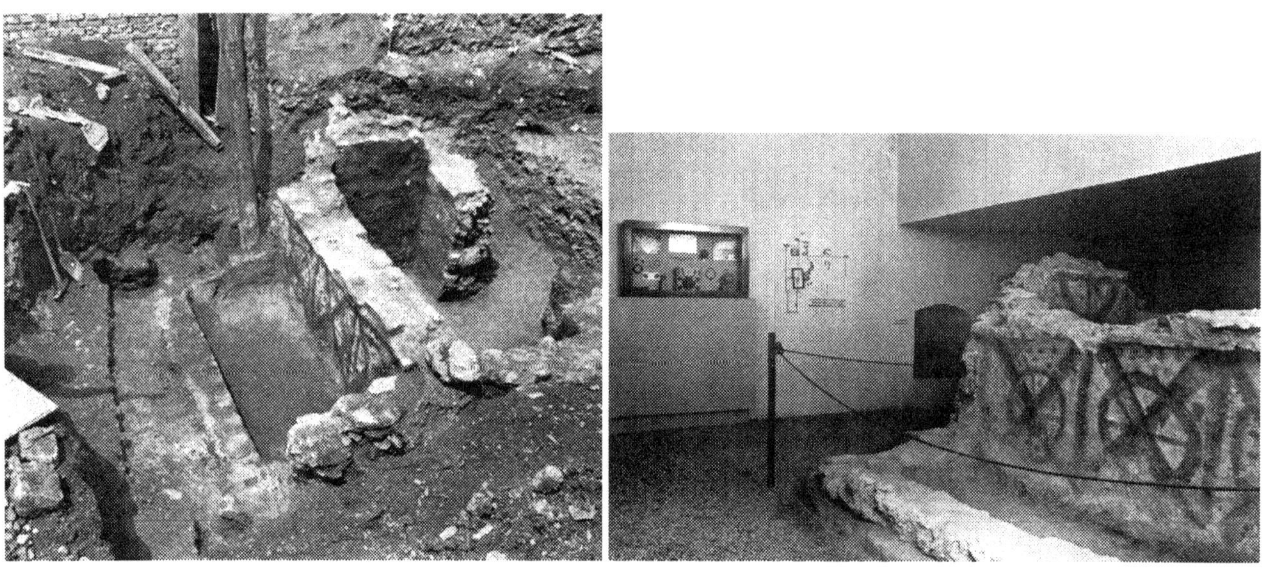

245a-b. Pécs, double grave (at 8 Geisler Eta Street), 32.II.i. Based on FÜLEP 1984.

246. Pécs, gem from grave G17, 32.II.j. Based on FÜLEP 1984.

247. Pécs, finger ring, 32.I.h. Based on SCHEIBER 1983.

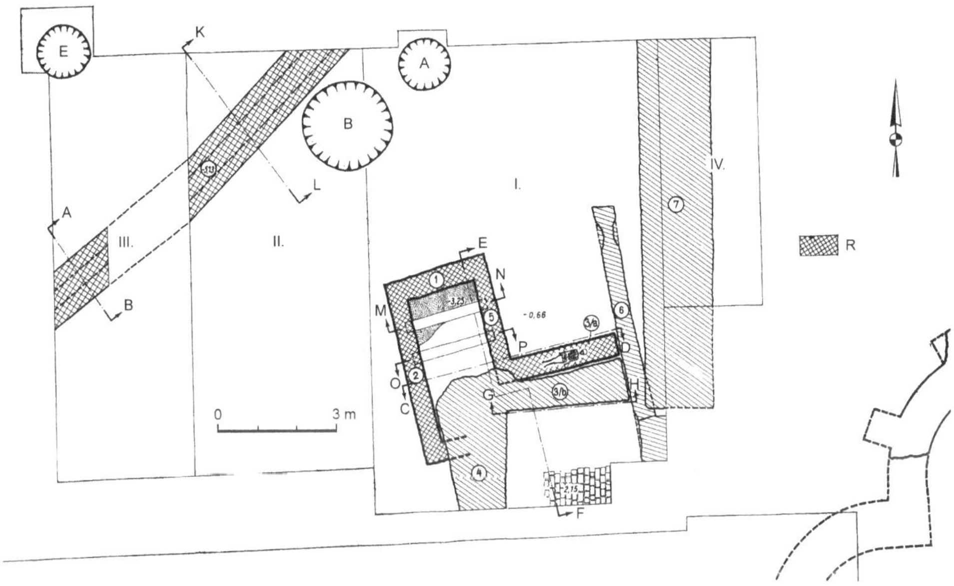

248. Pécs, general plan and ground plan of so-called tomb VII, 32.II.e. R: Roman walls, I, II, III: trail trenches, the conduit in III. A, B, E: digging-ins. The other walls are medieval. Based on FÜLEP 1962a.

249. Excavation photo. of 32.II.e, grave. FÜLEP 1962a.

250. Excavation plan of 32.II.e. **a:** grave under the wall, before opening, **b:** after opening. FÜLEP 1962a.

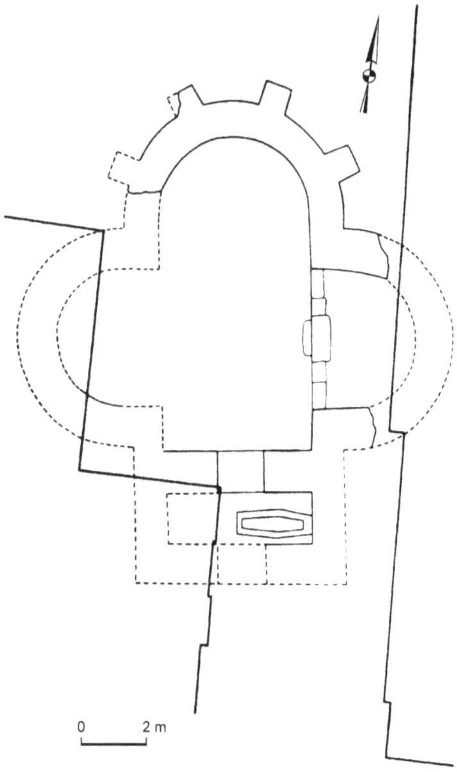

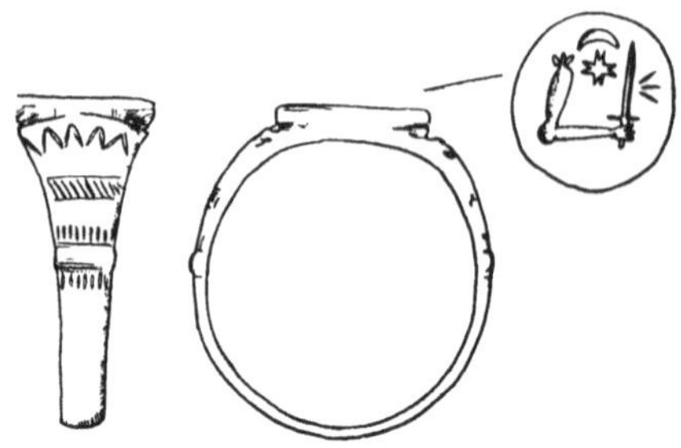

252. Medieval finger ring found at the excavation of 32.II.1.

251. Pécs, cella trichora, ground plan of 32.II.1. Based on GOSZTONYI 1943.

253. Fresco of 32.II.1. Based on FÜLEP 1984.

254. Fresco of 32.II.1. Based on FÜLEP 1984.

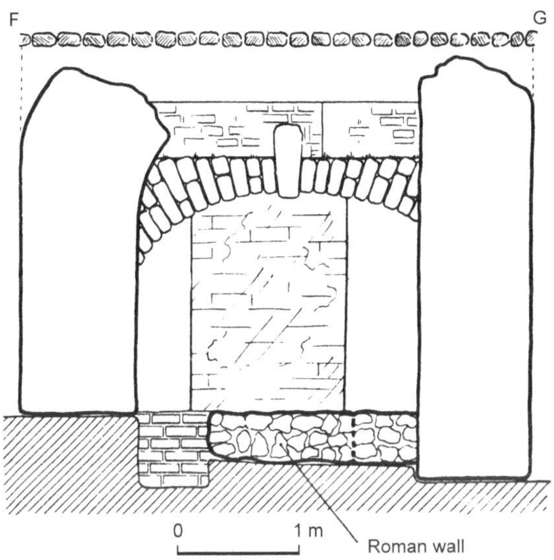

255. Section prepared during the excavation of 32.II.1. Section F-G. FÜLEP 1959a. R: Roman wall.

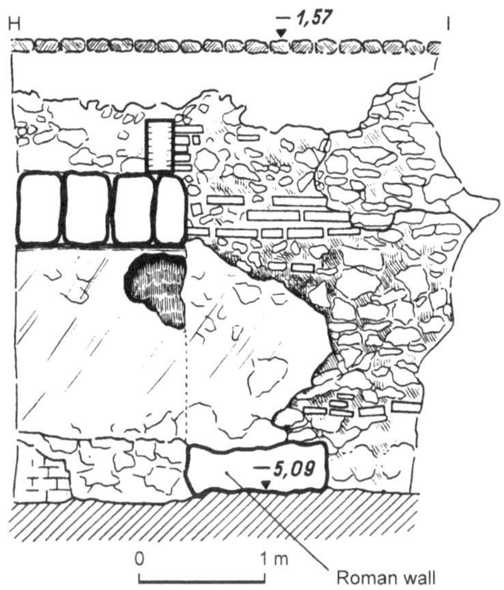

256. Section prepared during the excavation of 32.II.1. Section H-I. FÜLEP 1959a. R: Roman wall.

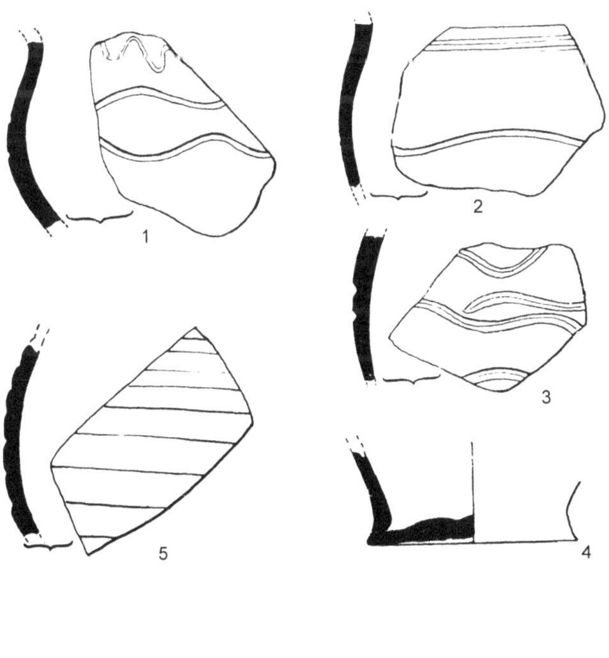

258. Tessera found during the excavation of 32.II.1. FÜLEP 1959a.

257. Drawing of the pottery fragments found during the excavation of 32.II.1. FÜLEP 1959a.

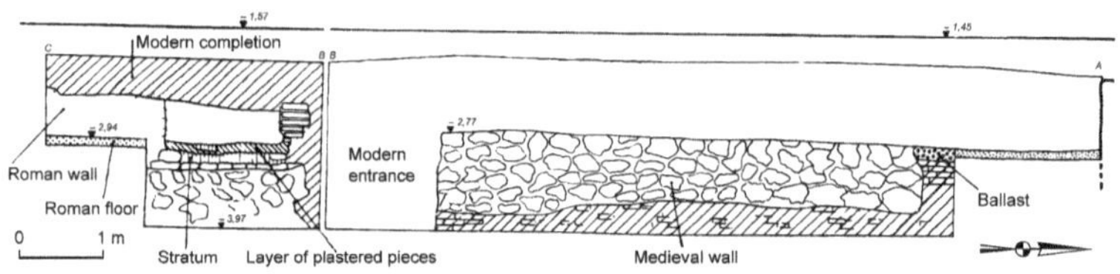

259. Section prepared during the excavation of 32.II.1. Sections A-B-C. FÜLEP 1959a.

261. Pécs, graffito found while excavating the gutter. Based on FÜLEP 1961b.

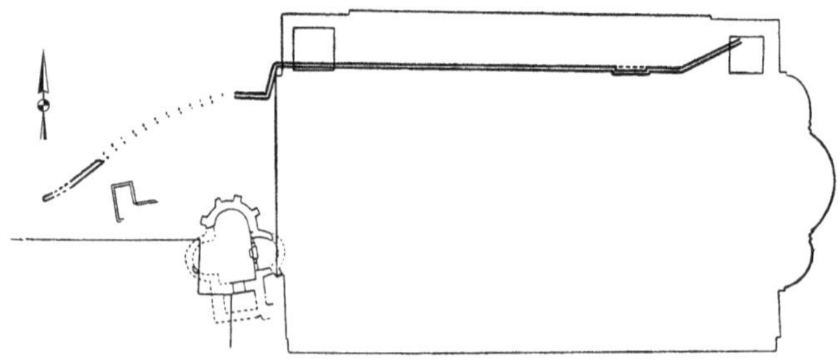

260. Pécs, canalization on the northern side of the cathedral. Based on FÜLEP 1961a.

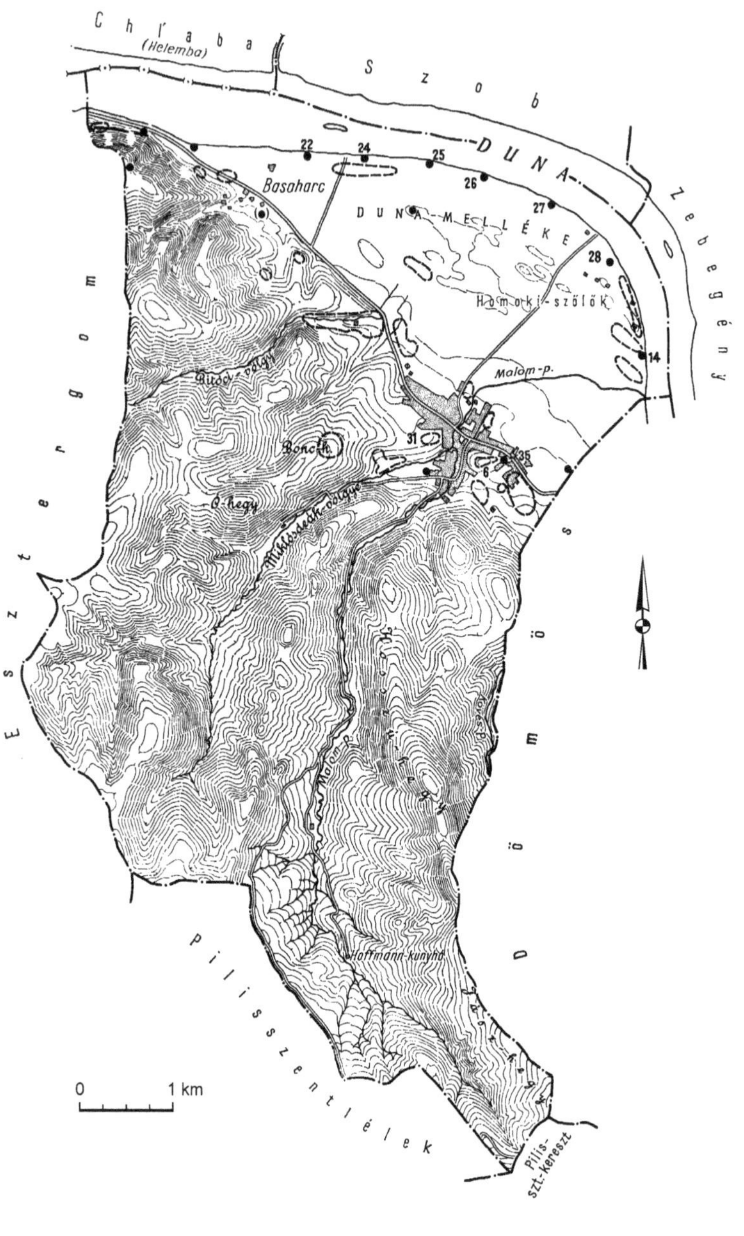

262. Pilismarót-Kishegy, fort, cemetery, 33.I.a. General plan based on MRT 5/17. No. 6 indicates the site of the fort, No. 31 indicates the cemetery. No. 35 is the hillside downwards from the fort, where also some graves were found. They are indicated by No. 35. Nos. 22, 24, 25, 26 and 27 are watchtowers along the Danube River.

263. Ground plan of the so-called fort and 33.I.a-c. Based on FINÁLY 1907. 1: excavated wall, 2: wall, took out, 3: supposed line of the wall.

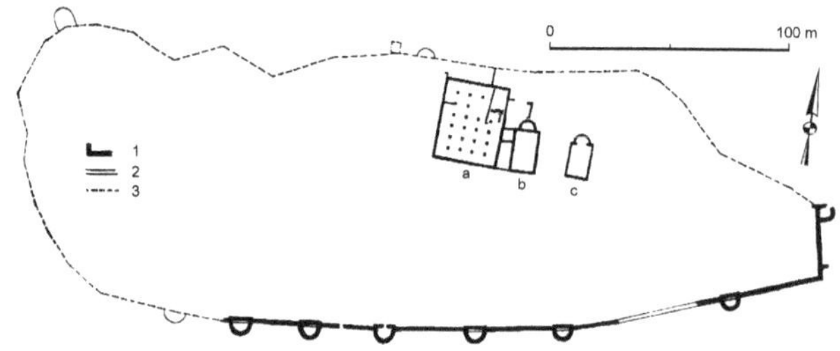

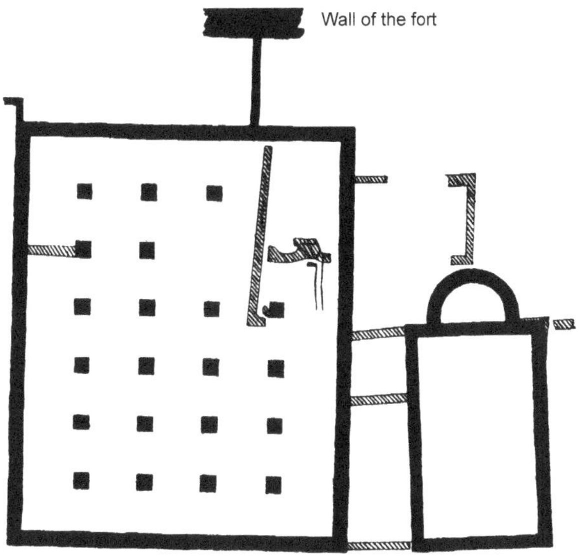

264. Ground plan of 33.I.a-b. Based on FINÁLY 1907.

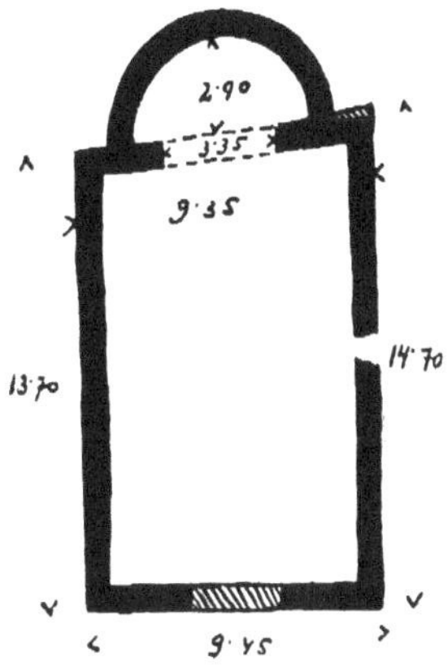

265. Ground plan of 33.I.c. Based on FINÁLY 1907.

266. Pilismarót, cup from the cemetery. BARKÓCZI 1960.

267a-b. Pölöske, dagger-sheath, 35.II.a. Photo: SUGÁR Lajos.

268. Vessel found next to 35.II.a. Based on Severin Katalog.

269. Detail of 35.II.a. Photo: SUGÁR Lajos.

270. Detail of 35.II.a. Photo: SUGÁR Lajos.

271. Detail of 35.II.a. Photo: SUGÁR Lajos.

272. Decoration of the side-plate of 35.II.a. Photo: SUGÁR Lajos.

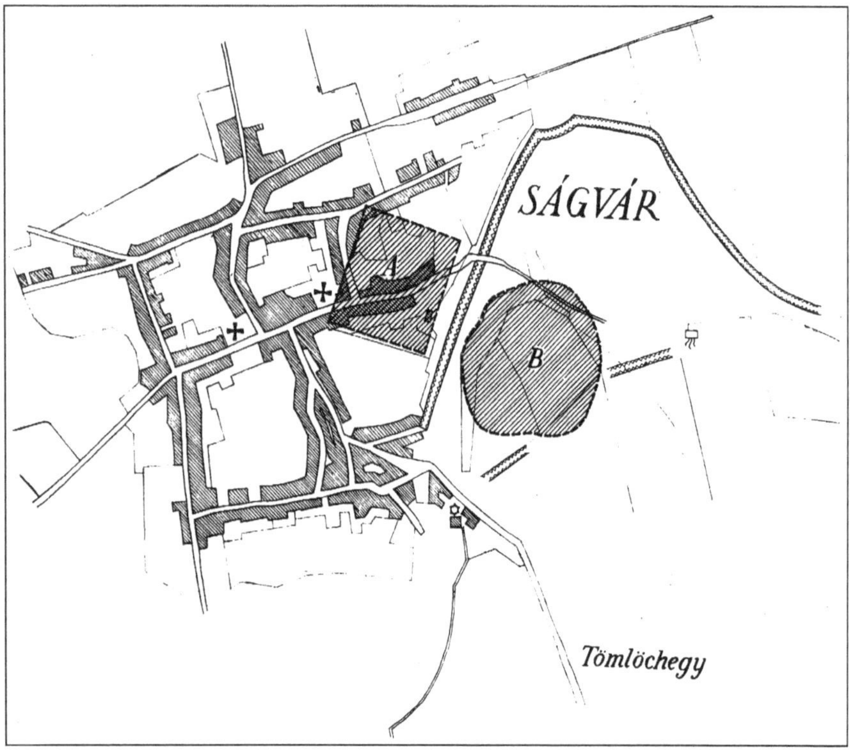

273. Ságvár, general plan. A: fortification, B: cemetery. Based on BURGER 1966.

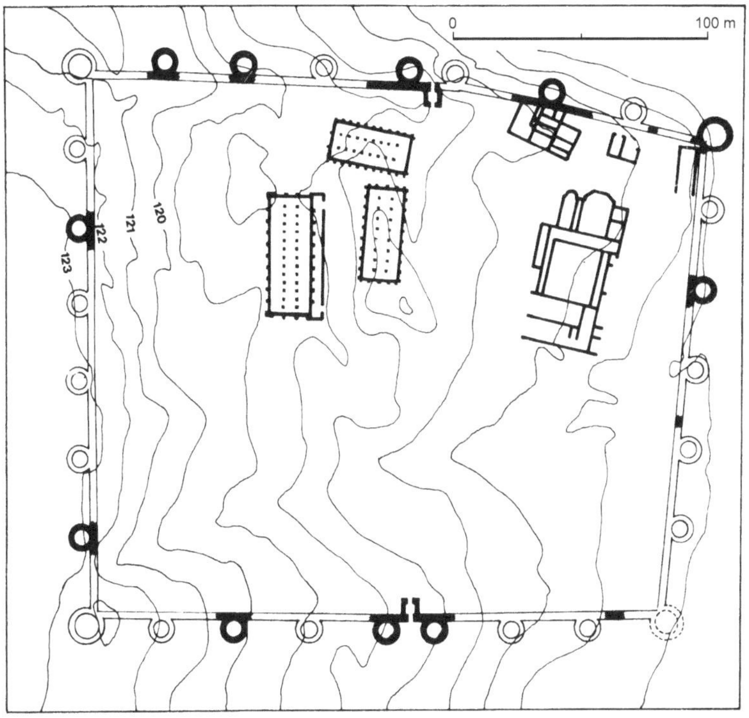

274. Ságvár, ground plan of the fortification. Based on TÓTH E 1985.

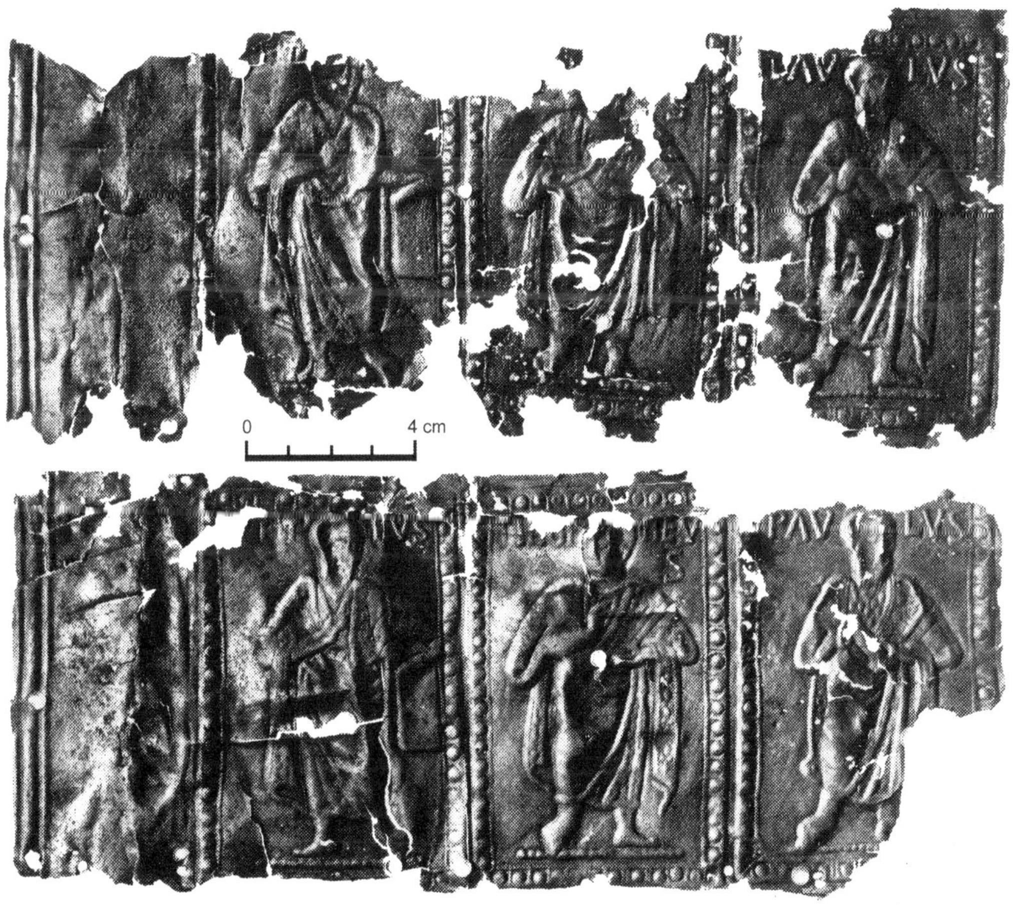

275. Ságvár, casket mount from the fort, 36.I.a. Photo: SUGÁR Lajos.

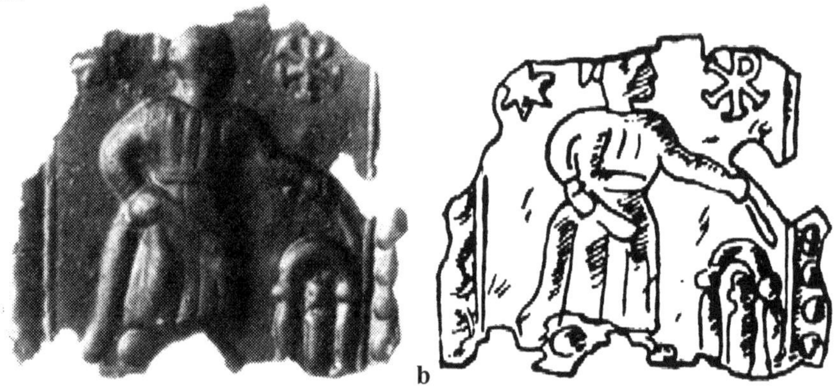

a b

276. Ságvár, casket mount from a grave, 36.I.b. **a:** photo: KÁDAS Tibor. **b:** the author's drawing.

277. Ságvár, grave of 36.I.b. Based on BURGER 1966.

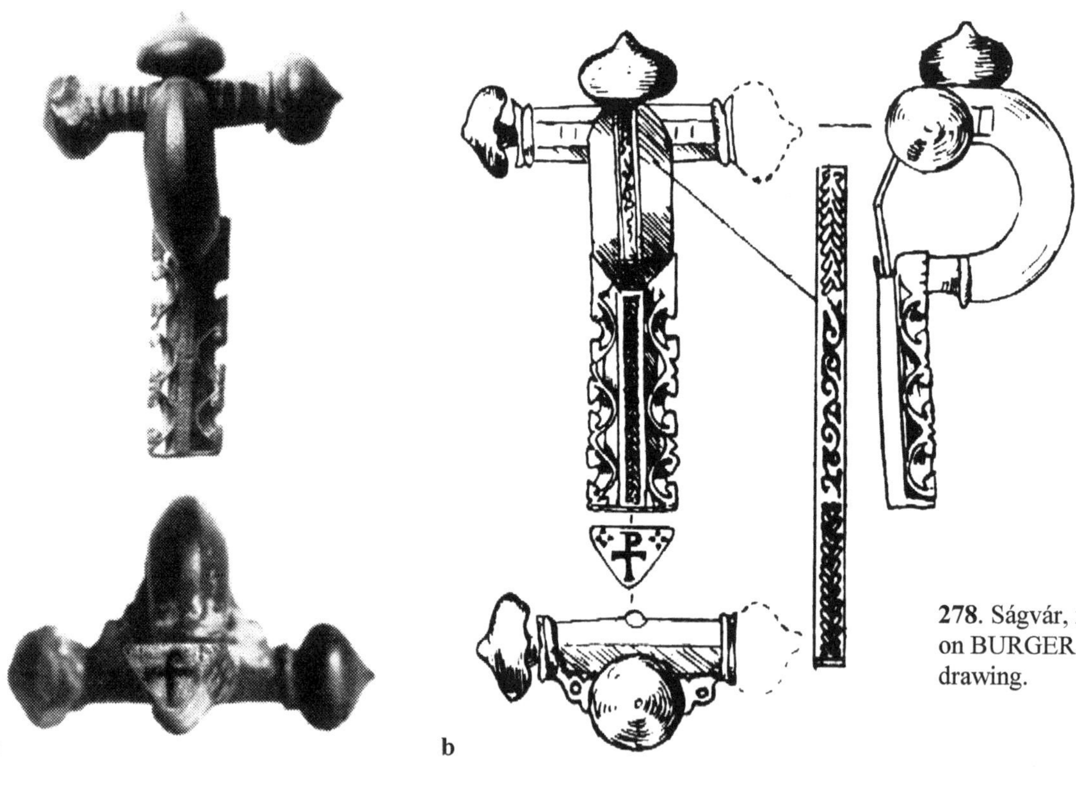

a b

278. Ságvár, fibula, 36.I.c. Based on BURGER 1966. **a:** photo. **b:** drawing.

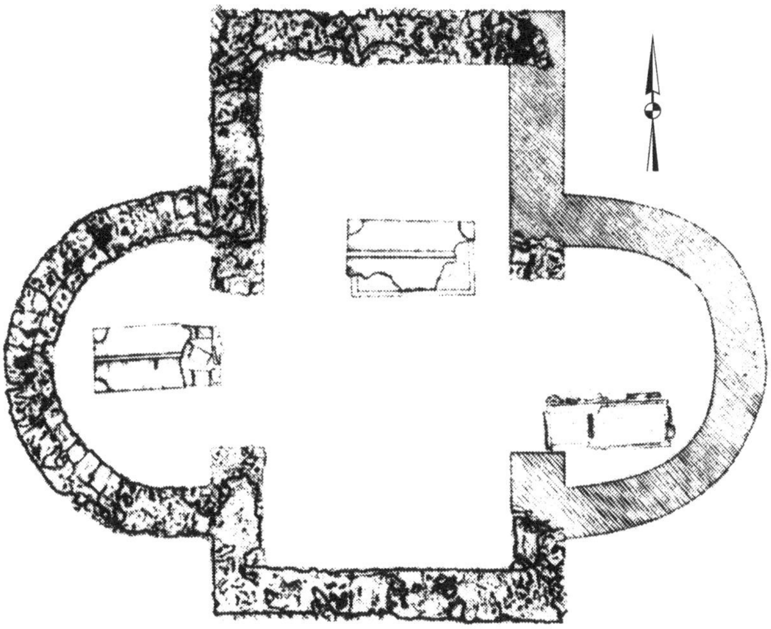

279. Ságvár, ground plan of cemetery-chapel with two apses, 36.II.a. Based on RADNÓTI 1939.

280. Excavation photos. of 36.II.a. Based on BURGER 1966. **a:** western apse and sarcophagus in the northern part, **b:** photo. of the western apse from the south, **c:** grave of the western apse - brick grave with a sarcophagus cover, **d:** eastern apse with a grave.

282. Ságvár, cella III, 36.II.c. Based on BURGER 1966.

281. Ságvár, cella II from two views, 36.II.b. Based on BURGER 1966.

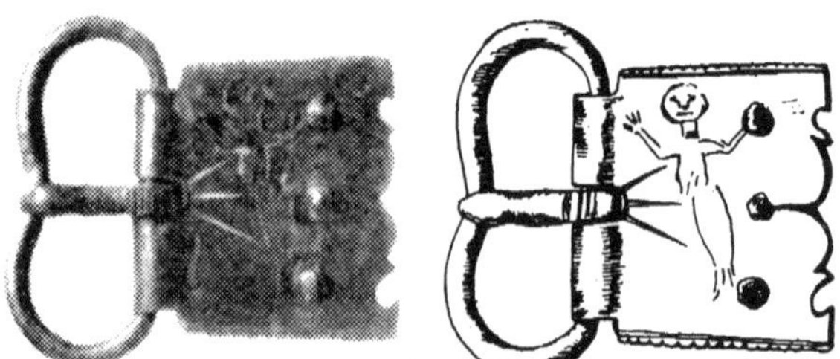

a **b**

283. Ságvár, buckle, 36.II.d. Based on BURGER 1966. **a:** photo. **b:** drawing.

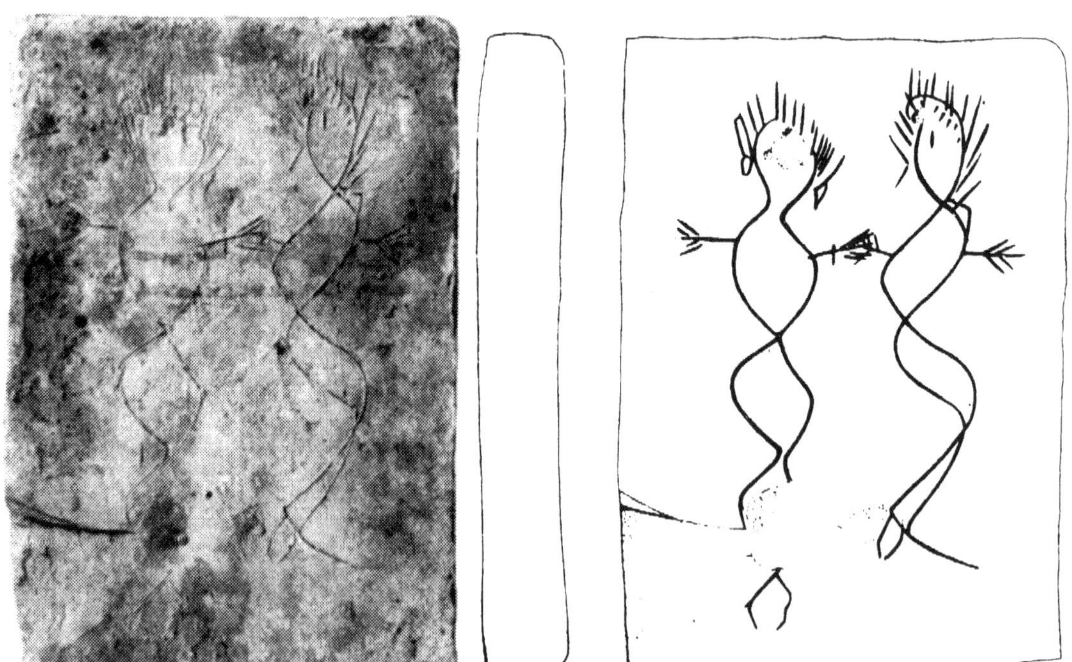

284. Ságvár, brick with incised pattern, 36.II.e. Based on BURGER 1966. **a:** photo. **b:** drawing.

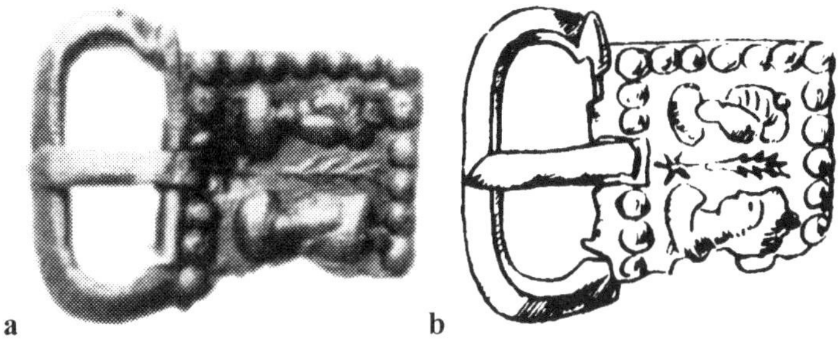

285. Ságvár, buckle. Based on BURGER 1966. **a:** photo. **b:** drawing.

286. Keszthely-Fenékpuszta, finger ring. Based on BARKÓCZI 1968.

287. Ságvár, brick with incised pattern, 36.II.f. Based on BURGER 1966. **a:** photo. **b:** drawing.

288. Darova (Daruvár), brick. Based on NAGY L 1938.

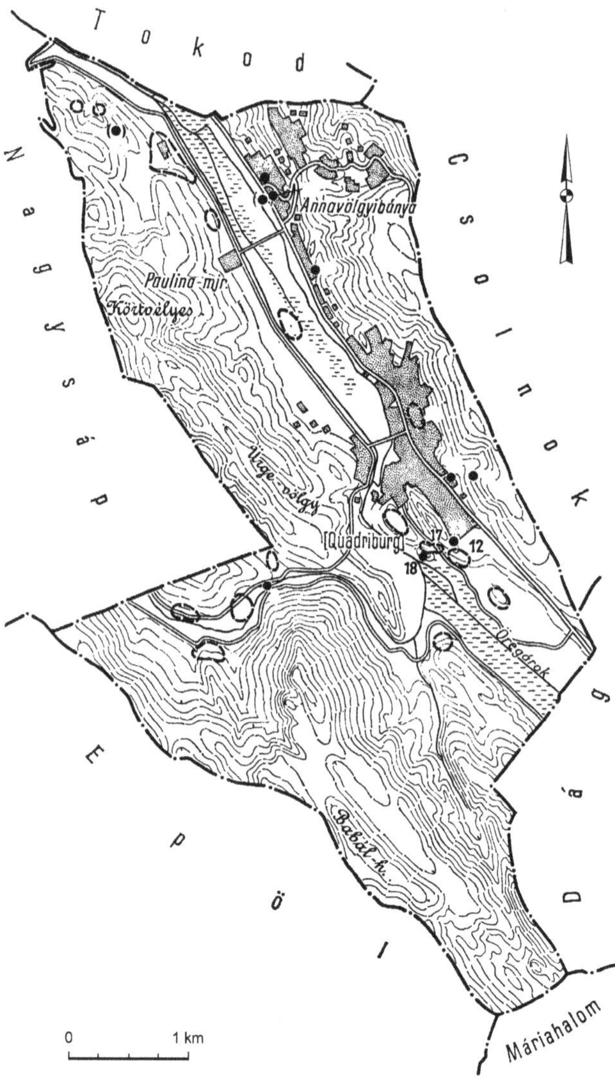

289. Sárisáp, general plan: 17: site of Roman settlement, 18: site of the so-called grave-chapel. Based on MRT 5, 19.

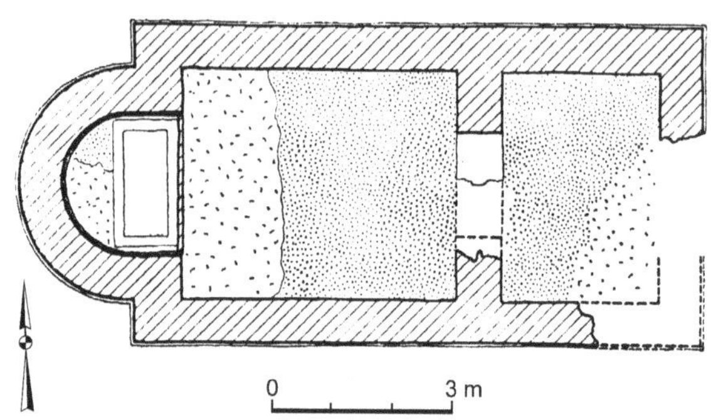

290. Sárisáp, ground plan of the so-called grave-chapel, 37.II.a. Based on MRT 5, 19.

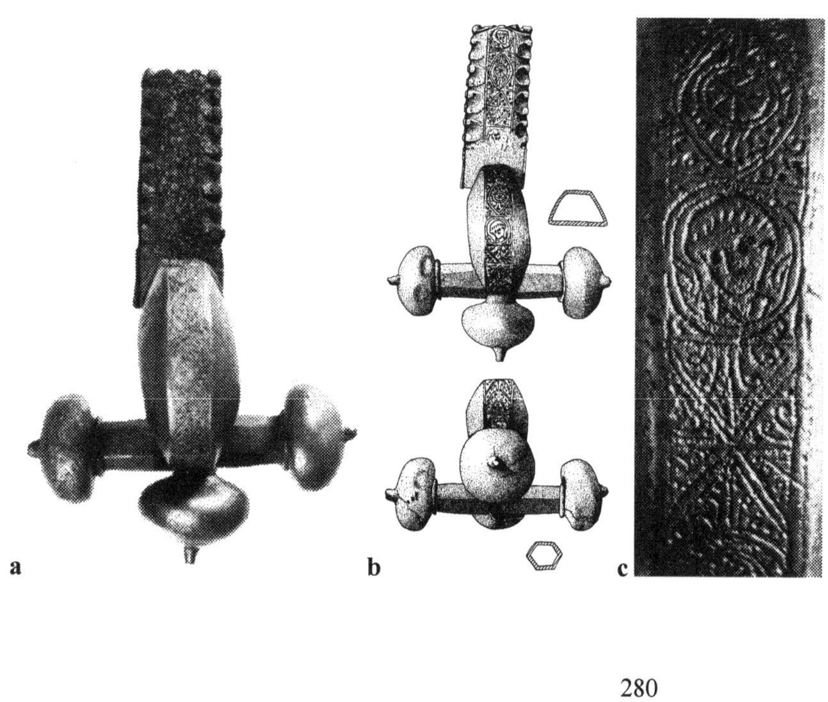

291. Somlóvásárhely, fibula, 40.I.a. Based on TÓTH E 1994a. **a:** photo. **b:** drawing, **c:** detail.

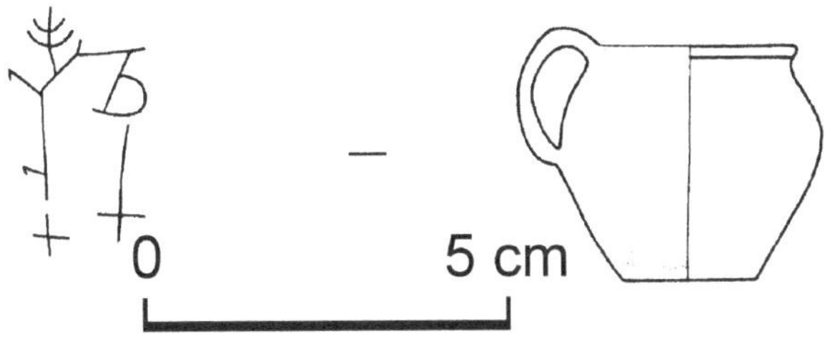

292. Somogyszil, pottery with incised pattern, 41.II.a. Drawn by DUKAY Bernadett.

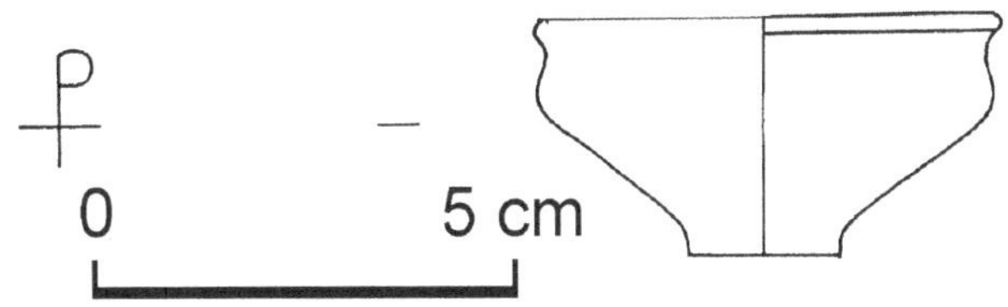

293. Somogyszil, pottery with incised pattern, 41.II.b. Drawn by DUKAY Bernadett.

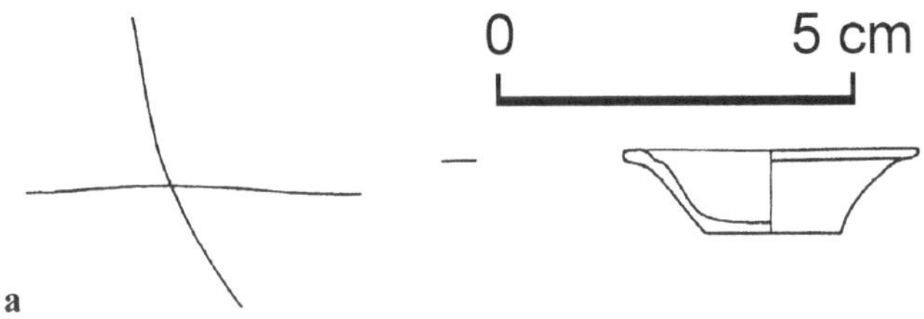

a

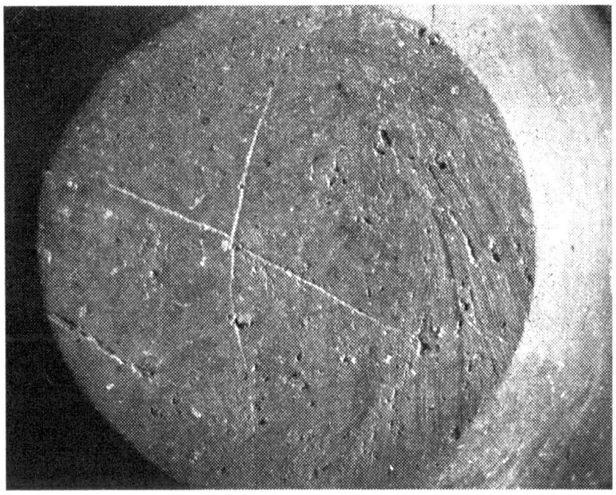

b

294. Somogyszil, pottery with incised pattern, 41.II.c. **a:** drawing by DUKAY Bernadett. **b:** photo: Mrs. GŐZSY.

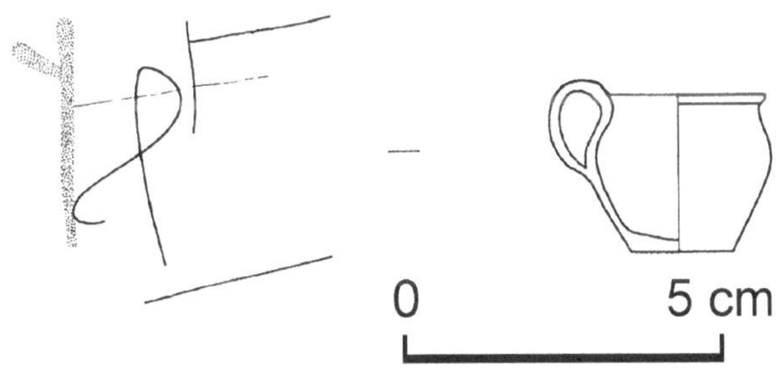

295. Somogyszil, pottery with incised pattern, 41.II.d. Drawn by DUKAY Bernadett.

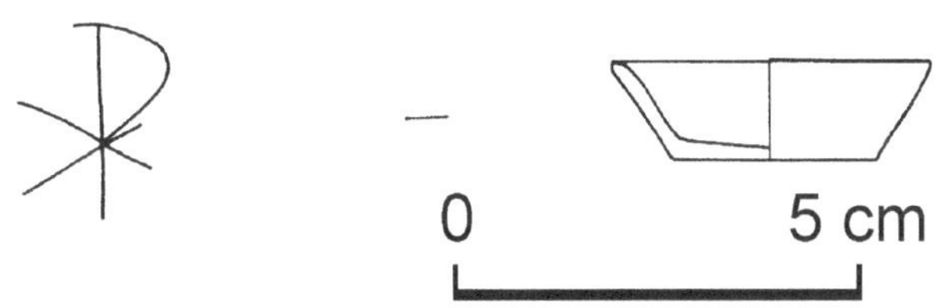

296. Somogyszil, pottery with incised pattern, 41.II.e. Drawn by DUKAY Bernadett.

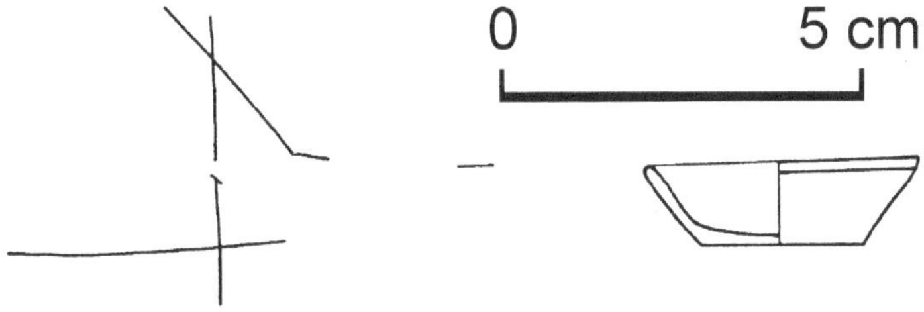

297. Somogyszil, pottery with incised pattern, 41.II.f. Drawn by DUKAY Bernadett.

a

b

298. Somogyszil, **a:** ichi smoothed in the bottom of an earthenware bowl, 41.II.g. Photo: Mrs. GŐZSY. **b:** Esztergom, altar. Based on MRT 5.

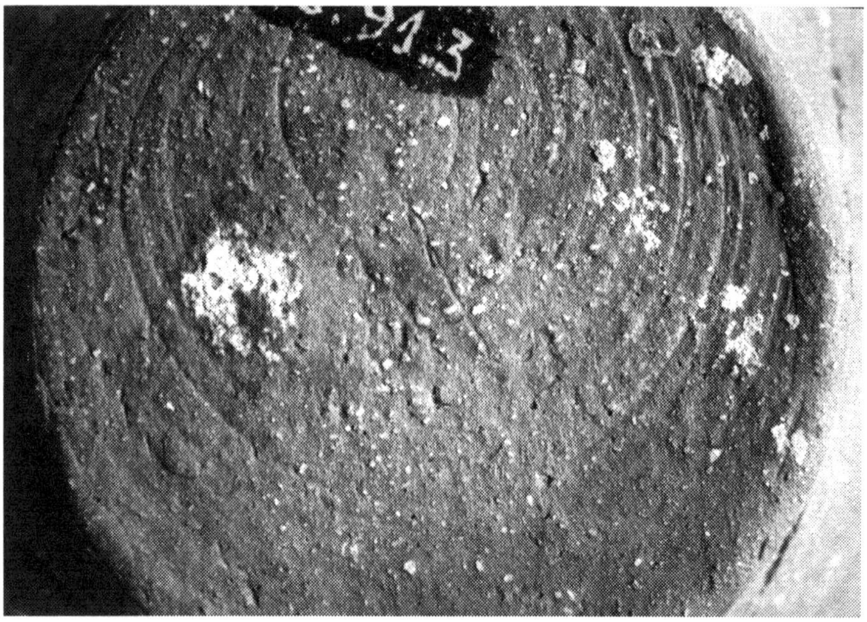

299. Somogyszil, pottery with incised pattern, 41.II.h. Photo: Mrs. GŐZSY.

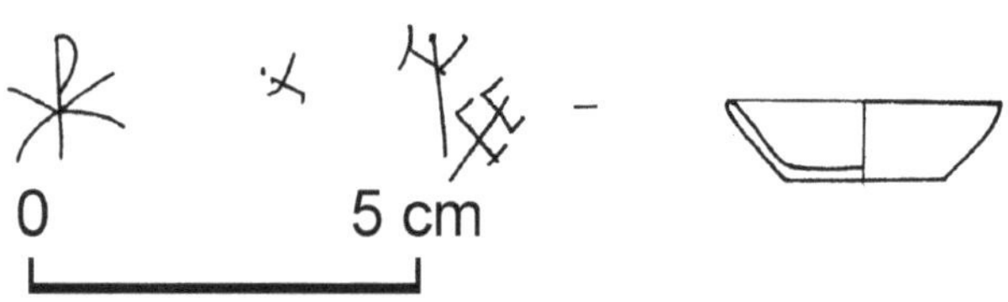

300. Somogyszil, pottery with incised pattern, 41.II.i. Drawn by DUKAY Bernadett.

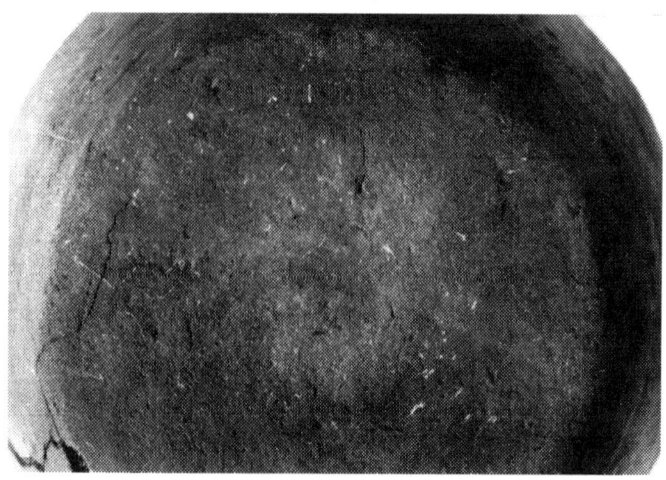

301. Somogyszil, pottery with incised pattern, 41.II.j. Photo: Mrs. GŐZSY.

302. Somogyszil, X smoothed in the bottom of an earthenware bowl, 41.II.k. Photo: Mrs. GŐZSY Gáborné.

303. Tác, drawings of earthenware incisions. Based on GABLER/KOCZTUR 1976.

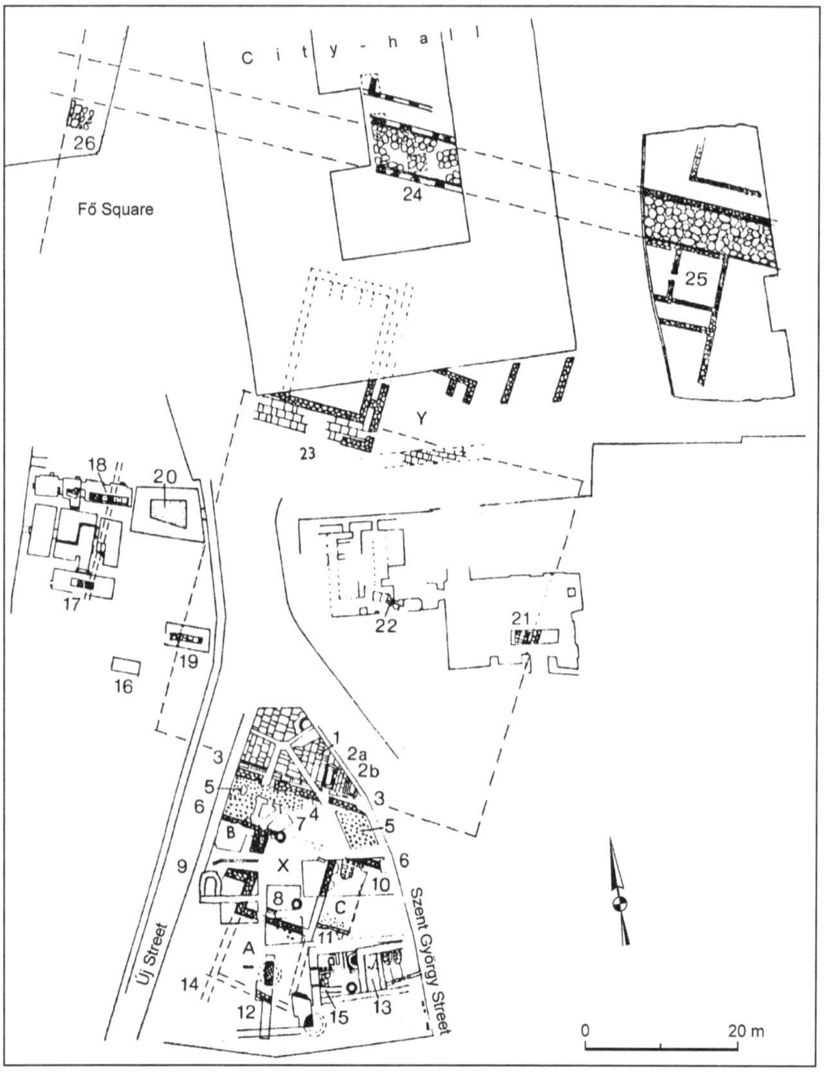

304. Sopron, general plan of the excavation of the forum. Based on GÖMÖRI 1986.

305. Sopron, jug fragment with inscription, 42.II.a. Based on GÖMÖRI 1986.

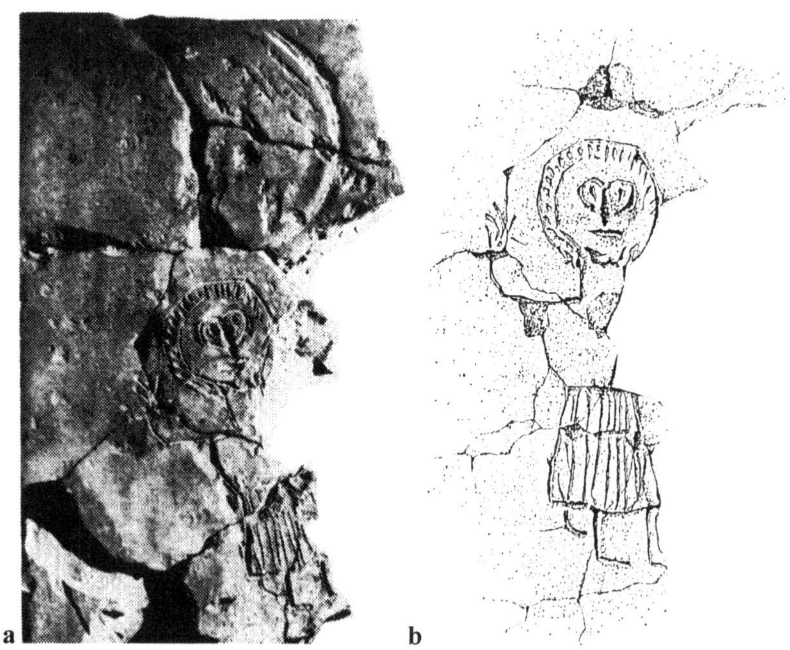

306. Sopron, brick sherd, 42.II.b. Based on GÖMÖRI. **a:** photo. **b:** drawing.

307. Sopron, marble slab, 42.II.c. Photo by ADORJÁN Attila.

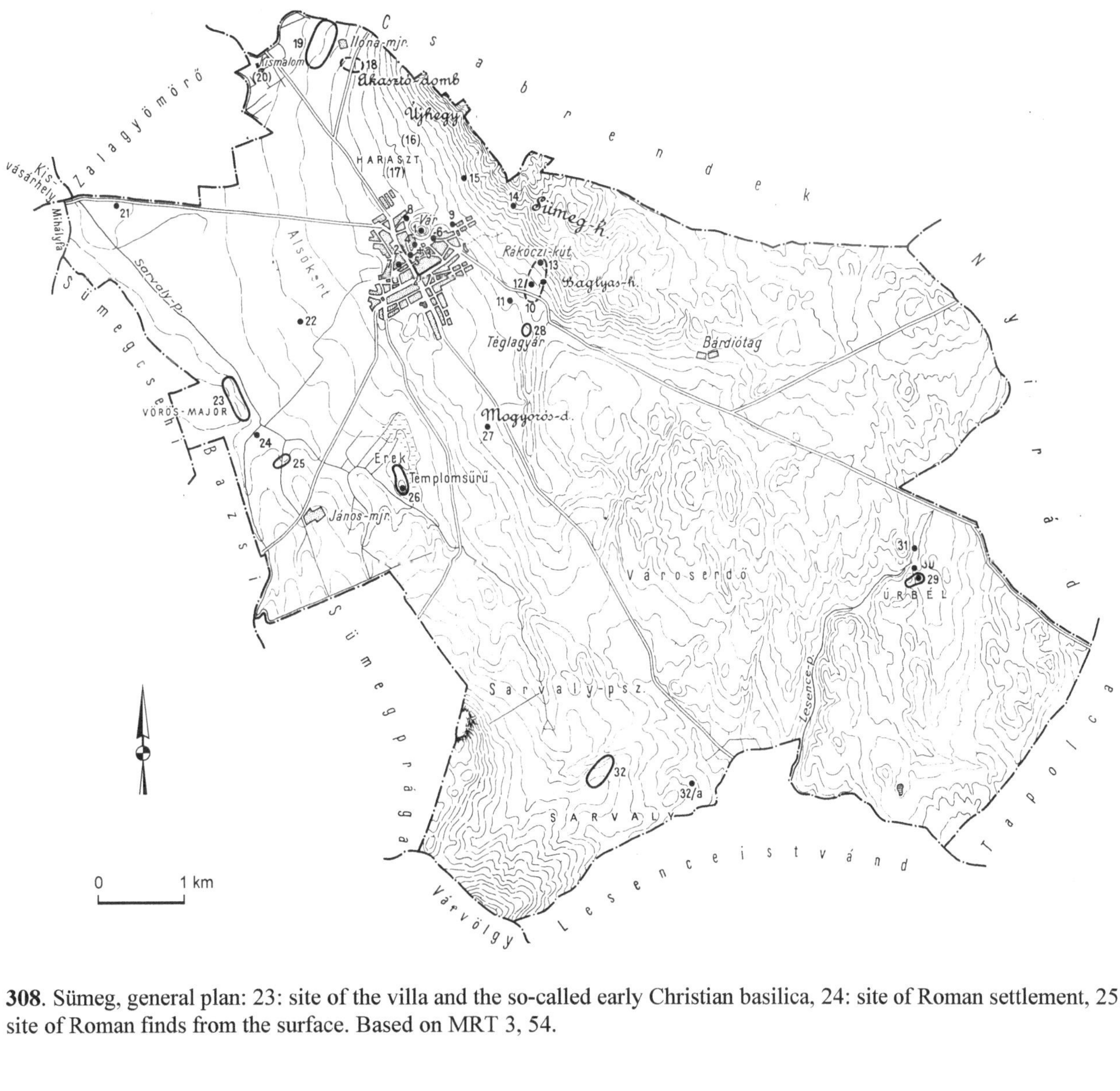

308. Sümeg, general plan: 23: site of the villa and the so-called early Christian basilica, 24: site of Roman settlement, 25: site of Roman finds from the surface. Based on MRT 3, 54.

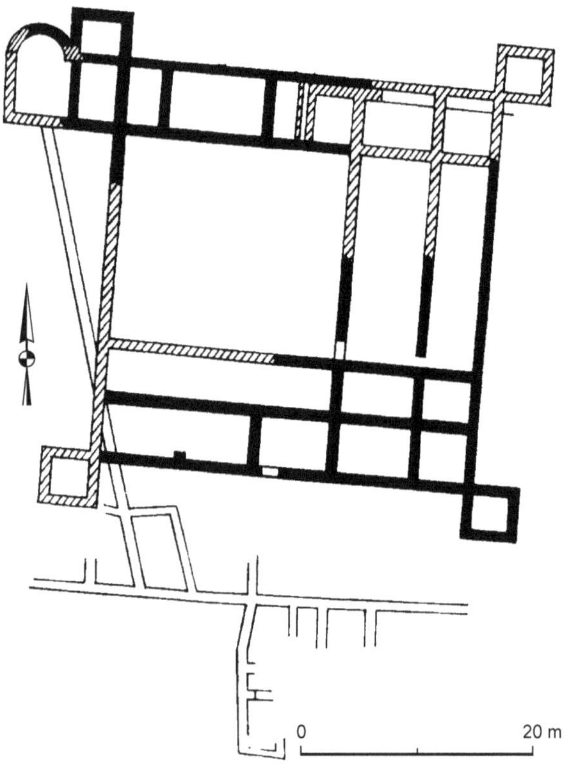

309. Sümeg, building I (villa), 43.I.a. Based on THOMAS 1964.

310. Sümeg, building II (so-called early Christian basilica), 43.II.a. Based on THOMAS 1964.

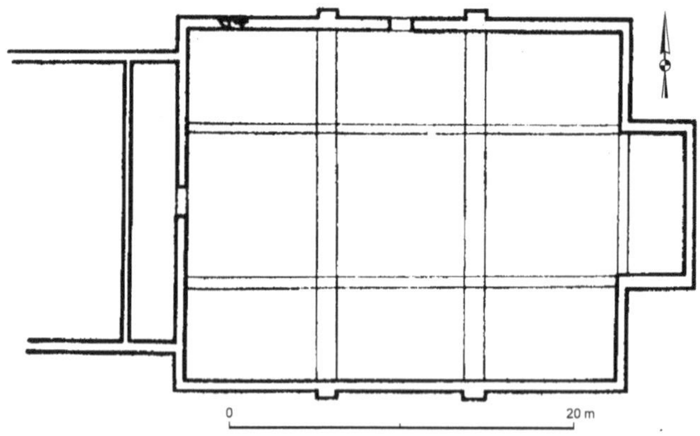

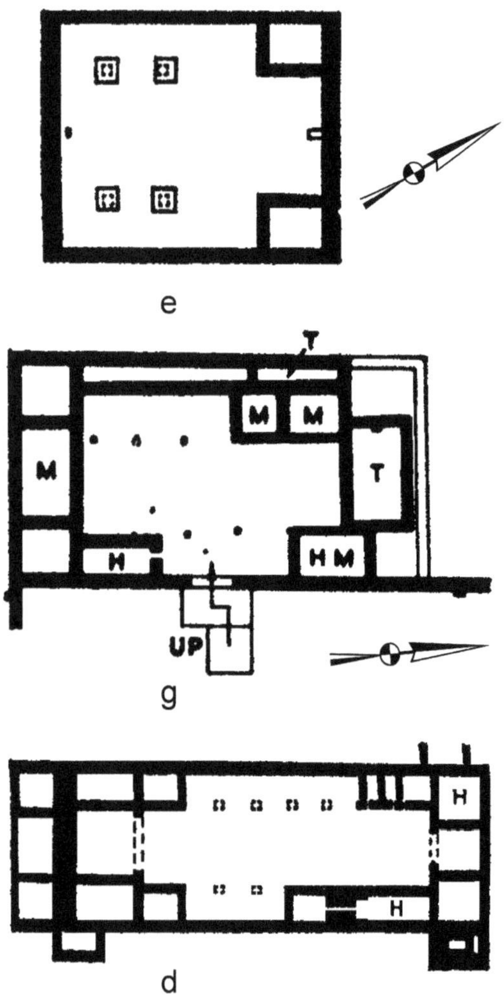

311. Parallels to building I. Based on RIVET 1969.

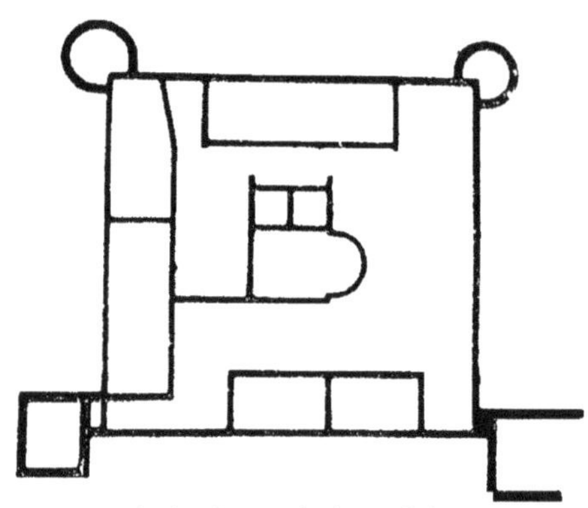

312. Ground plan of the Orlandovci villa. Based on SARNOWSKI 1978.

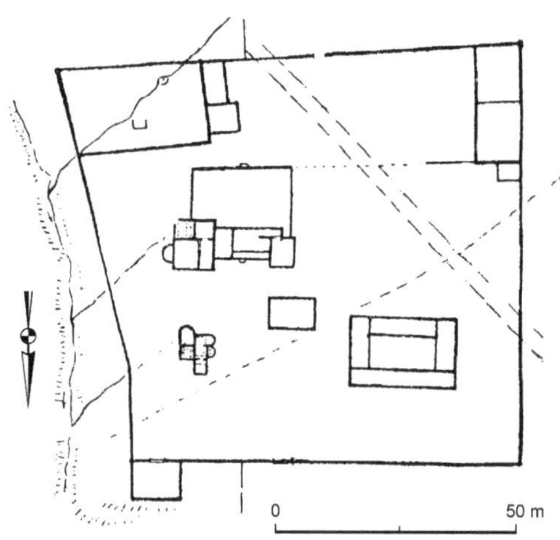

313. Ground plan of the Pforzheim villa. Based on SARNOWSKI 1978.

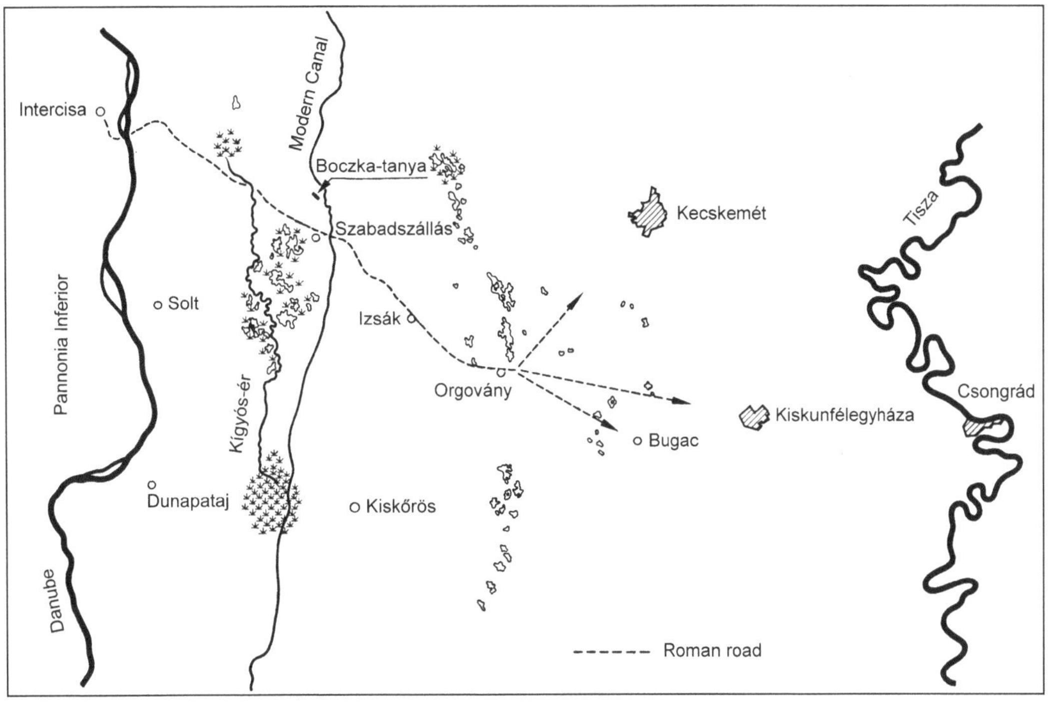

314. Szabadszállás, general plan. Based on H.-TÓTH E 1969.

315. Szabadszállás, glass cup, 44.II.a. Based on H.-TÓTH E 1971.

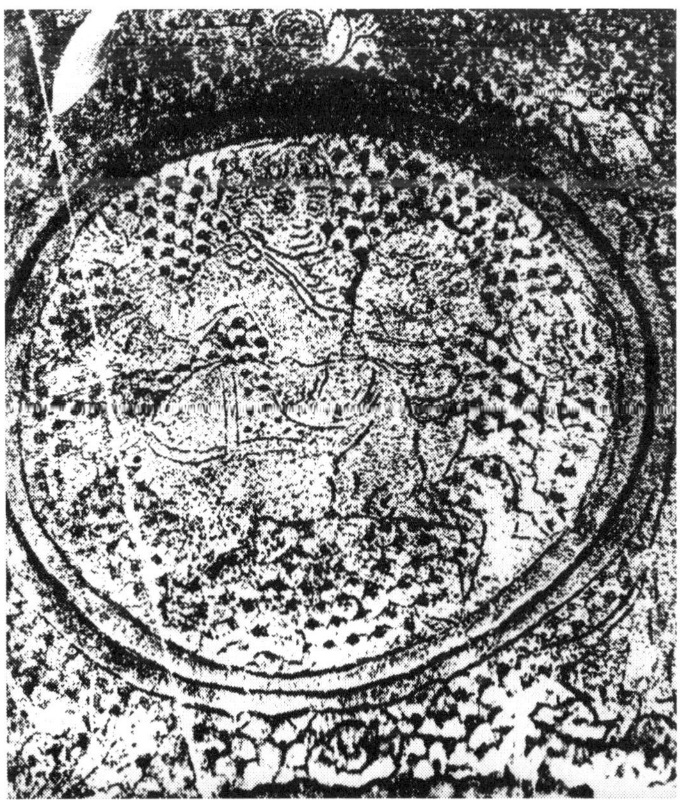

316. Part of 44.II.a. The first medallion. Based on H.-TÓTH E 1971.

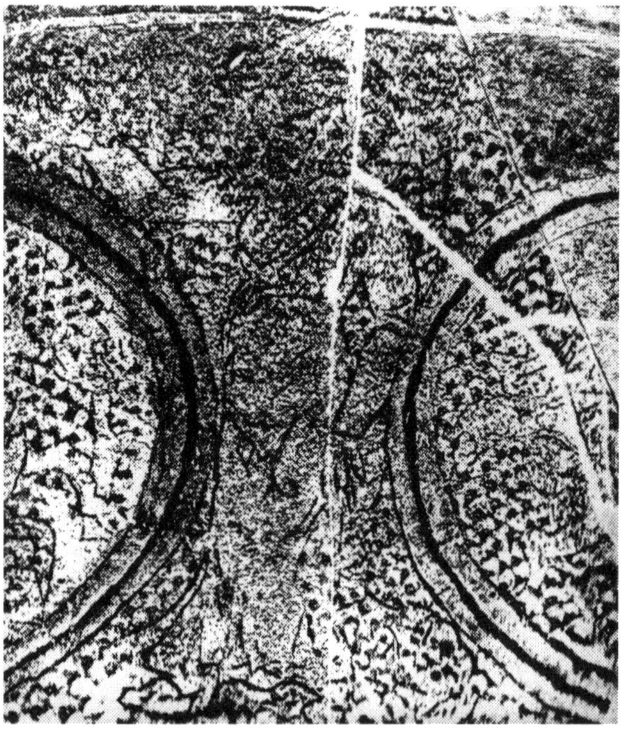

317. Part of 44.II.a. The woman figure between the first and the second medallions. Based on H.-TÓTH E 1971.

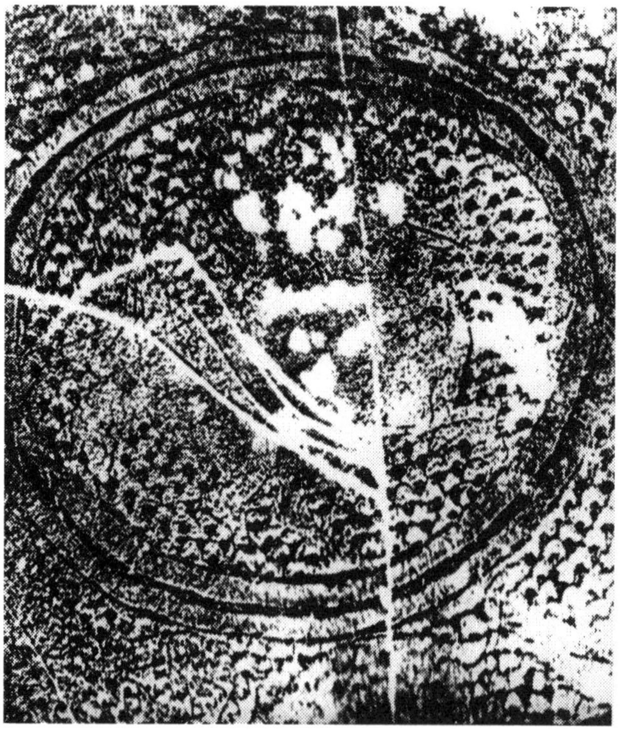

318. Detail of 44.II.a. Second medallion. Based on H.-TÓTH E 1971.

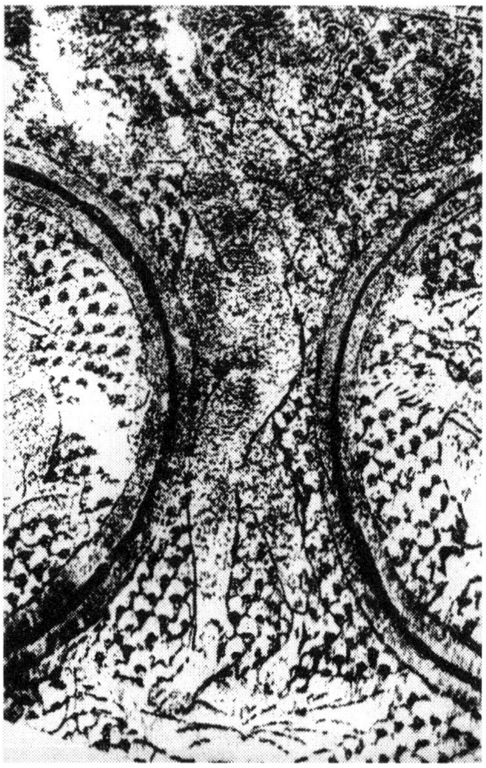

319. Detail of 44.II.a. Satyr between the medallions. Based on H.-TÓTH E 1971.

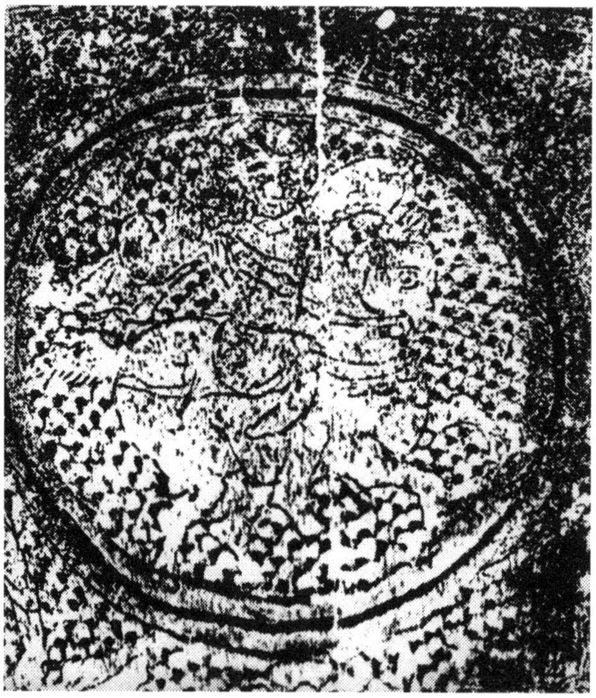

320. Detail of 44.II.a. Third medallion. Based on H.-TÓTH E 1971.

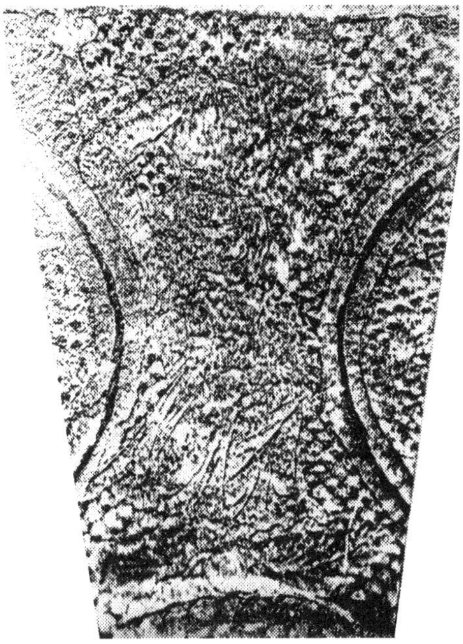

321. Detail of 44.II.a. Third figure between the medallions. Based on H.-TÓTH E 1971.

322. Superficies of 44.II.a. Drawing by Dr H.-TÓTH E.

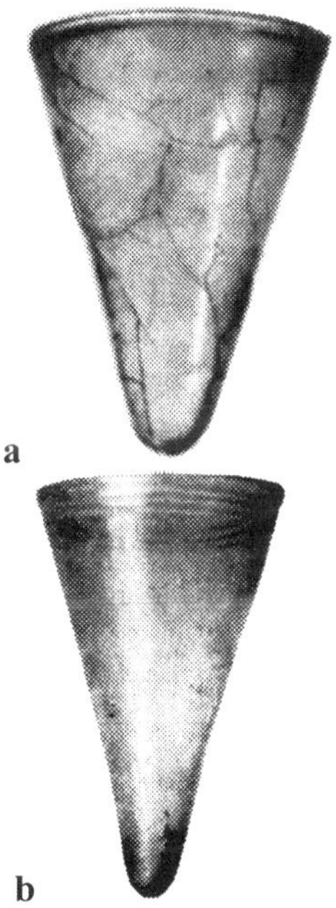

a

b

323. Parallels for the cup. **a:** Bátaszék-Kövesdpuszta, based on PÉTERFI 1993. **b:** Ságvár, grave 32, based on BURGER 1966.

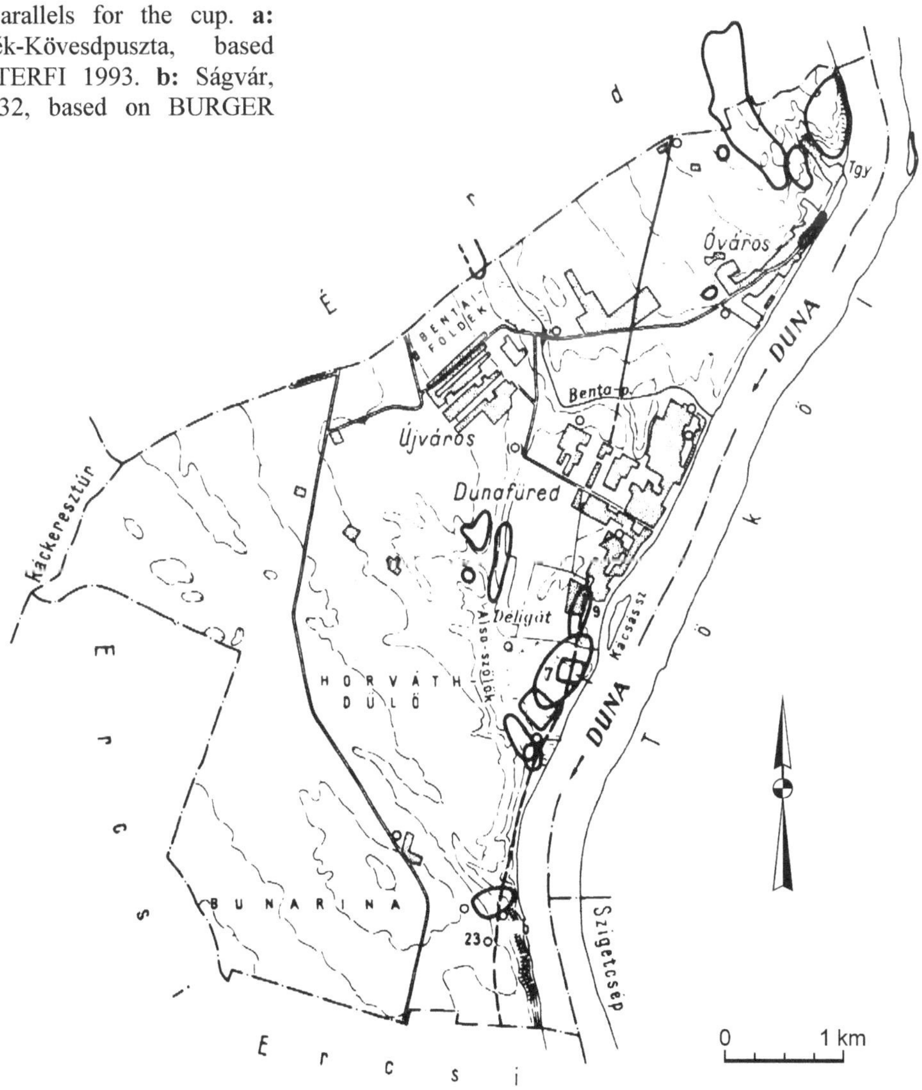

324. Százhalombatta, general plan: 6: fort, 7: vicus belonging to the fort, 8: southern cemetery, 9: northern cemetery, 23: watchtower, Based on MRT 7, 27.

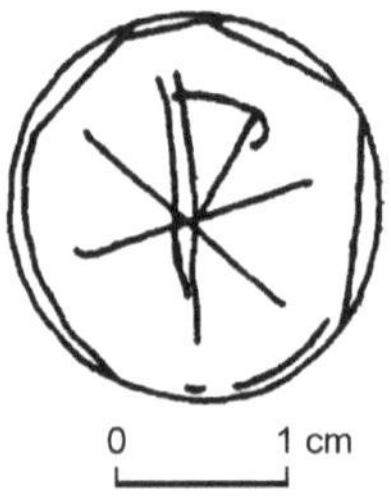

325. Százhalombatta-Dunafüred, bone tessera, 45.I.a. Based on MÓCSY 1955.

326. Székesfehérvár, screen, 46.I.a. as it was found, based on NAGY L 1938.

327. Szekszárd, vas diatretum, 47.II.b. Based on Severin Katalog.

328. Szekszárd, sarcophagus, 47.II.a. Based on Severin Katalog.

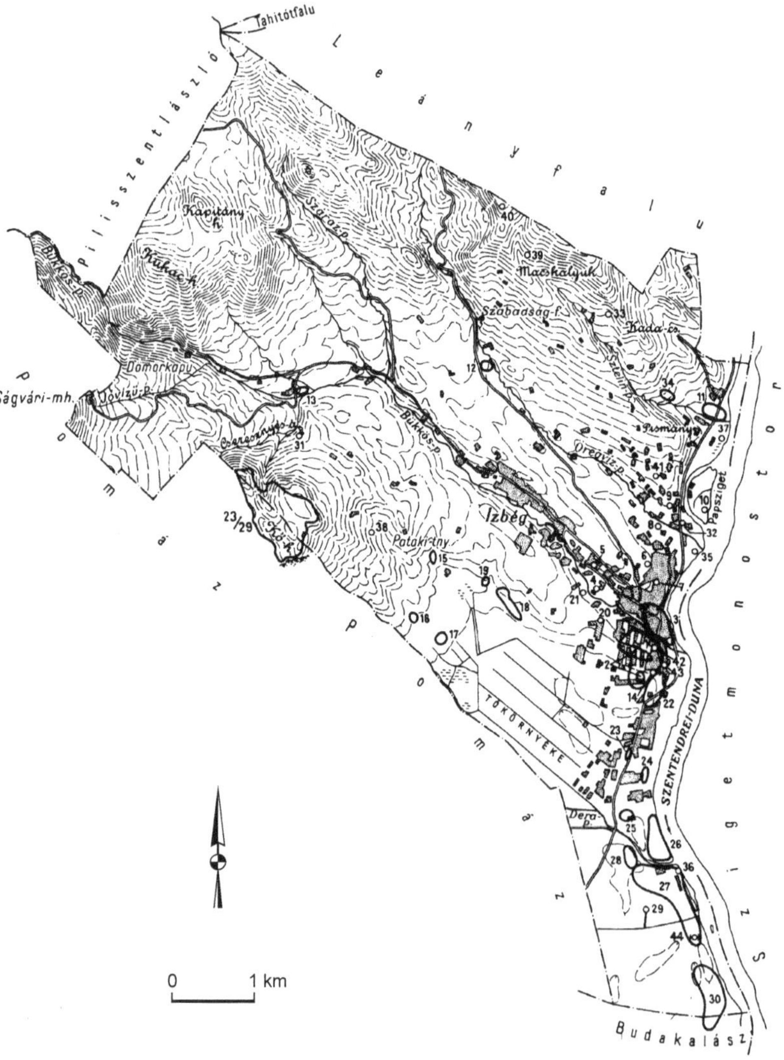

329. Szentendre, general plan. Based on MRT 7, 28.

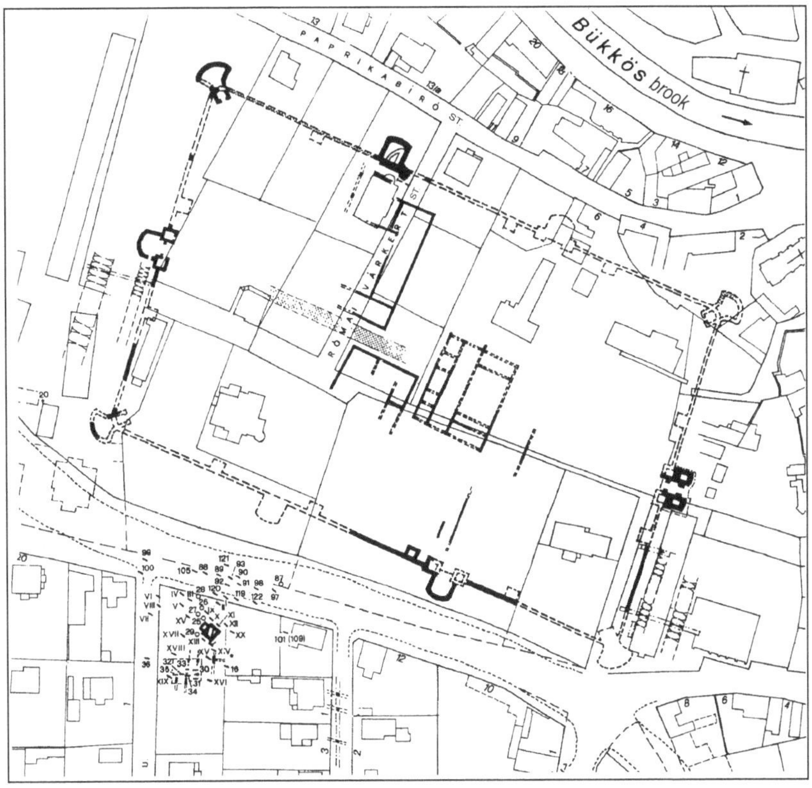

330. Szentendre, fortress and its environs, general plan. Based on MRT 7, 28.

331. Szentendre, casket mount, 48.I.a. Photo: KÁDAS Tibor.

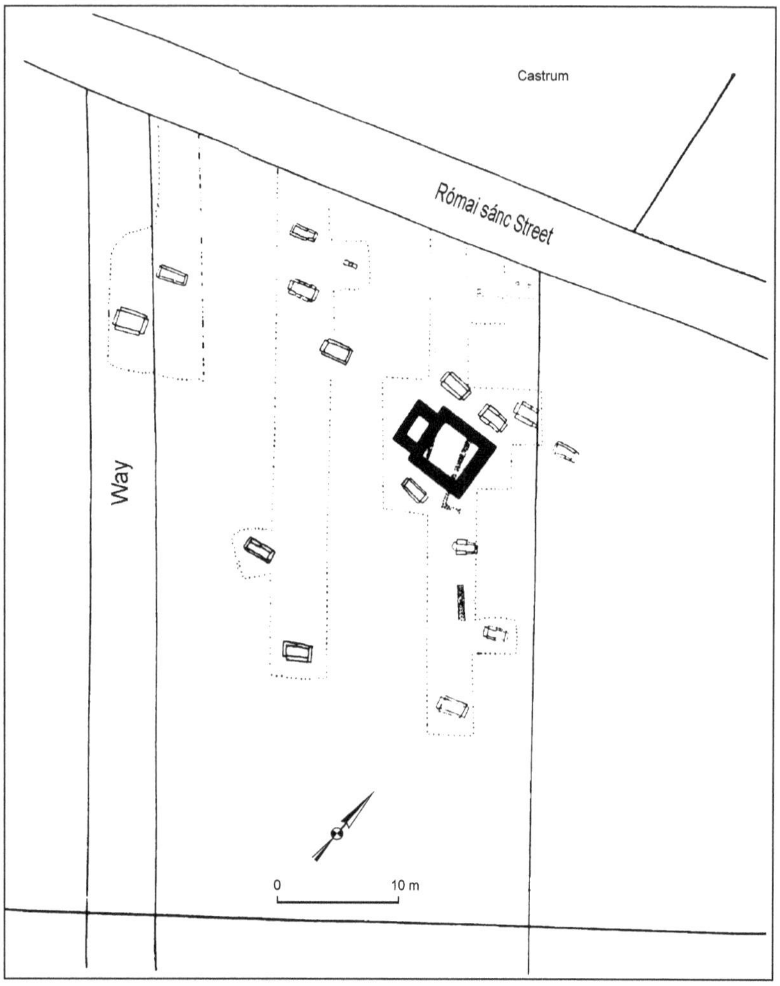

332. Szentendre, cemetery, 48.I.b. Based on NAGY L 1938. I-II. cemetery chapel.

a
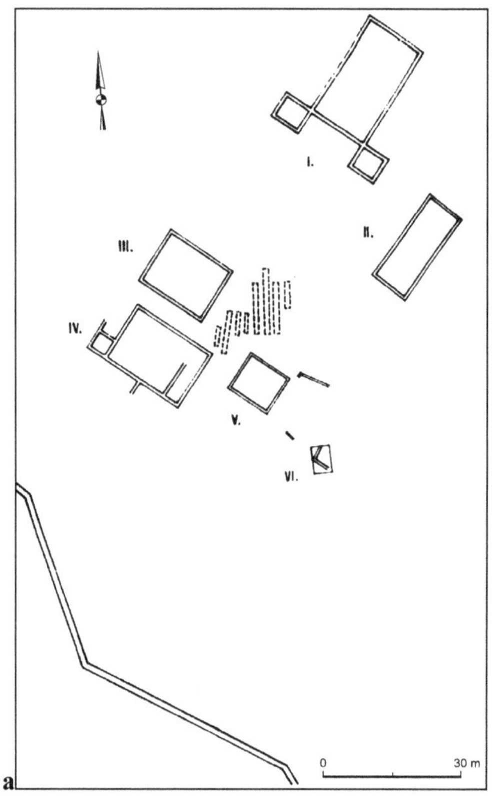

b
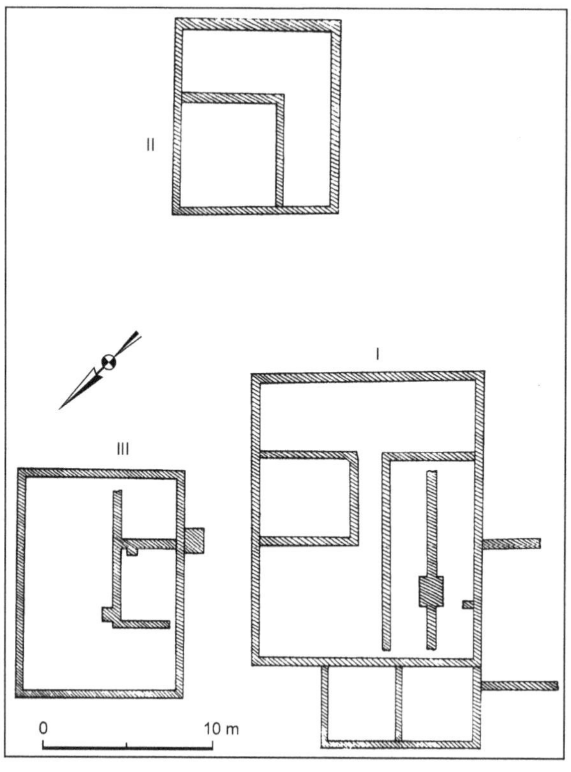

333. Szentkirályszabadja-Romkút, comprehensive drawing of the buildings, 49.II.a. **a:** based on THOMAS 1964. **b:** based on MRT 2, 44.

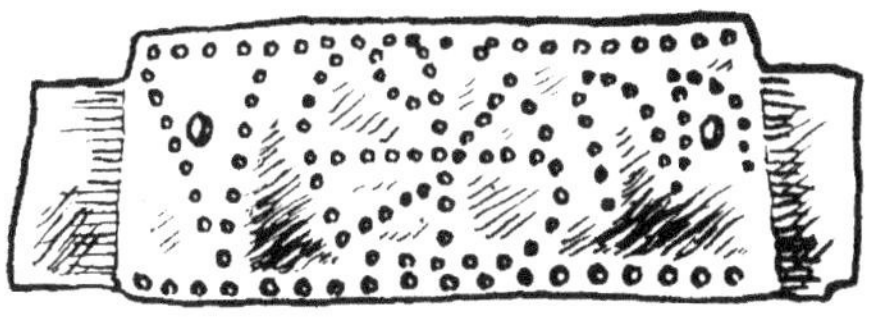

334. Szilágypuszta-Malomalja, bronze sheet, 50.II.a. Based on THOMAS 1964.

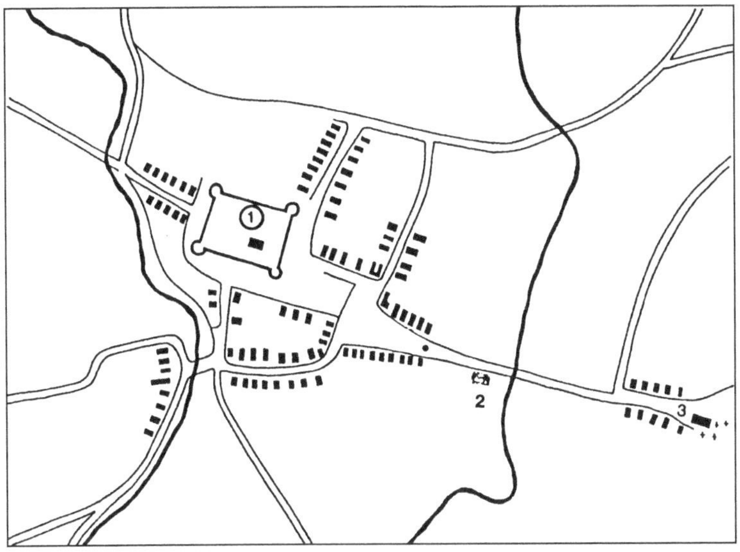

335. Szombathely, reconstruction of the 15th century situation: 1. castle and the Virgin Mary Church, 2. Franciscan Church, 3. Saint Martin's Church. Based on KISS G/ TÓTH E 1993.

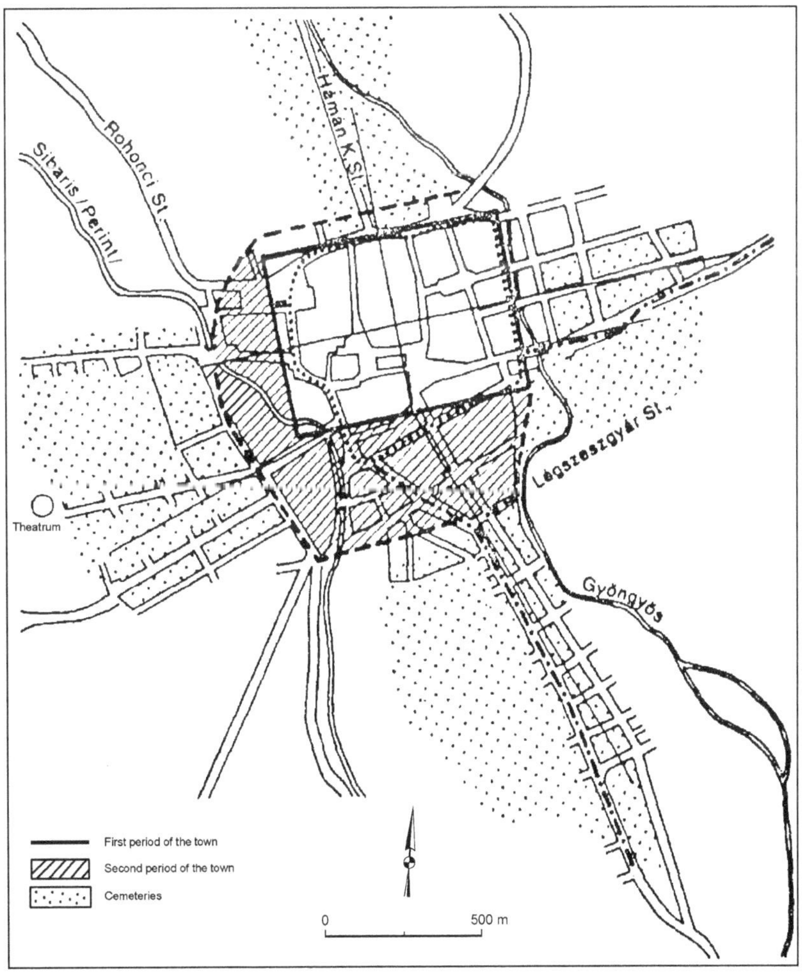

336. Szombathely's topography and periods according to earlier knowledge. Based on BUOCZ 1967.

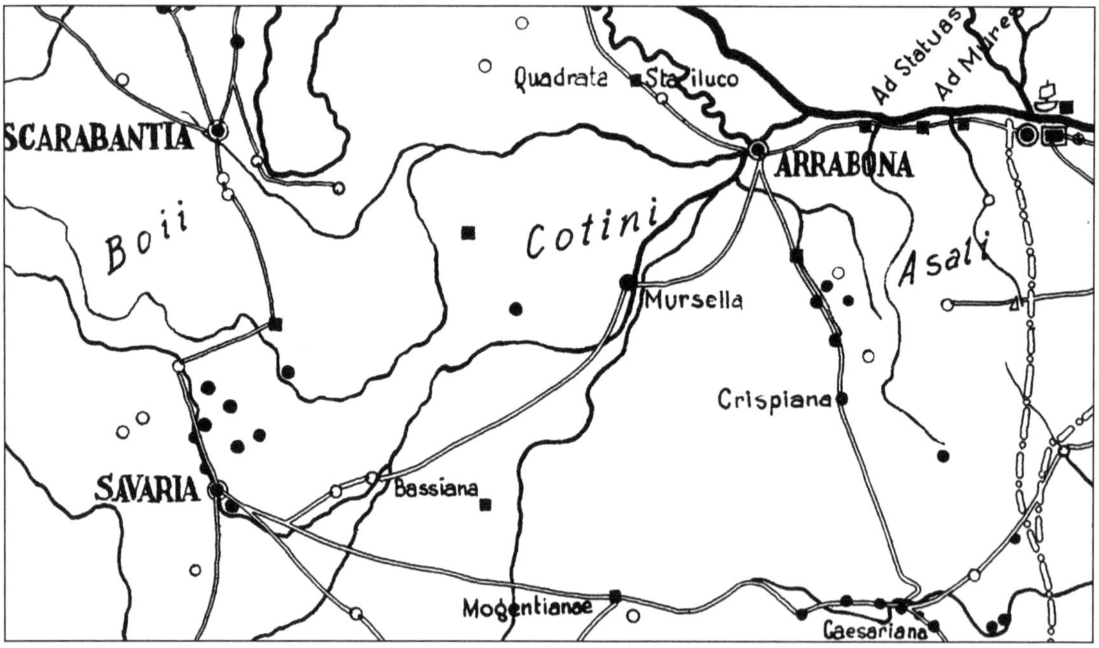

337. Link between Savaria and Arrabona. Based on GRAF 1936.

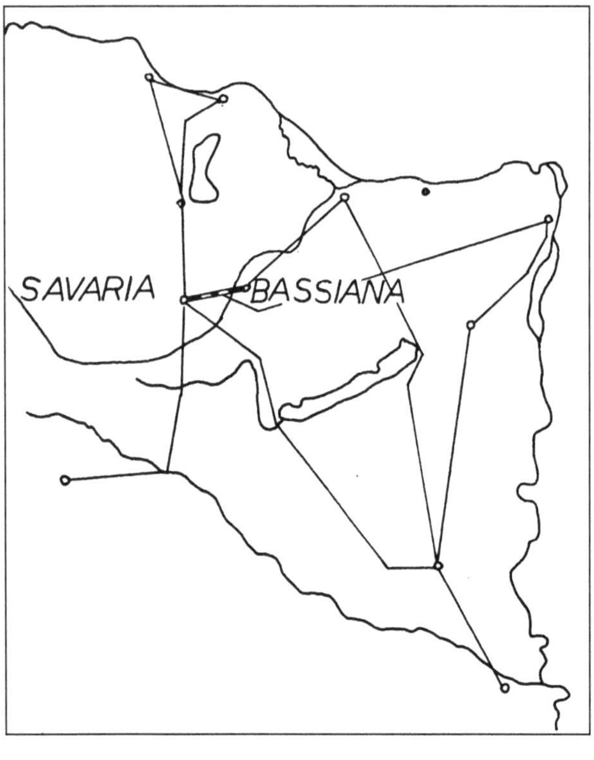

338. Link between Savaria and Bassiana. Based on TÓTH E 1977b.

339. Savaria's town wall and gates. Based on MEDGYES 1997.

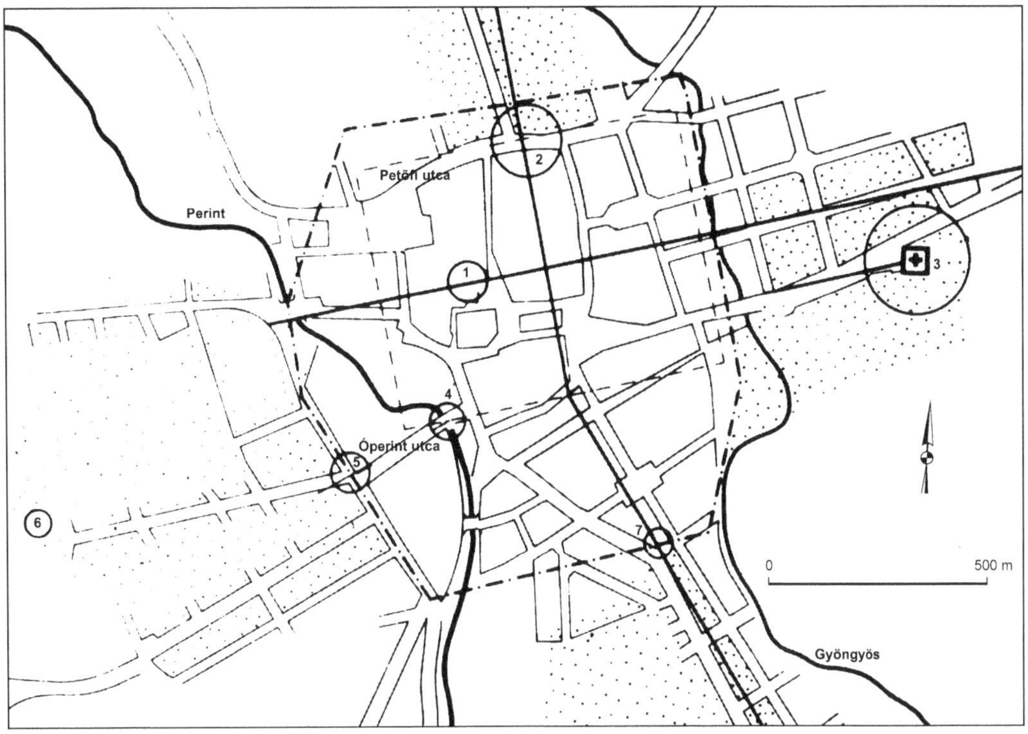

340. Szombathely, reconstructed places of the *passion*'s events based on TÓTH E 1973. 1. Imperial palace, 2. Sopron gate, 3. Saint Martin's supposed native house and supposed Christian cemetery, 4. place of Quirinus' martyrdom, 5. western gate, 6. amphitheatre, 7. southern gate.

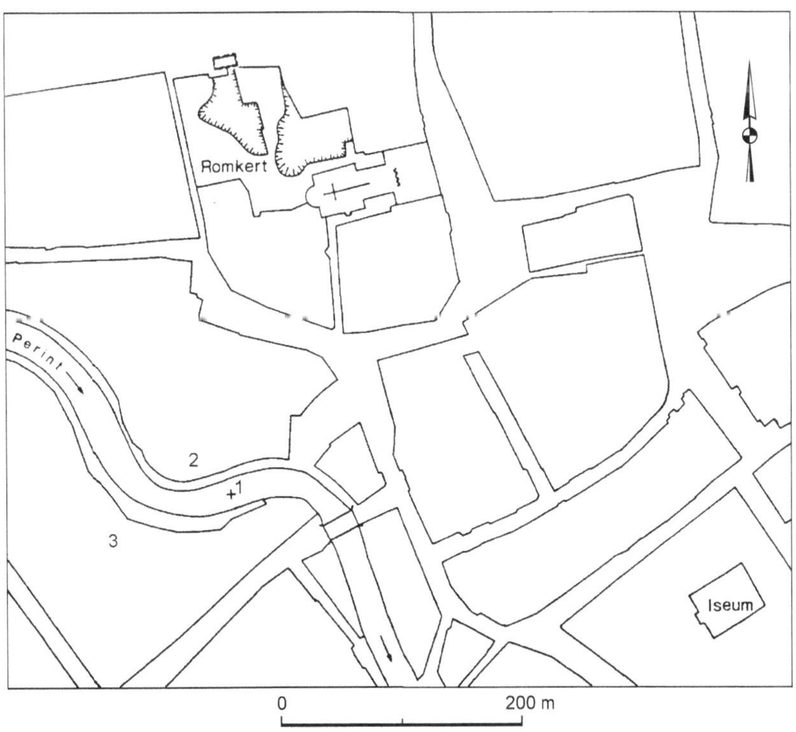

341. Szombathely, places of the events in the *passion,* author's reconstruction. 1: place where the body of the deceased was found, 2: *locus orationis*, area inhabited by Christians in the 3rd century, 3: area inhabited by Christians in the 4th century. Romkert = Garden of ruins.

342. Szombathely, tombstone of Aurr. Flavius et Leonis, 51.I.a. Based on THOMAS 1977b.

343. Szombathely, tombstone of Aurelius Iodorus, 51.I.b. Based on THOMAS 1977b.

344. Szombathely, tombstone of *pictures pelegrini*, 51.I.e. Based on THOMAS 1977b.

345. Szombathely, tombstone of L. Maximinus, 51.I.f. Based on THOMAS 1977b.

b

a

346a-b. Szombathely, bronze lamp, 51.I.g. Based on THOMAS 1977b.

347. Szombathely, Menas ampulla, 51.I.h. Based on THOMAS 1977b.

348. Szombathely, marble screen, 51.II.a. Based on Steindenkmäler.

349. Karatas, fragment of marble screen. Based on NIKOLAJEVIĆ 1978.

350. Szombathely, marble slab (cover *loculi*?), 51.II.b. Based on Steindenkmäler.

351. Szombathely, grave-inscription of Aelia Kalendina, 51.II.c. Based on Steindenkmäler.

352. Szombathely, tombstone of Aurelia Iustina, 51.II.d. Based on Steindenkmäler.

353. Szombathely, tombstone with *carmina*, 51.II.e. Based on Steindenkmäler.

354. Szombathely, tombstone of Namius Quintus, 51.II.f. Based on Steindenkmäler.

355. Szombathely, fragment of tombstone, 51.II.g. Based on Steindenkmäler.

356. Szombathely, roof tile, 51.II.h. Based on THOMAS 1977b.

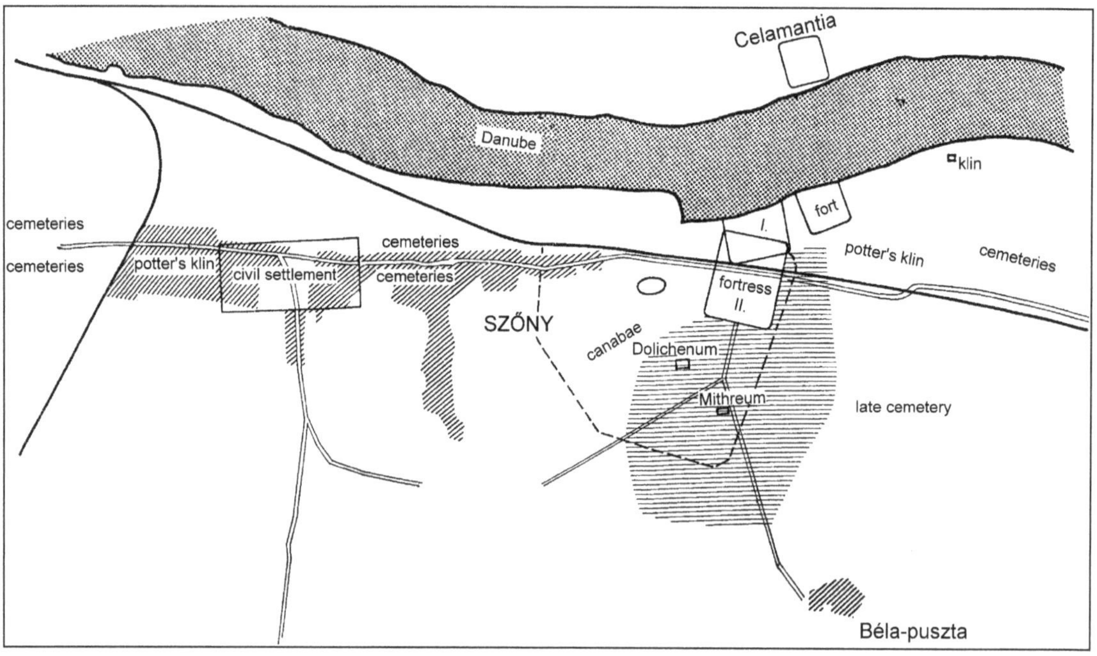

357. Szőny, general plan. Based on RIU 2.

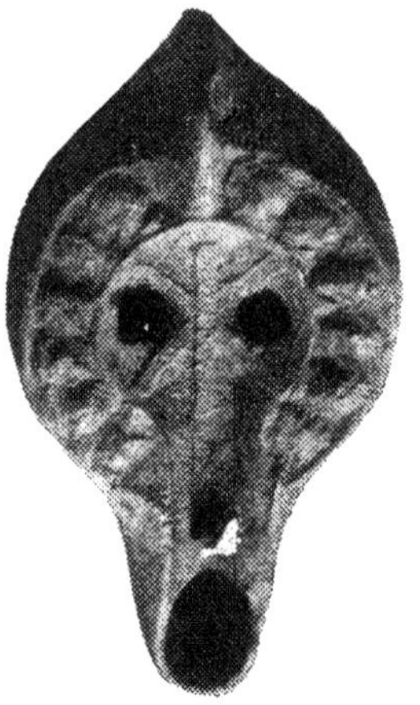

358. Szőny, clay lamp, 52.I.a. Based on IVÁNYI 1935.

359. Szőny, clay lamp, 52.I.b. Based on IVÁNYI 1935.

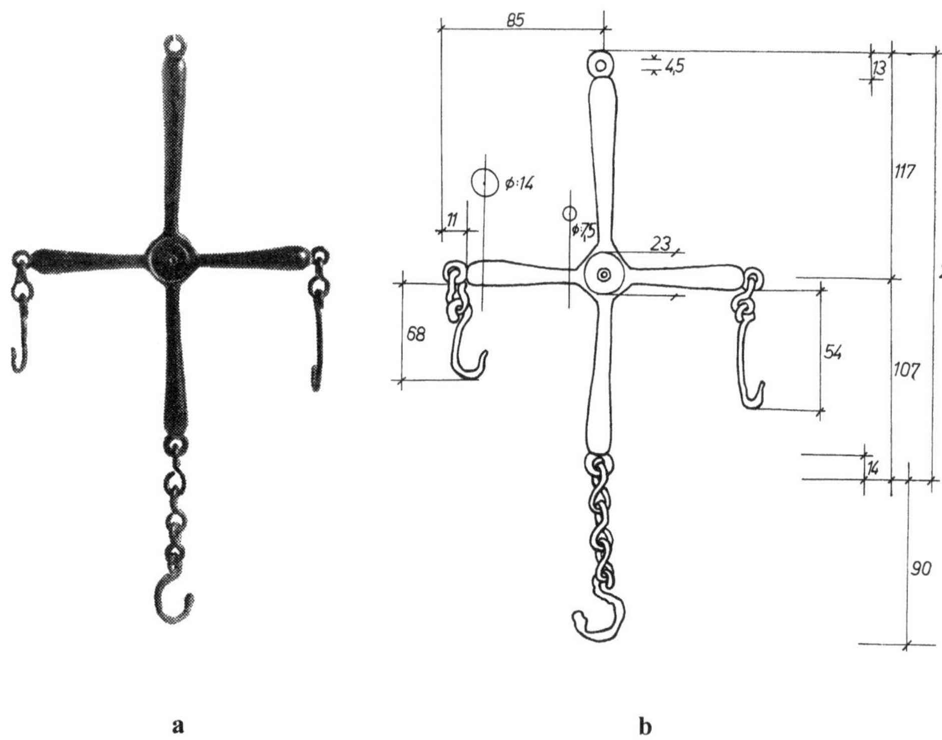

360. Szőny, bronze suspension for a lamp, 52.I.c. **a:** photo, **b:** drawing. Based on TÓTH E 1977a.

361. Szőny, silver finger ring, 52.I.d. Based on NAGY L 1938.

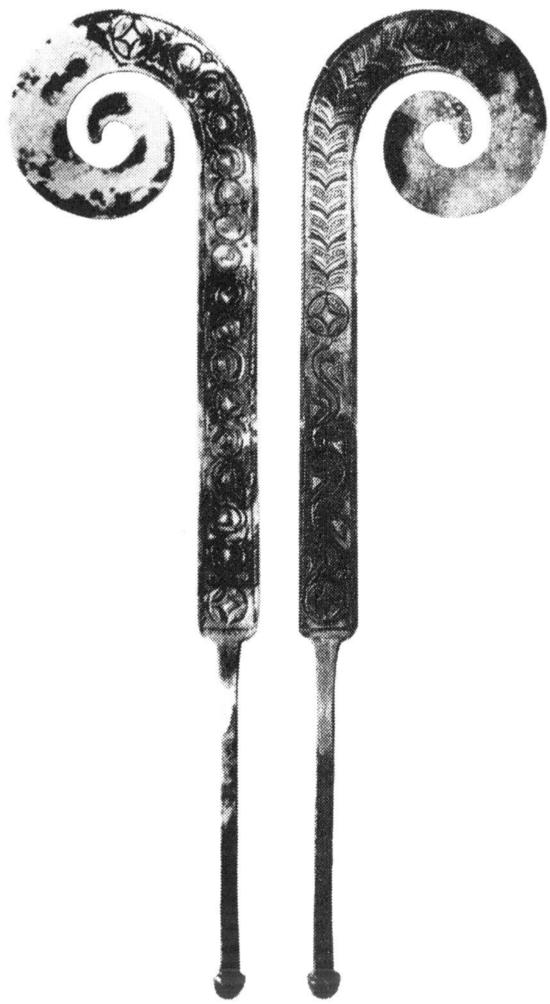

362. Szőny, so-called augur's staff / so-called crosier, 52.II.a.

a

b

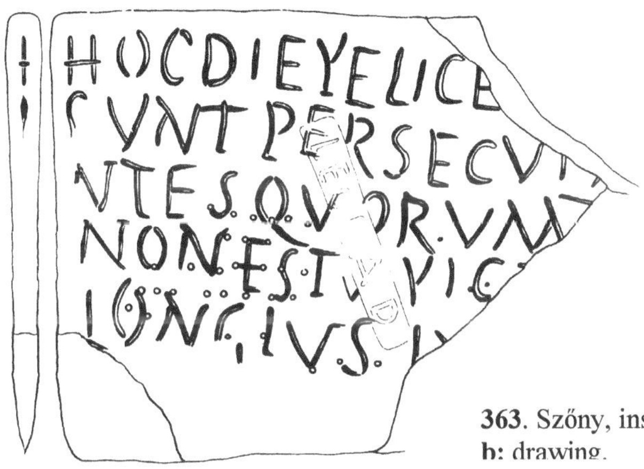

363. Szőny, inscribed roof tile, 52.II.b. Based on THOMAS 1974. **a:** photo, **b:** drawing.

a

b

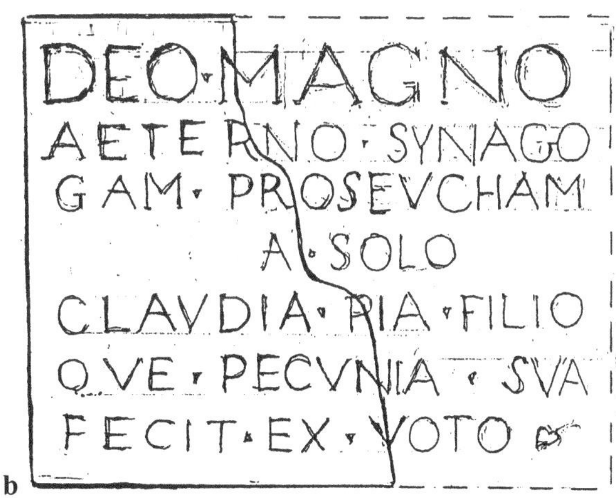

364. Szőny, votive plaque, 52.II.c. **a:** based on RIU 440. **b:** based on TÓTH I 1995.

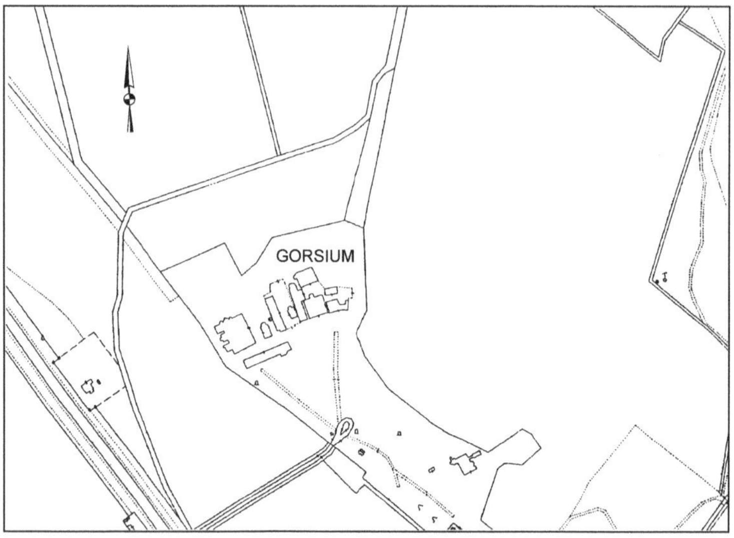

365. Tác, general plan based on the survey by the Geodéziai és Térképészeti Vállalat in 1989. Scale: 1:4000.

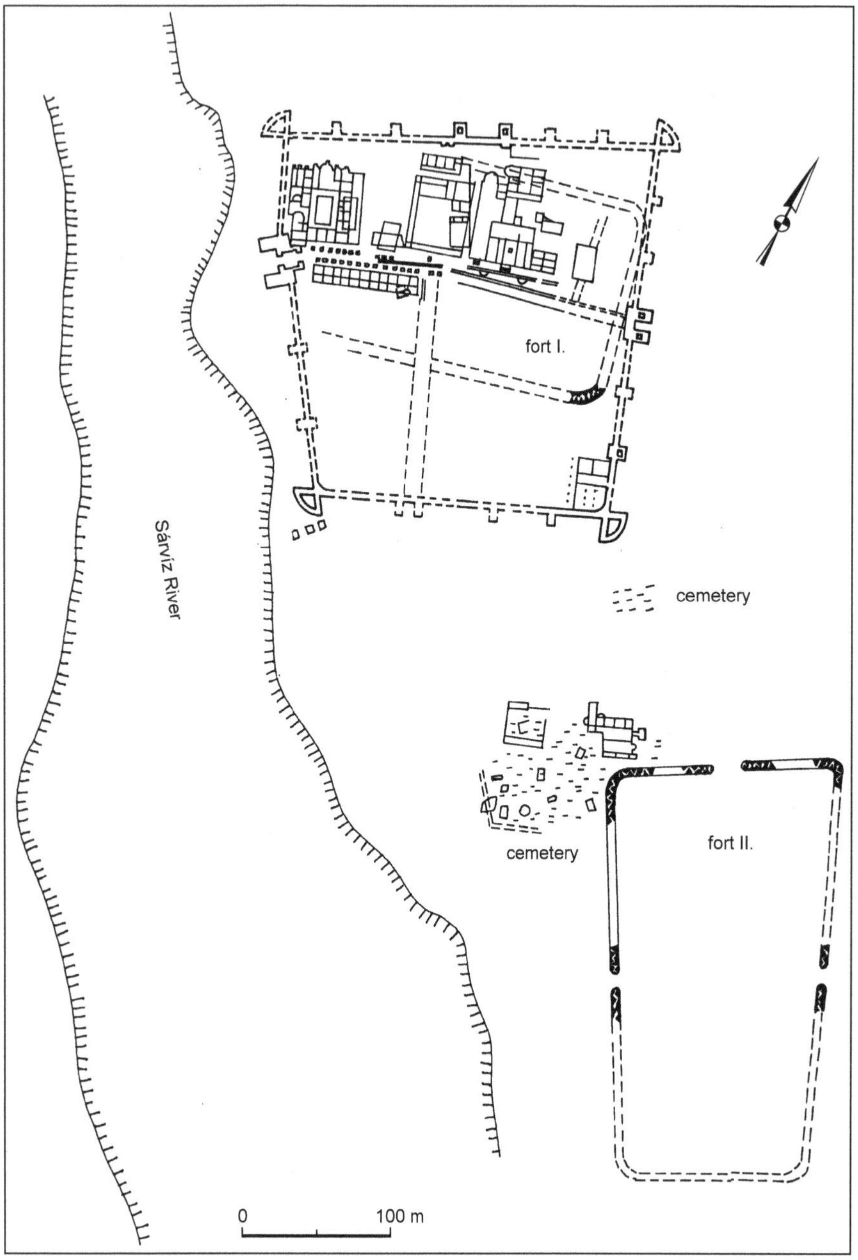

366. Tác, ground plan of excavations. Based on FITZ.

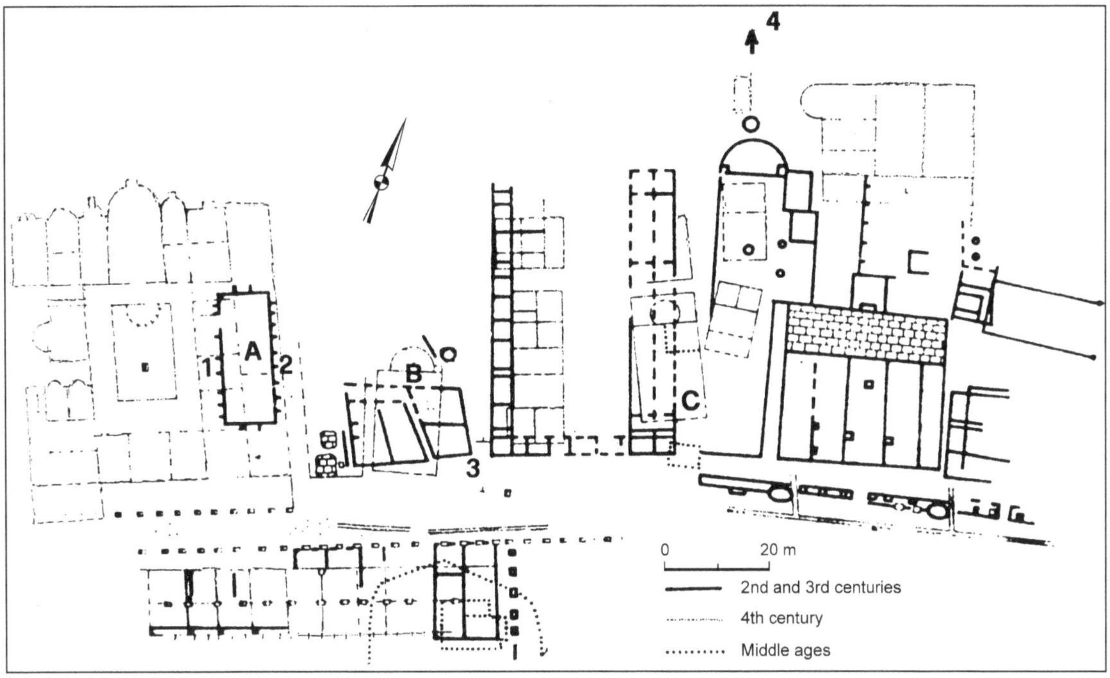

367. Tác, general ground plan of excavations. Based on BARKÓCZI/SALAMON 1984.

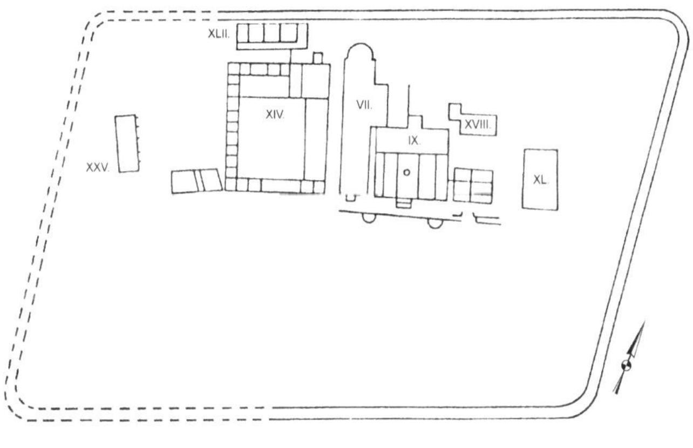

368. Tác, ground plan of buildings from the 2nd and 3rd centuries. Based on TÓTH E 1991.

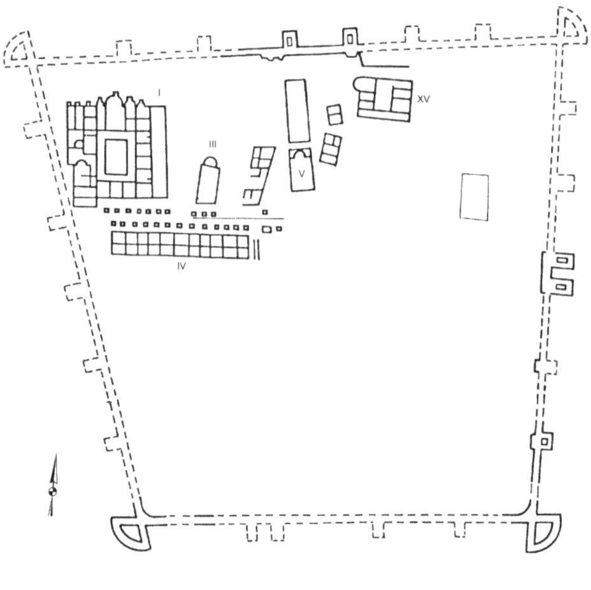

369. Tác, ground plan showing 4th century conditions. Based on TÓTH E 1991.

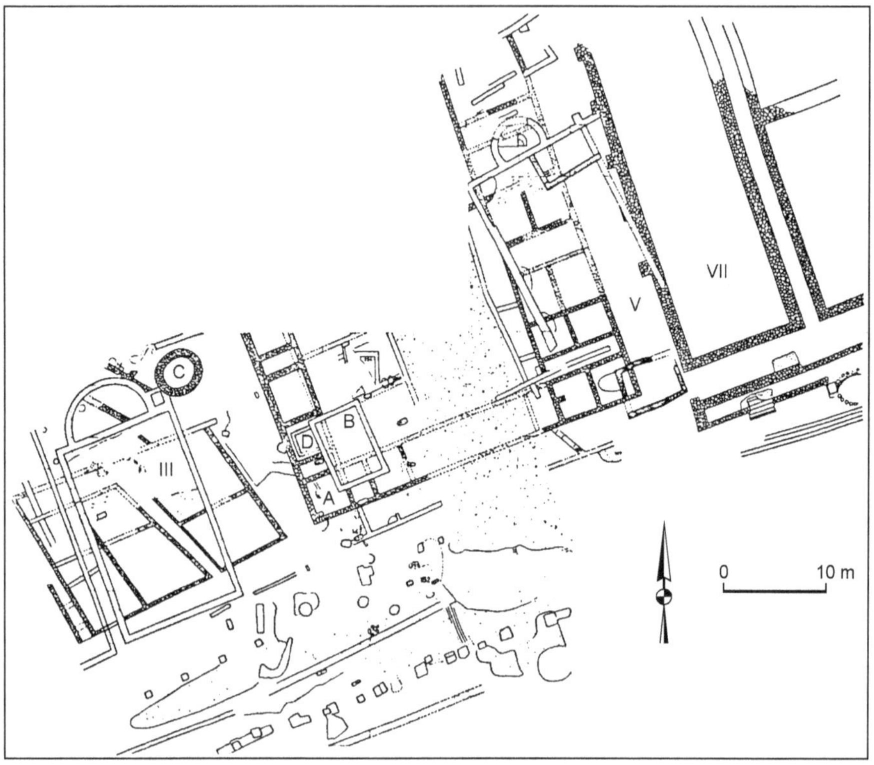

370. Tác, the environs of building III, 53.II.a. and of building V, 53.II.b. "C" indicates the well. Based on FITZ 1976b.

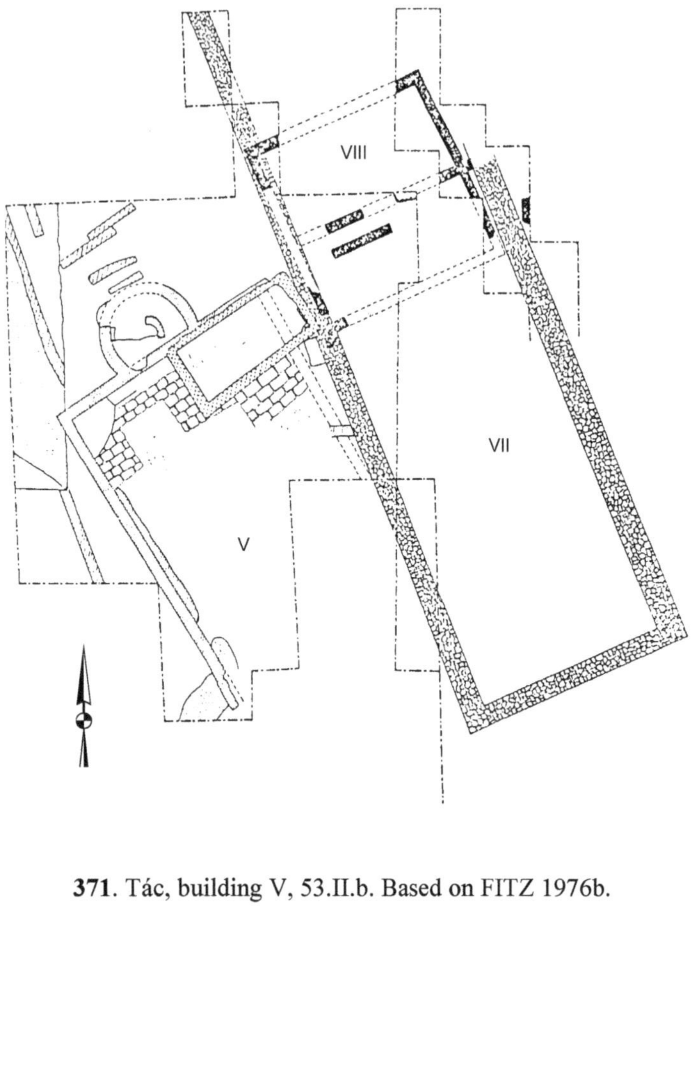

371. Tác, building V, 53.II.b. Based on FITZ 1976b.

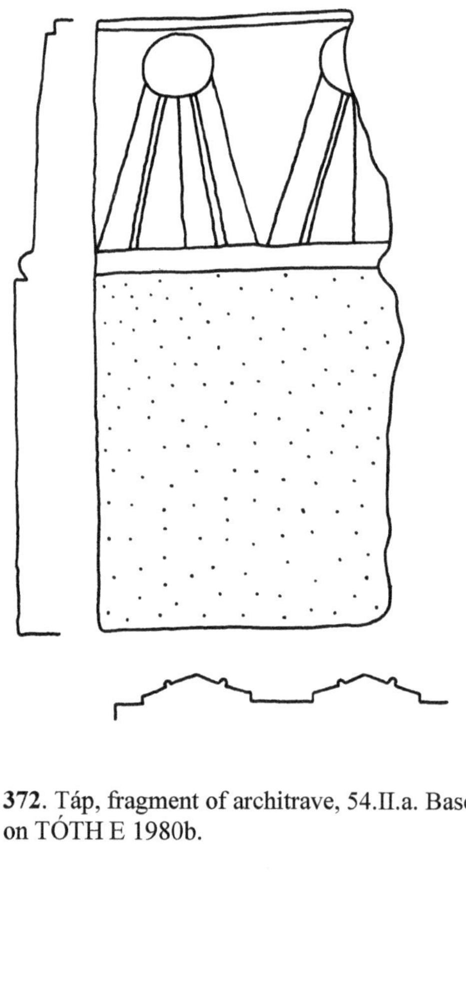

372. Táp, fragment of architrave, 54.II.a. Based on TÓTH E 1980b.

373. Tata, finger ring, 55.I.a. Based on TÓTH E 1979.

a

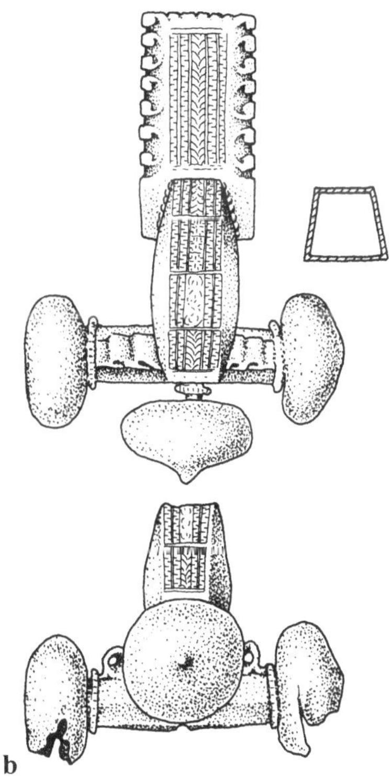

b

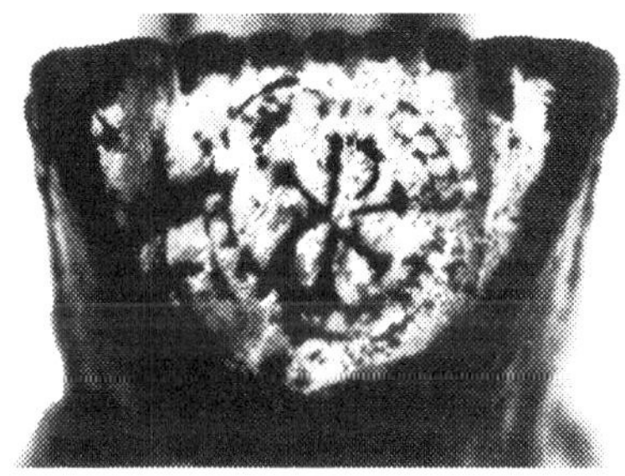

c

374. Tihany-Sajkód, fibula, 56.I.a. Based on TÓTH E 1994a. **a:** photo. **b:** drawing, **c:** detail.

375. Tokod, general plan: 15: Roman road, 16: settlement with workshops, 17: early cemetery, 18-19: late cemeteries, 20: so-called fort. Based on MRT 5, 22.

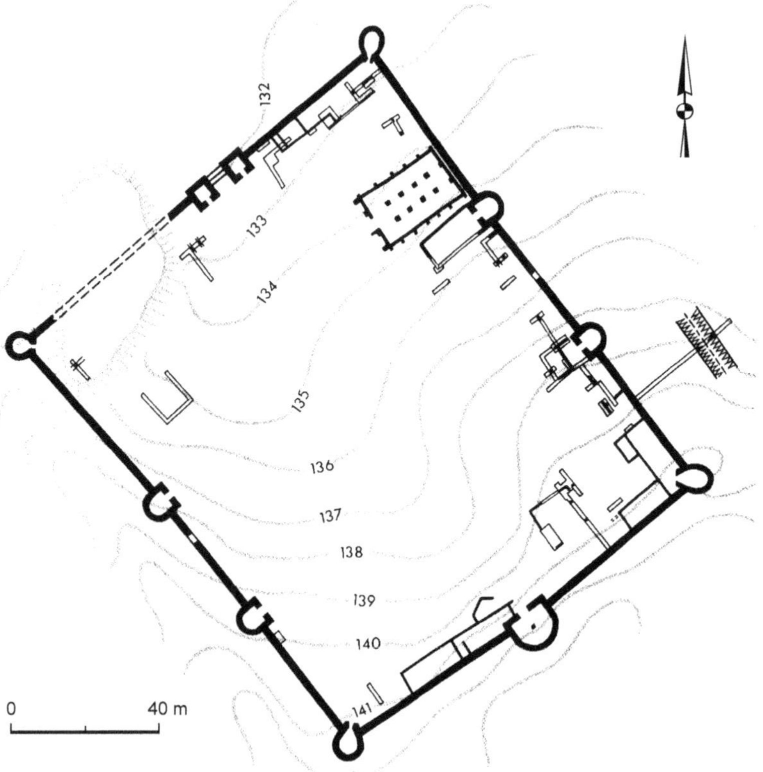

376. Tokod, ground plan of the so-called fort. Based on SOPRONI 1985.

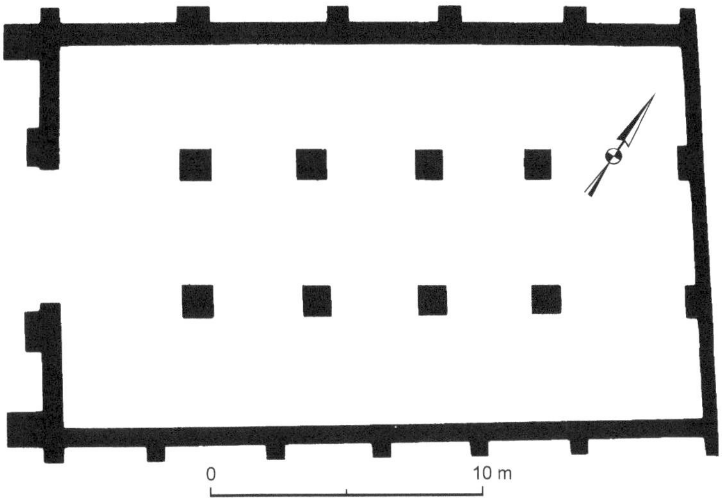

377. Tokod, ground plan of the so-called horreum. Drawing by ŐSI Sándor.

378. Tokod, pottery sherd, 57.I.a. Based on Severin Katalog.

379. Tokod, pottery sherd, 57.I.b. Based on Severin Katalog.

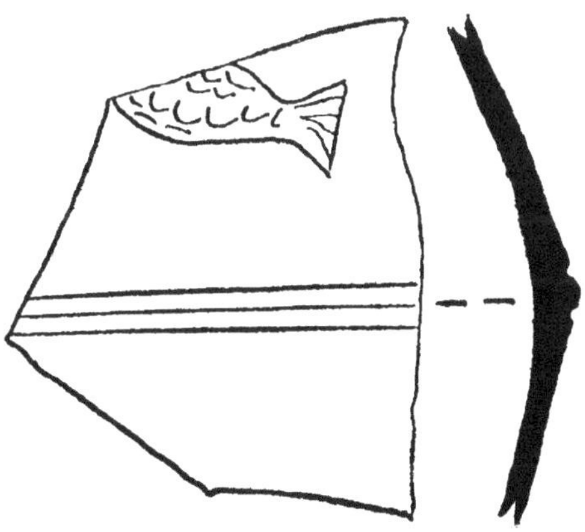

380. Tokod, pottery sherd, 57.I.c. Based on Tokod 1981 (LÁNYI).

381. Tokod, pottery sherd, 57.I.d. Based on Severin Katalog.

382. Tokod, casket mount, 57.I.e. Photo: SUGÁR Lajos.

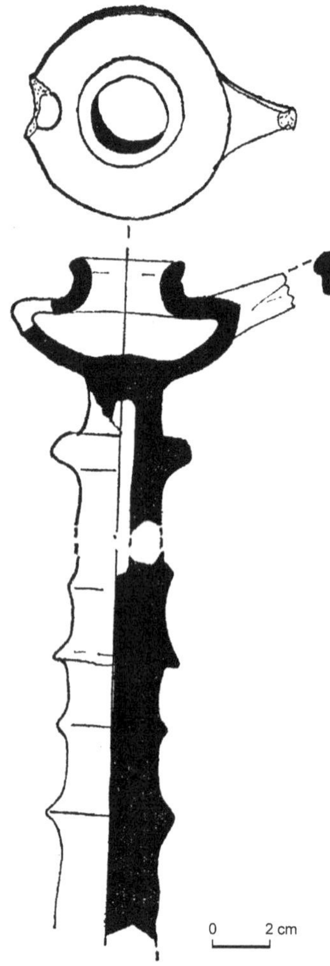

383. Tokod, clay lamp. Based on Tokod 1981 (LÁNYI).

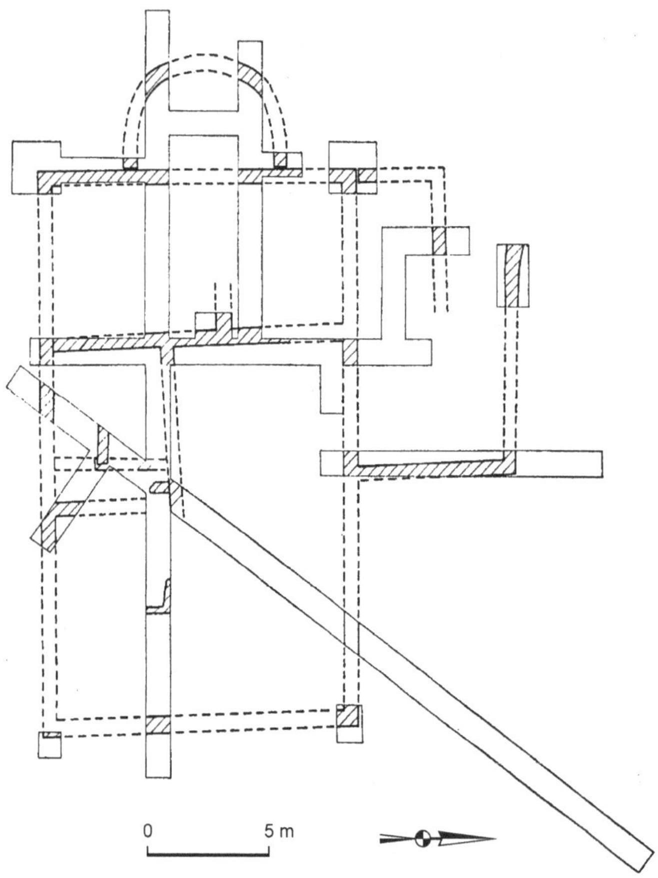

384. Ugod-Dióspuszta, so-called church, 59.II.a. Based on MRT 4, 75.

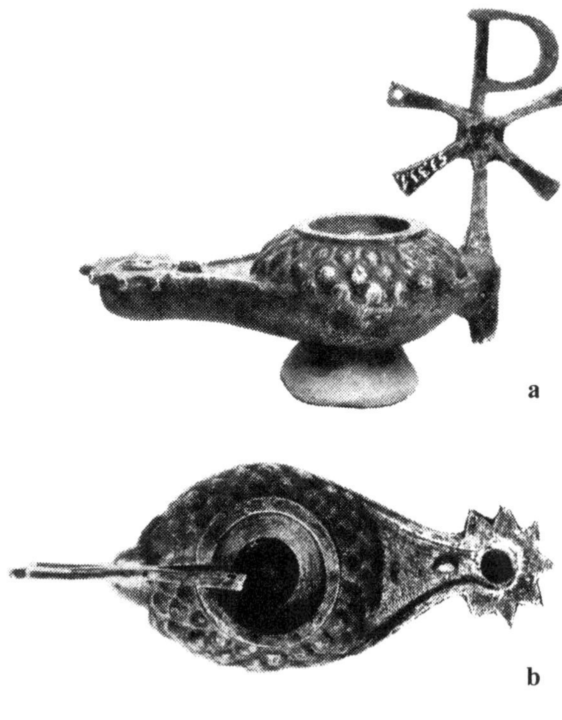

385a-b. Zalaegerszeg (Kaszaháza), bronze lamp, 60.I.a. Based on THOMAS 1977b.

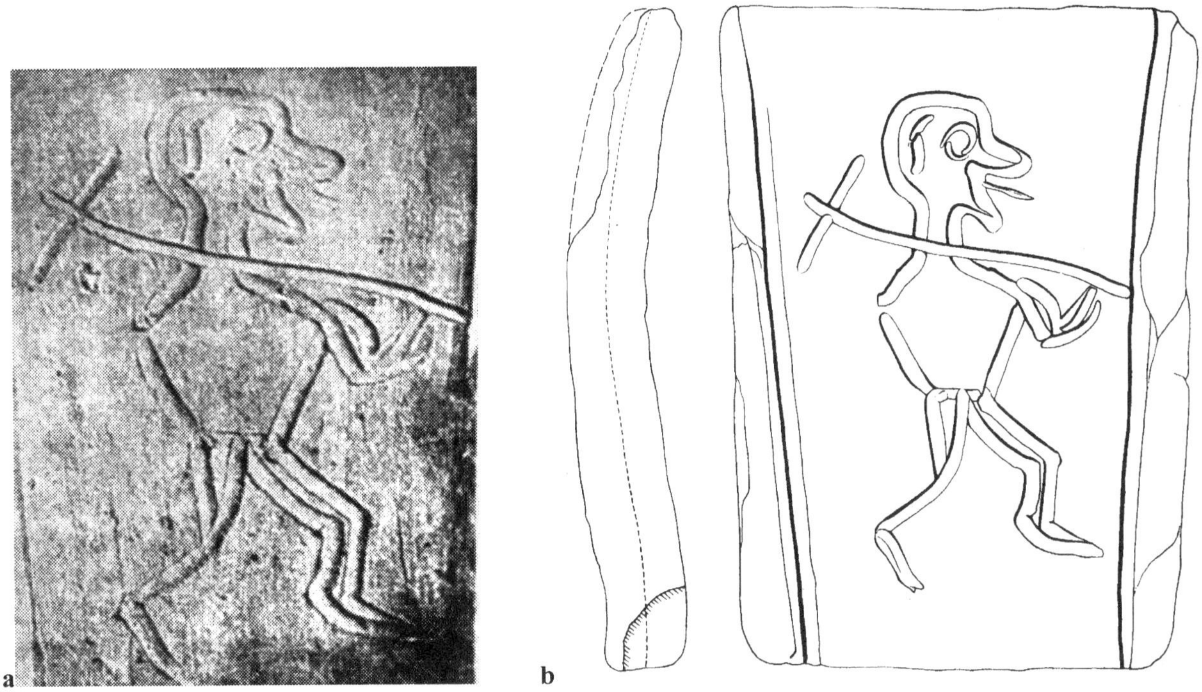

386. Rusovce (Oroszvár), roof tile, 61.II.a. Based on THOMAS 1980. **a:** photo. **b:** drawing.

387. Loc. unknown, suspension for a lamp, 62.I.a. Based on TÓTH E 1977a.

388. Loc. unknown, suspension for a lamp, 62.I.b. Based on TÓTH E 1977a.

389. Loc. unknown, suspension for a lamp, 62.I.c. Based on TÓTH E 1977a.

390. Reconstruction of 62.I.c. Drawn by ŐSI Sándor.

391. Biertan, suspension for a lamp. Based on RÖMER 1969.

392. Loc. unknown, finger ring, 62.II.a. Based on NAGY L 1938.

394. Loc. unknown, finger ring, 62.I.h. Based on NAGY L 1938.

393. Loc. unknown, clay lamp, 62.I.g. Based on Severin Katalog.

395. Loc. unknown, gem, 62.II.b. Based on THOMAS 1977.

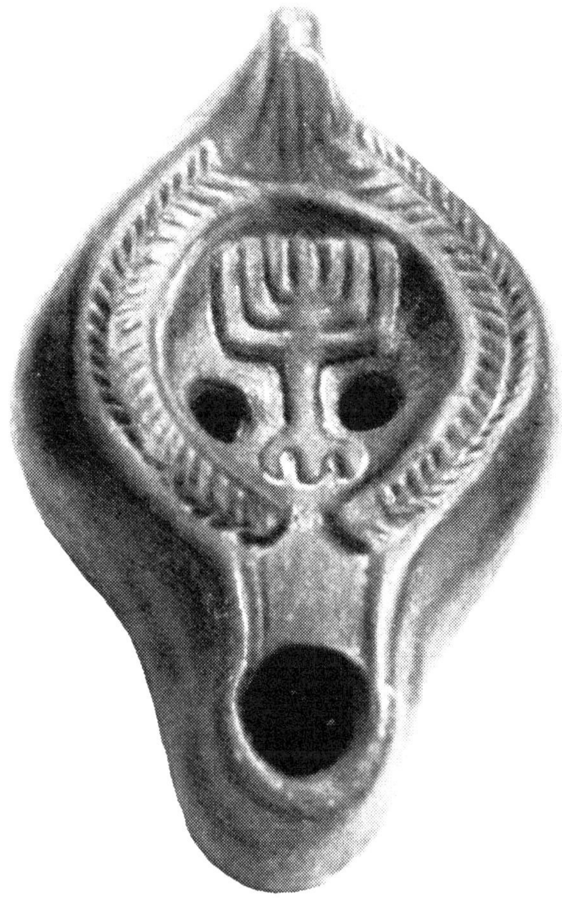

396. Loc. unknown, clay lamp, 62.I.i. Based on SCHEIBER1983.

397. Loc. unknown, bust of an emperor, 62.I.j. Based on THOMAS 1988.

www.ingramcontent.com/pod-product-compliance
Lightning Source LLC
LaVergne TN
LVHW071204210926
848482LV00015B/720